Methods in Stream Ecology

Methods in Stream Ecology

Volume 2: Ecosystem Function

Third Edition

Edited by

Gary A. Lamberti and F. Richard Hauer

ACADEMIC PRESS

An imprint of Elsevier

Academic Press is an imprint of Elsevier
125 London Wall, London EC2Y 5AS, United Kingdom
525 B Street, Suite 1800, San Diego, CA 92101-4495, United States
50 Hampshire Street, 5th Floor, Cambridge, MA 02139, United States
The Boulevard, Langford Lane, Kidlington, Oxford OX5 1GB, United Kingdom

Notices
Knowledge and best practice in this field are constantly changing. As new research and experience broaden our understanding, changes in research methods, professional practices, or medical treatment may become necessary.

Practitioners and researchers must always rely on their own experience and knowledge in evaluating and using any information, methods, compounds, or experiments described herein. In using such information or methods they should be mindful of their own safety and the safety of others, including parties for whom they have a professional responsibility.

To the fullest extent of the law, neither the Publisher nor the authors, contributors, or editors, assume any liability for any injury and/or damage to persons or property as a matter of products liability, negligence or otherwise, or from any use or operation of any methods, products, instructions, or ideas contained in the material herein.

Library of Congress Cataloging-in-Publication Data
A catalog record for this book is available from the Library of Congress

British Library Cataloguing-in-Publication Data
A catalogue record for this book is available from the British Library

ISBN: 978-0-12-813047-6

For information on all Academic Press publications visit our website at https://www.elsevier.com/books-and-journals

Working together
to grow libraries in
developing countries

www.elsevier.com • www.bookaid.org

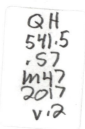
Publisher: Candice G. Janco
Acquisition Editor: Laura S. Kelleher
Editorial Project Manager: Emily Thomson
Production Project Manager: Mohanapriyan Rajendran
Designer: Matthew Limbert
Cover Photo: Gary Lamberti

Typeset by TNQ Books and Journals

Contents

List of Contributors

Numbers in the parentheses indicate the pages on which the authors' contributions begin.

DAVID G. ARMANINI (335) Department of Geography, Western University and Canadian Rivers Institute; Prothea, Via Manzoni 41, 20121 Milan, Italy

DONALD J. BAIRD (335) Environment and Climate Change Canada and Canadian Rivers Institute, Department of Biology, University of New Brunswick, Fredericton, NB, E3B 5A3 Canada

MICHELLE A. BAKER (129) Department of Biology and the Ecology Center, Utah State University, Logan, UT, 84322 USA

E.F. BENFIELD (71) Department of Biological Sciences, Virginia Polytechnic Institute and State University, Blacksburg, VA, 24061 USA

ARTHUR C. BENKE (235) Department of Biological Sciences, University of Alabama, Tuscaloosa, AL, 35487 USA

JONATHAN P. BENSTEAD (255) Department of Biological Sciences, University of Alabama, Tuscaloosa, AL, 35487 USA

KATHRYN BOYER (113) Department of Fisheries and Wildlife, Oregon State University, Corvallis, OR, 97331 USA

AMY J. BURGIN (173) Ecology and Evolutionary Biology and Environmental Studies, Kansas Biological Survey and The University of Kansas, Lawrence, KS, 66045 USA

JAMES L. CARTER (293) United States Geological Survey, Menlo Park, CA, 94025 USA

JOSEPH M. CULP (335) Environment and Climate Change Canada and Canadian Rivers Institute, Department of Biology, University of New Brunswick, Fredericton, NB, E3B 5A3 Canada

WALTER K. DODDS (173) Division of Biology, Kansas State University, Manhattan, KS, 66506 USA

SOLANGE DUHAMEL (197) Lamont-Doherty Earth Observatory, Columbia University, Palisades, NY, 10964 USA

SALLY A. ENTREKIN (55) Department of Biology, University of Central Arkansas, Conway, AR, 72035 USA

NATHAN T. EVANS (319) Southeast Environmental Research Center, Florida International University, Miami, FL, 33199 USA

MICHELLE A. EVANS-WHITE (255) Department of Biological Sciences, University of Arkansas, Fayetteville, AR, 72701 USA

STUART E.G. FINDLAY (21) Cary Institute of Ecosystem Studies, Millbrook, NY, 12545 USA

JACQUES C. FINLAY (3) Department of Ecology and Evolutionary Biology, University of Minnesota, St. Paul, MN, 55108 USA

KEN M. FRITZ (71) National Exposure Research Laboratory, U.S. Environmental Protection Agency, Cincinnati, OH, 45268 USA

CATHERINE A. GIBSON (255) The Nature Conservancy — New York, Albany, NY, 12205 USA

STANLEY V. GREGORY (113) Department of Fisheries and Wildlife, Oregon State University, Corvallis, OR, 97331 USA

NATALIE A. GRIFFITHS (55) Environmental Sciences Division, Oak Ridge National Laboratory, Oak Ridge, TN, 37831 USA

JACK W. GRUBAUGH (37) Department of Biological Sciences, University of Tennessee Martin, Martin, TN, 38237 USA

ANGELA GURNELL (113) School of Geography, Queen Mary University of London, London, E1 4NS United Kingdom

ROBERT O. HALL, JR. (219) Department of Zoology and Physiology, University of Wyoming, Laramie, WY, 82071 USA

MORGAN J. HANNAFORD (293) Department of Life Science, Industry and Natural Resources, Shasta College, CA, 96049 USA

F. RICHARD HAUER (83) Center for Integrated Research on the Environment and Flathead Lake Biological Station, University of Montana, Missoula, MT, 59812 USA

ANNE E. HERSHEY (3) Department of Biology, University of North Carolina at Greensboro, Greensboro, NC, 27402 USA

JAMES M. HOOD (255) Aquatic Ecology Laboratory Department of Evolution, Ecology, and Organismal Biology, The Ohio State University, Columbus, OH, 43210 USA

ERIN R. HOTCHKISS (219) Département des sciences biologiques, Université du Québec à Montréal; Department of Biological Sciences, Virginia Polytechnic Institute and State University, Blacksburg, VA, 24061 USA

ALEXANDER D. HURYN (235) Department of Biological Sciences, University of Alabama, Tuscaloosa, AL, 35487 USA

JOHN J. HUTCHENS, JR. (37) Department of Biology, Coastal Carolina University, Conway, SC, 29528 USA

GARY A. LAMBERTI (55) Department of Biological Sciences, University of Notre Dame, Notre Dame, IN, 46556 USA

AMY M. MARCARELLI (173) Department of Biological Sciences, Michigan Technological University, Houghton, MI, 49931 USA

G. WAYNE MINSHALL (83) Stream Ecology Center, Department of Biological Sciences, Idaho State University, Pocatello, ID, 83209 USA

ROBERT M. NORTHINGTON (3) Climate Change Institute and School of Biology and Ecology, The University of Maine, Orono, ME, 04469 USA

THOMAS B. PARR (21) Oklahoma Biological Survey, University of Oklahoma, Norman, OK, 73019 USA

BRUCE J. PETERSON (3) The Ecosystems Center, Marine Biological Laboratory, Woods Hole, MA, 02543 USA

HERVÉ PIÉGAY (113) National Center for Scientific Research, University of Lyon cedex 07, France

ALEXANDER J. REISINGER (147) Cary Institute of Ecosystem Studies, Millbrook, NY, 12545 USA

VINCENT H. RESH (293) Department of Environmental Science, Policy & Management, University of California, Berkeley, Berkeley, CA, 94720 USA

SCOTT L. ROLLINS (277) Department of Life Sciences, Spokane Falls Community College, Spokane, WA, 99224 USA

EMMA J. ROSI (147) Cary Institute of Ecosystem Studies, Millbrook, NY, 12545 USA

AMANDA T. RUGENSKI (83) Department of Ecology and Evolutionary Biology, Cornell University, Ithaca, NY, 14853 USA

THOMAS P. SIMON (319) School of Public and Environmental Affairs, Indiana University, Bloomington, IN, 47405 USA

ALAN D. STEINMAN (197) Annis Water Resources Institute, Grand Valley State University, Muskegon, MI, 49441 USA

R. JAN STEVENSON (277) Department of Integrative Biology, Michigan State University, East Lansing, MI, 48824 USA

ERIC A. STRAUSS (173) Department of Biology, University of Wisconsin — La Crosse, La Crosse, WI, 54601 USA

JENNIFER L. TANK (147) Department of Biological Sciences, University of Notre Dame, Notre Dame, IN, 46556 USA

SCOTT D. TIEGS (55,71) Department of Biological Sciences, Oakland University, Rochester, MI, 48309 USA

J. BRUCE WALLACE (37) Odum School of Ecology, University of Georgia, Athens, GA, 30602 USA

JACKSON R. WEBSTER (129) Department of Biology, Virginia Polytechnic Institute and State University, Blacksburg, VA, 24061 USA

ADAM G. YATES (335) Department of Geography and Canadian Rivers Institute, Western University, London, ON, N6A 5C2 Canada

Preface

When the first edition of *Methods in Stream Ecology* was published in 1996, we hoped that it would prove useful to practicing stream ecologists and perhaps as a supplementary textbook for aquatic ecology courses. We were surprised and delighted that the book was accepted worldwide as a basic text in stream ecology. While the first edition served well for 10 years as a reference for both instruction and research, stream ecologists from around the world continued to expand this important and dynamic research area. After a decade, the book was in need of modernization to keep pace with important methodological developments, as well as an expanded understanding of stream ecology. In 2007, we published the second edition of this book that not only stressed exercises that could generally be completed within a few hours or an afternoon of intensive field work, but also provided both classroom-style exercises and research-level methods appropriate for the most rigorous investigations. Since 2007, stream ecology has grown substantially in the breadth of inquiry and the complexity of technologies used in research, especially in the realm of molecular ecology. In this third edition of *Methods in Stream Ecology*, we have added content and research depth to previous chapters and developed new and exciting chapters in areas of inquiry not previously covered in prior editions.

As we pointed out in both previous editions, perhaps no other area of aquatic ecology requires a more interdisciplinary approach than stream ecology. Geology, geomorphology, fluid mechanics, hydrology, biogeochemistry, nutrient dynamics, microbiology, botany, invertebrate zoology, fish biology, food web analysis, bioproduction, biomonitoring, and molecular biology are but a few of the disciplines from which stream ecology draws. This integration is even more true today as advances in technology open opportunities not only for new types and sophistication of metrics, but also more importantly opportunities to ask new, exciting, and ecologically important questions. The science of stream ecology continues to advance at a remarkably rapid rate, as evidenced by the virtual explosion of publications in stream ecological research throughout the world. Along with the rapid increase in research activity, the teaching of stream ecology at the upper undergraduate and graduate levels at major colleges and universities has become foundational to comprehensive aquatic ecology programs. Today, more than ever, scientists, government agencies, resource managers, and the general public are increasingly aware of stream ecology as an integrative and holistic science that can help societies around the globe grapple with environmental degradation of their priceless running water resources. Because streams and rivers are fundamental to human existence, many organizations, both in government and in the private sector, have emerged to protect these unique habitats so vital to global biodiversity, complexity, sustainability, and human well-being.

Stream ecology has experienced many areas of rapidly advancing research, methodologies, and coupled technologies from advanced modeling to genomics. The serious student or researcher will find that the chapters in this third edition have been significantly updated to reflect these advancements. Our contributing authors have strived to provide the most comprehensive and contemporary series of methods in stream ecology to be used for teaching or conducting the most cutting-edge research. We trust that the book will be valuable to both the stream ecology student and the most seasoned scientist. Resource managers employed in the private sector or by federal or state agencies should continue to find this book an indispensable reference for developing monitoring approaches or evaluating the efficacy of their field and laboratory techniques.

We have added several topics not covered in either the first or second editions, reflecting the rapid growth and change occurring in the science of stream and river ecology. We have, by necessity, now split the book into two volumes. In *Methods in Stream Ecology: Volume 1—Ecosystem Structure*, we focus on the various structural components of streams. We have organized these chapters into three sections: (A) Physical Processes, (B) Stream Biota, and (C) Community Interactions. In *Methods in Stream Ecology: Volume 2—Ecosystem Function*, we focus on the various functional components of streams. We have organized these chapters also into three sections: (D) Organic Matter Dynamics, (E) Ecosystem Processes, and (F) Ecosystem Assessment. Importantly, to facilitate transition between the two volumes, we have sequentially numbered the chapters from 1 to 40, with *Volume 1* containing 22 chapters and *Volume 2* containing 18 chapters. Please note that individual chapters may cross-reference chapters in both volumes. Useful supplemental materials for both volumes will be posted to the same website: http://booksite.elsevier.com/9780124165588/index.php.

The first edition contained 31 chapters. In this third edition, the 40 chapters represent roughly a 25% increase in chapter numbers and significant updating of content. Each chapter consists of (1) an Introduction, (2) a General Design section, (3) a Specific Methods section, (4) Questions, (5) a list of needed Materials and Supplies, and (6) relevant References. The Introduction provides background information and a literature review necessary to understand the principles of the topic. The General Design presents the conceptual approach and principles of the methods. The Specific Methods generally begin with relatively simple goals, objectives, and techniques and increase in the level of difficulty and sophistication; Basic Methods are suitable for the classroom whereas Advanced Methods are applicable to high-end research projects. Each method is explained in step-by-step instructions for conducting either field or laboratory investigations. The methods presented are of research quality, and while it is not the book's goal to produce an exhaustive manual, the chapters present rigorous methods that provide sound underpinnings for both instruction and research purposes. In each case, the methods presented are used frequently by the authors in their personal research or instruction. The Questions listed at the end of each chapter are formulated to encourage critical evaluation of the topic and the methods that were used to address a particular stream ecology issue. The comprehensive list of Materials and Supplies itemizes equipment, apparatus, and consumables necessary to conduct each method and is generally organized by each specific method to allow simple checklists to be made. Finally, this third edition provides a Glossary at the end of each volume that defines the most common terms that appear in the chapters. We are including an accompanying website that can be easily accessed by a simple search. The website above contains supplementary materials, such as forms that can be printed and spreadsheets that can be downloaded onto field tablets and computers for field data entry and later data analysis. We have also posted downloadable figures for each chapter to the book website to aid in instruction.

If this book is being used for course instruction, we recommend that instructors carefully consider the chapters and methods that they wish to use and plan carefully to budget the necessary time for setup, sampling, and analysis to complete individual or group research projects. Generally, classes should begin with Basic Methods and then delve more deeply into Advanced Methods as time and resources allow. We hope that all of the chapters will enrich the field of stream ecology as a rigorous scientific discipline. We continue to encourage the use of this third edition to assist in the formulation of exciting ecological questions and hypotheses and, to that end, the chapters present sound methods for discovery. For course instruction, we recommend use of moderate-sized streams from 3 to 12 m wide that are easily waded. Smaller streams should be avoided by a large class, such as 10−20 students, because of the impacts incurred on a small environment. Large rivers are limiting to class instruction because of safety concerns and the inherent difficulties associated with sampling deep, flowing waters. Of course, research projects for graduate students or professionals will dictate the selection of stream size.

Reviewers and users of the previous editions found this book to be particularly "user friendly." We have maintained this tenor in the third edition as one of our primary goals. As in the previous editions, we have attempted to present a book with a logical flow of topics and a uniform chapter format and style, an approach that our authors embraced and implemented. We sincerely thank our contributing authors and coauthors from the previous editions, who once again gave their knowledge, talent, and time for the benefit of our science. We also welcome the new authors in this third edition and likewise thank them for their remarkable efforts on behalf of our science.

The inspiration for this book arose from our own research and teaching. Numerous colleagues and students have encouraged us through the first and second editions and now this third edition. We are thankful for their input. Our graduate and undergraduate students continue to be a source of inspiration and encouragement to us. We gratefully acknowledge the assistance and financial support of our outstanding home institutions, the University of Montana and the University of Notre Dame. Finally, and most importantly, we once again thank our families for their continued love and support. Our wives, Brenda Hauer and Donna Lamberti, energize and inspire us, not only throughout this endeavor, but in all of life. Our children, Andrew and Bethany Hauer and Sara and Matthew Lamberti, provide us with joy on a daily basis. We are forever grateful to them.

F. Richard Hauer
Gary A. Lamberti

Section D

Organic Matter Dynamics

Gary A. Lamberti and F. Richard Hauer

The input, distribution, and processing of organic matter in stream ecosystems have been a subject of great interest to stream ecologists for nearly half a century. The concept of external "fueling" of stream ecosystems by organic matter derived from the watershed was first expressed in Noel Hynes' seminal Baldi Memorial Lecture "The Stream and Its Valley" in 1974. This notion was further amplified in the mid-1970s with studies of the dominant role of allochthonous matter in the energy budget of streams of Hubbard Brook Experimental Forest, and then formally conceptualized in the River Continuum Concept—still the fundamental paradigm in stream ecology. The chapters in Section D detail standard methods to collect, characterize, and analyze the organic matter of streams, complemented with methods to manipulate organic matter to answer intriguing scientific questions. Chapter 23 presents the methods for using stable isotopes to trace the flow of organic matter, produced both internally and externally, through stream food webs and includes specific applications for impacted systems. Subsequent chapters in this section work through different forms of organic matter, historically based on dissolution or particle size. Chapter 24 describes the smallest fraction, dissolved organic matter, that often dominates total organic matter budgets of streams. Chapter 25 treats the next fraction, fine particulate organic matter, that is often overlooked but supports a myriad of filter feeders and other ecosystem processes. The visible coarse particulate organic matter (CPOM), such as leaves, is considered in Chapters 26 and 27. Chapter 26 describes methods for manipulating CPOM to study retention, carbon cycling, and the responses of food webs to CPOM loading. As leaf litter is so central to food webs in many streams, Chapter 27 presents both classical and modern approaches to study the decomposition and breakdown of leaf litter. Chapter 28 examines the riparian zone, a significant source of organic matter and habitat, and provides methods to characterize this critical ecotone affecting stream function including remote-sensing approaches. This section concludes with a treatment of large wood (LW) in Chapter 29, including modeling exercises for estimating LW abundance and dynamics during stream restoration efforts. Overall, these seven chapters present diverse and tested methods for measuring and manipulating the organic matter of streams and provide clear linkages to inform our knowledge of ecosystem structure and function.

Chapter 23

Stable Isotopes in Stream Food Webs

Anne E. Hershey[1], Robert M. Northington[2], Jacques C. Finlay[3] and Bruce J. Peterson[4]

[1]Department of Biology, University of North Carolina at Greensboro; [2]Climate Change Institute and School of Biology and Ecology, The University of Maine; [3]Department of Ecology and Evolutionary Biology, University of Minnesota; [4]The Ecosystems Center, Marine Biological Laboratory

23.1 INTRODUCTION

Stream *food webs* describe the trophic relationships among organisms in streams and integrate the dynamics of organic matter and nutrient processing with community interactions. Food webs differ in structure between stream types, although they all have some common elements (Cummins, 1973). Most streams have approximately three or four trophic levels with much connectivity, but disturbed streams typically have simplified food webs (e.g., Townsend et al., 1998). Allochthonous and autochthonous resources occupy the lowest trophic level, but the relative importance of these resources to the food web varies widely between streams. Furthermore, defining trophic relationships for consumers is more complex. Certain groups of macroinvertebrates and some vertebrates are characterized as either grazers or detritivores, apparently occupying the primary consumer trophic level. However, both the producers and especially detritus are intimately associated with heterotrophic microbes and microzoans, which grazers and detritivores also ingest. Thus, these consumers are functioning as both primary and secondary consumers (i.e., somewhere between trophic levels 2 and 3). Predators often have mixed diets including some combination of detritus, diatoms, animal prey, and other predators, placing them somewhere between trophic levels 3 and 5. In addition, several functional feeding groups can be found among detritivores (shredders, collector−filterers, collector−gatherers; see Chapter 20) and predators (Peckarsky, 1982). Many approaches have been used to characterize stream food webs, including gut analyses, functional feeding group analyses, observations, experiments, energy budgets, and others. More recently, *stable isotope* analysis has proven to be very powerful for characterizing resources and trophic structure in stream food webs; quantifying resource use; and testing hypotheses about the role of natural gradients, natural and anthropogenic stressors, and biotic interactions in controlling the food webs.

In recent years, the use of stable isotopes in stream ecology studies has become commonplace. Study of the ratio of heavy to light stable isotopes of carbon (^{13}C:^{12}C) and nitrogen (^{15}N:^{14}N) can be especially useful for following transfers of C and N from terrestrial, aquatic, and marine sources to primary and secondary consumers (see Peterson and Fry, 1987; Fry, 2006). Since carbon is the typical "currency" in ecosystems, having a tool to trace carbon flow through an ecosystem is essential. Furthermore, nitrogen is a key, sometimes limiting, plant, algal, and consumer nutrient, and, in excess, a serious pollutant. Thus, following the source and fate of nitrogen in a stream food web is also essential to understanding food web and ecosystem structure and function; stable isotopes of N are very useful for that purpose.

Stable isotopes of other elements also may be useful in stream food web studies (Table 23.1). For example, ^{34}S can be useful for discriminating marine versus terrestrial sources to consumers (e.g., Peterson and Fry, 1987), tracing pulp mill effluent (Wayland and Hobson, 2001), coal ash (Derda et al., 2006), or in discriminating among some terrestrial detrital sources (McArthur and Moorhead, 1996). The utility of stable isotopes of hydrogen (^{1}H and ^{2}H or D for deuterium) in plant physiology research has been well known for decades (Ehlerginger and Rundel, 1989), but only recently have hydrogen isotopes been examined for understanding aquatic food webs (Doucett et al., 2007; Finlay et al., 2010; Solomon et al., 2011; Karlsson et al., 2012). Although deuterium (D) comprises only about 0.015% of the hydrogen atoms in the biosphere (Ehlerginger and Rundel, 1989; Sulzman, 2007), its ability to bind with other atoms (e.g., oxygen and carbon) has made it a valuable tool for assessing hydrologic and ecological relationships. Additionally, the twofold greater mass of D relative to that of the more common hydrogen stable isotope, ^{1}H, makes it sensitive to physical processes such as condensation, evaporation, precipitation, and physiological processes within organisms' bodies (Gat, 1996; Sulzman, 2007). As a result,

TABLE 23.1 Potential sources of natural abundance stable isotope variation in organic matter sources that can be used to examine food web structure, diet sources, and energy flow in rivers and riparian zones.

Habitats or Sources	Stable Isotope	Primary Mechanism	Locations Observed	Examples
Longitudinal				
Pool–riffle	$\delta^{13}C$, $\delta^{15}N$	Fractionation	High algal growth, low CO_2	Finlay et al. (1999) and Trudeau and Rasmussen (2003)
River confluence	$\delta^{13}C$, $\delta^{15}N$, $\delta^{34}S$	Source	Mixing of chemically distinct waters	Finlay and Kendall (2007)
Upstream–downstream	$\delta^{13}C$, $\delta^{15}N$	Source and fractionation	Spring fed streams, point source inputs (e.g., sewage)	Kennedy et al. (2005)
Within Site				
Terrestrial–aquatic	$\delta^{13}C$, $\delta^{15}N$, $\delta^{34}S$, δD	Source and fractionation	Widespread	Finlay (2001), Doucett et al. (2007), Solomon et al. (2009), and Dekar et al. (2012)
Benthic–planktonic	$\delta^{13}C$	Fractionation	Rivers with attached and planktonic algae	Debruyn and Rasmussen (2002) and Delong and Thorp (2006)
River–riparian interface	$\delta^{13}C$, $\delta^{15}N$, $\delta^{34}S$	Source and fractionation	Large streams, rivers	Bastow et al. (2002) and Kato et al. (2004)
Marine–river	$\delta^{13}C$, $\delta^{15}N$	Source	Coastal streams and rivers	Chaloner et al. (2002) and MacAvoy et al. (2000)
Sewage effluent	$\delta^{13}C$, $\delta^{15}N$	Source	Downstream of wastewater treatment plants (most urban areas)	Ulseth and Hershey (2005) and Northington and Hershey (2006)
Methane	$\delta^{13}C$, δD	Source	Aquatic sediments	Deines et al. (2009), Jones and Gray (2011), and Hershey et al. (2015)
Pulp mill effluent	$\delta^{34}S$	Source	Downstream of pulp mills	Wayland and Hobson (2001)
Coal ash discharge	$\delta^{34}S$	Source	Downstream of coal ash spills	Derda et al. (2006)

Modified from Finlay and Kendall (2007).

many studies examine the relationship of D relative to other stable isotopes (e.g., ^{18}O, ^{13}C, ^{15}N, ^{34}S) to track its movement through food webs and ecosystems in general.

At this point, we introduce some isotope terminology and methodology to facilitate further discussion and use of stable isotopes in stream food webs. Isotope ratios in samples relative to the same ratios in standards can be measured very accurately using mass spectrometry. The accuracy of the measurement is typically much greater (~ 10-fold) than the natural variation in the environment. The standards used are materials that have very uniform distributions of the elemental isotopes of interest, and do not change through time. For carbon, the standard is carbonate rock from the Pee Dee Belemnite formation. The nitrogen standard is air. For hydrogen, it is standard mean ocean water. Because the light element is far more abundant than the heavy isotope in the standards and environmental samples, the deviation in the isotope ratio of the sample relative to the standard is expressed in parts per thousand (‰, also termed "per mil"). The value in ‰ is designated as the δ *value* (delta value, also referred to as del value, see http://wwwrcamnl.wr.usgs.gov/isoig/res/funda.html) as follows for carbon and nitrogen:

$$\delta^{13}C \text{ or } \delta^{15}N = [(R_{sample} - R_{standard})/R_{standard}] \times 1000 \tag{23.1}$$

where $R_{sample} = {}^{13}C{:}^{12}C$ or $^{15}N{:}^{14}N$ in the sample and $R_{standard} = {}^{13}C{:}^{12}C$ or $^{15}N{:}^{14}N$ in the standard.

The same equation can be applied to other elements with multiple stable isotopes. Samples *enriched* in the heavy isotope are said to be isotopically "heavy" and have higher δ values, whereas samples *depleted* in the heavy isotope are isotopically "light" (relatively rich in the lighter isotopes, e.g., ^{12}C or ^{14}N) and have lower δ values. Note that for carbon, most samples are ^{13}C depleted compared to the standard. Thus, the $\delta^{13}C$ of many organic and inorganic materials in the environment typically has a negative δ value. Whether sample δ values tend to be positive, negative, or variable in sign, they differ among elements depending on the respective standard reference materials used and environmental conditions.

As stable isotopes cycle in ecosystems the elements undergo shifts in the relative abundance of heavy and light isotopes during various reactions or processes, referred to as *fractionation* (Fry, 2006). Fractionation means that light and heavy isotopes move at slightly different rates as they undergo reactions (including metabolic processes) because they differ slightly in mass. The result is that the reactants and products (e.g., resource and consumer) end up with different isotope ratios. The common example for carbon is the approximately 20‰ fractionation in CO_2 from uptake through biomass incorporation in trees (see Peterson and Fry, 1987; Fry, 2006.). The $\delta^{13}C$ value for most trees is about −27‰ to −30‰, which is considerably less than the −8‰ $\delta^{13}C$ value for atmospheric CO_2. This large difference is due to the combined effect of slower diffusion of ^{13}C through the stomata and slower reaction rates of ^{13}C during photosynthesis. This fractionation accounts for the consistent large negative $\delta^{13}C$ value seen in biota compared to the atmosphere. In lakes and streams the $\delta^{13}C$ value of dissolved inorganic carbon (DIC), which is assimilated by algae, varies considerably because stream and lake waters are not usually in equilibrium with the atmosphere. Thus, algal $\delta^{13}C$ is also quite variable among and within streams (e.g., Finlay, 2004; Hill et al., 2008; Ishikawa et al., 2012). Often it is distinct from terrestrial leaf litter, but not always. In a hypothetical food web presented as an example later (see next section), we assign diatoms a $\delta^{13}C$ value of −35‰. This is not unrealistic for diatoms or the entire biofilm in a forested stream if the stream is supersaturated with CO_2 derived partially from the decomposition of terrestrial detritus ($\delta^{13}C \sim -28‰$). DIC in such a stream might average −15‰, rather than the atmospheric value of −8‰, reflecting both atmospheric and respiratory CO_2 sources. Several factors can affect algal fractionation of C (e.g., see Finlay et al., 1999; see *Advanced Method 2*), but if, as an example, algal fractionation of C is 20‰, the algal $\delta^{13}C$ value should be −15‰−20‰ = −35‰. In contrast, nitrogen fixation by microbes and plants often exhibits little fractionation and it is not uncommon for plants to have $\delta^{15}N$ values close to the 0‰ atmospheric value. However, microbial processes such as nitrification and denitrification, and animal metabolism, fractionate nitrogen isotopes sufficiently such that all food webs contain components with significant variation in N isotope ratios (>1‰, but often with 5‰−10‰).

Various organic matter sources in many ecosystems have different $\delta^{13}C$ and $\delta^{15}N$ ratios, and in those cases, diets of animals can be inferred from the $\delta^{13}C$ and $\delta^{15}N$ values in animal tissues. Fractionation varies among consumers and elements, but animal tissues are usually just slightly enriched (an average of 0.3−0.5‰) in ^{13}C relative to their food, but significantly enriched in ^{15}N (an average of 3.4‰) relative to their food (see Vander Zanden and Rasmussen 1999; Post, 2002; McCutchan et al., 2003). If algae with $\delta^{13}C = -32‰$ and $\delta^{15}N = 0‰$ are the only food of a grazer, then the isotopic composition of the animal is predicted to be $\delta^{13}C = -32‰$ to −31‰ and $\delta^{15}N = +2.5‰$ to +3.5‰. The trophic enrichment in ^{15}N is sufficiently large that it provides a useful tool for designation of trophic level if the δ values of food sources in the food web are known (Minagawa and Wada, 1984) or can be inferred (Vander Zanden and Rasmussen, 1999). For example, an animal with a $\delta^{15}N$ value of 6‰ would occupy one higher trophic level than an animal with a $\delta^{15}N$ value of 3‰. Since there is little trophic fractionation in carbon isotopes, but a relatively large and predictable fractionation in nitrogen isotopes, the combination of $\delta^{13}C$ and $\delta^{15}N$ isotopes is frequently used as an aid to determine both pathways of organic matter transfer and trophic structure in ecosystems. When carbon and nitrogen stable isotopes can resolve a food web sufficiently for the question being asked, combined use of $\delta^{13}C$ and $\delta^{15}N$ can be an ideal tool for studying stream food webs. Conveniently, analytical technology is such that labs that provide stable isotopes analysis services typically provide $\delta^{13}C$ and $\delta^{15}N$ from the same sample.

Carbon isotopes might be distinct, for example, in a stream ecosystem bordered by C_4 riparian grasses, such as found on a prairie. Primary consumers might be utilizing either terrestrial detritus or epilithic diatoms as food. Many grasses are C_4 plants, which are isotopically enriched in ^{13}C compared to C_3 plants, which include riparian trees (for discussion of why this occurs see Peterson and Fry, 1987; Fry, 2006). If C_4 grass is the primary terrestrial input and has a $\delta^{13}C$ value of −14‰, and the diatoms have a $\delta^{13}C$ value of −30‰, it will be easy to distinguish whether a consumer is a detritivore or a grazer (see *Advanced Method 1*). Some consumers, especially some collectors will have intermediate values, suggesting that they utilize food derived from both sources. Top predators frequently have $\delta^{13}C$ values that are intermediate between detritus and algae because they feed on both grazers and detritivores. However, if the same stream flows through a gallery forest, detrital sources to the stream may be dominated by litter from C_3 trees ($\sim -28‰$), which would be isotopically distinct from the C_4 grasses. Furthermore, algal $\delta^{13}C$ may vary along natural gradients such as current regimes, productivity, and stream size (Finlay, 2004; Ishikawa et al., 2012; see *Advanced Method 2*). In streams with anthropogenic influence, algal and detrital resources may be more distinct (e.g., Northington and Hershey, 2006, and see *Basic Method*).

Although $\delta^{13}C$ has most often been used to determine food sources for consumers and $\delta^{15}N$ has been used to delineate trophic levels, $\delta^{15}N$ can also serve as a tracer of food source under special conditions. A good example is the study of the importance of spawning salmon to stream food webs. ^{15}N derived from salmon carcasses is enriched compared to detrital resources because salmon are high trophic level predators in the ocean and some large lakes (e.g., Mathison et al., 1988; Kline et al., 1990; Schuldt and Hershey, 1995; Bilby et al., 1996; Fisher-Wold and Hershey, 1999; Chaloner et al., 2002; Wipfli et al., 2003). Carcasses can be eaten directly by consumers transferring salmon-derived N to consumer biomass, and, as the carcasses decompose, ^{15}N incorporated into their tissues becomes incorporated into microbial biomass and mineralized, becoming bioavailable to algae. These processes result in ^{15}N enrichment in the food web that reflects dependence on salmon nitrogen. Similarly stable isotopes of both C and S, and sometimes other elements, can be effective tracers of novel food sources (e.g., MacAvoy et al., 2000).

Anthropogenic N is commonly enriched in ^{15}N relative to natural sources, but may occasionally be depleted. Anthropogenic ^{15}N is reflected in the δ^{15}N values of stream organisms to the extent that food web components incorporate it. Municipal wastewater or manure from agriculture may have a δ^{15}N value as high as 15‰ above nonanthropogenic sources (McClelland and Valiela, 1998; Karr et al., 2003; Anderson and Cabana, 2005, Table 23.1). However, the degree of ^{15}N enrichment relative to "background" values depends on the extent of fractionation processes such as denitrification and volatilization that preferentially remove ^{14}N. For example, fertilizers are generally depleted in ^{15}N relative to other nitrogen sources because they are produced by chemical reactions that fix N from the atmosphere, which has a δ^{15}N of 0‰ (see Peterson and Fry, 1987). N from fossil fuel combustion and from N-fixing organisms such as legumes is similarly depleted in ^{15}N. Excess N can enhance fractionating N losses that leave the residual pools ^{15}N enriched. When enriched anthropogenic ^{15}N becomes incorporated into producers (McClelland and Valiela, 1998) and consumers (Fry, 1999), it provides a useful food web tracer of the anthropogenic influence on the food web (Ulseth and Hershey, 2005; Northington and Hershey, 2006; Bullard and Hershey, 2013).

Frequently, both detrital and algal organic matter sources have very similar C and N isotopic compositions, making the determination of sources assimilated by consumers from natural isotopic abundance of N and C impossible. In such cases, δD can be a very powerful tool (Doucett et al., 2007). In particular, δD can be useful for discriminating terrestrial versus aquatic resources in stream food webs. Aquatic consumers demonstrate very little fractionation in δD, and many times fall between δD of their food and their environmental water (Solomon et al., 2009). Thus, when using δD to estimate food resources in a consumer's diet, δD values must be corrected for the δD of environmental water (e.g., Solomon et al., 2009; Stenroth et al., 2015). Another option is to deliberately introduce an isotopic signal, such as ^{15}N-enriched NH_4^+ or NO_3^-, which also traces biogeochemical processing of N through the food web (e.g., Peterson et al., 2001; Mulholland et al., 2008). This approach is discussed in Chapter 31.

23.1.1 Hypothetical Food Web Using δ^{13}C and δ^{15}N

An example of a hypothetical stream food web analyzed using C and N stable isotopes is shown in Fig. 23.1. This hypothetical stream is known to receive large amounts of leaf detritus, but also has sufficient light input to support a benthic diatom community. Samples of detritus, epilithic algae (diatoms), insects (grazers, collectors, shredders, predatory insects), and fish have been collected and analyzed for C and N stable isotope ratios. As expected, the tree leaf detritus had a δ^{13}C value of -28‰ and a δ^{15}N value of 0‰. Pure diatom samples would not be collected in the field and are difficult to extract in the lab (see Hamilton et al., 2005) as the cells grow in an epilithic or epibenthic matrix of microbial slime and detritus, but for illustrative purposes we assign them a δ^{13}C value of -35‰ and δ^{15}N value of $+2$‰. Assuming a literature-based trophic transfer shift of $+0.4$‰ for C and $+3.4$‰ for N (e.g., Vander Zanden and Rasmussen 1999), the predicted values for insects with contrasting feeding modes and predators are shown in Fig. 23.1. Note the wide separation in δ^{13}C values for consumers feeding upon diatoms versus detritus, clearly indicating their organic matter sources. Also note that predators have higher δ^{15}N values than their prey. Insectivorous fish have a similar position in the web as

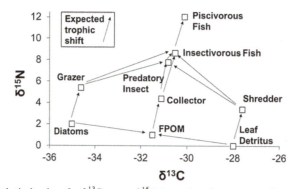

FIGURE 23.1 Sample biplot of hypothetical values for δ^{13}C versus δ^{15}N for various components of a typical stream food web. *Arrows* indicate hypothesized trophic transfers based on a fractionation of approximately 3.4‰ for N and 0.4‰ for C for each trophic level.

predatory insects, but are slightly more enriched in ^{15}N because they include predatory insects in their diet as well as the other insect groups. Fine particulate organic matter (FPOM) is derived from both algal and detrital components and has a δ^{13}C value that is intermediate between algae and detritus (Fig. 23.1). The system has about four trophic levels. While this example is oversimplified for illustrative purposes, it is not very different from what we see in many stream ecosystems (Fry, 1991). In real ecosystems, there may be more than two organic matter sources and many consumers are likely to have more generalized (mixed) diets, leading to less clear isotopic separation and more ambiguity in interpretation of organic matter transfer through the food web. Another potential problem with distinguishing consumer food resources using δ^{13}C is that methane-derived carbon may also become incorporated into food webs, via consumer consumption of methane-oxidizing bacteria; methane δ^{13}C in aquatic sediment (e.g., −50‰ to −110‰) is very negative compared to photosynthetically derived carbon (Whiticar, 1999; Jones and Gray, 2011; see Fig. 23.2C). Such limitations of natural abundance

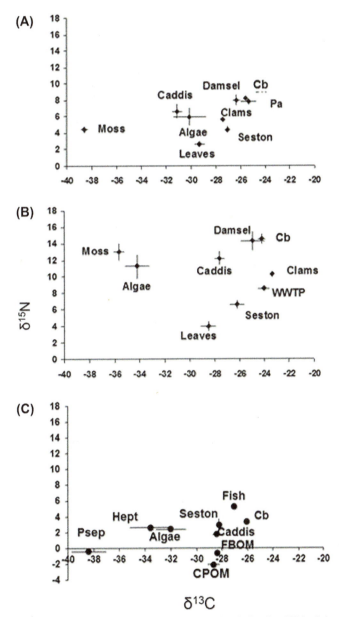

FIGURE 23.2 Isotope biplots of dominant food web components at three sites in two North Carolina, USA, piedmont streams: (A) urban North Buffalo Creek site not impacted by point source inputs; (B) North Buffalo Creek site ∼5 km downstream of a wastewater treatment plant; (C) Talbot's Branch, a relatively pristine forested stream. Comparison between biplots A and B illustrates that most components in North Buffalo Creek show shifts in both δ^{15}N and δ^{13}C indicating incorporation of sewage-derived N and C into the food web (from Ulseth and Hershey, 2005). Comparison of A and B with Talbot's Branch food web (from Rushforth, 2006) illustrates that the nonurban forested streams show no evidence of anthropogenic N (note compressed δ^{15}N axis). Very negative δ^{13}C value of Psephenid beetles (Psep) in Talbot's Branch compared to particulate carbon sources illustrate incorporation of methane-derived carbon via methane-oxidizing bacteria. *Caddis*, Hydropsychidae caddisflies; *Damsel*, damselflies; *clams*, Asian clams; *Cb*, *Cambarus* crayfish, *Pa*, *Procambarus* crayfish; *Psep*, Psephenidae water penny beetle larvae; *Fish*, *Notropis* shiners. *Modified from Ulseth and Hershey (2005) and Rushforth (2006).*

isotope studies emphasize the importance of having additional information from gut contents, feeding studies, morphological analyses, stable isotope data for other elements, and/or tracer experiments. Generally, an explicit hypothesis will govern whether or what types of supporting data or experiments would be of most value.

23.1.2 Overview of Basic and Advanced Methods

In this chapter, our objectives are to illustrate a generalized method (*Basic Method*) for characterizing and comparing stream food webs using $\delta^{13}C$ and $\delta^{15}N$ and three more advanced experimental approaches (*Advanced Methods 1−3*) for examining specific questions or hypotheses about resource use in stream food webs using $\delta^{13}C$, $\delta^{15}N$, and δD. Collectively, these approaches will allow you to (1) identify the principle sources of organic matter for a stream; (2) assign consumers or consumer groups to trophic levels within the food web; (3) identify specific consumer food sources; (4) determine the influence of anthropogenic factors on consumer resource use; (5) evaluate impact of flow or geomorphic factors on basal resource stable isotope signatures; and (6) evaluate incorporation of novel resources or changes in resource availability.

23.2 GENERAL DESIGN

As discussed throughout this book, the economy of a stream is influenced by a combination of allochthonous and autochthonous resources utilized by a community of consumers, superimposed on a geomorphic and land use template and constrained by hydrologic conditions. We provide a few examples of how we can use stable isotope analyses to explore these various influences on stream food webs by comparing consumer resources and resource use under contrasting conditions that influence the resource quality or quantity. Although the examples relate to particular questions, the approaches are easily generalizable and can be readily modified to address other questions or compare other stream types than those described.

23.2.1 General Site Selection Considerations

For the methods presented below that investigate stream food webs using stable isotope analyses, it makes the most sense to choose sites for which prior information is available. Food web studies in such streams may be more meaningful than in streams not previously studied because such studies would complement existing data, contribute to an overall improved understanding of the ecosystem, and potentially reveal the need for additional studies to improve that understanding. Knowledge of the taxonomy and life history of the organisms at your sites would facilitate development of a hypothesized food web based on perceived food sources and functional feeding groups (Merritt et al., 2008) that could be compared with the food web developed using stable isotope data. If comprehensive taxonomic and functional feeding group information is available, it would be useful to construct a conventional food web diagram for your sites that includes four trophic levels. Be sure to represent major organic matter resources, dominant consumers at each trophic level and functional feeding group, and linkages between food resource and consumer components. Examples of stream food web diagrams are provided in Cummins and Klug (1979), Power (1990), and Benke and Wallace (1997).

The success of a food web study using natural abundance of stable isotopes depends upon finding a site or sites where components of the food web have enough differences in the stable isotope or isotopes of interest to address the question being asked. A preliminary isotopic analysis of a few samples of a few dominant consumers can help guide the site selection process. There is also considerable information in the literature regarding variation in stable isotope signatures of resources (see Table 23.1 for some examples) and consumers. But seemingly similar ecosystems may be poised very differently with respect to isotope values due to variations in local biogeochemistry. At the very least, you should do a pilot investigation to identify the dominant consumers and resources at your prospective sites to inform your selection of consumers to use for your stable isotope investigation. Identifying consumers to genus or species level, if possible, is preferable since even closely related species sometimes have quite different diets. The number of consumers or consumer groups to be included will depend on what is present at your sites as well as availability of time and funding. You will learn more by including more consumers, but at a minimum you should be sure to include the potential basal resources that are present in your stream (e.g., leaf litter, periphyton, FPOM, filamentous algae, bryophytes), representative primary consumers (suspected grazers, collectors and shredders, see Chapter 20), and suspected predator trophic levels. If you include vertebrates, most of which are predators, you must have the necessary collecting permits. Identify the consumers you have chosen to the lowest practical taxonomic level. Be sure to include any organic matter sources from your sites that are likely to have distinct $\delta^{15}N$ or $\delta^{13}C$ values.

23.2.2 General Procedures

23.2.2.1 Influence of Urbanization on Food Webs

Urban development can have significant impacts on the food webs and trophic dynamics of streams (see Paul and Meyer, 2001; Walsh et al., 2005). One of the potential impacts of urbanization on stream food webs is the introduction of anthropogenically derived N and C (McClelland and Valiela, 1998; Ulseth and Hershey, 2005; Northington and Hershey, 2006). These elements may come from point and nonpoint source inputs (e.g., wastewater treatment and industrial effluents, pet waste, sewage leaks, storm water runoff, which carries fertilizer and other contaminants, etc.). Since anthropogenic sources of N and, to a lesser extent, C often have stable isotopic signatures that are distinct from natural sources, they can be detected in food webs using $\delta^{15}N$ and $\delta^{13}C$ (Fig. 23.2 A and B). It is not unusual for $\delta^{15}N$ to be enriched in food web components by 10‰ or so due to wastewater input compared to more pristine streams, but seston or fine benthic organic matter (FBOM) may also be enriched in ^{13}C (comparison of Fig. 23.2B and C). Point sources, such as wastewater effluent, often provide strong isotope signatures, but nonpoint and small point source inputs generally also produce detectable isotopic changes in stream food web components. As a result, stream food webs from urban streams, among many other changes (Walsh et al., 2005), are often isotopically distinct from food webs in forested or other natural stream ecosystems (e.g., Ulseth and Hershey, 2005; Rushforth, 2006, Fig. 23.2A−C).

23.2.2.2 Gradients in $\delta^{13}C$ and $\delta^{15}N$ in Basal Resources

Methods using $\delta^{13}C$ and $\delta^{15}N$ have been the most widely used natural abundance stable isotope tracers of stream food web studies over the past ∼35 years. Natural variations in isotope ratios of potential food sources have been used to quantify diet sources and energy flow pathways, and trophic enrichment in ^{15}N has been used to estimate trophic position and food chain length. These applications require information about the range and variability of stable isotopes at the base of food webs. Large differences in stable isotope values of basal resources arising from natural processes or human actions can be used to examine energy flow through streams food webs. Differences in water velocity, carbon or nutrient sources, primary production, or mixing of chemically distinct water masses may produce stable isotope variation that can be used as intrinsic tracers of diet sources. Pool−riffle sequences, headwater springs, tributary junctions, estuarine transition zones, lake outlets, and locations of point source pollution may produce useful variation in basal resources that can be leveraged to examine consumer resource use over a variety of spatial scales. As noted above, inputs of sewage to a stream can provide a natural label of productivity that may be traced into food webs downstream of the sewage point source input, often for many kilometers downstream (e.g., Ulseth and Hershey, 2005). At a more local scale, variation in water velocity may provide a natural label of algal production from adjacent pool and riffle habitats (e.g., Finlay et al., 1999, Fig. 23.3). Some other examples of useful natural abundance variation are provided in Table 23.1. The presence of such variation can

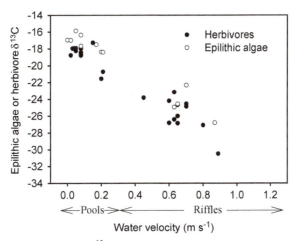

FIGURE 23.3 Relationship of epilithic algal and herbivore $\delta^{13}C$ to water velocity during midsummer in the South Fork Eel River, an open canopied river in northern California, USA. Herbivore taxa sampled were primarily Ephemeroptera and Trichoptera except in fast water velocities where blepharicerid larvae were also collected. Unlike other riffle herbivores, blepharicerids graze on rock tops in fast-flowing water. Each point represents a composited sample of epilithic algae or herbivores within a 1-m^2 patch with similar water velocity within a pool or riffle. *Figure modified from Finlay et al. (2002).*

sometimes be predicted from relatively simple measurements and observations such as water velocity, stream size, and pH. Some preliminary sampling is almost always necessary for verification and establishment of a specific study design. A small pilot study can determine if useful isotopic tracers occur at the selected site. A pilot study can also help identify the tracers that are most likely to be helpful to address the questions of interest, minimizing costs through identification of the combination of stable isotopes most likely to yield the necessary isotopic separation between sources in space and time.

23.2.2.3 Use of δD in Stream Food Web Studies

Stable isotope ratios of hydrogen (δD) can be used to investigate hydrological, biochemical, and ecological relationships in streams. In combination with $\delta^{18}O$, δD can be used to establish home ranges or sources of aquatic organisms (West et al., 2006; Solomon et al., 2009; Myers et al., 2012) due to regional and latitudinal differences in isotopic composition of the water in which they developed (Ehlerginger and Rundel, 1989; Gat, 1996; Rubenstein and Hobson, 2004). Differences in physiological processes drive major differences in D accumulation in tissues of primary producers. Freshwater algae have lower δD relative to terrestrial plants due to lipid biosynthesis processes (Zhang and Sachs, 2007). Terrestrial plants, on the other hand, become more D enriched due to the multiple fractionation events involved in photosynthesis (Fogel and Cifuentes, 1993) and the specific pathway used (CAM > C4 > C3; reviewed in Marshall et al., 2007). Building structural components in terrestrial plants (e.g., leaves, branches) leads to additional increases in δD (Dawson et al., 2002).

Differences in δD of allochthonous and autochthonous basal resources have been noted in both lentic (Cole et al., 2011; Solomon et al., 2011; Karlsson et al., 2012) and lotic (Doucett et al., 2007; Finlay et al., 2010; Dekar et al., 2012) ecosystems. In stream food webs spanning a wide geographic range, autochthonous resources are consistently more depleted (~50‰–100‰) compared to terrestrial sources (Doucett et al., 2007; Finlay et al., 2010; Dekar et al., 2012). In some cases, δD differences among groups of autochthonous primary producers (e.g., diatoms, filamentous algae, and cyanobacteria) are significant and clearly reflected in consumers (Finlay et al., 2010). Terrestrial resource groups also tend to be distinct, with grasses generally having lower δD values than tree leaves (Doucett et al., 2007; Dekar et al., 2012). Nontraditional basal resources such as methane can also be differentiated using δD as well as $\delta^{13}C$ (e.g., Kohzu et al., 2004; Deines et al., 2009; Hershey et al., 2015). These features of δD make it an appealing alternative to $\delta^{13}C$ in studies of resource utilization, as the variation in and overlap of $\delta^{13}C$ in allochthonous and autochthonous basal resources often make source partitioning difficult (Ishikawa et al., 2013).

Understanding ecological interactions in the face of climate change will be an important avenue of research for future studies using δD. Given the importance of δD in understanding hydrological dynamics (Gat, 1996), regions undergoing drought conditions will likely demonstrate very enriched values of δD due to evaporative concentration and lack of precipitation recharge while wetter regions will likely become even more depleted by greater dilution of the δD signal by isotopically lighter forms of precipitation (Gat, 1996). Anthropogenic changes in landscapes around streams and rivers brought about by agriculture or urbanization will be reflected in δD of waters (Kendall and Coplen, 2001; Hibbs et al., 2012). In turn, these landscape changes will influence the balance of terrestrial and aquatic basal resources (Bartels et al., 2012; Griffiths et al., 2013) and types of aquatic consumers (Stenroth et al., 2015) comprising the stream food web.

23.2.2.4 Use of Mixing Models to Quantify Food Web Transfers

When a consumer has an intermediate isotope value between two sources, a simple two-source *mixing model* can be constructed to calculate the relative contribution of each source to consumer diets (e.g., Kline et al., 1990). An equation for a simple two-source mixing model for $\delta^{15}N$ of a consumer feeding on a mixture of algae and detritus, corrected for a trophic shift of 3.4, is as follows:

$$\delta^{15}N_{consumer} = \left[s\delta^{15}N_{algae} + (1-s)\delta^{15}N_{detritus} \right] + 3.4 \tag{23.2}$$

where s = proportion of the diet derived from an algal source and $(1-s)$ is the proportion derived from a detrital source. This can be rearranged to calculate the proportion of the diet derived from algae:

$$s = \left(\delta^{15}N_{consumer} - \delta^{15}N_{detritus} - 3.4 \right) / \left(\delta^{15}N_{algae} - \delta^{15}N_{detritus} \right) \tag{23.3}$$

Commonly, there are multiple basal sources for consumers in a food web such that a two-source mixing model is not suitable. Phillips and Koch (2002) developed a mixing model that evaluates contributions of three sources when signals for two isotopes are available, which also accounts for differences in proportional contribution of each isotope. Their model can also be generalized to $n+1$ sources, when δ values for n isotopes are available. In addition, for cases with $>n+1$ sources, Phillips and Gregg (2003) have developed a method (IsoSource) which gives the range of possible contributions from the different source materials for which isotopic data are available. More recently Bayesian approaches to stable isotope mixing models, including an application in R (SIAR model, Parnell et al., 2010), provide probability distributions of multiple source contributions (see reviews Parnell et al., 2013; Phillips et al., 2014). Although these approaches are becoming widely used and provide appealing solutions, like all models, they depend on the quality of the data, the question(s) being asked, and the underlying assumptions. It is important to remember that diet composition varies over time and space, and trophic fractionation is often unknown or variable within a consumer group or with resource quality; these and other factors need to be considered in model application and interpretation.

23.3 SPECIFIC METHODS

23.3.1 Basic Method: Comparison of Stream Food Webs Using Stable Isotopes of C and N

This exercise is designed to compare food webs in two streams or stream reaches, one of which is influenced by urbanization. The underlying principle, discussed above, is that a shift in the isotopic signature of stream food web components will result from changes in N or C sources having unique stable isotopic signatures. Two options for study of urbanization are investigated in this *Basic Method*: (1) comparison of food webs from an urban and a forested stream on the same landscape and (2) evaluation of incorporation of effluent from an urban wastewater treatment plant in a stream food web. For option 1, we might hypothesize that urbanization shifts the invertebrate consumer community away from terrestrial leaf litter resources toward algal resources, reflecting increased nutrient loading and reduced canopy cover. For option 2, we might hypothesize that sewage effluent represents a novel and bioavailable resource that has a strong influence on stream food webs. Note that the method outlined is easily generalizable to alternative stream food web comparisons that might be of local interest.

23.3.1.1 Site Selection

For option 1, choose mid-order urban and forested stream sites that are accessible and have a representative range of stream habitats for your region. Select a study reach (or multiple reaches) which contains distinct habitat types, such as pool and riffle units. Ideally, you should take samples from multiple (minimum n = 3) habitats of each type, but number of habitats and samples of each type will be dictated by the specific questions of interest, as well as resources and time. For option 2, choose sites on an urban stream that are upstream and downstream of a wastewater treatment plant. You might use the same urban stream site as in option 1 for your upstream site. Depending on the size of the city and the stream receiving its effluent, the signature of the plant will continue to be seen for several kilometers downstream. For example, a wastewater treatment plant draining the northern half of Greensboro, NC, USA (population 279,639; www.google.com/webhp?sourceid=chrome-instant&ion=1&espv=2&ie=UTF-8#q=city+of+greensboro+nc+population&*), discharges into a stream to a base flow discharge of ~ 0.3 m³/s. The δ^{15}N and δ^{13}C signatures of the wastewater treatment plant are clearly apparent ~ 7 stream km downstream (Ulseth and Hershey, 2005). The δ^{13}C signal of the plant attenuates to background but the δ^{15}N signal is still evident ~ 14 km downstream (Bullard and Hershey, 2013). Accordingly, your site does not have to be immediately below the plant. See Section 23.2.1 above for discussion of preliminary data that you should collect before initiating your stable isotope study, as described below.

1. Collect samples of food resources and consumers. In each representative stream habitat, sample available primary producer groups (see Chapters 11−13), detrital coarse particulate organic matter (CPOM) and FPOM (including benthic and suspended) components (see Chapters 25−26), invertebrate macroconsumers or groups that you selected based on your preliminary studies (see *General Site Selection Considerations*), and, if permits are in place, fish (see Chapter 16). Sort consumers in the field, placing groups or taxa in separate containers, then store on ice for transport to the laboratory.
2. A few representatives from each consumer taxon should be preserved in ethanol for taxonomic verification.
3. Prepare samples of components for stable isotope analyses, and send to a university or commercial laboratory for δ^{13}C and δ^{15}N analysis (unless your own institution has the capability to analyze stable isotopes). Look at the website for the laboratory you are using for specific instructions on sample preparation. Some general guidelines are as follows:
   a. Sample mass required for stable isotope analyses depends on the elemental mass concentration for the stable isotope ratio of interest. As noted above, you should review the sample preparation instructions for the lab where you are sending your samples. As a general guideline you should expect to send a dry mass of about 1 mg, 2−3 mg, and 4−6 mg for consumer, algae, and detritus samples, respectively, for combined δ^{15}N and δ^{13}C analyses. Filter water for suspended FPOM (seston) onto glass fiber filters, clogging the filter during the filtering process to increase sample mass relative to filter mass, then trim the edges (that contain no sample material). After trimming, you

can usually send all of a 25-mm filter or half of 47-mm filter, although specific labs may vary in their requirements for filter samples. Animals should be held in filtered stream water in a cool place for at least several hours, or overnight, to help clear their guts (the contents might bias the isotopic signal). You can start this step by filtering the stream water you use to transport your animals through a syringe filter. Following gut clearance, crustaceans, snails, or bivalves should be removed from carbonate shells (be sure to remove all of the shell material!) because carbonate, which is not assimilated into the food web, has a highly enriched $\delta^{13}C$ value compared to animal body tissue that will distort your data. Depending on the size of the consumer, you likely will need to pool a few to several individuals of a species for a single analysis to have enough mass. For larger organisms, it is possible to analyze specific tissues, which often differ within a single organism due to fractionation during metabolism (see Peterson and Fry, 1987). However, in this exercise we will use either whole body analysis or muscle tissue (for fish or crayfish). For each consumer or resource, you should try to process a minimum of three replicates. You may have resources for more replication, but you may also want to consider additional consumers. In any case, the value of more replication should be weighed against cost, as well as the risk of depleting the local abundance of the consumers of interest in your reach (especially important for small streams where you plan to sample repeatedly through time).

b. Place the clean samples in microfuge tubes and dry in an oven at 60°C, or use a freeze drier, if available. Dried samples can be stored indefinitely in microfuge tubes once they are dried and snapped closed. Alternatively, samples can be frozen for later dissection and drying. Ethanol-preserved or formalin-preserved samples can be used, but the preservation may result in some alteration of both C and N δ-values (Hobson et al., 1997; Arrington and Winemiller, 2001; Heidrun and Gray, 2003).

c. Pack dry samples into individual tin capsules (4×6 mm or 9×10 mm) for stable isotope analysis and arrange in 96-well trays. Always wear gloves to avoid contaminating samples. For samples involving pooling of several individual consumers, several animals can be packed whole into a single preweighed capsule, then weighed. Dried samples such as leaf fragments and larger consumers should be ground to a powder, weighed, then packaged in preweighed tin capsules. Samples should be thoroughly homogenized because different parts of a leaf fragment or animal tissues (if not using whole animals) may vary in isotopic composition by several parts per thousand. Tin capsules should be arranged into well plates with sample ID information on a spreadsheet that must be submitted with the samples. Again, the stable isotope lab website will have instructions on sample packaging and shipping as well as spreadsheet requirements. If you plan to do serious work, send a known blind sample as a standard with each shipment to increase your confidence in the resulting data.

4. Your data will arrive as a spreadsheet from the isotope laboratory. In addition to the $\delta^{13}C$ and $\delta^{15}N$ values, the spreadsheet will contain the C/N ratio of your samples, along with other information and any comments. Plot the data for $\delta^{13}C$ and $\delta^{15}N$ values of each consumer along with the respective standard errors or standard deviations to look for the range of values and the pattern of grouping of $\delta^{13}C$ and $\delta^{15}N$ values (see Fry, 1991 for examples). A rule of thumb is that values that are different by 1‰ or less should not be considered different unless you have enough true field replicates (not multiple subsamples from the same pooled sample) to compute an accurate variance. While the precision of laboratory analyses may be better than 0.1‰, the variability in most ecosystem components and in sampling/processing is almost always larger (Fry, 1991; see Fig. 23.2).

5. Using the $\delta^{13}C$ and $\delta^{15}N$ values of your basal resource and consumer samples, construct food web diagrams in the form of biplots (see Fig. 23.2) for each stream or reach. What can you learn about urban influence (or other contrast you chose) on your food web from comparing the two biplots? How do your biplots compare with the conventional food web diagram depicting energy flow based on functional feeding groups or other information? Do the stable isotopic distributions support your functional feeding group classifications? Compare the $\delta^{13}C$ and $\delta^{15}N$ values between reaches with paired t-tests, assuming that you sampled similar components at each site. Calculate the proportion of urban or wastewater N or C in consumer diets using one or more of the mixing model approaches described (see Section 23.2.2.4).

23.3.2 Advanced Method 1: Tracing Use of a Novel Terrestrial Particulate Organic Matter Source in a Stream Food Web

This exercise will investigate use of a novel particulate organic matter (POM) source, corn detritus, in a stream food web by replacing or supplementing naturally occurring accumulations of terrestrial detritus, much of which originates from C_3 plants, with senescent corn stalks. Corn is a C_4 plant, similar to many riparian grasses and is isotopically heavy ($\delta^{13}C \sim -14$ to $-10‰$) compared to C_3 plants ($^{13}C \sim -28‰$ to $-26‰$), such as most trees and some grasses (discussed in Section 23.1). The approach described later labels the components of the ecosystem most heavily dependent on terrestrial litter because those components will incorporate the $\delta^{13}C$ signature of a C_4 plant. Corn stalks are readily available during fall in areas that are near an agricultural region that grows corn. The method is designed to be used in a low order stream reach or reaches with debris dams of a size that may be suitable for manipulation. Such sites often have organic resources (i.e., tree leaves, algae) with $\delta^{13}C$ values that are highly ^{13}C depleted compared to C_4 plants. The experiment involves the following steps:

1. Obtain the upper meter of corn stalks that include most or all of the leaves from up to 25 senescent corn plants from a local farmer or your garden. There should be multiple stems per plant. Precondition them for several days in an aerated trough or flow-through system in the laboratory to initiate microbial conditioning (Cummins, 1973) of the corn detritus.
2. Select a stream reach or multiple reaches where you can identify at least 10 small debris dams and 10 riffle areas which do not contain debris dams, and mark these along the bank with flagging tape or stakes. For each pair of debris dams, randomly select (e.g., coin toss) one member of the pair to serve as a reference dam and the other to serve as the treatment dam.
3. Collect supporting data on stream geometry, temperature, substrate characteristics, canopy cover, and discharge.
4. Collect initial samples of periphyton, CPOM, FPOM, and consumers for isotope analyses as in the *Basic Method* (above), being sure to represent the functional diversity of primary consumers and predators available from each of the debris dams and riffle areas. Prepare these samples for $\delta^{13}C$ stable isotope analyses as soon as you can after you set up your experiment (steps 5–6).
5. Prepare five bundles of corn stalks by wrapping five plants together with strong cord or securing with cable ties. For very small or larger streams, you might want to use fewer or more stalks in each bundle. Tie them in a few places near the central portion of the stalk segments, but be sure to allow the ends of the bundles to remain loose for water flow and consumer habitat.
6. Remove the five debris dams (or the litter and FBOM associated with woody debris dams) that are designated as the treatment dams and replace them with corn stalk bundles, staking them in place with rebar or other secure mechanism, or attaching them to large, well anchored woody debris. Be sure to retain replicate samples of corn stalk material for stable isotope analysis. A variation of the study design is also described (see *Advanced Method 1 Alternate*) involving addition of supplemental corn stalk debris dams.
7. Collect weekly samples of a few consumers from each pair of the debris dams and from the corresponding riffle areas, taking care to minimize disturbance of the entire dam. In the treatment dams, do this carefully by clipping part of one corn stalk from each treatment dam and place the sample in a sorting pan. In the reference dams, carefully remove a handful of leaves from each dam and place it in the pan with water. Save some of the CPOM from each sample for stable isotope analyses. Sort samples in the field to be sure that you have enough material for isotope analyses. Also collect samples of FPOM from debris dam samples by filtering FPOM that collects in your sorting pan through a glass fiber filter. If necessary to obtain enough mass or enough consumers, place a fine mesh net in the field around part the bundle or debris dam or immediately downstream and gently shake to dislodge FPOM and consumers, taking care not to disturb the entire dam. Weekly sampling will serve to help you evaluate the rate that consumers shift toward isotopic equilibrium during the experiment and will also serve as a safeguard so that you will have some data in the unfortunate (but all too common) possibility that the experiment gets washed out in a flood! Check security of corn debris dams at the weekly sampling visits and adjust as necessary to reduce the possibility that they will be dislodged.
8. Terminate the experiment after 4–6 weeks. The decision about how long the experiment should run might be site specific, depending on length of the growing season or other constraints on field work. Resample food resources and consumers from control and corn litter debris dams and riffle areas as in initial sampling (step 4), collecting the same components if possible. Prepare these samples for stable isotope analyses as before and send your samples to the isotope lab that you are using.
9. Examine and analyze your data. For each resource and consumer, graph $\delta^{13}C$ response to weekly sampling. Did any of the consumers show a shift toward the $\delta^{13}C$ of corn detritus, and had they reached isotopic equilibrium? For each consumer or consumer group, compare the final $\delta^{13}C$ values in riffle, control debris dams, and corn debris dams graphically by plotting mean $\delta^{13}C$ ($\pm SD$, $\pm SE$, or $\pm 95\%$ CI), then analyze your data with a one-way ANOVA [response variable $=\delta^{13}C$ for each consumer or consumer group, treatment = habitat (riffle, control debris dam, and corn debris dam habitats]. For consumers that showed significant incorporation of corn litter, calculate the proportion of corn litter in their diet using one or more of the mixing model approaches described (see Section 23.2.2.4). Did FPOM collected from the corn debris dams show a $\delta^{13}C$ shift relative to that associated with control debris dams? Did the corn detritus shift its $\delta^{13}C$ value with conditioning?

Alternate experimental design approach: Instead of removing debris dams, supplemental debris dams can be added to the stream by staking the bundles in secure locations along the edge of the channel (see also Chapter 26). Be sure to place them where they will remain wet under base flow conditions.

23.3.3 Advanced Method 2: Tracing Spatial Variation in Diet Sources for Food Webs

This method uses spatial differences in algal $\delta^{13}C$ in response to water velocity as a natural label of algal productivity in adjacent or nearby habitat patches in a pool–riffle sequence (Fig. 23.3). Slow diffusion of CO_2 through water creates conditions where discrimination against $^{13}CO_2$ is reduced (a physiological effect) or where expression of reduced discrimination is not realized because much of the available CO_2 is used (a physical effect) (Finlay, 2004). Because water velocity determines the extent of diffusive boundary layer conditions near substrates; autotrophic $\delta^{13}C$ may be negatively affected by water velocity; labeling $\delta^{13}C$ of stream production and local consumers in ways that can be used to ask questions about the effects of consumer and resource movement, diet sources, and trophic interactions in stream food webs (Table 23.1). Settings with strong contrasts in flow, moderate to high demand for CO_2, and stable flow conditions are conducive to strong $\delta^{13}C$ contrasts between pool and riffle production. Other ideas for situations that are likely to generate useful stable isotope variation in an ecosystem are presented in Table 23.1 and may be substituted here with appropriate modifications. This method involves the following steps:

1. Identify useful natural variation in algal (or biofilm) $\delta^{13}C$ in response to water velocity in a pool–riffle sequence. Conditions of stable flows for >2 weeks will be helpful to reduce variation, and more productive streams may show greater differentiation with water velocity. Measurement of flow velocity is required for the method (below) and is useful for planning purposes. Flow can be measured 5 cm above the substrate with a standard water velocity meter (e.g., propeller or magnetic types). More detailed information can be obtained with an acoustic doppler velocimeter, which may allow more detailed characterization of fluid flow conditions in the near bed environment, at the cost of slower data collection.
2. Preliminary sampling will likely be highly informative for design of the study. Sample grazers or benthic algae along a water velocity gradient in an adjacent pool–riffle sequence using the *Basic Method*. Sampling at 5–10 locations that vary in water velocity (minimum range of 0.3 m/s) may be sufficient to provide information on study design. Once these data are available, examine the relationships between $\delta^{13}C$ and water velocity with bivariate regression analyses. Is there a relationship between flow and $\delta^{13}C$?
3. Using information gained from your preliminary data, select 10–15 points to sample for algae or biofilms, benthic invertebrates, and vertebrates (if proper permits are available). At each point, measure flow and collect cobbles or other hard substrate into a wide sorting pan. Collect periphyton by removing macroscopic biomass to a container or scraping epilithic algae from surfaces into a small pan. The sample can be washed into a 50-cc centrifuge tube for processing in the lab. Sort and select invertebrate taxa based on abundance and functional feeding group for stable isotope analyses. If possible, include representatives of common functional feeding groups at each point. Composite at least two individuals for a single sample of large bodied taxa (e.g., late instar caddisflies), and 5–20 individuals for smaller taxa (e.g., baetid mayflies and black flies). Note the location of any grazers on cobbles (e.g., rock surface, side, or bottom). Other potential resources such as conditioned leaves or macrophytes may also be important to sample where they occur; it is unlikely that they will occur along the flow gradient, so a smaller number of samples will likely be adequate.
4. Process your algal and invertebrate samples according to procedures described in the *Basic Method* section. It may be necessary to examine algal samples via microscopy to confirm that your sample is mostly algal in nature. Procedures exist for purifying samples with significant presence of other materials (detritus, bacteria); such methods require special equipment (see Hamilton et al., 2005).
5. Prepare samples for stable isotope analyses. If necessary carbonates may be removed from dried samples of algae or organic matter using the acid fumigation technique described by Harris et al. (2001). This method avoids use of acid washing to remove carbonates which may alter isotope values (e.g., Bunn et al., 1995).
6. Examine and analyze your data. For each resource and consumer, graph $\delta^{13}C$ as a function of water velocity. Do algae and scraper $\delta^{13}C$ values decrease with increasing velocity? Do other consumers respond similarly? Do filter feeding insect larvae appear to be using the same resources as grazers or collector gatherers? If substantial $\delta^{13}C$ variation exists, it will be informative to create a biplot of $\delta^{13}C$ (x-axis) and $\delta^{15}N$ (y-axis) identifying resources and consumers in pools and riffles with different symbols. Are predators tracking variation in local benthic resources or are they integrating resources over larger spatial gradients? If adequate isotopic separation exists, use mixing models to estimate contribution of resources to consumer diets as described above. If algal $\delta^{13}C$ varies across a wide range, algal values may

overlap with terrestrial $\delta^{13}C$ values, so it may be helpful to examine other isotopes. If other stable isotopes were measured, create similar graphs with these data, and using a more complex (3+ source) mixing model.

7. Stable isotopes may vary with season or flow conditions. If possible, evaluate whether isotopic values are stable over time by repeating sampling at a subset of points on a biweekly basis.

23.3.4 Advanced Method 3: Tracing Shifts in Basal Resources of Stream Food Webs Using δD

This exercise examines stream food web dynamics in response to basal resources derived from either terrestrial leaf matter or autochthonous production by algae. This experiment simulates how resources (and food webs) can change as heavily canopied streams reliant upon allochthonous inputs transition toward open-canopy, autochthonous systems. Here, the experimental treatments (i.e., leaves) applied will act as a proxy for darkened streams in heavily forested catchments. By establishing a gradient of terrestrial input, we could hypothesize that changes in light penetration translate to altered basal resources (Fig. 23.4) with isotopic incorporation of the altered resources by multiple levels of consumer. In this exercise, the investigator would be able to examine food web structure as a response to proportional changes in basal resources over time. Using leaves such as maple would release highly colored leachate into the water, changing the optical properties of the water, and thereby affecting biological interactions and food web components. Altering the light environment within the experimental units will be a primary control on the development of algal resources for the food webs in the experiment. We suggest using δD isotopes to examine food web components, given its ability to consistently differentiate between allochthonous and autochthonous resources in a variety of stream ecosystems (e.g., Doucett et al., 2007; Finlay et al., 2010).

This experiment would best be carried out in recirculating artificial stream channels, where each channel would act as an independent experimental unit (see Chapter 19). To establish a series of experimental streams with differing optical properties, we suggest using different masses of leaves in each channel to shift the resource contribution from mostly allochthonous to autochthonous, while creating a gradient of change in the color of water in each treatment. In this way, investigators can mimic more natural stream conditions. The different leaf masses for a given treatment could be set up as categories related to the color of the resulting treatment (e.g., dark, moderate, clear), which would be based on the total mass of leaves added to each channel. Enough stream channels should be constructed to have adequate replication for each treatment level. We suggest at least four replicates of each leaf/color treatment. This method uses the following steps:

1. Leaf packs should be constructed with dried leaves as described in Chapter 27, using onion bags to allow for organisms to colonize. For small, drain-pipe style recirculating channels (described below), we suggest the following treatments: high color (highest leaf mass), moderate color (intermediate leaf mass), and no color (no leaves). Investigators may also choose to include other treatment levels (using different masses) to establish a more varied gradient of light availability if resources are available. Be sure to make enough leaf packs for all treatments and replicates. A subset of dried leaves should be ground and processed for initial δD using techniques for dry sample preparation described in the *Basic Method*.

2. Select a source of water and substrates from a stream nearby for establishing the experimental channels. Take samples of the supply water for determination of δD, which will be important for isotopic food web analysis later (Stenroth et al., 2015).

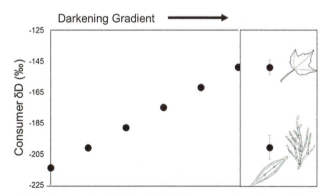

FIGURE 23.4 Possible results from a darkening experiment that would alter the resource base for aquatic consumers. Allochthonous (e.g., leaf) and autochthonous (e.g., algal) end member δD are represented to the right, and predicted consumer isotopic values relative to the darkening treatments are presented to the left. The treatments will become darker (due to higher leaf mass) as you move from the left to the right of the x-axis. *The basal resource data used are based on Doucett et al. (2007), Finlay et al. (2010), and Dekar et al. (2012).*

3. We suggest using an experimental stream setup similar to those described by Kominoski et al. (2015), where arrays of aluminum stream channels ($0.15 \times 0.15 \times 4$ m) run at a continuous discharge (e.g., 0.1 L/s) to maintain a constant water depth (e.g., 0.1 m). In this type of channel array, treatments may be randomly spaced, and still independent from one another statistically. For a system such as this, we suggest the following leaf masses for each treatment: high color: 20 g maple leaves, moderate color: 10 g maple leaves, and no color: 0 g maple leaves. The masses of leaves suggested for the three treatments would be an ideal mass: volume ratio to elicit the distinct color changes inherent to this experimental design. These masses can be adjusted based on the type of artificial stream setup available, taking care to maintain a similar water volume/leaf mass ratio and similar water depth.

4. Add leaf packs to the channels in the quantities necessary for each darkening treatment. Place unglazed tiles in the experimental streams as collection substrates for algal growth. These tiles will be collected at the end of the experiment in all treatments, scraped, and processed for δD. Leave the channels open at the top in order for insect colonization to occur throughout the course of the experiment.

5. Once the streams have been set up and the leaf packs added, the experiment should be allowed to run for at least one month, although this time may be adjusted based on the specific experimental questions and the time of the year in which the experiment is conducted. Establishment of treatment water coloration should occur within the first 24 h of the experiment due to the rapid leaching of labile leaves such as maple. Basic stream measurements should be taken over the course of the experiment, including temperature, pH, dissolved oxygen, and conductivity. Light measurements should be taken in all of the treatments using a light meter to verify differences in experimental conditions.

6. At the completion of the experiment, all food web components should be collected from each replicate treatment stream. For each stream channel, invertebrates should be identified and individuals pooled for sufficient mass for δD isotope analysis. Invertebrates could then be grouped by trophic level, where primary consumers and predators for each treatment are identified and processed for δD. Solid samples (e.g., invertebrates, algal scrapings, leaf material) should be dried prior to grinding and processing for the isotope laboratory, as described above in the *Basic Method* except that the samples would be analyzed for δD. Water from each stream should also be collected to represent the final δD for each treatment stream. Liquid samples should be collected and sent for analysis based on protocols defined by the isotope lab that you have chosen.

23.3.5 Data Analysis

1. For all treatments and replicates, δD values returned from the isotope lab should be sorted by data types such as consumers, basal resources, and treatment water. Due to δD as a component of water molecules, all stream consumer isotope data should then be corrected for the treatment water from which it was collected:

$$\delta D_{corr} = \frac{(\delta D_{cons} - (\omega \times \delta D_{water}))}{(1 - \omega)} \tag{23.4}$$

where the corrected value of δD_{corr} is related to the consumer isotopic content (δD_{cons}), corrected for water (δD_{water}) and the dietary contribution of hydrogen ($\omega = 0.185$, Stenroth et al., 2015).

2. Mixing models (see above) may be employed to determine relative contributions from terrestrial versus aquatic resources along the gradient of light availability (or leaf quantity). These proportional changes may then be examined statistically using one-way ANOVA if the color levels are treated as categories (high, moderate, low). If leaf masses are used as the independent variable, analyze your data using regression analysis.

3. Consumer δD may be plotted to visually demonstrate isotopic shifts resulting from the treatments established at the beginning of the experiment (e.g., Fig. 23.4).

23.4 QUESTIONS

1. Consider the food web diagrams you constructed in the *Basic Method*. Do your food web diagrams suggest differences in food web structure or resource use between your streams or stream reaches? Do your stable isotope studies lead you to additional questions or hypotheses about food web structure or dynamics?

2. Do dominant consumers show considerable overlap in resource use or minimal overlap? Do the same consumer or consumers shift resources between streams or stream reaches?

3. Does the food web you constructed using stable isotope data differ from those that might be constructed using other methods?

5. Did stable isotope biplots constructed in the *Basic Method* reveal any point or nonpoint sources of ^{15}N or ^{13}C, and if so, to what extent? Did the anthropogenic source affect both δ^{13}C and δ^{15}N? Design an experiment using stable isotopes to further explore the importance of the anthropogenic source to the food web.

6. If you conducted *Advanced Method 1*, consider that corn is foreign to the system and may not cycle like riparian leaf litter. Do your data provide any insight into that issue? Explain.

7. If you used *Advanced Method 2*, or measured water velocity when you sampled for the *Basic Method*, did you observe any flow sensitivity of δ^{13}C for any food web components? If not, why might this be the case?

8. If you conducted *Advanced Method 3*, how well did δD differentiate between the basal resources in your different treatments? Were there significant changes in the isotopic signatures of primary consumers and predators across your treatments? Did the different treatments change the structure of your stream food webs? How much did the δD of the treatment water influence the signatures of your different food web components?

23.5 MATERIALS AND SUPPLIES

General (see also Chapters 12 and 15 for algae and invertebrate protocols, respectively)
   Field supplies for collecting/field sorting aquatic insects and basal resources
   Laboratory supplies and taxonomic keys for identifying consumers or consumer groups
   Glass fiber filters with appropriate filter funnel for filtering seston (SPOM) from water samples
   Filter forceps
All isotope studies
   Drying oven or freeze dryer
   Glass vials (scintillation vials work well)
   Microfuge tubes
   4×6 mm or 9×10 mm tin capsules
   96-well plates for small tin capsules, 24- or 48-well plates for larger capsules
   Stream samples of organic matter components and macroconsumers
   Glass rods to grind up invertebrates in microfuge tubes
   Grinder (e.g., coffee grinder or tissue grinder) for coarse organic matter
Additional supplies for specific methods
   Advanced Method 1
      Corn stalks
      Rebar (if using)
      Twine to secure corn stalks (if not using rebar)
   Advanced Method 2
      Current meter
   Advanced Method 3
      Recirculating streams
      Leaf packs
      Unglazed ceramic tiles
      Light meter
      Multiprobe sonde or handheld probes to measure physicochemical stream variables (temperature, pH, DO, conductivity)

REFERENCES

Anderson, C., Cabana, G., 2005. δ^{15}N in riverine food webs: effects of N inputs from agricultural watersheds. Canadian Journal of Fisheries and Aquatic Sciences 62, 333−340.

Arrington, D.A., Winemiller, K.O., 2001. Preservation effects on stable isotope analysis of fish muscle. Transactions of the American Fisheries Society 131, 337−342.

Bartels, P., Cucherousset, J., Steger, K., Eklov, P., Tranvik, L.J., Hillebrand, H., 2012. Reciprocal subsidies between freshwater and terrestrial ecosystems structure consumer resource dynamics. Ecology 93, 1173−1182.

Bastow, J.L., Sabo, J.L., Finlay, J.C., Power, M.E., 2002. A basal aquatic-terrestrial trophic link in rivers: algal subsidies via shore-dwelling grasshoppers. Oecologia 131, 261−268.

Benke, A.C., Wallace, B.J., 1997. Trophic basis of production among riverine caddisflies: implications for food web analysis. Ecology 78, 1132−1145.

Bilby, R.E., Fransen, B.R., Bisson, P.A., 1996. Incorporation of nitrogen and carbon from spawning coho salmon into the trophic system of small streams: evidence from stable isotopes. Canadian Journal of Fisheries and Aquatic Sciences 53, 164–173.

Bullard, A.E., Hershey, A.E., 2013. Impact of *Corbicula fluminea* (Asian clam) on seston in an urban stream receiving wastewater effluent. Freshwater Science 32, 976–990.

Bunn, S.E., Loneragan, N.R., Kempster, M.A., 1995. Effects of acid washing on stable isotope ratios of C and N in penaeid shrimp and seagrass: implications for food-web studies using multiple stable isotopes. Limnology and Oceanography 40, 622–625.

Chaloner, D.T., Martin, K.M., Wipfli, M.S., Ostrom, P.H., Lamberti, G.A., 2002. Marine carbon and nitrogen in southeastern Alaska stream food webs: evidence from artificial and natural streams. Canadian Journal of Fisheries and Aquatic Sciences 59, 1257–1265.

Cole, J.J., Carpenter, S.R., Kitchell, J., Pace, M.L., Solomon, C.T., Weidel, B., 2011. Strong evidence for terrestrial support of zooplankton in small lakes based on stable isotopes of carbon, nitrogen, and hydrogen. Proceedings of the National Academy of Sciences of the United States of America 108, 1975–1980.

Cummins, K.W., 1973. Trophic relations of aquatic insects. Annual Review of Entomology 18, 183–206.

Cummins, K.W., Klug, M.J., 1979. Feeding ecology of stream invertebrates. Annual Review of Ecology and Systematics 10, 147–172.

Dawson, T.E., Mambelli, S., Plamboeck, A.H., Templer, P.H., Tu, K.P., 2002. Stable isotopes in plant ecology. Annual Review of Ecology and Systematics 33, 507–559.

deBruyn, A.M.H., Rasmussen, J.B., 2002. Quantifying assimilation of sewage-derived organic matter by riverine benthos. Ecological Applications 12, 511–520.

Deines, P., Wooller, M.J., Grey, J., 2009. Unravelling complexities in benthic food webs using a dual stable isotope (hydrogen and carbon) approach. Freshwater Biology 54, 2243–2251.

Dekar, M.P., King, R.S., Back, J.A., Whigham, D.F., Walker, C.M., 2012. Allochthonous inputs from grass-dominated wetlands support juvenile salmonids in headwater streams: evidence from stable isotopes of carbon, hydrogen, and nitrogen. Freshwater Science 31, 121–132.

Delong, M.D., Thorp, J.H., 2006. Significance of instream autotrophs in trophic dynamics of the Upper Mississippi River. Oecologia 147, 76–85.

Derda, M., Chmielewski, A.G., Licki, J., 2006. Stable isotopes of sulphur in investigating pollution sources. Environment Protection Engineering 32, 63–68.

Doucett, R.R., Marks, J.C., Blinn, D.W., Caron, M., Hungate, B.A., 2007. Measuring terrestrial subsidies in aquatic food webs using stable isotopes of hydrogen. Ecology 88, 1587–1592.

Ehlerginger, J.R., Rundel, P.W., 1989. Stable isotopes: history, units, and instrumentation. In: Rundel, P.W., Ehleringer, J.R., Nagy, K.A. (Eds.), Stable Isotopes in Ecological Research. Springer-Verlag, New York, NY, pp. 1–15.

Finlay, J.C., Power, M.E., Cabana, G., 1999. Effects of water velocity on algal carbon isotope ratios: implications for river food web studies. Limnology and Oceanography 44, 1198–1203.

Finlay, J.C., 2001. Stable-carbon-isotope ratios of river biota: implications for energy flow in lotic food webs. Ecology 82, 1052–1064.

Finlay, J.C., Khandwala, S., Power, M.E., 2002. Spatial scales of carbon flow in a river food web. Ecology 83, 1845–1859.

Finlay, J.C., 2004. Patterns and controls of lotic algal stable carbon isotope ratios. Limnology and Oceanography 49, 850–861.

Finlay, J.C., Kendall, C., 2007. Stable isotope tracing of organic matter sources and food web interactions in watersheds. In: Lajtha, K., Michener, R. (Eds.), Stable Isotopes in Ecology and Environmental Science. Blackwell, Oxford, UK, pp. 283–333.

Finlay, J.C., Doucett, R.R., McNeely, C.M., 2010. Tracing energy flow in stream food webs using stable isotopes of hydrogen. Freshwater Biology 55, 941–951.

Fisher-Wold, A.K., Hershey, A.E., 1999. The effects of salmon carcass-derived nutrients on periphyton growth, aufwuchs growth, and wood decomposition. Canadian Journal of Fisheries and Aquatic Sciences 18, 2–16.

Fogel, M.L., Cifuentes, L.A., 1993. Isotope fractionation during primary production. In: Engel, M.H., Macko, S.A. (Eds.), Organic Geochemistry. Plenum Press, New York, NY, pp. 73–98.

Fry, B., 1991. Stable isotope diagrams of freshwater food webs. Ecology 72, 2293–2297.

Fry, B., 1999. Using stable isotopes to monitor watershed influences on aquatic trophodynamics. Canadian Journal of Fisheries and Aquatic Sciences 56, 2167–2171.

Fry, B., 2006. Stable Isotope Ecology. Springer, New York, NY.

Gat, J.R., 1996. Oxygen and hydrogen isotopes in the hydrologic cycle. Annual Review of Earth and Planetary Sciences 24, 225–262.

Griffiths, N.A., Tank, J.L., Royer, T.V., Roley, S.S., Rosi-Marshall, E.J., Whiles, M.R., Beaulieu, J.J., Johnson, L.T., 2013. Agricultural land use alters the seasonality and magnitude of stream metabolism. Limnology and Oceanography 58, 1513–1529.

Hamilton, S.K., Sippel, S.J., Bunn, S.E., 2005. Separation of algae from detritus for stable isotope or ecological stoichiometry studies using density fractionation in colloidal silica. Limnology and Oceanography: Methods 3, 149–157.

Harris, D., Horwáth, W.R., van Kessel, C., 2001. Acid fumigation of soils to remove carbonates prior to total organic carbon or CARBON-13 isotopic analysis. Soil Science Society of America Journal 65, 1853–1856.

Heidrun, F., Gray, J., 2003. Effect of preparation and preservation procedures on carbon and nitrogen stable isotope determinations from zooplankton. Rapid Communications in Mass Spectrometry 17, 2605–2610.

Hershey, A.E., Northington, R.M., Hart-Smith, J., Bostick, M., Whalen, S.C., 2015. Methane efflux and oxidation, and use of methane-derived carbon by larval Chironomini, in arctic lake sediments. Limnology and Oceanography 60, 276–285.

Hibbs, B.J., Hu, W., Ridgway, R., 2012. Origin of stream flows at the wildlands-urban interface, Santa Monica Mountains, California, U.S.A. Environmental and Engineering Geoscience 18, 51–64.

Hill, W.R., Fanta, S.E., Roberts, B.J., 2008. C-13 dynamics in benthic algae: effects of light, phosphorus, and biomass development. Limnology and Oceanography 53, 1217−1226.

Hobson, K.A., Gibbs, H.L., Gloutney, M.L., 1997. Preservation of blood and tissue samples for stable-carbon and stable-nitrogen isotope analysis. Canadian Journal of Zoology 75, 1720−1723.

Ishikawa, N.F., Doi, H., Finlay, J.C., 2012. Global meta-analysis for controlling factors on carbon stable isotope ratios of lotic periphyton. Oecologia 170, 541−549.

Ishikawa, N.F., Hyodo, F., Tayasu, I., 2013. Use of carbon-13 and carbon-14 natural abundance for stream food web studies. Ecological Research 28, 759−769.

Jones, R.I., Gray, J., 2011. Biogenic methane in freshwater food webs. Freshwater Biology 56, 213−229.

Karlsson, J., Berggren, M., Ask, J., Bystrom, P., Jonsson, A., Laudon, H., Jansson, M., 2012. Terrestrial organic matter support of lake food webs: evidence from lake metabolism and stable isotopes of consumers. Limnology and Oceanography 57, 1042−1048.

Karr, J.D., Showers, W.J., Jennings, G.D., 2003. Low-level nitrate export from confined dairy farming detected in North Carolina streams using delta N-15. Agriculture Ecosystems and Environment 95, 103−110.

Kato, C., Iwata, T., Wada, E., 2004. Prey use by web-building spiders: stable isotope analyses of trophic flow at a forest-stream ecotone. Ecological Research 19, 633−643.

Kendall, C., Coplen, T.B., 2001. Distribution of oxygen-18 and deuterium in river waters across the United States. Hydrological Processes 15, 1363−1393.

Kennedy, T.A., Finlay, J.C., Hobbie, S.E., 2005. Eradication of invasive *Tamarix ramosissima* along a desert stream increases native fish density. Ecological Applications 15, 2072−2083.

Kline, T.C., Goering, J.J., Mathisen, O.A., Poe, P.H., 1990. Recycling of elements transported upstream by runs of pacific salmon: 1. ^{15}N and ^{13}C evidence in Sashin Creek, southeastern Alaska. Canadian Journal of Fisheries and Aquatic Sciences 47, 136−144.

Kohzu, A., Kato, C., Iwata, T., Kishi, D., Murakami, M., Nakano, S., Wada, E., 2004. Stream food web fueled by methane-derived carbon. Aquatic Microbial Ecology 36, 189−194.

Kominoski, J.S., Rosemond, A.D., Benstead, J.P., Gulis, V., Maerz, J.C., Manning, D.W.P., 2015. Low-to-moderate nitrogen and phosphorous concentrations accelerate microbially driven litter breakdown rates. Ecological Applications 25, 856−865.

MacAvoy, S.E., Macko, S.A., McInich, S.P., Garman, G.C., 2000. Marine nutrient contributions to freshwater apex predators. Oecologia 122, 568−573.

Marshall, J.D., Brooks, J.R., Lajtha, K., 2007. Sources of variation in the stable isotopic composition of plants. In: Michener, R., Lajtha, K. (Eds.), Stable Isotopes in Ecology and Environmental Science, second ed. Blackwell Publishing, Malden, MA, pp. 22−60.

Mathison, O.S., Parker, P.L., Goering, J.J., Kline, T.C., Poe, P.H., Scalan, R.S., 1988. Recycling of marine elements transported into freshwater by anadromous salmon. Verhandlungen der Internationalen Vereinigung für Theoretische und Angewandte Limnologie 23, 2249−2258.

McArthur, J.V., Moorhead, K.K., 1996. Characterization of riparian species and stream detritus using multiple stable isotopes. Oecologia 107, 232−238.

McClelland, J.W., Valiela, I., 1998. Linking nitrogen in estuarine producers to land-derived sources. Limnology and Oceanography 43, 577−585.

McCutchan, J.H., Lewis, W.M., Kendall, C., McGrath, C.C., 2003. Variation in trophic shift for stable isotope ratios of carbon, nitrogen, and sulfur. Oikos 102, 378−390.

Merritt, R.W., Cummins, K.W., Berg, M.B., 2008. An Introduction to the Aquatic Insects of North America, fourth ed. Kendall/Hunt Publishers, Dubuque, IA.

Minagawa, M., Wada, E., 1984. Stepwise enrichment of ^{15}N along food chains: further evidence and the relation between δ^{15}N and animal age. Geochimica et Cosmochimica Acta 48, 1135−1140.

Mulholland, P.J., Helton, A.M., Poole, G.C., Hall, R.O., Hamilton, S.K., Peterson, B.J., Tank, J.L., Ashkenas, L.R., Cooper, L.W., Dahm, C.N., Dodds, W.K., Findlay, S.E.G., Gregory, S.V., Grimm, N.B., Johnson, S.L., McDowell, W.H., Meyer, J.L., Valett, H.M., Webster, J.R., Arango, C.P., Beaulieu, J.J., Bernot, M.J., Burgin, A.J., Crenshaw, C.L., Johnson, L.T., Niederlehner, B.R., O'Brien, J.M., Potter, J.D., Sheibley, R.W., Sobota, D.J., Thomas, S.M., 2008. Stream denitrification across biomes and its response to anthropogenic nitrate loading. Nature 452, 202−205.

Myers, D.J., Whitledge, G.W., Whiles, M.R., 2012. Evaluation of δD and δ^{18}O as natural markers of invertebrate source environment and dispersal in the middle Mississippi River-floodplain ecosystem. River Research and Applications 28, 135−142.

Northington, R.M., Hershey, A.E., 2006. Effects of stream restoration and wastewater treatment plant effluent on fish communities in urban streams. Freshwater Biology 51, 1959−1973.

Parnell, A.C., Inger, R., Bearhop, S., Jackson, A.L., 2010. Source partitioning using stable isotopes: coping with too much variation. PLoS One 5, e9672.

Parnell, A.C., Phillips, D.L., Bearhop, S., Semmens, B.X., Ward, E.J., Moore, J.W., Jackson, A.L., Grey, J., Kelly, D.J., Inger, R., 2013. Bayesian stable isotope mixing models. Environmetrics 24, 387−399.

Paul, M.J., Meyer, J.L., 2001. Streams in the urban landscape. Annual Review of Ecology and Systematics 32, 333−365.

Peckarsky, B.L., 1982. Aquatic insect predator-prey relations. BioScience 32, 261−266.

Peterson, B.J., Fry, B., 1987. Stable isotopes in ecosystem studies. Annual Review of Ecology and Systematics 18, 293−320.

Peterson, B.J., Wollheim, W.M., Mulholland, P.J., Webster, J.R., Meyer, J.L., Tank, J.L., Marti, E., Bowden, W.B., Valett, H.M., Hershey, A.E., McDowell, W.B., Dodds, W.K., Hamilton, S.K., Gregory, S., D'Angelo, D.J., 2001. Control of nitrogen export from watersheds by headwater streams. Science 292, 86−90.

Phillips, D.L., Gregg, J.E., 2003. Source partitioning using stable isotopes: coping with too many sources. Ecosystems 136, 261−269.

Phillips, D.L., Koch, P.L., 2002. Incorporating concentration dependence in stable isotope mixing models. Oecologia 130, 114−125.

Phillips, D.L., Inger, R., Bearhop, S., Jackson, A.L., Moore, J.W., Parnell, A.C., Semmens, B.X., Ward, E.J., 2014. Best practices for use of stable isotope mixing models in food-web studies. Canadian Journal of Zoology 92, 823–835.

Post, D.M., 2002. Using stable isotopes to estimate trophic position: models, methods, and assumptions. Ecology 83, 703–718.

Power, M.E., 1990. Effects of fish in river food webs. Science 250, 811–814.

Rubenstein, D.R., Hobson, K.A., 2004. From birds to butterflies: animal movement patterns and stable isotopes. Trends in Ecology and Evolution 19, 256–263.

Rushforth, H.M., 2006. Nitrogen Uptake, Food Web Nutrient Transfer and Community Structure in a Restored Urban Stream Compared to a Pristine and an Unrestored Urban Stream (M.S. thesis). University of North Carolina at Greensboro.

Schuldt, J.A., Hershey, A.E., 1995. Impact of salmon carcass decomposition on Lake Superior tributary streams. Journal of the North American Benthological Society 14, 259–268.

Solomon, C.T., Carpenter, S.R., Clayton, M.K., Cole, J.J., Coloso, J.J., Pace, M.L., Vander Zanden, M.J., Weidel, B.C., 2011. Terrestrial, benthic, and pelagic resource use in lakes: results from a three-isotope Bayesian mixing model. Ecology 92, 1115–1125.

Solomon, C.T., Cole, J.J., Doucett, R.R., Pace, M.L., Preston, N.D., Smith, L.E., Weidel, B.C., 2009. The influence of environmental water on the hydrogen isotope ratio in aquatic consumers. Oecologia 161, 313–324.

Stenroth, K., Polvi, L.E., Faltstrom, E., Jonsson, M., 2015. Land-use effects on terrestrial consumers through changed size structure of aquatic insects. Freshwater Biology 60, 136–149.

Sulzman, E.W., 2007. Stable isotope chemistry and measurement: a primer. In: Michener, R., Lajtha, K. (Eds.), Stable Isotopes in Ecology and Environmental Science, second ed. Blackwell Publishing, Malden, MA, pp. 1–21.

Townsend, C.R., Thompson, R.M., McIntosh, A.R., Kilroy, C., Edwards, E., Scarsbrook, M.R., 1998. Disturbance, resource supply, and food-web architecture in streams. Ecology Letters 1, 200–209.

Trudeau, W., Rasmussen, J.B., 2003. The effect of water velocity on stable carbon and nitrogen isotope signatures of periphyton. Limnology and Oceanography 48, 2194–2199.

Ulseth, A.J., Hershey, A.E., 2005. Stable isotope natural abundances trace anthropogenic nitrogen and carbon in an urban stream. Journal of the North American Benthological Society 24, 270–289.

Vander Zanden, M.J., Rasmussen, J.B., 1999. Primary consumer $\delta^{13}C$ and $\delta^{15}N$ and the trophic position of aquatic consumers. Ecology 80, 1395–1404.

Walsh, C.J., Roy, A.H., Feminella, J.W., Cottingham, P.D., Groffman, P.M., Morgan II, R.P., 2005. The urban stream syndrome: current knowledge and the search for a cure. Journal of the North American Benthological Society 24, 706–723.

Wayland, M., Hobson, K.A., 2001. Stable carbon, nitrogen, and sulfur isotope ratios in riparian food webs on rivers receiving sewage and pulp-mill effluents. Canadian Journal of Zoology 79, 5–15.

West, J.B., Bowen, G.J., Cerling, T.E., Ehleringer, J.R., 2006. Stable isotopes as one of nature's ecological recorders. Trends in Ecology and Evolution 21, 408–414.

Whiticar, M.J., 1999. Carbon and hydrogen isotope systematics of bacterial formation and oxidation of methane. Chemical Geology 161, 291–314.

Wipfli, M.S., Hudson, J.P., Caouette, J.P., Chaloner, D.T., 2003. Marine subsidies in freshwater ecosystems: salmon carcasses increase the growth rates of stream-resident salmonids. Transactions of the American Fisheries Society 132, 371–381.

Zhang, Z., Sachs, J.P., 2007. Hydrogen isotope fractionation in freshwater algae: I. variations among lipids and species. Organic Geochemistry 38, 582–608.

Chapter 24

Dissolved Organic Matter

Stuart E.G. Findlay[1] and Thomas B. Parr[2]

[1]Cary Institute of Ecosystem Studies; [2]Oklahoma Biological Survey, University of Oklahoma

24.1 INTRODUCTION

Dissolved organic matter (DOM) is a central component of aquatic ecosystem structure and function (Lindeman, 1942) providing energy and critical building blocks for organismal growth and function. It consists of the elements C, H, O, N, P, and S configured into millions of different organic molecules and occurs in every compartment of the hydrologic cycle from oceans and rain to surface water and groundwater. Its quantity and composition reflects specific biotic and abiotic ecosystem functions and their relative rates (e.g., gross primary productivity and ecosystem respiration). The stoichiometric characteristics (e.g., molar ratio of C:N:P) of the DOM pool vary broadly (e.g., C:N ratios ranging from 10 to 50; Kaushal and Lewis, 2003) and depend on the source of DOM (see also Chapter 36).

An ever-proliferating vocabulary is used to describe the different types of DOM. When communicating your research to the general public, many people are familiar with the term "tannins" (a type of humic acid) and the brown color they confer on water. Researchers commonly refer to the major DOM components in stream water as "humic" and "fulvic" materials. These are broad classes of organic acids containing aromatic rings (Thurman, 1985) typically derived from microbial degradation of vascular plant material. Identifiable monomers, such as simple sugars and amino acids, are readily detectable with current techniques (Kaplan and Newbold, 2003) and although present in low concentrations may be particularly important in supporting heterotrophic activity (see Chapters 9 and 10). More complex carbohydrates, peptides, proteins, and nucleic acids are all present, but in highly variable concentrations. Modern perspectives on DOM increasingly avoid the terms "humic" and "fulvic," but have yet to replace them with more descriptive terms. New terms are continually evolving out of studies investigating the molecular structure and functional activity of DOM at different resolutions (e.g., Cory and McKnight, 2005; Stubbins et al., 2014).

In stream ecosystems, DOM is often the largest proportion of the total organic matter standing stock and dominates the carbon flux downstream during hydrologic low (or base) flows (Webster and Meyer, 1997). Typically, its abundance is expressed as a concentration of an organic element of interest. For example, the concentration of DOC in headwater streams ranges from 1 to 5 mg/L of carbon, while streams draining high-organic soils or vegetated wetlands may have concentrations of 20 or even 50 mg/L (Fig. 24.1A). Standing stock of DOC in headwater streams, assuming an average stream depth of 0.5 m, translates into 0.5–2.5 g of C/m^2 in headwater streams and 10–25 g of C/m^2 in wetland streams. Typically, this organic matter standing stock is small relative to particulate matter in debris dams, but may be large relative to fine particles in transport or algal biomass. In fact, the ratio of DOC to particulate organic carbon (POC) in transport is almost always >1 (Meybeck, 1982) and in many systems, DOC is 95% of the total organic carbon (DOC + POC) in baseflow transport. Indeed, the quantity of DOC passing a given point in a stream is typically several fold greater than for POC, except under the temporally pulsed high flow conditions where POC can dominate.

It is important to keep in mind that, while DOM includes only ~6 different elements in > trace quantities (C, H, O, N, P, and S), these elements are arranged into a staggering array of configurations spanning a range of molecular sizes. The most common definition of DOM is the organic material that passes through a given filter size—typically this is 0.45 μm, but depending on the study intent, may range from 0.2 to 0.7 μm, and thus, to some degree is researcher defined. This DOM can be split into the dissolved organic form of each element, typically DOC (carbon), DON (nitrogen) and sometimes DOP (phosphorus). However, for most systems, carbon dominates the dissolved organic pool (>50%) and its

Methods in Stream Ecology. http://dx.doi.org/10.1016/B978-0-12-813047-6.00002-4

FIGURE 24.1 Examples of dissolved organic matter in (A) stream water samples from seven third- to fifth-order streams in the Upper Peninsula of Michigan, USA, ranging in dissolved organic matter concentration from ~5 to 40 mg C/L and (B) leachates of fallen leaves from five different tree species found in the upper Midwest, USA. *(A) Photo by J. Larson. (B) Photo by E. Strauss.*

measurement is easily performed and analytically reliable. As such we here, and in the literature, refer interchangeably between DOM and DOC. Because measurement of DOM is derived from biomass, it contains carbohydrates, aromatics, lipids, amino acids, nucleotides, and other compounds derived from growth and decomposition processes. The stoichiometric ratios of these elements or the dominance of a particular molecular constituent can be linked to rates of both heterotrophic metabolism and nutrient transformation (see Chapters 9, 32–34, and 36).

DOM derives from both autochthonous (within the stream; e.g., algae, macrophytes) and from allochthonous (outside the stream; e.g., soils, wetlands, fallen leaves). The contributions from these sources can follow a variety of natural cycles. At the daily time scale, Kaplan and Bott (1989) demonstrated that diel fluctuation in White Clay Creek (Pennsylvania, USA) DOC concentration characterized by higher DOC during the day as compared to night was driven by exudates from autotrophic (or autochthonous) production during the day. At longer time scales, annual input of leaves or reconnection of disconnected soils (e.g., soils of small wetlands and seeps) may also increase stream water DOC. DOM derived from autochthonous autotrophic production at a variety of time scales may be a critical source of C in ecosystems either because autochthonous supply dominates the pool via less direct mechanisms such as "priming" (Hotchkiss et al., 2014). These autochthonous sources of DOC may be the more biologically available portion despite occurring at a lower concentration than the larger, and arguably more stable pool of terrestrially derived DOM. Compounds released from living plants and fresh litter are generally of lower molecular weight (MW) and have fewer aromatic moieties and thus more metabolically active (bioavailable or labile) than compounds leached from soils and aged litter (McKnight et al., 2001). On the other hand, "aged" terrestrial DOM can also support metabolism (McCallister and del Giorgio, 2012); although likely dependent on different pathways of breakdown. Thus, DOM origin and its links to composition have a strong effect on decomposition rates, residence times, and other aspects of DOC physicochemistry.

Earlier views on the bioavailability of DOM tended toward the opinion that because DOM was exported in stream water, most of the compounds must be refractory to biotic degradation in a sense representing the "leftovers" from decomposition in terrestrial systems (cf. Kaplan et al., 1980). More recently it has become clear that the compositional complexity of DOM also represents a continuum of lability with degradation rates from minutes to years (Cory and Kaplan, 2012; Koehler et al., 2012) and will be removed/consumed/metabolized within streams and rivers given a sufficient opportunity to interact with biotic (benthic and hyporheic biofilms; see Chapters 8−11) and abiotic (photodegradation, sorption) mechanisms (Battin et al., 2008) and processes such as sorption on mineral surfaces or in biofilms (Freeman and Lock, 1995). Sorption of DOM to microbial biofilms may well be the first step in its eventual metabolism and can serve as a scavenging and buffering mechanism enhancing retention of DOM in a stream reach. Deposition on mineral surfaces is particularly important during movement of DOM through soil profiles, certain minerals have a great potential for removal and protection of DOM (McCracken et al., 2002) leading to changes in concentration and composition.

Aside from biotic assimilation within sediments, DOM is also susceptible to direct photomineralization and alteration in bioavailability following exposure to full sunlight (Wiegner and Seitzinger, 2001). The energy in solar radiation interacts particularly strongly with certain classes of compounds (e.g., complex phenolics) resulting in direct release of CO_2, small organic molecules, and often a decrease in light-absorbing capability (i.e., photobleaching) (Moran and Zepp, 1997). These chemical transformations are often, but not always, accompanied by an increase in the ability of a DOM mixture to support bacterial growth, presumably due to "priming" of complex compounds for biotic metabolism or release of specific, directly assimilable compounds (Tranvik and Bertilsson, 1999). Photolytic effects on DOM in streams and rivers are diverse with some reports of strong positive effects of light exposure on DOM degradation with others showing neutral or negative effects (e.g., Findlay et al., 2001; Wiegner and Seitzinger, 2001; Brisco and Ziegler, 2004; Cory et al., 2014).

As the flux of DOM is often the largest component in stream material budgets and represents a complex pool of millions of molecules, one of the challenges facing stream ecologists has been detecting relatively small rates of DOM addition or removal given a large and typically variable background flux. Investigators seeking to disentangle this challenge have found interpretable shifts in DOM amount and composition along a stream's length apparently connected to changes in both sources and sinks throughout the drainage network (Vannote et al., 1980; Creed et al., 2015; Mineau et al., 2016).

Many studies have shown the importance of organic and/or poorly drained soils in a catchment contributing to stream DOC loads. For example, the annual flux of DOC can be predicted reasonably well by the quantity of wetlands in a catchment (e.g., Dillon and Molot, 1997). Soil properties often contribute substantially to relationships between land cover and stream DOM yield (Aitkenhead-Peterson et al., 2003). On smaller temporal or spatial scales, short duration hydrologic events (e.g., snowmelt, storms) can lead to near-surface flow through high organic soil layers with a disproportionate influence on annual DOC export. For instance, it has been estimated that a large proportion of the annual yield of DOC from Rocky Mountain streams occurs during a brief period in the early spring (Brooks et al., 1999). Thus, DOM concentrations can vary dramatically over the span of a few hours (Pacific et al., 2010), and one of the difficulties in constructing mass balances is quantifying episodic export particularly when trying to balance these fluxes against in situ retention or metabolism process discussed previously.

The methods and analyses used in measuring and understanding DOM processes through quantity and composition have evolved over time (Sharp et al., 2002). Bulk methods that look at the elemental composition, UV/Vis optical absorbance, or fluorescence are widely used. Increasingly, researchers are investigating the ecological effects of natural and anthropogenic processes on DOM composition at near-molecular levels using advanced instrumental methods. Characterization of individual molecular formulae can be obtained from Fourier transform ion cyclotron resonance mass spectrometry, but does not provide definitive structural information. Characterization of the relative abundance of specific structures (e.g., aromatic rings or carboxylic acid groups) can be achieved through nuclear magnetic resonance (NMR) spectroscopy (e.g., ^{13}C NMR; cf. Brown et al., 2004). Such methods bring us closer to ecometabolomics or understanding ecosystem and individual organism level metabolic processes recorded by DOM composition. However, a complete analysis of the elemental and structural composition of DOM, down to the level of single molecules, has never been completely described for any streams or other ecosystems.

From an anthropocentric point of view, DOM abundance and characteristics affect both contaminant transport and drinking water quality. DOM interacts with organic contaminants acting as binding sites for some less soluble organic compounds or interaction with metals (e.g., Chin, 2003; McKnight et al., 1992). Methyl mercury concentrations in an array of surface waters were positively correlated with bulk DOC (Dennis et al., 2005). The quantity of DOM is also relevant for drinking water supplies since excess chlorination of high DOM waters can result in the formation of halogenated compounds with carcinogenic properties (Rook, 1977; Rodriguez and Sérodes, 2001).

The methods described in this chapter will be useful to students and researchers seeking to understand biotic and abiotic ecosystem functions and their feedbacks on DOM pools and flows. Changes in the composition, quality, and quantity of DOM in stream ecosystems across temporal and spatial scales may be used to identify and quantify changes in stream ecosystem structure or function produced by natural phenomena or anthropogenic disturbances. Such approaches may represent a next generation approach to ecosystem analysis and management.

24.2 GENERAL DESIGN

24.2.1 Site Selection and Study Design

Broadly, studies employing DOM analysis in stream ecology are focused on understanding how ecosystem structure and function create and respond to variability over space and time in the quantity, composition, production, and degradation of DOM. Study designs typically consist of end-member characterization, before-after-control-impact (BACI), factorial designs, and gradients. A study focused on understanding characteristics of different sources of DOM to a stream may be designed to sample all the potential sources (end members) in a watershed. In contrast, a gradient study may forego a detailed, watershed level understanding and instead seek to describe variability across watersheds or streams with different treatments (e.g., land use change, effects of nutrients, climatic gradients). When an anthropogenic or natural perturbation is anticipated, a BACI experiment may be particularly useful. However, the duration and frequency of before and after sampling may be critical in capturing changes in long- and short-term dynamics—2–3 years of pre- and postsampling may not be adequate to characterize the true impacts. A factorial design may be employed in the field or laboratory to understand mechanisms producing and transforming DOM by comparing similarly sized watersheds differing in the characteristic of interest, manipulate conditions in similar watersheds, and manipulate hypothesized mechanisms affecting DOM in laboratory mesocosms. So, in short, the selection of sites and sampling frequency will be question dependent. Below, we describe some of the more common approaches spanning a range in both complexity and the type of information generated.

24.2.2 Sample Collection and Analysis

As a functional definition, DOM ranges from truly dissolved to almost colloidal material. The most widely accepted pore size for DOM is defined as 0.45 μm, but a variety of filter materials and a range of filter sizes are used to collect DOM samples. A range of pore sizes from 0.2 to 0.7 μm are available for plastic filters [e.g., nylon, nitrocellulose, polytetrafluoroethylene (PTFE)], while most glass fiber filters (GFFs) are only available in 0.7 μm or greater (smaller pore sizes are available but less common). Prior to use, residual organics should be removed from GFFs by combusting them at ∼500°C, but plastic synthetic filters can only be rinsed or leached. Because of this, GFFs with a nominal pore size of 0.7 μm are most widely used.

Many methods are available for filtering DOM, which may be broadly categorized as positive and negative pressure methods. The lowest cost options for positive pressure filtration are typically syringe-driven reusable filter holders. Negative pressure (vacuum) filtration can be performed in the field using a vacuum hand pump or in the lab using a vacuum system. It is often desirable to use vacuum filtration for samples requiring filtration to 0.22 μm due to the amount of time it requires. Vacuum filtration may also degas the sample lowering the O_2 concentration. If using these samples for biotic assays later, it is advisable to check the O_2 in the sample and reaerate it by shaking. Similarly, syringe filtration may infuse additional gasses into the sample if the syringe contains headspace.

The bottles that samples are collected in should be organic free. Acid-washed and DI-rinsed amber glass bottles combusted at 500–550°C are preferred. In addition, 40-mL amber glass "VOC vials" with PTFE-lined silicon septum caps are often desirable as these can be placed directly into many DOC and N analyzers. Septa may be cleaned with hot persulfate. High density polyethylene bottles may be cleaned by acid washing (10% v/v HCl) and DI rinsing, although these still may leach plastic (i.e., DOC). Amber-colored glass bottles are recommended to reduce photodegradation of the DOM or photosynthetic production of new DOM if samples are unfiltered.

Samples can be preserved with refrigeration over the short term (e.g., hours to days). While freezing inorganic nutrients for longer term storage (i.e., months) is common, freezing DOM samples can lead to irreversible precipitation of less soluble compounds and alter optical properties (Fellman et al., 2008). Alternatively, a variety of preservatives (e.g., acidification, mercuric chloride) can be used for longer term storage (Kaplan, 1994), but they may interfere with subsequent optical and compound specific analyses.

Numerous instruments are available for carbon concentration analysis, and most are automated. Several of these instruments will concurrently analyze nitrogen. The primary methods of C and N analysis involve high temperature or pressure and a catalyst that converts organic and inorganic C to CO_2 and N to NO, which is subsequently measured by infrared and chemiluminescence detectors, respectively. DOC can be measured directly by first acidifying samples and purging the inorganic C as CO_2 or it may be measured indirectly as the difference between total and inorganic C (see Findlay et al., 2010). Similarly, DON and DOP can be obtained by subtraction of inorganic from total dissolved nitrogen or total dissolved phosphorus (see Chapters 32 and 33).

The composition of DOM is complex, and while progress has been made toward a complete chemical characterization of freshwater organic constituents, it is neither practical nor necessary for most ecological studies. Given the analytical complexity, a host of coarser-resolution characterization schemes have been used and shown to offer ecologically relevant information about DOM sources and dynamics. For bulk DOM, many of the constituents are defined functionally; for instance, fulvic and humic acids are separated based on solubility at low pH (Thurman, 1985). Several selective resins, which target the humic acid fraction, have been used for both characterization and preparative isolation. Gross MW distribution can be determined with a series of nominal MW cut-off filters (Leff and Meyer, 1991). Optical analyses based on UV/Vis absorption and fluorescence can provide further information on composition.

Because of their relative ease of use and wide availability, these optical properties have become popular in understanding DOM dynamics in stream ecosystems. The distinction between the two is reasonably simple. Most, but not all, molecules contain structures which absorb UV/Vis light at particular wavelengths. A fraction of these molecules contain structures such as aromatic rings or heteroatoms (e.g., N, O, or S replacing a C in a ring) in organic acids and proteins that reemit light at longer wavelengths. This is known as fluorescence. UV/Vis indices such as the specific ultraviolet absorbance at 254 nm ($SUVA_{254}$), the slope of the UV/Vis spectrum (S_e), and the slope ratios (S_r) in different portions of the spectrum have been related to the structure and MW of DOM (Weishaar et al., 2003), photodegradation (Helms et al., 2008), and other processes. Indices from fluorescence measurements are developed from ratios of fluorescence intensities (I) at specific wavelengths (λ). The fluorescence index (FI) indicates terrestrial vascular plant or more microbial sources (McKnight et al., 2001), the humification index is an indicator of the degree of aromatic structure (Ohno, 2002), and the biological index indicates the degree of recent autochthonous production (Huguet et al., 2009). Fluorescence approaches also produce excitation emission matrices (EEMs), which are collections of all the emission responses across a range of excitation wavelengths for a sample (i.e., several thousand points). Parallel Factor Analysis (PARAFAC) enables researchers to deconvolute these complex EEMs into statistically and chemically distinct components (Stedmon and Bro, 2008; Murphy et al., 2013). In practice, the relative abundance of EEM components is typically highly correlated with fluorescence indices. It is important to note that the interpretations of fluorescence indices and PARAFAC results are not universal (Ishii and Boyer, 2012). Rather they are developed and interpreted from a specific ecosystem or set of systems and do not preclude a similar signal with a very different interpretation in a different ecosystem. The appropriateness of the index for the data should always be assessed.

24.3 SPECIFIC METHODS

24.3.1 Basic Method 1: A Stream and Its Multiple Sources of Dissolved Organic Matter

How do different biotic and hydrologic watershed compartments interact to shape the quantity and composition of DOM in a watershed? Potential sources of DOM to a water body include inflowing surface waters, throughfall, pore water in local wetlands or high organic soils, and groundwater. Depending on the system under consideration, these samples are easy to collect although the timing and frequency of collection may be critical during episodic inputs. Sampling sites and timing should be established based on research questions about DOM dynamics of interest before embarking on a sampling program, such as the one described next.

1. DOM from different pathways may vary in its quantity and composition. The chemical signature of each pathway can be investigated using UV/Vis absorbance or fluorescence.
   a. Identify the major sources of DOM contributing to the stream.
   b. Optionally, install wells or lysimeters in subsurface hydrologic compartments (e.g., hyporheic or riparian zones, deep and shallow groundwater, throughfall, etc.).
   c. Iron can strongly interfere with DOM optical properties and its abundance should be assessed before selecting your method of DOM composition analysis.
   d. If possible, characterize the hydrologic contribution of each source (e.g., discharge gage, groundwater depth, rainfall gage, etc.).
2. Collect water from the relevant hydrologic compartments. Filter and preserve water either in the field or immediately upon return to the laboratory. Collect the necessary volume of filtrate in organic carbon and nitrogen free bottles.
3. Determine concentrations of C and N with an automated total organic carbon/total nitrogen (TOC/TN) analyzer.
4. Measure absorbance at 254 and 280 nm in a DOC analyzer using a spectrophotometer (e.g., a scanning spectrophotometer or a single wavelength spectrophotometer like the one used to measure chlorophyll-a). Calculate $SUVA_{254}$ and slope indices if you collected the full spectrum (Table 24.1).
5. To augment field sampling of aqueous sources, various potential sources can be examined with lab-based leaching experiments. For example, different leaf types (separated by species or apparent degree of decay) can be leached under standard conditions to determine their potential contribution to surface water DOM (Fig. 24.1B). Similarly, different

TABLE 24.1 Summary of ultraviolet and visible spectrum (UV/Vis) optical indicators frequently used to characterize dissolved organic matter (DOM). Lambda (λ) is the specific wavelength of interest. It is important to note that UV/Vis optical metrics are not universal. In particular, metrics developed to describe DOM in a "normal" stream may not be appropriate to use in streams with novel DOM sources (e.g., sewage discharge or large quantities of fresh leaf leachate). It is always necessary to verify that the selected metric fits or describes the features in your data accurately.

Measure	Equation	Units	Indication
UV absorbance	A_λ	cm^{-1}	DOC concentration
Specific UV absorbance at 254 nm (SUVA$_{254}$)	$\dfrac{A_{254\ nm}}{[DOC]} \times 100$	L mg-C^{-1} m^{-1}	Composition, molecular weight, and aromaticity (Weishaar et al., 2003)
Napierian absorption coefficient (a_λ)	$= \dfrac{A_\lambda}{l} \times 2.303$ where l is the measurement path length in meters; a 1 cm cuvette would be 0.01	m^{-1}	DOC concentration
Spectral slope (S_e)	$a_\lambda = a_{\lambda_{ref}} \times e^{-S_e(\lambda - \lambda_{ref})}$ Common ranges for which S_e is calculated are 275–295 nm or 350–400 nm; λ_{ref} is the lowest wavelength in the range	nm^{-1}	Both S_e and S_r related to DOM molecular weight, photodegradation, and a variety of other environmental processes (Twardowski et al., 2004; Helms et al., 2008)
Slope ratio (S_r)	$\dfrac{S_{e_1}}{S_{e_2}}$	Dimensionless	

soil types representing a range of organic content can be subjected to a standard leaching treatment to determine their relative potential as a source of DOM. The "standard conditions" used will to some extent determine the quantity and composition of organic matter extracted from the solid source samples and their comparability among studies. In many cases stream water or synthetic rainwater may be the most ecologically meaningful leaching agents.

6. Optionally and slightly more advanced, determine the importance of different sources by mass balance. Using the concentrations of DOC and DON measured above, estimates of water entering via different pathways, and unique optical fingerprints of DOM, one can roughly estimate which pathways are important sources of DOM at different times of the year and hydrologic stages.

24.3.2 Advanced Method 1: Heterotrophic Activity of Dissolved Organic Matter

With some more time, the samples/sources of water in Basic Method 1 can also be analyzed for the bioavailability of DOM to address the question: How do different sources and their respective compositions affect DOM bioavailability? DOM bioavailability is a continuum from the most to the least reactive species, but for convenience may be parsed into highly, semi, and nonlabile components (Cory and Kaplan, 2012). If direct analysis of DOC/N is available, a range of experiments can reveal differences in heterotrophic DOC/N removal or production of DON under either field or laboratory conditions. The simplest approach is to collect water samples from the sources as illustrated by Fig. 24.1A or streams before and just after leaf fall, riparian wetland versus open channel, or groundwater seeps versus open channel. For a classic example see Qualls and Haines (1992).

24.3.2.1 Dark Incubation to Assess Bioavailability

1. Most waters contain sufficient bacterial cells to degrade DOM over 28 days without an added inoculum. Collect DOM from sources of interest and filter to 0.7 μm—this will remove grazers while allowing most bacteria to pass through. You can either collect water in 500-mL amber bottles and remove an aliquot at each time interval or collect water in replicate 40-mL vials sacrificed at each interval. Either approach is appropriate but the latter may better account for bottle effects, be less susceptible to contamination during the course of the experiment, and may be easier to manage in a small space despite the large number of vials that may be required.
2. Degradation of DOM may be measured as first order decay of the DOM. Typical measurement intervals might be 0, 1, 3, 7, 14 days. More frequent measurements may allow you to model highly and semilabile pools of DOM using a double exponential model (McDowell et al., 2006). Loss of DOM from the most labile sources should be observable within

TABLE 24.2 Common equations for modeling.

Model	Equation
First order	$[C_t] = [C_0] \cdot e^{-k \cdot t}$
Double exponential	$[C_t] = [C_0] \cdot a \cdot e^{-k_1 \cdot t} + [C_0] \cdot (100 - a) \cdot e^{-k_2 \cdot t}$

hours to days. Twenty-eight days is typically sufficient for complete bioavailability and is in general agreement with results from more aggressive bioreactor approaches. Longer incubations may reveal insight into degradation of non-labile pools (Koehler et al., 2012).

3. Simple dissolved oxygen measurements should also be made to (1) ensure aerobic (or anaerobic) conditions are maintained, (2) infer differences in metabolic pathways utilizing DOM, and (3) with appropriate controls, to account for the possibility that different rates of DOC loss were simply due to precipitation/sorption.
4. Determine net loss of bulk DOC by measurement of concentration following procedures above as appropriate for the available instrumentation. If absorbance is used, it must be kept in mind that absorbance may change due to shifts in relative abundance of chromogenic compounds without a concomitant decline in bulk DOC concentration. Parallel samples to follow biological oxygen demand (see Chapters 5 and 14) may help resolve this question.

24.3.2.2 Data Analysis

5. Decay is frequently modeled as a first-order chemical reaction (Table 24.2). If a sufficient number of time series measurements have been made, a double exponential may be helpful in resolving the decay kinetics of fast and slow decay pools within the sample. The double exponential will require parameter estimation in a program such as Matlab or [R].
6. If feasible, particularly in conjunction with hyporheic water sampling (see Chapter 8), a set of water samples from different locations along hyporheic flow paths can be used to track changes in DOC and dissolved oxygen.

24.3.3 Advanced Method 2: Enzymatic Characterization of the Dissolved Organic Matter Demand

Quantifying the components of DOM that actually support heterotrophic metabolism remains one of the most complex and exciting challenges in the field. With the diversity of compounds (many highly complex and poorly characterized); the diversity of microbes and their tremendous metabolic capacity, the possible interactions and variability are daunting. Enzymes represent a reasonably simple experiment for unraveling the composition of DOM that microbes are utilizing. The most conservative interpretation of enzyme data is that different patterns, "fingerprints," of allocation of activity among enzymes represents some biologically meaningful difference in biologically available components of DOM. In a more mechanistic view, enhancement of particular enzymes is driven by changes in the target substrate although it must be recognized that the model compounds used in the assays cannot represent the full range in a class of organic compounds. For example, leucine aminopeptidase is the most common peptidase, but in principle it is not reactive to other terminal peptides.

24.3.3.1 Enzyme Collection

1. We have used a series of extracellular enzyme assays to describe differences in DOM components undergoing degradation among streams (cf. Findlay et al., 2001). The presumption is that changes in the units of DOM being degraded are accompanied by shifts in allocation among enzymes targeting different classes of compounds. For example, if proteins become relatively more important as a growth substrate for stream bacteria, one would predict increased peptidase activity.
2. It is important to remember that the products of enzymatic hydrolysis of polymeric carbon compounds (DOM) may be used for either catabolism (energy) or anabolism (biomass) and so even if the monomer is present there is still some benefit from polymer degradation. Therefore, high abundances of polymers are generally expected to induce the appropriate enzyme.
3. Interpretation of the enzymes degrading organic carbon compounds differs from enzymes targeting specific elements. For instance, high phosphatase activity is often viewed as indicative of P limitation since the cells have essentially "switched" to an organic form of P. For organic carbon degrading enzymes, the analogy would be that absence of

monomers triggers degradation of the polymeric form. However, enzymes targeting organic complexes may also be induced by small quantities of the product (the monomer) and so, in contrast to the phosphate example, enzyme response to monomer could be either positive or negative.

4. Microbial enzyme activity may vary depending on location in the stream and time of year. Depending on the sampling location the enzymes collected will represent a mixed community of autotrophs and heterotrophs (C acquiring enzymes being more characteristic of heterotrophs). Bed sediments, epilithic biofilms, and the water column are all potential sources of samples for extracellular enzyme activity (EEA) analysis.

5. Sample preservation is a challenge for EEA (Smucker et al., 2009). Freezing enzymes may change their activity and the impacts on your samples should be first characterized. Refrigeration may be acceptable if samples are being run within a few days. Immediate analysis may give the most unimpeachable results, but may require you to limit the scope of your study.

6. a. For **bed sediments**, collect $\sim$100 mL of sediment into a suitable container. Sieve through a mesh (e.g., 2 mm) to remove large debris and invertebrates. Homogenize by vortexing.

   b. For **epilithic biofilms**, loosen a known area of biofilm (e.g., 25 cm^2 or more for less productive streams) with a toothbrush and wash into a suitable container (see Chapter 12). Homogenize by vortexing.

   c. For **water column bacterioplankton**, collect an unfiltered water sample (e.g., 250 mL)

24.3.3.2 Enzyme Activity Measurement

7. Enzyme activities can be measured with either spectrophotometric or fluorometric approaches, and the approach you choose may depend on the samples you wish to analyze. Additionally, the pH of the solution affects both the enzyme activity and ability to detect the fluorescent signal; so care is required when comparing across water sources. Bringing all samples to a common pH gets around this but obviously alters ambient conditions. Both approaches use the cleavage of a fluorescent or UV/Vis absorbing molecule. Fluorometric approaches use 4-methylumbelliferyl- or 7-amino-4-methyl-coumarin-linked fluorescing substrates and standard and require smaller volumes of sample and substrates, but requires a microplate-reading fluorometer (Table 24.3). The spectrophotometric methods use *p*-nitrophenol (PNP)-linked substrates, and PNP is released as the model substrates are cleaved. Bacterial films growing in the dark on inorganic substrates (buried stones, shaded ceramic tiles) are a useful model community since they rely on supply of organic carbon from the water column. For the PNP substrates it is possible to incubate actual stones in solutions of stream water and substrate although this generates a fair amount of waste and requires larger quantities of the substrates. The procedure can be scaled down by scrubbing the biofilm off the stones/tiles and resuspending in stream water or buffer to which small quantities of substrate are added. If a plate reader is available multiple assays are possible on a single plate greatly expediting sample processing. Also, the plate can be read repeatedly allowing better resolution of the time course of color or fluorescence development.

TABLE 24.3 Preparation of 100 mL stock of methylumbelliferyl (MUB) or 7-amido-4-methyl-coumarin-linked substrates and interpretation.

Enzyme	Substrate	Quantity of Substrate[a] (mg)
Phosphatase	4-MUB-phosphate	25.62
Leucine-aminopeptidase	L-leucine 7-amido-4-methyl-coumarin	32.48
Endopeptidase	4-MUB-P-guanadinobenzoate	37.33
β-N-acetylglucosaminidase	4-MUB-N-acetyl-β-glucosaminide	37.94
α-Glucosidase	4-MUB-α-D-glucoside	33.83
β-Glucosidase	4-MUB-β-D-glucoside	33.83
Xylosidase	4-MUB-β-D-xyloside	30.83
Esterase	4-MUB-acetate	31.82
	4-MUB-propionate	23.22
	4-MUB-butyrate	24.66

[a]*Grams of substrate added to 100 mL of 5 mM bicarbonate buffer to prepare a 1 mM stock of substrate.*

8. Rates of enzyme activity are affected by pH. Some enzymes have higher activity at low pH, which decreases as pH increases, and others behave oppositely. For example, two C and N acquiring enzymes, *N*-acetylglucosaminidase (NAG) and L-leucine-aminopetidase, are most active at pH ~4–5 and pH ~7–8, respectively. Running samples at ambient pH measures the rates at which the enzymes are working in the environment and may elucidate cross-site differences in the rates at which microbes are utilizing DOM. Running samples at a constant pH may better elucidate differences in the quantities of enzymes produced among sites.

9. To measure enzyme activities, complete the following protocol:
    a. *Bicarbonate buffer*: Make a 100-mM stock solution by dissolving 8.4 g NaHCO₃ in 1.0 L of deionized water. Stock solution pH is 8.2.
    b. *Enzyme substrates*: Dilute 50 mL of stock buffer solution to 1.0 L to make 5 mM solution for substrates. It is advisable to optimize the substrate solution so that demand is saturated for the duration of the experiment. This may be done by performing a series of experiments (following this protocol) using a representative sample at different substrate concentrations (German et al., 2011) or different sample concentrations; 1 mM is frequently sufficient to saturate biofilm demand. Note that some substrates (Leu-aminopeptidase, esterase) are unstable in buffer and should be made up in autoclaved deionized water.
    c. *Blanks*: Blanks consisting of substrate with no biofilm suspension, biofilm with no substrate, or even autoclaved biofilm suspension are run in parallel to detect any instability in substrates.
    d. M*icroplate setup*: Each standard 96-well microplate has 8 rows and 12 columns. It should be set up in a logical manner that contains reference standards, substrate controls, sample controls, quench controls, and the samples for analysis. Once set up with appropriate replication, approximately five samples can be analyzed per plate. See Online Worksheet 24.1 for an example setup and reagent and volume.
    e. Warm up microplate reader for 1 h and perform any necessary calibrations and checks. Set excitation wavelength at 365 nm and emission wavelength at 450 nm.
    f. Dispense water sample and blanks first (Load 1), and substrate solutions last (Load 2). Load 2 starts the reaction. Immediately place the plate into the plate reader and read it. This is your Time 0 measurement.
    g. Read again at 0.1–1.0 h intervals, depending on activity. Some enzymes (esterases, proteases) consistently develop color (or fluorescence) faster than some of the hydrolases. Measurement interval and duration should be adjusted accordingly.
    h. If working in multiple streams spanning a range of pH, it may be desirable to conduct the analysis at the ambient pH. The cleaved fluorophores and reference standards fluoresce most strongly at a pH 8–9. So, streams at different pH may have the same activity, but the cleaved substrates may exhibit different fluorescence. To correct this, raise the pH. After confirming that reaction rates were linear, and fluorescence has increased 10–100 (or more) times over background, end the experiment by adding ~10 μL of 0.5 N NaOH to raise the pH above 9 and make a final read. The plate cannot be used for further reads.

24.3.3.3 Data Analysis

Please see Online Worksheet 24.1 for excel calculations facilitating QA/QC and data analysis.

10. Once the data have been collected, it is necessary to perform a quality control check to ensure that the collected data are "good." Typically, this is done by analyzing the coefficient of variance (CV) in the assayed sample replicates [(standard deviation/mean of replicates) × 100]. The CV allows you to detect samples that may contain outliers. Samples with a CV > ~10 should be inspected for "runaway" or "dead" wells that had inordinately high or low reaction rates—it happens sometimes. With eight replicates in your assayed samples you can safely remove several "outlier" wells to bring your CV closer to 10. Exceptionally high or low readings occur less frequently in the quenches and controls, but should also be checked.

Quench coefficient for a sample is:

11. Calculate the EEA as follows:

$$Q_{coeff} = \frac{\overline{QC}}{\overline{RC}} \tag{24.1}$$

where $\overline{QC}$ = average response (fluorescence standard units, FSU) of the quench control for a sample and $\overline{RC}$ = average response (FSU) of the reference control.

Emission coefficient for the reference control is:

$$E_{\text{coeff}} = \frac{\overline{RC}}{\text{Refquantity}} = \frac{\text{FSU}}{\text{nmol}} \tag{24.2}$$

where 0.5 is the quantity of reference standard in the well in nanomole.
The activity at the final read time (t_f) may be calculated as:

$$\text{Activity} = \frac{\left(\dfrac{AS_{t_f}}{Q_{\text{coeff}}} - \text{SaC}_{t_f} - \text{SuC}_{t_f} \right)}{E_{\text{coeff}} \times 0.2\ \text{mL} \times (t\ \text{h}) \times N} = \frac{\text{nmol}}{\text{h} \times (\text{mL} \times N)} \tag{24.3}$$

where 0.2 mL is the volume of sample in the well and N is an optional unit of normalization. For example, if you count the number of cells in the water sample you are assaying for enzymes, you might normalize to cell/mL and the resulting activity would be in units of $\text{nmol} \cdot \text{hr}^{-1} \cdot \text{cell}^{-1}$. You may also wish to normalize to organic content in your sample (mg AFDM/mL) or area scrubbed for the sample (area per mL sample collected).

12. Data are typically analyzed with multivariate techniques (e.g., principal components analysis) that reduce the multiple enzyme assays to two or three dimensions. Ratios of C, N, and P acquiring enzymes may also be calculated to indicate the nutrient status of the community sampled (Sinsabaugh et al., 2009).

24.3.4 Advanced Method 3: Fluorescent Analysis of Dissolved Organic Matter Composition

DOM fluorescence is one way of gaining reasonably detailed information about the DOM pool and should be considered complementary to enzyme analysis. Although the entire DOM pool does not fluoresce, behavior of the fluorescent fraction may be indicative of processes affecting underlying nonfluorescing fractions. Fluorescence contained in EEMs integrates a variety of ecosystem processes. Advanced statistical methods such as PARAFAC, enable us to statistically separate independent components of the DOM pool. Unique responses of components to ecosystem processes can then be used as a tracking tool. This section will focus on the basic approaches to collecting high-quality fluorescence data. For PARAFAC modeling of that data we refer the reader to several published tutorials with actively maintained software packages (http://www.models.life.ku.dk/dreem; Murphy et al., 2013). For a thorough text on organic matter fluorescence in aquatic systems, we refer the reader to Coble et al. (2014).

1. Collect filtered samples in C- and N-free amber glass vials. Some plastics, particularly fresh plastics, will have a fluorescence signal.
2. Measure fluorescence; properly setting up and calibrating the instrument is essential for reliable data from run to run.
   a. Each instrument has an instrument-specific bias in the excitation (the light source) and the emission (the detector) which must be corrected. Most new instruments will do this automatically. Once corrected fluorescence results are more comparable between different instruments (Cory et al., 2010; Murphy et al., 2010).
      i. Understanding the unique response of your instrument is important for interpreting your data. For example, an FI of 1.2 on one instrument may be an FI of 1.3 or 1.4 on another. Standard reference materials, such as those produced by the International Humic Substances Society, can help.
   b. Fluorescence is also subject to an "inner-filter effect" where the light emitted by fluorescence may be reabsorbed by the sample. This can be achieved by diluting samples and applying a mathematical correction based on UV/Vis absorbance measurements (Ohno, 2002). So, any fluorescence EEM you collect should also have a UV/Vis spectrum that spans the relevant wavelengths. The mathematical correction is only appropriate when the absorbance in a 1 cm path length cuvette is less than 0.2 (scaled to cuvette path length, dilution of the fluorescence sample, and the UV/Vis sample should match).
   c. Instrument settings vary from instrument to instrument. In general, EEMs are measured between excitation wavelengths of 250−550 nm in 5-nm increments (generally, noise increases below 250 and there is little fluorescence above 450 nm). A smaller increment is acceptable but not necessary. Emission is measured in 2−5 nm increments between 250 and 600 nm. Longer wavelengths typically have less fluorescence. The scan rate or integration time is how quickly your instrument makes a measurement. Faster scans can produce noisier data, while slower scans will produce better data but take longer. With most commercially available instruments, an adequate fluorescence scan takes 15−20 min. Next generation fluorometers can produce high-quality EEMs in 2−3 min.

d. Once the instrument is configured, collect a water blank with DI water. This will be used later to remove background noise, Rayleigh scattering, water Raman peaks, and to normalize the data to the water Raman peak.

e. Warm samples to room temperature and read them in the fluorometer. Take care that you wipe the outside of the cuvette with a clean Kimwipe or other low lint tissue to remove any moisture or particles.

f. Several factors may cause "noisy" EEMs. If the noise cannot be eliminated by improving laboratory methods or instrument maintenance, it can be reduced by increasing the integration time (slower scan rates) or collecting multiple EEMs (same sample different aliquots). Both will add time to your analysis.

g. Each EEM will be stored as a separate file or collection of files depending on the instrument manufacturer.

3. You now must perform several steps to obtain fully "corrected" data. If the instrument does not automatically perform the corrections, then they must be performed in a program like [R] or Matlab (a manual approach is not advisable). Fortunately, detailed tutorials and "workflows" with code are available. The two most popular packages in Matlab are "DOMFluor" and its successor "DREEM." The latter builds upon DOMFluor's PARAFAC analysis algorithm adding significant functionality for correcting EEMs and validating models. The need for correction and preprocessing depends on the model of fluorometer and settings used. Some common corrections include the following:

a. Blank subtraction: Subtract the DI EEM (2.d) from each sample EEM.

b. Inner-filter correction: Apply $I = I_0 \left(10^{-(b_{ex} * A_{ex} + b_{ex} * A_{em})} \right)$ where I = corrected intensity, I_0 = measured intensity, $b_{ex \text{ or } em}$ are 1/2 the corresponding cuvette internal diameters (path lengths), A_{ex} and A_{em} are the blank corrected UV/Vis absorbances at the corresponding ex and em wavelengths. For a "standard" $\sim$4-mL square cuvette with a 1 cm width, $b_{ex} = b_{em} = 0.5$.

c. Raman normalization: From an emission scan at ex 348 nm, integrate the area under the water Raman peak centered at em 396 nm and divide the EEM by this value. This helps make runs on different days and instruments more comparable. Other methods using model compounds may also be employed (e.g., quinine sulfate). Units will then be Raman units (RU) or quinine sulfate units (QSU).

4. Fluorescence indices are an option for rapidly gaining insight into the processes shaping the DOM pool. Three indices are readily calculated from an EEM or from emission scans of the relevant wavelengths. These indices often show a high degree of correlation to PARAFAC components. If you are calculating indices for more than a few samples you may wish to develop a script to automate the calculation of these indices from multiple files (see Appendix Tables 1 and 2).

5. PARAFAC is the current "state of the art" in analyzing EEMs. There are two approaches to using PARAFAC; one is to fit your data to an existing PARAFAC model and the other is to develop your own model (Fellman et al., 2009). There is no universal PARAFAC model or interpretation of components, although some generalities may exist (Ishii and Boyer, 2012). Data fitting may be appropriate if the data set and the model it is being fit to were generated on similar instruments and collected using similar methods and experimental treatments. If your study design intends to capture novel sources or processes, model fitting may poorly represent them and result in high residuals. If residual analysis reveals clear peaks that are higher than $\sim$10% of the original data, you should consider developing a new model. For developing a new model (and more details on collecting fluorescence data and fitting models) we refer the reader to Stedmon and Bro (2008) or Murphy et al. (2013).

6. Once you have a valid model, you need to interpret your "components."

a. If the "end members," were sampled and included, you can interpret the component sources relative to their abundance in those end members.

b. If your design involved experimental treatments (e.g., biodegradation), you can interpret the ecological function of the components form their responses.

c. With some caution, you may consider comparing your components to ones previously published (Fellman et al., 2009; Murphy et al., 2014; Parr et al., 2014).

7. You will receive several pieces of information from the PARAFAC modeling process. We are most interested in the maximum fluorescence of each component (F_{max}). Conceptually, these are a fluorescence concentration proportional to the absolute molecular concentration, that proportionality factor is never known. These data should be interpreted as follows:

a. Among samples, if component 1 has twice the fluorescence in *sample 1* compared to *sample 2* it is reasonable to conclude it is twice as abundant (assuming all other water chemistry variables are approximately equal).

b. Within a sample, if *component 2* has twice the fluorescence compared to *component 1*, it would be incorrect to assume that it is more abundant. Different molecules have different fluorescence intensities when exposed to the same light (quantum yield).

c. Because of this, these data are generally reported in two ways. For investigating the *abundance* of a single component, F_{max} can be treated as a concentration. For investigating changes in a single component's *contribution to composition*, percent F_{max} ($\%F_{max}$) may be calculated as F_{max} (for a given component) divided by the sum of F_{max} values measured for that sample. Both may be appropriate for a given study, but they will inform different questions.

24.3.4.1 Data Analysis

8. It is important to note that some caution should be exercised when interpreting changes in the percent of composition of DOM. For example, if F_{max} for most components changes little across sites, but one or two change significantly, it can change the $\%F_{max}$ for all components.
9. Statistical analysis of DOM composition may be done using a variety of approaches. Ordinary least square regressions may be useful in relating F_{max} values to individual water chemistry parameters or ecosystem rates. If you have collected data on multiple factors potentially influencing DOM composition or abundance (e.g., GIS data about land cover, geology, census, etc.), then a multivariate ordination approach (RDA, variance partitioning) may be the best way of analyzing and graphically presenting the relationships.

24.3.5 Advanced Method 4: Limitations to Degradation

Biotic and abiotic factors may limit DOM degradation. Using the basic DOM biodegradation method described in Section 24.3.2.1 (1), you can assess the effects of nutrients and light/photolysis. You can further expand these two experiments to look at their effects on individual components of the DOM pool by integrating fluorescence measurements as described in Section 24.3.2.1 (3). You can also cross-check nutrient and light treatments to test the interaction of nutrient availability and photodegradation of DOM.

24.3.5.1 Nutrients

The availability of inorganic nutrients can limit degradation. DOM bioavailability is an intrinsic property of the individual molecules driven by the elemental stoichiometry and structural characteristics. In simple terms, the compound may or may not contain the right balance of elements to support metabolism. On the other hand, microbes can acquire elements from the external environment to "supplement" the DOM in question and allow metabolism of an otherwise unsuitable material. By amending DOM samples in with DIN, DIP, or both you can infer whether the intrinsic composition of the DOM in question was or was not adequate.

24.3.5.2 Photolysis/Photobleaching

To what extent does heterotrophic activity depend on photobleaching? This method can be conducted even if direct DOC analysis is not available. Some DOC samples will show obvious declines in absorbance, which may even be detectable with the naked eye. Often a simple spectrophotometric absorbance measurement will show changes. Changes in these optical properties suggest that there has been a change in mass or composition of DOM. These measurements can be coupled with measures of oxygen consumption in sealed bottles of treated water to determine whether exposure to sunlight alters DOM biodegradability.

Experimental manipulations should include exposure to full sun in an open pan or quartz glass bottle, selective filtering with UV-blocking materials (e.g., Mylar), and, of course, a dark control. Ideally the exposure is related to real-world conditions. For instance, one might estimate the length of time that a water mass requires to travel through an open canopy reach of stream (see also Chapters 4 and 7). As above, the required measurements are simply the change in DOM over time estimated via DOC quantification or change in absorbance.

24.3.6 Advanced Method 5: Sorption of Dissolved Organic Matter in Soils

How do the source sink dynamics of soils interact to control the DOM composition observed in streams? Different soil types and horizons can variably act as sources or sinks for DOC moving in soil water. Small columns of organic (with or without leaf litter) or mineral soil can be perfused with solutions of DOC (or distilled water) to determine whether they are net contributors or sinks for DOC. As for the photolysis experiment, it may be possible to detect changes in DOC by simple spectrophotometric measurements if an analyzer is not available. It is also feasible to use the effluents from these columns as source waters for the bioavailability/metabolism assays described above.

24.4 QUESTIONS

1. Based on DOC concentrations and volumes of water entering a stream reach (or other water body), what are the largest sources of DOC? Are the sources with highest concentrations the largest overall contributor?
2. What DOC sources are most variable over various time scales?
3. What litter types (or ages) have greatest potential to contribute DOC?
4. What DOC sources are most susceptible to heterotrophic metabolism? What sources are most susceptible to photolysis?
5. If a change in absorbance occurs after exposure to sunlight how large an effect will this have on light penetration? Are there differences in enzyme "fingerprints" among sources, streams? Which enzymes are most sensitive?
6. What soil types or layers show greatest capacity to supply versus remove DOC from solution? Is DOM from various sources equally susceptible to sorption to mineral soils?
7. After DOC passes through sorptive soil layers can it support heterotrophic activity? Are soils in your region likely to be net sources or sinks for DOM?

24.5 MATERIALS AND SUPPLIES

Filters ≤ 0.7 μm
Filtration apparatus
$NaHCO_3$
Carbon-free 18.2 mΩ deionized water
Microplate
5 mM bicarbonate buffer solution
Leu-aminopeptidase substrate
Esterase substrate
Microfluorometer (microplate reader)
UV/Vis spectrophotometer
Kimwipes
Scanning fluorometer
Assorted laboratory glassware

REFERENCES

Aitkenhead-Peterson, J.A., McDowell, W.H., Neff, J.C., 2003. Sources, production, and regulation of allochthonous dissolved organic matter inputs to surface waters. In: Findlay, S.E.G., Sinsabaugh, R.L. (Eds.), Aquatic Ecosystems: Interactivity of Dissolved Organic Matter. Academic Press, New York, NY, pp. 25—70.

Battin, T.J., Kaplan, L.A., Findlay, S.E.G., Hopkinson, C.S., Marti, E., Packman, A.I., Newbold, J.D., Sabater, F., 2008. Biophysical controls on organic carbon fluxes in fluvial networks. Nature Geoscience 1, 95—100.

Brisco, S., Ziegler, S., 2004. Effects of solar radiation on the utilization of dissolved organic matter (DOM) from two headwater streams. Aquatic Microbial Ecology 37, 197—208.

Brooks, P.D., McKnight, D.M., Bencala, K.E., 1999. The relationship between soil heterotrophic activity, soil dissolved organic carbon (DOC) leachate, and catchment-scale DOC export in headwater catchments. Water Resources Research 35, 1895—1902.

Brown, A., McKnight, D.M., Chin, Y.P., Roberts, E.C., Uhle, M., 2004. Chemical characterization of dissolved organic material in Pony Lake, a saline coastal pond in Antarctica. Marine Chemistry 89, 327—337.

Chin, Y.P., 2003. The speciation of hydrophobic organic compounds by dissolved organic matter. In: Findlay, S.E.G., Sinsabaugh, R.L. (Eds.), Aquatic Ecosystems: Interactivity of Dissolved Organic Matter. Academic Press, New York, NY, pp. 161—184.

Coble, 2007. Marine optical biogeochemistry: the chemistry of ocean color. Chemical Reviews 107, 402—418.

Coble, P.G., Lead, J., Baker, A., Reynolds, D.M., Spencer, R.G.M. (Eds.), 2014. Aquatic Organic Matter Fluorescence. Cambridge University Press, New York, NY.

Cory, R.M., Kaplan, L.A., 2012. Biological lability of streamwater fluorescent dissolved organic matter. Limnology and Oceanography 57, 1347—1360.

Cory, R.M., McKnight, D.M., 2005. Fluorescence spectroscopy reveals ubiquitous presence of oxidized and reduced quinones in dissolved organic matter. Environmental Science and Technology 39, 8142—8149.

Cory, R.M., Miller, M.P., McKnight, D.M., Guerard, J.J., Miller, P.L., 2010. Effect of instrument-specific response on the analysis of fulvic acid fluorescence spectra. Limnology and Oceanography: Methods 8, 67—78.

Cory, R.M., Ward, C.P., Crump, B.C., Kling, G.W., 2014. Sunlight controls water column processing of carbon in arctic fresh waters. Science 345, 925—928.

Creed, I.F., McKnight, D.M., Pellerin, B.A., Green, M.B., Bergamaschi, B.A., Aiken, G.A., Burns, D.A., Findlay, S.E.G., Shanley, J.G., Striegl, R.G., Aulenbach, B.T., Clow, D.W., Laudon, H., McGlynn, B.L., McGuire, K.J., Smith, R.A., Stackpoole, S.M., 2015. The river as a chemostat: fresh perspectives on dissolved organic matter flowing down the river continuum. Canadian Journal of Fisheries and Aquatic Sciences 72, 1272—1285.

Dennis, I.F., Clair, T.A., Driscoll, C.T., Kamman, N., Chalmers, A., Shanley, J., Norton, S.A., Kahl, S., 2005. Distribution patterns of mercury in lakes and rivers of northeastern North America. Ecotoxicology 14, 113—123.

Dillon, P.J., Molot, L.A., 1997. Effect of landscape form on export of dissolved organic carbon, iron and phosphorous from forested stream catchments. Water Resources Research 33, 2591—2600.

Fellman, J.B., D'Amore, D.V., Hood, E., 2008. An evaluation of freezing as a preservation technique for analyzing dissolved organic C, N and P in surface water samples. The Science of the Total Environment 392, 305—312.

Fellman, J.B., Miller, M.P., Cory, R.M., D'Amore, D.V., White, D., 2009. Characterizing dissolved organic matter using PARAFAC modeling of fluorescence spectroscopy: a comparison of two models. Environmental Science and Technology 43, 6228—6234.

Findlay, S.E.G., Quinn, J., Hickey, C., Burrell, G., Downes, M., 2001. Land-use effects on supply and metabolism of stream dissolved organic carbon. Limnology and Oceanography 46, 345—355.

Findlay, S., McDowell, W.H., Fischer, D., Pace, M.L., Caraco, N., Kaushal, S.S., Weathers, K.C., 2010. Total carbon analysis may overestimate organic carbon content of fresh waters in the presence of high dissolved inorganic carbon. Limnology and Oceanography: Methods 8, 196—201.

Freeman, C., Lock, M.A., 1995. The biofilm polysaccharide matrix — a buffer against changing organic substrate supply? Limnology and Oceanography 40, 273—278.

German, D.P., Chacon, S.S., Allison, S.D., 2011. Substrate concentration and enzyme allocation can affect rates of microbial decomposition. Ecology 92, 1471—1480.

Helms, J.R., Stubbins, A., Ritchie, J.D., Minor, E.C., Kieber, D.J., Mopper, K., 2008. Absorption spectral slopes and slope ratios as indicators of molecular weight, source, and photobleaching of chromophoric dissolved organic matter. Limnology and Oceanography 53, 955—969.

Hotchkiss, E.R., Hall Jr., R.O., Baker, M.A., et al., 2014. Modeling priming effects on microbial consumption of dissolved organic carbon in rivers. Journal of Geophysical Research 119, 982—995.

Huguet, A., Vacher, L., Relexans, S., Saubusse, S., Froidefond, J.M., Parlanti, E., 2009. Properties of fluorescent dissolved organic matter in the Gironde Estuary. Organic Geochemistry 40, 706—719.

Ishii, S.K.L., Boyer, T.H., 2012. Behavior of reoccurring PARAFAC components in fluorescent dissolved organic matter in natural and engineered systems: a critical review. Environmental Science and Technology 46, 2006—2017.

Kaplan, L.A., Newbold, J.D., 2003. The role of monomers in stream ecosystem metabolism. In: Findlay, S.E.G., Sinsabaugh, R.L. (Eds.), Aquatic Ecosystems: Interactivity of Dissolved Organic Matter. Academic Press, New York, NY, pp. 97—119.

Kaplan, L.A., Bott, T.L., 1989. Diel fluctuations in bacterial activity on streambed substrata during vernal algal blooms: effect of temperature, water chemistry and habitat. Limnology and Oceanography 34, 718—733.

Kaplan, L.A., Larson, R.A., Bott, T.L., 1980. Patterns of dissolved organic carbon in transport. Limnology and Oceanography 25, 1034—1043.

Kaplan, L.A., 1994. A field and laboratory procedure to collect, process and preserve freshwater samples for dissolved organic carbon analysis. Limnology and Oceanography 39, 1470—1476.

Kaushal, S.S., Lewis, W.M., 2003. Patterns in the chemical fractionation of organic nitrogen in Rocky Mountain streams. Ecosystems 6, 483—492.

Koehler, B., von Wachenfeldt, E., Kothawala, D., Tranvik, L.J., 2012. Reactivity continuum of dissolved organic carbon decomposition in lake water. Journal of Geophysical Research 117, 1—14.

Leff, L.G., Meyer, J.L., 1991. Biological availability of dissolved organic carbon along the Ogeechee River. Limnology and Oceanography 36, 315—323.

Lindeman, R., 1942. The trophic-dynamic aspect of ecology. Ecology 23, 399—417.

McCallister, S.L., del Giorgio, P.A., 2012. Evidence for the respiration of ancient terrestrial organic c in northern temperate lakes and streams. Proceedings of the National Academy of Sciences 109, 16963—16968.

McCracken, K.L., McDowell, W.H., Harter, R.D., Evans, C.V., 2002. Dissolved organic carbon retention in soils: comparison of solution and soil measurements. Soil Science Society of America Journal 66, 563—568.

McDowell, W.H., Zsolnay, A., Aitkenhead-Peterson, J.A., Gregorich, E.G., Jones, D.L., Jödemann, D., Kalbitz, K., Marschner, B., Schwesig, D., 2006. A comparison of methods to determine the biodegradable dissolved organic carbon from different terrestrial sources. Soil Biology and Biochemistry 38, 1933—1942.

McKnight, D.M., Bencala, K.E., Zellweger, G.W., Aiken, G.R., Feder, G.L., Thorn, K.A., 1992. Sorption of dissolved organic carbon by hydrous aluminum and iron oxides occurring at the confluence of Deer Creek with the Snake River, Summit County, Colorado. Environmental Science and Technology 26, 1388—1396.

McKnight, D.M., Boyer, E.W., Westerhoff, P.K., 2001. Spectrofluorometric characterization of dissolved organic matter for indication of precursor organic material and aromaticity. Limnology and Oceanography 46, 38—48.

Meybeck, M., 1982. Carbon, nitrogen and phosphorous transport in world's rivers. American Journal of Science 282, 401—450.

Miller, M.P., McKnight, D.M., Cory, R.M., Williams, M.W., Runkel, R.L., 2006. Hyporheic exchange and fulvic acid redox reactions in an alpine stream/wetland ecosystem, Colorado front range. Environmental Science and Technology 40, 5943—5949.

Mineau, M.M., Wolheim, W.M., Buffam, I., Findlay, S.E.G., Hall Jr., R.O., Hotchkiss, E.B., Koenig, L.E., McDowell, W.H., Parr, T.B., 2016. Dissolved organic carbon uptake in streams: a review and assessment of reach-scale measurements. JGR Biogeosciences 121, 2019—2029.

Moran, M.A., Zepp, R.G., 1997. Role of photoreactions in the formation of biologically labile compounds from dissolved organic matter. Limnology and Oceanography 42, 1307—1316.

Murphy, K.R., Butler, K.D., Spencer, R.G.M., Stedmon, C.A., Boehme, J.R., Aiken, G.R., 2010. Measurement of dissolved organic matter fluorescence in aquatic environments: an interlaboratory comparison. Environmental Science and Technology 44, 9405—9412.

Murphy, K.R., Stedmon, C.A., Graeber, D., Bro, R., 2013. Fluorescence spectroscopy and multi-way techniques: PARAFAC. Analytical Methods 5, 6557.

Murphy, K.R., Stedmon, C.A., Wenig, P., Bro, R., 2014. Openfluor — an online spectral library of auto-fluorescence by organic compounds in the environment. Analytical Methods 6, 658.

Ohno, T., 2002. Fluorescence inner-filtering correction for determining the humification index of dissolved organic matter. Environmental Science and Technology 36, 742–746.

Pacific, V.J., Jencso, K.G., McGlynn, B.L., 2010. Variable flushing mechanisms and landscape structure control stream DOC export during snowmelt in a set of nested catchments. Biogeochemistry 99, 193–211.

Parr, T.B., Ohno, T., Cronan, C.S., Simon, K.S., 2014. ComPARAFAC: a library and tools for rapid and quantitative comparison of dissolved organic matter components resolved by parallel factor analysis. Limnology and Oceanography: Methods 12, 114–125.

Qualls, R.G., Haines, B.L., 1992. Biodegradability of dissolved organic matter in forest throughfall, soil solution, and stream water. Soil Science Society of America Journal 586, 578–586.

Rodriguez, M.J., Sérodes, J.B., 2001. Spatial and temporal evolution of trihalomethanes in three water distribution systems. Water Research 35, 1572–1586.

Rook, J.J., 1977. Chlorination reactions of fulvic acids in natural waters. Environmental Science and Technology 2, 478–482.

Sharp, J.H., Carlson, C.A., Peltzer, E.T., Castle-Ward, D.M., Savidge, K.B., Rinker, K.R., 2002. Final dissolved organic carbon broad community intercalibration and preliminary use of DOC reference materials. Marine Chemistry 77, 239–253.

Sinsabaugh, R.L., Hill, B.H., Follstad Shah, J.J., 2009. Ecoenzymatic stoichiometry of microbial organic nutrient acquisition in soil and sediment. Nature 462, 795–798.

Smucker, N.J., DeForest, J.L., Vis, M.L., 2009. Different methods and storage duration affect measurements of epilithic extracellular enzyme activities in lotic biofilms. Hydrobiologia 636, 153–162.

Stedmon, C.A., Bro, R., 2008. Characterizing dissolved organic matter fluorescence with parallel factor analysis: a tutorial. Limnology and Oceanography: Methods 6, 1–6.

Stubbins, A., Lapierre, J.F., Berggren, M., Prairie, Y.T., Dittmar, T., Del Giorgio, P.A., 2014. What's in an EEM? Molecular signatures associated with dissolved organic fluorescence in boreal Canada. Environmental Science and Technology 48, 10598–10606.

Thurman, E.M., 1985. Organic Geochemistry of Natural Waters. Martinus Nijhoff/Dr. W. Junk, Dordrecht, the Netherlands.

Tranvik, L., Bertilsson, S., 1999. Contrasting effects of solar UV radiation on dissolved organic sources for bacterial growth. Ecology Letters 4, 458–463.

Twardowski, M.S., Boss, E., Sullivan, J.M., Donaghay, P.L., 2004. Modeling the spectral shape of absorption by chromophoric dissolved organic matter. Marine Chemistry 89, 69–88.

Vannote, R.L., Minshall, G.W., Cummins, K.W., Sedell, J.R., Cushing, C.E., 1980. The river continuum concept. Canadian Journal of Fisheries and Aquatic Sciences 37, 130–137.

Webster, J.R., Meyer, J.L., 1997. Stream organic matter budgets – a synthesis. Journal of the North American Benthological Society 16, 141–161.

Weishaar, J.L., Aiken, G.R., Bergamaschi, B.A., Fram, M.S., Fujii, R., Mopper, K., 2003. Evaluation of specific ultraviolet absorbance as an indicator of the chemical composition and reactivity of dissolved organic carbon. Environmental Science and Technology 37, 4702–4708.

Wiegner, T.N., Seitzinger, S.P., 2001. Photochemical and microbial degradation of external dissolved organic matter inputs to rivers. Aquatic Microbial Ecology 24, 27–40.

Wilson, H.F., Xenopoulos, M.A., 2009. Effects of agricultural land use on the composition of fluvial dissolved organic matter. Nature Geoscience 2, 37–41.

APPENDIX 24.1

Summary of fluorescence indicators frequently used to characterize DOM. Lambda (λ) is the specific wavelength of interest. These equations are expressed as the ratio of emission intensities measured at a given excitation wavelength. I is the fluorescence intensity measured at a given combination of excitation and emission wavelengths. The units of I may be in RU or QSU, but the units of the calculated index are always dimensionless. It is important to note that fluorescence indices are not universal. Metrics developed to describe DOM in a "normal" stream may not be appropriate to use in streams with novel DOM sources (e.g., sewage discharge or large quantities of fresh leaf leachate). It is always necessary to verify that the selected metric fits or describes the features in your data accurately.

Fluorescence	Excitation (λ, nm)	Emission Intensity Equation	Indication
Fluorescence Index (FI)	370	$\dfrac{I_{470\,nm}}{I_{520\,nm}}$	Indicator of microbial ($> \sim 1.5$) or terrestrial ($< \sim 1.3$) source (Cory et al., 2010).
Humification Index (HIX)	254	$\dfrac{\sum_{435\,nm}^{480\,nm} I}{\sum_{435\,nm}^{480\,nm} I + \sum_{300\,nm}^{345\,nm} I}$	Indicator of degree of "humic" character (Ohno, 2002).
Biological Index (BIX, β/α)	310	$\dfrac{I_{380\,nm}}{I_{430\,nm}}$ or $\dfrac{I_{380\,nm}}{\text{max value between } I_{420\,nm} \text{ and } I_{435\,nm}}$	An indicator of autochthonous inputs, higher values indicate greater input (Wilson and Xenopoulos, 2009; Huguet et al., 2009).

APPENDIX 24.2

Summary of fluorescence excitation emission matrix (EEM) methods frequently used to characterize DOM.

EEMs	Description	Implementation	Units	Source
Peak picking	Maximum peak intensity within a defined area	Manual or computer scripted selection of the maximum values in predefined area	RU, QSU	Coble (2007)
Parallel factor analysis (PARAFAC)	Statistically distinct patterns of fluorescence	Accomplished in Matlab with free toolbox (DREEM, see http://www.models.life.ku.dk/dreem)	RU, QSU	Stedmon and Bro (2008) and Murphy et al. (2013)
Redox index	Derived from reduced and oxidized PARAFAC components	$\dfrac{\sum F_{max}\ \text{reduced}}{\sum F_{max}\ \text{reduced} + \sum F_{max}\ \text{reduced}}$		Miller et al. (2006)

Chapter 25

Transport and Storage of Fine Particulate Organic Matter

John J. Hutchens, Jr.[1], J. Bruce Wallace[2] and Jack W. Grubaugh[3]

[1]*Department of Biology, Coastal Carolina University;* [2]*Odum School of Ecology, University of Georgia;* [3]*Department of Biological Sciences, University of Tennessee Martin*

25.1 INTRODUCTION

Fine particulate organic matter (FPOM) includes particles in the size range of >0.45 to <1000 μm (1.0 mm) that are either suspended in the water column or deposited within lotic habitats. Size fractions of FPOM can be further divided into the categories of medium large (250−1000 μm), small (100−250 μm), fine (45−100 μm), very fine (25−45 μm), and ultrafine (0.45−25 μm). Suspended fine particulate material, also referred to as *seston*, includes all living (e.g., bacteria, algae, protozoans, invertebrates, etc.) and nonliving (amorphous organic matter, detritus, as well as suspended inorganic sediment) material within the size range from 0.45 μm to 1 mm. Seston can originate from many sources, including the breakdown of larger particles by physical forces, animal consumption, microbial processes, flocculation of dissolved substances, and terrestrial inputs (Wotton, 1984, 1990). Transported loads of seston vary greatly among lotic systems, from micrograms in some small streams to metric tons in larger streams and rivers (see Chapter 5).

Seston functions as an important food resource for many filter-feeding invertebrates (Wallace and Merritt, 1980; Benke et al., 1984), as well as for some vertebrates in large rivers, such as paddlefish (*Polyodon spathula*) endemic to the Mississippi River Basin, USA. In some situations, such as below the outflow of dams or lake outlets, filter-feeding populations of benthic invertebrates (see Chapter 15) can remove large portions of suspended seston from the water column within a few kilometers (Maciolek and Tunzi, 1968; Voshell and Parker, 1985). Dense populations of filter-feeding black flies can also transform large quantities of dissolved organic matter and FPOM in rivers into much larger fecal pellets (Wotton et al., 1998; Malmqvist et al., 2001), which increases the rate of particle deposition. Aggregations of mussels not only transfer seston to benthic habitats by filter feeding and subsequently producing feces and pseudofeces, but this deposited material can alter benthic macroinvertebrate community structure (Howard and Cuffey, 2006). Bioturbation of stream substrates by the benthos, in turn, was hypothesized to be the cause of persistent nighttime peaks in seston concentrations (80% higher) compared to daytime values during baseflow of a piedmont stream (Richardson et al., 2009). The downstream transport of seston is also important to the theme of conceptualizing streams as longitudinally linked systems (Vannote et al., 1980; Minshall et al., 1985) and the concept of material spiraling in stream ecosystems (Newbold et al., 1982; Webster, 2007). Indeed, seston is important to many stream ecosystem processes and represents a major pathway of organic matter transport, deposition, and export. FPOM is thus an important consideration in ecosystem organic matter budgets (e.g., Fisher and Likens, 1973; Cummins et al., 1983; Golladay, 1997; Webster and Meyer, 1997; Tank et al., 2010).

FPOM occurs not only in the water column as seston, but is also found deposited in lotic habitats as fine benthic organic matter (FBOM). FBOM standing crops are rarely adequately assessed in stream research due, in part, to its high spatial variation. Sometimes FBOM is ignored completely, or measurements are done in conjunction with benthic sampling for macroinvertebrates with a relatively large mesh size (e.g., 250 μm) that underestimates the total stored FBOM. For example, Minshall et al. (1982) found that standing crops of benthic organic matter may be underestimated by as much as 65% when sampling devices with 250-μm meshes are used. Additionally, standing crops of organic matter may vary greatly between erosional (e.g., riffles and outcrops) and depositional areas (e.g., pools) of streams. Debris dams, for example, often are sites of high FBOM storage (Bilby, 1980; Smock et al., 1989). FBOM and associated microbes serve as an important resource for animals adapted for deposit feeding (collector—gatherers; see Chapter 20), which includes a wide assortment of invertebrates as well as some collecting—gathering fishes (e.g., sturgeon, suckers). Many deposit-feeding animals have low assimilation efficiencies, and the ingestion and reingestion of FBOM and associated microbes may occur many times in longitudinally linked systems. Unfortunately, only a few studies have attempted to quantify the turnover of FPOM. Fisher and Gray (1983) estimated that fine particle feeders ingested over four times their weight per day, and the entire standing crop of FPOM in Sycamore Creek, Arizona, USA, was ingested and egested every 2—3 days. Larval midges (Diptera: Chironomidae), an abundant collector—gatherer in many streams, contributed 12%—74% of total FPOM export in two southern Appalachian Mountain, USA, streams, with higher contributions occurring during a year of lower discharge (Romito et al., 2010). The role of chironomids in FPOM turnover was estimated by combining measurements of gut passage time using seasonal applications of fluorescent pigments to in-stream detritus with annual measurements of assimilation efficiency and secondary production. FBOM storage varies greatly within heterogeneous stream environments. In small headwater streams, the highest standing crops of FBOM are usually associated with pools and wood debris dams (Bilby and Likens, 1980; Huryn and Wallace, 1987; Smock et al., 1989). In large river systems, slack water habitats such as sloughs and backwaters are repositories for large amounts of FBOM; during high flow conditions floodplains adjacent to large rivers can serve as both source and sink of seston and FBOM (Grubaugh and Anderson, 1989).

A number of approaches have been used to estimate FPOM quality and will only be mentioned here. The organic:inorganic matter ratio is simply an estimate of the relative amount of organic and inorganic matter in seston and can be easily determined from procedures described below for seston sampling (see Basic Method 1). This ratio often varies greatly for different size fractions of seston, with smaller size classes having a greater proportion of inorganic material (ash) than larger size fractions. Some studies (e.g., Lamberti and Resh, 1987; Angradi, 1993a; Bukaveckas et al., 2011) have examined the organic constituents of seston such as chlorophyll a (see Chapter 12), while others have examined other organic material and microbial activity such as respiration (Peters et al., 1989; Bonin et al., 2000, 2003; Yoshimura et al., 2008; Richardson et al., 2013; see also Chapters 27 and 34 for examples of respirometry techniques). Edwards (1987) evaluated the importance of bacteria in seston and in the growth of filter-feeding black fly larvae (Edwards and Meyer, 1987). Carlough and Meyer (1991) found sestonic protozoans to be an important component of seston in a low-gradient, blackwater river. Voshell and Parker (1985) used microscopy to examine directly the frequency and type of particles in various size categories. The amounts in each category (e.g., animals, diatoms, other algae, vascular plants, and amorphous detritus) are estimated by the areal standard unit method used in phytoplankton studies as described by Welch (1948). Wallace et al. (1987) used a microscope and digitizer interfaced with a computer for similar analyses; however, these latter methods are not appropriate for bacteria and protozoans.

The movement of FPOM downstream should be viewed as dynamic, involving respiration as well as movement from the substratum to water column and vice versa. Recent methods have been developed to assess this movement as carbon spiraling (also see Chapter 26). For example, radioactively tagged particles have been used to study movement and deposition of seston (Cushing et al., 1993; Minshall et al., 2000; Monaghan et al., 2001; Thomas et al., 2001; Newbold et al., 2005). Hall et al. (1996) estimated transport distances of bacteria by tagging them with fluorescent markers. Fluorescently labeled latex particles (Harvey et al., 2012) and titanium dioxide particles (Karwan and Saiers, 2009) have been used to investigate deposition and entrainment of ultrafine particles. Several seston analogs have also been used to examine rates of transport and deposition including corn pollen and glass beads (Miller and Georgian, 1992; Webster et al., 1999; Georgian et al., 2003), *Lycopodium* spores (Wanner and Pusch, 2000), and yeast (Paul and Hall, 2002). Stable isotope analysis of carbon and nitrogen has been used to study origin and movement of seston and FBOM (Angradi, 1993b; McConnachie and Petticrew, 2006; Schindler Wildhaber et al., 2012; Marko et al., 2013). Raymond and Bauer (2001) used ^{14}C to find that several rivers transport very old (often >1000 years) seston originating from terrestrial soils. The origin of FPOM from potential sources both in and out of the stream channel can be identified using fluorescent spectroscopy, as seen in a study of the efficacy of urban stream restoration (Larsen et al., 2015).

In the following exercises, we describe seston sampling procedures for streams and rivers of various sizes and describe techniques to assess FPOM concentration, size distribution, and instantaneous estimates of total seston export. We also provide sampling techniques for FBOM and emphasize the relative importance of depositional and erosional habitats in the assessment of stream FBOM standing crops. The final exercise examines direct linkages between sestonic FPOM and filter-feeding biota. The specific objectives of these exercises are to (1) introduce the researchers and students to the importance and magnitude of FPOM transport in streams; (2) demonstrate techniques for collecting and analyzing seston and FBOM; (3) demonstrate the importance of hydrologic events in seston transport compared to baseflow conditions; (4) compare the relative importance of erosional and depositional habitats in FBOM storage in streams; and (5) illustrate direct consumption of suspended particles by filter-feeding larvae of black flies (Diptera: Simuliidae). The reader should gain an appreciation for the methods involved in assessing FPOM transport, storage, and use in streams.

25.2 GENERAL DESIGN

25.2.1 Seston

Instantaneous seston concentrations (e.g., mass per volume of water; mg/L) can be easily measured by filtering known volumes of water through preashed and preweighed glass fiber filters (GFFs). This simple approach can be used to compare seston concentrations during baseflow and short-term hydrologic events (e.g., storms), or for comparing seston concentrations among streams of various sizes. Percent ash in such measurements has been related to long-term watershed disturbance (Webster and Golladay, 1984). In some cases, seston particle size has been shown to vary with stream size, as smaller headwater streams draining forested areas have larger median particle sizes than larger rivers downstream (Wallace et al., 1982). However, with few exceptions, the majority of the particles transported by most streams during baseflow conditions are <50 μm in diameter (Sedell et al., 1978; Naiman and Sedell, 1979a, 1979b; Wallace et al., 1982).

A two-part sampling approach generally is adequate for sampling seston in small rivers and streams. The first part consists of collecting a 20- to 30-L grab sample for measuring concentrations of finer seston particles (i.e., <250 m). Under most conditions, seston particle size distributions are strongly skewed toward smaller size fractions; therefore, the second part of sampling uses a collection net to filter a large quantity of water to obtain reasonable concentration estimates for larger seston particle sizes (i.e., >250 μm). One of the best devices for this purpose is a Miller plankton tow net fitted with a 250-μm mesh collection net and flowmeter (Fig. 25.1).

Particle size separation requires a wet filtration system consisting of a series of stacked sieves of Nitex or bolting cloth netting of various sizes. Sieves can be constructed with short (4–5 cm) sections of PVC pipe with netting glued over one end and joined with connectors to form a stackable series of sieves. More elaborate wet filtration systems are constructed of threaded, stainless steel tubes fitted with stainless steel bolting cloth filters of various dimensions. The individual filters with Teflon gaskets are inserted between threaded sections of each tube to form a series of stackable sieves with a large funnel at the top of the apparatus (Fig. 25.2). A water sample is poured into the funnel and through the sieves under vacuum. Seston particle sizes are thus separated by sieve sizes and water passing though the smallest filter can be retained for the ultrafine fraction.

A complementary method to separating particles into size fractions as described above is to count fine particles in different size classes, which is very time consuming if manual light microscopy is used. Particle size fractionation can be facilitated by automated means using different technologies, including Coulter counters, digital imaging flow cytometers, digital holographic microscopy, and laser diffraction particle–size counters. Coulter counters can rapidly and accurately count and measure the size of individual particles, but are unable to distinguish between inorganic and organic particles (Kim and Menden-Deuer, 2013). Similarly, digital imaging flow cytometers rapidly count and size particles (Spaulding et al., 2012), but also can measure fluorescent qualities. Another automated imaging system that can measure particle size is digital holographic microscopy, which also can classify particles based on algorithms (Zetsche et al., 2014). Laser diffraction instruments have been used successfully in situ and in the laboratory with field-collected samples of riverine seston (Czuba et al., 2015). The ability of laser diffraction to distinguish organic particles from inorganic ones relies on data processing routines that vary with how light from the laser interacts with particles of different composition (Andrews et al., 2010).

Large rivers (>7th order) present considerable difficulties for seston sampling. Most of these rivers are nonwadeable, and sampling can only be conducted from bridges or boats. Furthermore, as a result of differential settling rates and lower

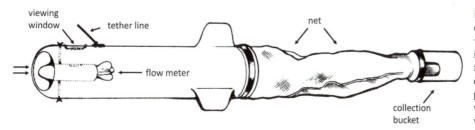

FIGURE 25.1 A Miller-type tow net, equipped with a flowmeter to record velocity of water filtered. The flowmeter is used to calculate distance of water filtered over the time interval the net is deployed (see seston concentration protocols and Table 25.1). A Plexiglas viewing window (optional) is ideal for viewing the dial on the current meter.

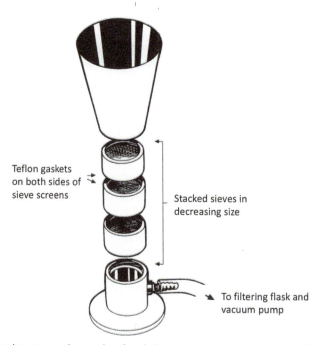

Teflon gaskets
on both sides of
sieve screens

Stacked sieves in
decreasing size

To filtering flask and
vacuum pump

FIGURE 25.2 Example of a wet filtration system using a series of stacked sieves in order of decreasing size. Note: vacuum line is connected to a large filtering flask to retain filtrate for ultrafine seston and to avoid pumping water into the vacuum pump (see particle size separation protocols).

current velocities near the water/substratum interface, seston concentrations and particle size distributions can vary greatly with depth as well as with sampling location relative to the thalweg (i.e., middle of the main channel). Adequate sampling of suspended material in large rivers requires depth-integrated, velocity-weighted samples taken at multiple depths along a transect. A variety of devices have been designed to collect integrated and weighted samples of total suspended sediments, and these lend themselves well to seston sampling in large rivers (e.g., Grubaugh and Anderson, 1989; see also Chapter 5). Drawbacks to their general use are that these samplers are expensive and require a trained operator. The reader should consult Edwards et al. (1988) for a discussion of such devices and sampling designs for use in large rivers.

Point samplers, although considerably less accurate than depth-integrating devices, are much less expensive, easier to operate, and can also be used in large rivers. The protocol provided herein for large river seston collection employs point-sampling techniques. Examples of point samplers include Kemmerer or Van Dorn bottles fitted with *line depressors* or fluked weights to facilitate a vertical descent of the sampler in high-velocity rivers. To estimate seston concentration using point samplers, samples need to be collected at several locations along a given vertical gradient, with the depth and number of samples dependent on the total depth of the water column.

Traditional methods of FPOM collection are limited in their ability to capture the spatial variability of FPOM in large rivers. Newer methods relying on remote sensing techniques are likely to become an important tool for quantifying FPOM loads in large rivers. These techniques take advantage of the colors associated with suspended matter. For example, turbidity in the Great Miami River, Ohio, USA, was related to specific wavelengths measured by an airborne hyperspectral sensor (Shafique et al., 2001). Using this relationship, the authors produced maps displaying the relative distribution of turbidity for 80 km of river. These maps showed plumes of clear water (i.e., low turbidity) associated with a wastewater treatment plant. Similar relationships with total suspended solids (TSS) have been found in other freshwater systems (e.g., Oron and Gitelson, 1996; Dekker et al., 2001). Although turbidity and TSS describe both inorganic and organic particulate matter, FPOM alone can be calculated using the proportion of organic matter estimated from traditional sampling (Dekker et al., 2001). Additional applications of remote sensing techniques in lotic systems are discussed by Johnson and Gage (1997) and Mertes (2002).

25.2.2 Fine Benthic Organic Matter

Laboratory protocols similar to those employed for seston can be used to determine stored FBOM concentrations and particle size distributions if samples are resuspended in a known volume of water prior to analysis. Sampling procedures, however, can be more complicated for FBOM storage, since stream characteristics such as current velocity, substrate particle size, and the presence or absence of retention devices must also be taken into consideration. These factors influence the physical storage of FBOM as well as the structure of the benthic community (e.g., Huryn and Wallace, 1987), and hence they affect transport and use of FPOM in lotic systems. To assess these influences, it is necessary to measure standing stocks of FBOM among stream habitats with differing morphological characteristics. For the purposes of this chapter, FBOM will be quantified in three size fractions: 0.45−250, 250−500, and 500−1000 μm (or 1.0 mm). Particle sizes that are greater than 1.0 mm constitute coarse particulate organic matter (CPOM), which is considered elsewhere in this book (see Chapter 26).

25.2.3 Linkages

Collector−gatherer and collector−filterer organisms use FBOM and seston, respectively, as food resources (see Chapter 20). One collector organism is the larval black fly, which uses cephalic fans to filter and remove small, suspended particles (<300 μm diameter) from the water column. Black flies and other collector−filterers are important in energy trans- formations in streams and are examples of direct linkages between seston and the biota (Wallace and Webster, 1996). Larval black flies also make good study organisms since they are often very abundant in many lotic habitats and have rapid (<1 h) gut passage times (e.g., Wotton, 1978, 1980).

In Advanced Method 1 (below), we describe exposure of larval black flies to a dense concentration of trackable FPOM (powdered charcoal) for a brief time, which will produce a distinct band in the larval guts. Representative larvae (preferably of different instars or sizes) will be collected and preserved at 10-, 20-, and 30-min intervals following particle release. In the laboratory, larvae from each time interval will be separated by size class (body length) and their guts dissected. Using a dissecting microscope equipped with an ocular micrometer, we will make two measurements: the distance from the posterior end of the head to the band of charcoal and the distance from the posterior end of the head to the tip of the abdomen. The ratio of these two (ratio W; Wotton, 1978) gives a measure of gut passage time based on the distance the band has moved.

25.2.4 Site Selection

25.2.4.1 Seston

Methods described herein are designed for both small streams and rivers; however, if used for class purposes, we recommend that they be restricted to lotic reaches that are safely wadeable. Application of this protocol to larger rivers requires working from either bridges or boats, both of which carry inherent risks. If a bridge site is used, the bridge should be close enough to the water surface, and water depth should be shallow enough that sampling equipment can reach the river bottom. The bridge should have sufficiently wide shoulders and limited automotive traffic to facilitate safety; sampling crews are strongly urged to exercise extreme caution and to wear "blaze-orange" garments and deploy road cones to enhance their visibility. Boat sampling requires that attention be given not only to sampling equipment and procedures, but also to boat maintenance, proper safety equipment, and safe boating practices. Because of the inherent dangers of sampling during high flow conditions, we recommend boat sampling be conducted only at low- to mean-flow stage conditions. Seston sampling during storm events can be dangerous even in small streams, and adequate precaution is again recommended.

Management of the sampling site is as important as site selection; disturbance of upstream substratum should be avoided when collecting water for seston analysis. Even minor disturbances such as wading across a stream can dislodge sufficient amounts of FPOM to greatly increase seston concentrations well above baseflow conditions. Be especially careful when working with a field team consisting of several individuals; the person or persons actually collecting seston samples should be upstream of other team members. Finally, the volume of water filtered or collected with the seston is crucial information in seston sampling. Care should be taken in both the field and laboratory when determining and recording volume measurements.

Frequently, especially when assessing the effects of ecosystem-level disturbances, it is desirable to know how FPOM transport responds to individual storms for disturbed and reference conditions, because many studies have shown much greater concentrations and transport of seston during storm discharges. Hewlett and Hibbert (1967) have provided a method for determining what constitutes a storm for catchments of various sizes. Many investigators have noted that the concentration of particulates increases rapidly during the rising limb of the hydrograph (i.e., hysteresis), and that peak concentrations are somewhat unpredictable, usually occurring before peak discharge (Fisher and Likens, 1973; Bilby and Likens, 1979; Gurtz et al., 1980; Fisher and Grimm, 1985; Webster et al., 1990; Wallace et al., 1991). Automated sampling device such as those manufactured by ISCO (ISCO, Inc., Lincoln, Nebraska, USA) can be used to obtain samples of up to 1 L at various intervals during individual storms. Storm sampling, including a sequence of samples taken over rising and falling hydrographs, can be initiated by wetness (rainfall) sensors of individual storms or sufficient rise in stream hydrographs (Wallace et al., 1991; Harmel et al., 2003). ISCO samplers, as well as several other commercially available devices, offer somewhat similar options, such as programmable operation and memory, water-level or stage recorders, water-level collection devices—including sample collection pump and storage—and discrete versus composite samples. Some negative aspects of such devices include high initial costs and high maintenance requirements. Various automated collection devices, settings, and sampling strategy considerations are discussed by Harmel et al. (2003).

25.2.4.1 Fine Benthic Organic Matter

In this method, logistical considerations are again of primary concern. The protocols for this method are most easily and safely accomplished in relatively shallow streams (<0.75 m depth) that are readily wadeable. Furthermore, site selection should be focused toward stream reaches with clearly heterogeneous channel features consisting of zones of contrasting current velocities, such as pools and other clearly depositional reaches, versus cobble riffles or bedrock outcrops. If the site selected is in a nonwadeable, medium-sized stream or river that is too deep for devices described in the FBOM method, SCUBA gear, in combination with elaborate sampling devices such as dome samplers, can be used to sample benthos and FBOM (Gale and Thompson, 1975; Platts et al., 1983). Such devices are expensive and will require trained operators. In backwaters or other slow current habitats, an Eckman dredge can be used, but this device often has a high loss of FBOM. In habitats with slow to moderate currents (e.g., channel borders and sloughs), a Ponar or petite Ponar dredge can be used; in moderate to fast currents (e.g., main channel) Peterson dredges are preferred. None of these dredges will function properly if the site selected has coarse substrata.

25.2.4.2 Linkages

This method is most easily performed in small streams less than 0.25 m in depth. The site should have a reasonably abundant black fly population; shallow outflow streams from lakes and small reservoirs are ideal locations because black fly larvae often form dense aggregations at such sites. It is also important to select areas where the charcoal slurry can be released immediately upstream of the black flies so that the slurry plume flows directly over the larvae. Sampling site management is again important; use caution to minimize disturbing black fly aggregations prior to, during, and following release of the charcoal slurry. Larvae should be collected, following the charcoal release, by standing to one side of the charcoal plume and reaching into the area covered by the plume.

25.3 SPECIFIC METHODS

25.3.1 Basic Method 1: Seston Concentration

25.3.1.1 Protocol for Seston Sampling in Streams and Small Rivers

1. For obtaining a *carboy* sample (Steps 1—3): cover the opening of a clean, 20- to 30-L carboy with a 250-μm mesh sieve or bolting cloth.
2. Using another carboy, bucket, or other vessel, collect a grab sample of stream water. Care should be taken not to disturb the substratum and collect resuspended benthic FPOM with the sample. Pour the water through the mesh and into the carboy until filled. This sample will be used to estimate FPOM concentrations of <250-μm particle size. If the sample is to be used for particle size analysis and seston concentration likely is low (e.g., during winter sampling or in streams with little allochthonous inputs), it is advisable to fill a second or even third carboy for sample processing.
3. Label the carboy(s) by sampling site and sample number and transport to the laboratory for processing. If the sample is to be used for particle size analysis, laboratory filtration should be completed within a few hours following collection of field samples.
4. For the *Miller net* sample (Steps 4—11): record the initial reading of the flowmeter in the Miller net prior to putting the sampler into the stream. Record start time of sampling (optional).
5. Suspend the Miller net in the water column with the front opening completely submersed. Use a tether line on the front of the sampler to secure the sampler in place (e.g., tied to a bridge rail, overhanging limb, or held by the operator if stream velocity is not prohibitive).

6. Sampling time will vary depending on the amount of suspended material. Generally, 10–30 min is an adequate sampling period, but more time might be needed if seston concentrations are low and little or no material is readily visible in the collection net at the end of sampling. Conversely, less time is needed if seston loads are heavy. If larger materials are present in the water column (i.e., leaves or sticks), check to make sure the opening of the Miller net or flowmeter is not obstructed.
7. Upon the completion of sampling, record the final flowmeter reading and (optionally) stop time of sampling.
8. Wash material from the collection net into a 1-mm mesh sieve nested over a 250-μm mesh sieve. This separates out CPOM, which would otherwise result in an overestimate of FPOM during sample processing.
9. Wash material retained on the 250-μm sieve into a sample bag or sample jar. This collection will be used to estimate FPOM concentrations of the particle size class >250 μm. Label the bag or jar by sampling site and sample number, and transport to the laboratory for processing. If the sample is to be used for particle size analysis, laboratory filtration should be completed within a few hours following the collection of field samples.
10. To estimate the volume of water filtered through the Miller net, the following information is needed: radius (r in m) of front opening and distance (d in m) of water filtered as measured by the flowmeter.
11. The volume (V in L) filtered is calculated by the equation:

$$V = r^2 d \times 1000 \qquad (25.1)$$

12. If elapsed sampling time (t) was recorded in seconds, a measure of velocity (v in m/s) can also be made:

$$v = d \div t \qquad (25.2)$$

Although velocity is not crucial to estimating seston concentration, it is easy by this method to calculate it as an additional physical parameter. A data sheet for recording and calculating water volume and velocity information using Miller-type tow nets is presented (see Table 25.1).

TABLE 25.1 Field collection of seston using Miller-type tow nets (see also Online Worksheet 25.1).

Date_____   Observers_____

Stream and site_____   Sampling location_____

_____   Stream stage condition_____

Net mesh size_____μm   **(A)** Diameter of net opening _____m   **(B)** Radius of net opening (A ÷ 2)_____cm   **(C)** pi = 3.1416

(D) Flowmeter conversion to meters _____   **(E)** Volume conversion (B^2 × C × 1000) _____

Sample & Filter Number	**(F)** Initial Flowmeter Reading	**(G)** Final Flowmeter Reading	**(H)** Elapsed Time (sec)	**(I)** Elapsed Flow (G − F)	**(J)** Water Filtered (m) (I × D)	**(K)** Volume Filtered (L) (J × E)	**(L)** Velocity (m sec^{-1}) (J ÷ H)

25.3.1.2 Protocol for Seston Sampling in Large Rivers

1. The protocol described is for bridge sampling. Measure river width below the bridge with a tape measure.
2. Divide the river width measurement by 11 and use the result to determine 10 equidistant points across the river. Using an erasable marking pen, clearly mark and number these 10 points on the bridge railing. These will be the locations of the sampling verticals.
3. Measure the distance from bridge rail to the water surface and from bridge rail to river bottom at each vertical; the difference between these distances is the total depth at each vertical.
4. The number and depth of collections for point samples at each vertical are determined from the water depth of individual verticals as follows:

Water Depth (m)	Sampling Depth (measured from surface)
≤1	60% of water depth
1−3	20% and 80% of water depth
>3	20%, 60%, and 80% of water depth

5. Lower a Kemmerer or Van Dorn bottle to an appropriate depth, and then close and retrieve the sampler. Filter samples through a 1-mm mesh sieve or bolting cloth into milk jugs to remove CPOM. Mark each jug as to collection site, vertical, and sample depth.
6. Transport samples to the laboratory to determine seston concentrations (see *Standard Processing Protocols* below).
7. To determine the mean seston concentration, first calculate the mean concentration for each vertical and then calculate the mean of all 10 verticals.

25.3.1.3 Standard Processing Protocols

1. Set up a microfiltration unit consisting of a filter holder, base, and funnel that can accommodate 47- or 50-mm-diameter filters. The microfiltration unit is seated on a 2- to 4-L capacity filtering flask connected to a vacuum pump with vacuum tubing. Filters are 47- or 50-mm-diameter GFFs, preashed, and preweighed[1].
2. For carboy and jug samples, vigorously shake the carboy or jug to resuspend seston. Pour a 1- to 4-L aliquot of the sample into a graduated cylinder. Record the volume of sample used.
3. Pour the aliquot into the microfiltration funnel and draw down onto a GFF under vacuum. Volume required will vary depending upon seston concentration in the aliquot. In general, volume should be sufficient to produce a clearly visible layer of seston on the GFF.
4. Rinse the microfiltration funnel with distilled/deionized, prefiltered water to ensure seston particles are not adhering to funnel. Remove GFF from the microfiltration unit with blunt forceps and return to its labeled aluminum square.
5. Repeat Steps 2−4 until three replicate aliquots have been filtered for each sample.
6. For Miller net samples, thoroughly wash seston out of the sample bag or sample jar and into the microfiltration funnel using distilled/deionized, prefiltered water and draw down onto a GFF under vacuum. Rinse microfiltration funnel and remove GFF as above.
7. To determine dry mass, seston samples and GFFs should be oven-dried (50°C for 24 h), desiccated (24 h), and weighed on an analytical balance.
8. To determine ash mass, dry-weighed seston samples and GFFs should be ashed in a muffle furnace (500°C for 0.5−1 h), rewetted with distilled/deionized water to restore waters of hydration (Weber, 1973), oven-dried (50°C for 24 h), desiccated (24 h), and weighed on an analytical balance (see also protocol in Chapter 12).
9. Masses obtained provide measures of ash-free dry mass (AFDM) or organic seston (dry mass/ash mass) and inorganic seston (ash mass). Masses from >250 to <250 μm fractions need to be mathematically combined if samples were collected using Protocol 1 (above). Seston concentrations may be reported directly as milligram or gram of seston per sample volume. However, it is preferable to standardize units to either mg/L or g/m^3.

---

1. Note: Filters should be free of binders such as Gelman type A/E or equivalent. Labeled squares of aluminum foil are useful for maintaining individual preashed and preweighed GFFs.

25.3.1.4 Particle Size Separation Protocols

1. Set up the wet filtration system consisting of a funnel, a series of stacked sieves in decreasing size order, a base to attach the sieve stack to a filtering flask, and an electric vacuum pump connected to the flask with vacuum tubing (Fig. 25.2). A large capacity (≥ 4 L) vacuum flask should be connected between the stacked sieves and the vacuum. For carboy samples, the largest sieve size should be 250 μm, which should be the smallest sieve size for Miller net samples.

2. For carboy samples (Steps 2−8), vigorously shake the carboy to resuspend seston. Slowly pour the sample from the carboy into the funnel of the filtration system. The volume of water required will vary depending upon seston concentration in the sample. In general, it will take the entire volume of the carboy, but somewhat less if seston loads are high. Under conditions of low seston concentrations, several carboys may be needed to obtain adequate samples.

3. Filtration will require turning off the vacuum periodically to empty the filtering flask to avoid pulling water into the vacuum pump. Carefully disconnect the stack and base from the flask and *record the volume of water* in the flask prior to discarding. Make sure at least 3 L of filtrate is retained to measure ultrafine seston. Reassemble the system and continue filtration.

4. When filtration is complete, compute and record the total volume of water filtered through the system. Disassemble the wet filtration system, arranging sieves such that size fractions are clearly denoted.

5. Set up a microfiltration unit as described in Step 1 of *Standard Processing Protocols*.

6. Starting with the largest sieve (i.e., 250 μm), wash retained material into the funnel of the microfiltration unit with distilled/deionized, prefiltered water and draw down onto a preashed, preweighed GFF under vacuum. Rinse the funnel to ensure seston particles are not adhering. Remove the GFF from the microfiltration unit with blunt forceps and return it to its labeled aluminum square.

7. Repeat Step 6 for the next smaller sieve size, carefully recording sieve sizes and corresponding GFF identification numbers. Continue until material from all sieves has been drawn down onto separate GFFs.

8. Filter reserved ultrafine seston filtrate onto a GFF. Volume required will vary depending upon seston concentration in the aliquot. In general, volume should be sufficient to produce a clearly visible layer of seston covering on the GFF. Record the volume of filtrate used.

9. Process GFFs and seston samples as described in Steps 7 and 8 of *Standard Processing Protocols*.

10. For Miller net samples (Steps 10−15), resuspend sampled material in distilled/deionized, prefiltered water and pour into the funnel of the wet filtration system. Carefully wash out sample bag or jar into funnel to ensure all material is recovered.

11. Draw down material into the stacked sieve column under a vacuum while rinsing the funnel with distilled/deionized, prefiltered water to ensure seston particles are not adhering.

12. When filtration is complete, disassemble the wet filtration system, arranging sieves such that size fractions are clearly denoted.

13. Set up a microfiltration unit as described in Step 1 of *Standard Processing Protocols*.

14. Starting with the largest sieve, wash retained material into the funnel of the microfiltration unit with distilled/deionized, prefiltered water and draw down onto a preashed, preweighed GFF under a vacuum. Rinse the microfiltration funnel and remove the GFF as above.

15. Repeat Step 13 for the next smaller sieve size, carefully recording sieve sizes and corresponding GFF identification numbers. Continue until material from all sieves has been drawn down onto separate GFFs.

16. Process GFFs and seston samples as described in Steps 7 and 8 of *Standard Processing Protocols*.

17. Seston concentrations can now be determined as milligram per sample volume for individual particle sizes. Since water volume filtered to obtain samples differs between carboy and Miller net samples, concentrations must be converted to a standard unit (e.g., milligram of seston/L) prior to comparison.

25.3.1.5 Optional Experiment A: Seston Export

1. Export, or total transport of seston, requires knowledge of stream discharge at the time of seston sampling (see Chapter 3 for methods of determining discharge). Estimates of total export are made by weighing seston concentration (mass per unit volume) by discharge (volume per unit time) to determine export or total transport (mass per unit time). Provided you have the necessary data, this is easily accomplished by multiplying total seston concentration (mg AFDM/L) $\times$ $1000 = $ mg AFDM/m^3. The product is multiplied by discharge (m^3/s) to estimate mg of FPOM exported per second.

2. One advantage to sampling large rivers is that these systems are routinely gauged, and information such as mean daily discharge (in ft^3/s or m^3/s) is easily obtainable. In the United States, such data are available in the US Geological

Survey's *Water Resources* data book published annually for each state (usually found in government publications sections of most libraries, or Internet at http://waterdata.usgs.gov/nwis). Similar systems exist in countries around the world. Daily loads of transported seston (seston export) for large rivers can then be estimated by adjusting for units of measure and multiplying seston concentration by mean daily discharge.

3. Other methods of estimating export include the use of rating curves (Cummins et al., 1983; Webster et al., 1990; see also Chapter 5), thus incorporating some aspect of discharge to estimate particulate organic matter (POM) concentrations. However, discharge and POM concentrations are generally poorly related (Bilby and Likens, 1979; Gurtz et al., 1980; Cuffney and Wallace, 1988). These studies indicate that infrequent sampling and poor rating curves are not good predictors of POM export.

4. Another method for continuous export measurements involves Coshocton proportional samplers, which are only suitable for small streams and require more elaborate instrumentation (Cuffney and Wallace, 1988; Wallace et al., 1991).

25.3.1.6 Optional Experiment B: Seston Sampling During Storms

1. In small streams with quickly fluctuating ("flashy") discharge, the bulk of the total suspended material is carried during the rising hydrograph of storms (e.g., Gurtz et al., 1980; Webster et al., 1990; Wallace et al., 1991). Sampling these events can be difficult due to their unpredictable timing and short duration. Although seston sampling for particle size analysis can be conducted under such conditions, it is extremely labor intensive, and for our purposes we will only examine total seston concentrations under conditions of baseflow and rising and falling hydrographs. As storms are largely unpredictable, this will require access to a stream located near your laboratory that can be readily sampled. Small, gauged streams are ideal for this purpose. If none are available, see Chapter 3 for stream gauging methods.

2. In some cases, a meter stick anchored vertically to an area where the cross-sectional profile can be measured will suffice as a gauge. Record the water height on the meter stick with each sample taken during the storm. Use standard processing procedures to calculate total dry mass, ash, and AFDM concentrations for these samples (see Chapter 12). You may wish to repeat these measurements over a several-day period if no storms occur. You should have a series of 10−15 clean bottles with caps (1- to 2-L capacity) for sampling as storms approach, as well as a supply of preashed and preweighed GFFs. Start your storm sampling sequence prior to the first rainfall if possible. Clearly record the time and water height for each subsequent sample as stream turbidity increases on the rising hydrograph during the storm. Brief and intense summer thundershowers are ideal for this purpose; however, severe electrical storms can be dangerous. Be sure *not* to seek shelter under tall trees between sampling intervals. Although timing is tricky for such storms, you should attempt to sample over a period that provides a series of samples taken over both the rising and falling hydrographs. In the laboratory, process each sample separately, clearly labeling each filter from the sequential samples. Determine total dry mass, ash, and AFDM (mg/L) for each sample in the storm sequence.

25.3.2 Basic Method 2: Fine Benthic Organic Matter

25.3.2.1 Protocols for Field Collection of Fine Benthic Organic Matter

1. Prepare a substantial amount of filtered stream water: pour stream water into a carboy or other large, clean vessel through a 250-μm mesh sieve.
2. Select sites that are characteristic of either depositional or erosional stream habitats (see Site Selection above). Place a graduated barrel (or large bucket) and paddle in close proximity to the sampling site.
3. With minimal disturbance to the substratum, force the sampling corer into the substratum. The core should be ≤22 cm in diameter and made of steel or PVC pipe. For cobble-riffle and bedrock-outcrop habitats, wrap a cloth towel around the outside base of the corer once it is in place to form an effective seal with the substratum.
4. Remove material from within the corer with either a plastic cup or hand-powered diaphragm pump (the latter works more efficiently on hard-bottomed substrata). Pass removed material through nested 1-mm and 250-μm mesh sieves that are positioned over the graduated barrel to retain water passing through the smaller mesh.
5. In riffle areas, cobbles inside the corer should be thoroughly brushed and disturbed while pumping; bedrock substratum also should be thoroughly brushed.
6. Once water has been removed from inside the coring device and the substratum cleaned of fine particles, thoroughly wash the sieves with filtered stream water, retaining material passing through the bottom sieve in the graduated barrel. Measure and record water volume in the barrel.

7. For the <250-μm fraction, stir water in the barrel thoroughly with the paddle and remove a subsample of the agitated water (0.2−1 L depending on the concentration of particles). Store subsamples in either separate bottles or large, self-closing plastic bags. Three replicate subsamples (stirring before each subsample) are desirable for each sample.
8. For the >250-μm fractions, discard material retained on the 1-mm mesh sieve, which is the CPOM fraction of the sample. Wash material retained on the 250-μm mesh sieve with filtered stream water into a suitable container (e.g., large plastic bag or wide-mouth bottle), and clearly label this container and the subsamples.
9. Repeat the sampling procedure for all targeted habitats (i.e., erosional and depositional areas).

25.3.2.2 Fine Benthic Organic Matter Processing Protocols

1. For the >250-μm fractions (Steps 1−5), wash contents of the sample container with tap water into a large pail and resuspend in water.
2. Pour the resuspended material through nested 500- and 250-μm mesh sieves. Allow time to drain samples thoroughly and transfer material to separate, labeled paper bags.
3. Oven-dry material and bags at 50°C to a constant weight (24 h to several days, depending on sample size). Place bags in a desiccator for 24 h.
4. Remove material from bags and weigh on a top-loading balance to determine dry mass.
5. Ash material at 500°C (small, heavy-gauged, aluminum baking pans work well for this purpose), and reweigh to obtain AFDM for the 250−500 and 500 μm to 1.0 mm size fractions.
6. For the <250-μm fractions (Steps 6−11), set up a microfiltration unit as described in Step 1 of *Standard Processing Protocols*.
7. Individually pour each of the three replicate subsamples into separate 1-L graduated cylinders and record the subsample volumes.
8. Pour the first subsample into the funnel of the filtration unit. Wash any material clinging to the subsample bag or graduated cylinder into the funnel with distilled/deionized, prefiltered water. Draw material down onto a GFF, washing sides of funnel with distilled/deionized, prefiltered water. Remove filter with blunt forceps and return to aluminum square.
9. Repeat Steps 7 and 8 for remaining replicates.
10. Dry, weigh, ash, and reweigh FBOM samples and GFFs following Steps 7 and 8 of *Standard Processing Protocols*.
11. AFDM of 0.45- to 250-μm-sized fraction is estimated as the mean of the following quantity calculated for each of the three subsamples:

$$AFDM = (\text{barrel volume} \div \text{subsample volume}) \times \text{subsample AFDM} \qquad (25.3)$$

12. FBOM quantity is normally expressed as g AFDM/m^2 of stream bottom. This requires you to know the area of your sampling device (in cm^2). Use the following equation for FBOM standing stocks estimated for each size fraction to express your results:

$$g\,AFDM/m^2 = \text{mg AFDM} \div 1000 \times 10,000 \div cm^2 \text{ of area sampled} \qquad (25.4)$$

The g AFDM/m^2 for each size fraction is summed to obtain total FBOM standing crop (in AFDM) in your sample.

25.3.3 Advanced Method 1: Linkages of Sestonic Fine Particulate Organic Matter to the Biota

25.3.3.1 Field Release and Larval Collection

1. In the field, thoroughly mix charcoal with stream water in one or two large pails until no more charcoal remains on the surface of the water to form a dense slurry of suspended charcoal (some continuous stirring even during release may be required to ensure suspension).
2. Position members of the team on either side of the black fly aggregation, being careful to minimize disturbance.
3. At a location 1−2 m upstream of the black fly aggregation, slowly pour the slurry back and forth across a 0.5−1.0 m width of stream, ensuring that the water passing over the larval aggregation is darkly stained with charcoal particles.

Pour slowly to ensure that the contents of the pail are not released as a massive instantaneous dosage. (A beaker can be used for removing the slurry from the bucket and releasing the mixture in the stream.)

4. The release should take 1—2 min. Record the starting and ending time of the slurry release. Note the width of the slurry passing over the aggregation and the lateral boundaries of the slurry (flagging attached to wire stakes may be useful for this purpose), and keep larval collections within boundaries.

6. Collect larvae at 10-, 20-, and 30-min intervals following slurry release.

7. Use collecting forceps to pick larvae and place in a vial prelabeled with the appropriate time interval and half-filled with 70% ethanol. Collectors should strive to sample a range of larval sizes at each period. Sampling should continue for about 1 min after each 10-min interval. Following each collection period, check all vials to ensure that the time interval is correctly indicated.

25.3.3.2 Laboratory Analysis

1. Separate vials into specific time intervals, and work with larvae from only one interval at a time to avoid confusion.

2. Starting with larvae collected 10 min after the charcoal release, use a dissecting microscope fitted with an ocular micrometer to divide black flies into size classes to the nearest 0.5 mm. Keep size classes separate.

3. For each size class, use the ocular micrometer to measure the distance from the posterior end of head to the tip of the abdomen on each larva. Record data as Distance x.

4. Using the point of the jeweler's forceps, carefully split open the larval integument from below the head to the tip of the abdomen.

5. With two pairs of jeweler's forceps, gently tease the gut out of the body cavity, keeping the head attached to the gut.

6. With the ocular micrometer, measure the distance from the posterior end of the head to the charcoal band in the gut. Record data as Distance y.

7. Repeat this procedure until all larvae from each size class and all three time intervals have been measured.

8. Upon completion of measurements, you should have recorded the following information for each larva collected:
   a) Collection interval (10, 20, or 30 min)
   b) Larval length (mm)
   c) Distance y in mm (posterior end of head to charcoal band)
   d) Distance x in mm (posterior end of head to tip of abdomen)
   e) Ratio W (Distance y ÷ Distance x).
   f) Plot the ratio of W (y-axis) against the larval length (x-axis) for each larva examined at the 10-min interval. Repeat this process for each larva measured for the 20-min interval, then the 30-min interval. You should be able to regress the values for the ratio of W and larval length for each time interval.

25.4 QUESTIONS

25.4.1 Seston

1. What is the total organic seston (in mg AFDM/L) concentration in your stream?

2. Based on your measurements of individual size classes, what sizes are the most abundant in terms of total organic seston in transport?

3. Suppose you repeat these measurements in smaller headwater streams or a larger downstream river. Would you expect the same results? Why or why not?

4. How does seston concentration vary with stream depth? With distance from the thalweg?

5. Does seston quality (in terms of organic:inorganic ratio) change with distance from the thalweg? If so, can you hypothesize as to why this change occurs?

6. If seston concentrations are available for two rivers or sites, or the same river in different seasons, compare estimates of seston export between rivers, sites, or seasons. How do they compare, and can you suggest any mechanism to account for differences?

7. (*Optional Experiment A*) Convert seston concentration data into an estimate of total seston export. What source did you use for discharge data? Can you predict seasonal patterns of seston export for your system based on what information you have on discharge and seston concentrations?

8. (*Optional Experiment B*) How did the dry mass, AFDM, and ash concentration change over the rising and falling hydrograph of the storm? If you are working in a gauged stream, it will be useful to plot each sample concentration

against discharge at the time the sample was collected. If not, you can plot each sample against water depth measured on the meter stick as a rough estimate of relative discharge for each sample.

9. (*Optional Experiment B*) At what stage of the storm sampling sequence were maximum and minimum seston concentrations reached? How do you explain your results?

10. (*Optional Experiment B*) Based on your sampling results during the storm, what problems do you see with calculating organic matter export for stream ecosystems? How does this influence organic matter budgets for a given stream reach?

25.4.2 Fine Benthic Organic Matter

11. How do FBOM particle size distributions and total FBOM standing crops compare between erosional and depositional habitats?

12. Hypothesize as to the specific physical characteristics in each habitat that may account for differences in FBOM standing crops.

13. Given differences in FBOM particle size distribution and standing crops, what are your hypotheses concerning the relative functional structure of the benthic macroinvertebrate community in each habitat? (See Chapters 15 and 20 for information concerning benthic community functional structure.)

25.4.3 Linkages of Sestonic Fine Particulate Organic Matter to the Biota

14. Do black fly larvae display any tendency to select food particles based on type of food available? Give reasons for your answer.

15. Black fly larvae have been described as feeding nonselectively on particles <300 μm in diameter. Based on your analyses of seston particle sizes, what significance do you attach to this observation with respect to particle size availability in lotic habitats?

16. For a specific time interval, i.e., 10, 20, or 30 min following charcoal exposure, is there any difference in gut passage times for larvae of different size classes? If so, what differences did you detect? What does the ratio of W versus larval length illustrate about gut passage times?

17. What is your best estimate of gut passage time for black fly larvae of different size classes? Did you notice any difference in charcoal bands after 30 min? How do you account for differences after longer time intervals (see Wotton, 1980)?

18. What do you see as the "ecological role" microfiltering collectors such as black flies play in stream ecosystems? Explain your answer.

25.5 MATERIALS AND SUPPLIES

Letters in parentheses indicate in which Method (1, 2, or 3 = Advanced Method 1) the item is used.

Aluminum squares (1, 2), approximately 60 mm side length, and numbered to facilitate filter identification
Balance (1, 2), analytical
Balance (2), top-loading
Bags, paper (2)
Bags, plastic (1, 2), self-closing (e.g., Ziploc or Whirl-Pak)
Bottles (2), wide-mouth, capped, 1- to 2-L capacity
Buckets, 10- to 15-L capacity (3)
Charcoal, fine-powered (3). *Optional*: fine-powdered florescent pigments can be substituted for charcoal (e.g., Miller et al., 1998). These are more expensive than powdered charcoal but easier to locate in the gut, especially for black fly larvae with heavily pigmented integuments. They also glow when exposed to a black light source, such as a mineral light used by geologists. One source of such pigments is Radiant Color, 2800 Radiant Ave., Richmond, CA 94804. Type P-1600 (average particle size = 5 μm) manufactured by Radiant Color have the added advantage that specimens can be mounted on glass slides without interference from the many solvents used in mounting.
Corer (2), hand-held or stove pipe with inside diameter ≥22.6 cm, made of steel or PVC pipe that can be forced into the substratum
Cup, plastic (2), for sampling depositional areas
Desiccator (1, 2), with $CaSO_4$ desiccant

Filters (1, 2), 47- or 50-mm GFFs without binder (e.g., Gelman type A/E, Whatman GFF, or equivalent). Prior to use, filters are ashed in a muffle furnace (500°C for 0.5–1 h), rewetted with distilled/deionized water to restore waters of hydration, oven-dried (50°C for 24 h), desiccated (24 h), and preweighed on an analytical balance. Store GFFs on labeled aluminum squares in a desiccator.

Flags (3) (optional), attached to wire stakes

Forceps, jewelers (3), blunt (1, 2)

Furnace, muffle (1, 2)

Graduated container (2); large pail or vinyl trash can marked for the volume of water at various depths

Graduated cylinders, 1-L capacity (1, 2)

Jugs, approximately 1-gal (3.8 L) capacity (2); milk jugs are good for this purpose as they are inexpensive and many are needed.

Microfiltration unit (1, 2), including:

    47- or 50-mm filter base, holder, and funnel

    filtering flask, 2- to 4-L capacity

    vacuum pump

    vacuum tubing

    blunt forceps

Marker, permanent ink (1, 2, 3)

Microscope, dissecting, binocular (3) fitted with an ocular micrometer

Notebook, field (1, 2, 3) with waterproof pages

Oven, drying (1, 2)

Paddle, canoe (2, 3)

Pump (2); hand-held, diaphragm-type, for sampling erosional areas

Sampler, point (1), including:

    Kemmerer or Van Dorn sampling bottle

    weighted messenger

    tether line marked in 0.5-m increments

    line depressor

Sampler, Miller-type tow net (1), including:

    sampler body with slightly tampered front (reduction fitting)

    collecting net, 250 μm

    catch bucket

    flowmeter

    tether line

Sieves, standard testing (1, 2); nestable, with mesh sizes of 250 μm, 500 μm, and 1.0 mm

Stopwatch (1, 3)

Tape, measuring (1); 10–50 m marked in 0.5-m increments

Wash bottles (1, 2)

Wet filtration unit (2); includes:

    top funnel

    nestable sieves of decreasing mesh sizes; examples: 500, 250, 100, 50, and 25 μm

    filtering flask, 2- to 4-L capacity

    Teflon gaskets

    vacuum pump

    vacuum tubing

Vials, 1 dram with stoppers (3); vials should be half-filled with 70% ethanol and prelabeled to indicate 10-, 20-, and 30-min collection intervals

Water, distilled/deionized and prefiltered through GFFs (1, 2)

REFERENCES

Andrews, S., Nover, D., Schladow, S.G., 2010. Using laser diffraction data to obtain accurate particle size distributions: the role of particle composition. Limnology and Oceanography: Methods 8, 507–526.

Angradi, T.R., 1993a. Chlorophyll content of seston in a regulated Rocky Mountain River, Idaho, USA. Hydrobiologia 259, 39–46.

Angradi, T.R., 1993b. Stable carbon and nitrogen isotope analysis of seston in a regulated Rocky Mountain River, USA. Regulated Rivers: Research and Management 8, 251–270.

Benke, A.C., Van Arsdall Jr., T.C., Gillespie, D.M., Parrish, F.K., 1984. Invertebrate productivity in a subtropical blackwater river: the importance of habitat and life history. Ecological Monographs 54, 25–63.

Bilby, R.E., 1980. Role of organic debris dams in regulating the export of dissolved and particulate matter from a forested watershed. Ecology 62, 1234–1243.

Bilby, R.E., Likens, G.E., 1979. Effects of hydrologic fluctuations on the transport of fine particulate organic carbon in a small stream. Limnology and Oceanography 24, 69–75.

Bilby, R.E., Likens, G.E., 1980. Importance of organic debris dams in the structure and function of stream ecosystems. Ecology 61, 1107–1113.

Bonin, H.L., Griffiths, R.P., Caldwell, B.A., 2000. Nutrient and microbial characteristics of fine benthic organic matter in mountain streams. Journal of the North American Benthological Society 19, 235–249.

Bonin, H.L., Griffiths, R.P., Caldwell, B.A., 2003. Nutrient and microbial characteristics of fine benthic organic matter in sediment settling ponds. Freshwater Biology 48, 1117–1126.

Bukaveckas, P.A., MacDonald, A., Aufdenkampe, A., Chick, J.H., Havel, J.E., Schultz, R., Angradi, T.R., Bolgrien, D.W., Jicha, T.M., Taylor, D., 2011. Phytoplankton abundance and contributions to suspended particulate matter in the Ohio, Upper Mississippi and Missouri Rivers. Aquatic Sciences 73, 419–436.

Carlough, L.A., Meyer, J.L., 1991. Bacterivory by sestonic protists in a southeastern blackwater river. Limnology and Oceanography 36, 873–883.

Cuffney, T.F., Wallace, J.B., 1988. Particulate organic matter export from three headwater streams: discrete versus continuous measurements. Canadian Journal of Fisheries and Aquatic Sciences 45, 2010–2016.

Cummins, K.W., Sedell, J.R., Swanson, F.J., Minshall, G.W., Fisher, S.G., Cushing, C.E., Petersen, R.C., Vannote, R.L., 1983. Organic matter budgets for stream ecosystems: problems in their evaluation. In: Barnes, J.R., Minshall, G.W. (Eds.), Stream Ecology: Application and Testing of General Ecological Theory. Plenum Press, New York, NY, pp. 299–353.

Cushing, C.E., Minshall, G.W., Newbold, J.D., 1993. Transport dynamics of fine particulate organic matter in two Idaho streams. Limnology and Oceanography 38, 1101–1115.

Czuba, J.A., Straub, T.D., Curran, C.A., Landers, M.N., Domanski, M.M., 2015. Comparison of fluvial suspended-sediment concentrations and particle-size distributions measured with in-stream laser diffraction and in physical samples. Water Resources Research 51, 320–340.

Dekker, A.G., Vos, R.J., Peters, S.W.M., 2001. Comparison of remote sensing data, model results and in situ data for total suspended matter (TSM) in the southern Frisian lakes. The Science of the Total Environment 268, 197–214.

Edwards, R.T., 1987. Seasonal bacterial biomass dynamics in the seston of two southeastern blackwater rivers. Limnology and Oceanography 32, 221–234.

Edwards, R.T., Meyer, J.L., 1987. Bacteria as a food source for black fly larvae in a blackwater river. Journal of the North American Benthological Society 6, 241–250.

Edwards, T.K., Glysson, G.D., Guy, H.P., Norman, V.W., 1988. Field Methods for Measurement of Fluvial Sediment. Department of the Interior, US Geological Survey, pp. 9–32.

Fisher, S.G., Gray, L.J., 1983. Secondary production and organic matter processing by collector macroinvertebrates in a desert stream. Ecology 64, 1217–1224.

Fisher, S.G., Grimm, N.B., 1985. Hydrologic and material budgets for a small Sonoran Desert watershed during three consecutive cloudburst floods. Journal of Arid Environments 9, 105–118.

Fisher, S.G., Likens, G.E., 1973. Energy flow in Bear Brook, New Hampshire: an integrative approach to stream ecosystem metabolism. Ecological Monographs 43, 421–439.

Gale, W.F., Thompson, J.D., 1975. A suction sampler for quantitatively sampling benthos on rocky substrates in rivers. Transactions of the American Fisheries Society 104, 398–405.

Georgian, T., Newbold, J.D., Thomas, S.A., Monaghan, M.T., Minshall, G.W., Cushing, C.E., 2003. Comparison of corn pollen and natural fine particulate matter transport in streams: can pollen be used as a seston surrogate? Journal of the North American Benthological Society 22, 2–16.

Golladay, S.W., 1997. Suspended particulate organic matter concentration and export in streams. Journal of the North American Benthological Society 16, 122–131.

Golladay, S.W., Webster, J.R., Benfield, E.F., Swank, W.T., 1992. Changes in stream stability following forest clearing as indicated by storm nutrient budgets. Archive für Hydrobiologie Supplement 90, 1–33.

Grubaugh, J.W., Anderson, R.V., 1989. Upper Mississippi River: seasonal and floodplain forest influences on organic matter transport. Hydrobiologia 174, 235–244.

Gurtz, M.E., Webster, J.R., Wallace, J.B., 1980. Seston dynamics in southern Appalachian streams: effects of clear-cutting. Canadian Journal of Fisheries and Aquatic Sciences 37, 624–631.

Hall, R.O., Peredney, C.L., Meyer, J.L., 1996. The effect of invertebrate consumption on bacterial transport in a mountain stream. Limnology and Oceanography 41, 1180–1187.

Harmel, R.D., King, K.W., Slade, R.W., 2003. Automated storm water sampling on small watersheds. Applied Engineering in Agriculture 19, 667–674.

Harvey, J.W., Drummond, J.D., Martin, R.L., McPhillips, L.E., Packman, A.I., Jerolmack, D.J., Stonedahl, S.H., Aubeneau, A.F., Sawyer, A.H., Larsen, L.G., Tobias, C.R., 2012. Hydrogeomorphology of the hyporheic zone: stream solute and fine particle interactions with a dynamic streambed. Journal of Geophysical Research 117, G00N11.

Hewlett, J.D., Hibbert, A.R., 1967. Factors affecting the response of small watersheds to precipitation in humid areas. In: Sopper, W.E., Lull, H.W. (Eds.), Forest Hydrology. Pergamon Press, Oxford, UK, pp. 275—290.

Howard, J.K., Cuffey, K.M., 2006. The functional role of native freshwater mussels in the fluvial benthic environment. Freshwater Biology 51, 460—474.

Huryn, A.D., Wallace, J.B., 1987. Local geomorphology as a determinant of macrofaunal production in a mountain stream. Ecology 68, 1932—1942.

Johnson, L.B., Gage, S.H., 1997. Landscape approaches to the analysis of aquatic ecosystems. Freshwater Biology 37, 113—132.

Karwan, D.L., Saiers, J.E., 2009. Influences of seasonal flow regime on the fate and transport of fine particles and a dissolved solute in a New England stream. Water Resources Research 45, W11423.

Kim, H., Menden-Deuer, S., 2013. Reliability of rapid, semi-automated assessment of plankton abundance, biomass, and growth rate estimates: Coulter Counter versus light microscope measurements. Limnology and Oceanography: Methods 11, 382—393.

Lamberti, G.A., Resh, V.H., 1987. Seasonal patterns of suspended bacteria and algae in two northern California streams. Archives für Hydrobiologie 110, 45—57.

Larsen, L., Harvey, J., Skalak, K., Goodman, M., 2015. Fluorescence-based source tracking of organic sediment in restored and unrestored urban streams. Limnology and Oceanography 60, 1439—1461.

Maciolek, J.A., Tunzi, M.G., 1968. Microseston dynamics in a simple Sierra Nevada lake-stream ecosystem. Ecology 49, 60—75.

Malmqvist, B., Wotton, R.S., Zhang, Y., 2001. Suspension feeders transform massive amounts of seston in large northern rivers. Oikos 92, 35—43.

Marko, K.M., Rutherford, E.S., Eadie, B.J., Johengen, T.H., Lansing, M.B., 2013. Delivery of nutrients and seston from the Muskegon River Watershed to near shore Lake Michigan. Journal of Great Lakes Research 39, 672—681.

McConnachie, J.L., Petticrew, E.L., 2006. Tracing organic matter sources in riverine suspended sediment: implications for fine sediment transfers. Geomorphology 79, 13—26.

Mertes, L.A.K., 2002. Remote sensing of riverine landscapes. Freshwater Biology 47, 799—816.

Miller, J., Georgian, T., 1992. Estimation of fine particle transport in streams using pollen as a seston analog. Journal of the North American Benthological Society 11, 172—180.

Miller, M.C., Kurzhals, M., Hershey, A.E., Merritt, R.W., 1998. Feeding behavior of black fly larvae and retention of fine particulate organic matter in a high-gradient blackwater stream. Canadian Journal of Zoology 76, 228—235.

Minshall, G.W., Brock, J.T., LaPoint, T.W., 1982. Characterization and dynamics of benthic organic matter and invertebrate functional feeding group relationships in the Upper Salmon River, Idaho (U.S.A.). International Revue Gesamten Hydrobiologie 67, 793—820.

Minshall, G.W., Cummins, K.W., Petersen, R.C., Cushing, C.E., Bruns, D.A., Sedell, J.R., Vannote, R.L., 1985. Developments in stream ecosystem theory. Canadian Journal of Fisheries and Aquatic Science 42, 1045—1055.

Minshall, G.W., Thomas, S.A., Newbold, J.D., Monaghan, M.T., Cushing, C.E., 2000. Physical factors influencing fine organic particle transport and deposition in streams. Journal of the North American Benthological Society 19, 1—17.

Monaghan, M.T., Thomas, S.A., Minshall, G.W., Newbold, J.D., Cushing, C.E., 2001. The influence of filter-feeding benthic macroinvertebrates on the transport and deposition of particulate organic matter and diatoms in two streams. Limnology and Oceanography 46, 1091—1099.

Naiman, R.J., Sedell, J.R., 1979a. Characterization of particulate organic matter transported by some Cascade Mountain streams. Journal of the Fisheries Research Board of Canada 36, 17—31.

Naiman, R.J., Sedell, J.R., 1979b. Benthic organic matter as a function of stream order in Oregon. Archive für Hydrobiologie 97, 404—422.

Newbold, J.D., Mulholland, P.S., Elwood, J.W., O'Neill, R.J., 1982. Organic carbon spiralling in stream ecosystems. Oikos 38, 266—272.

Newbold, J.D., Thomas, S.A., Minshall, G.W., Cushing, C.E., Georgian, T., 2005. Deposition, benthic residence, and resuspension of fine organic particles in a mountain stream. Limnology and Oceanography 50, 1571—1580.

Oron, G., Gitelson, A., 1996. Real-time quality monitoring by remote sensing of contaminated water-bodies: waste stabilization pond effluent. Water Research 30, 3106—3114.

Paul, M.J., Hall, R.O., 2002. Particle transport and transient storage along a stream-size gradient in the Hubbard Brook Experimental Forest. Journal of the North American Benthological Society 21, 195—205.

Peters, G.T., Benfield, E.F., Webster, J.R., 1989. Chemical composition and microbial activity of seston in a southern Appalachian headwater stream. Journal of the North American Benthological Society 8, 74—94.

Platts, W.S., Megahan, W.F., Minshall, G.W., 1983. Methods for Evaluating Stream, Riparian, and Biotic Conditions. General Technical Report INT-138. United States Department of Agriculture, Forest Service, Intermountain Forest and Range Experiment Station, Ogden, UT.

Raymond, P.A., Bauer, J.E., 2001. Riverine export of aged terrestrial organic matter to the North Atlantic Ocean. Nature 409, 497—500.

Richardson, D.C., Newbold, J.D., Aufdenkampe, A.K., Taylor, P.G., Kaplan, L.A., 2013. Measuring heterotrophic respiration rates of suspended particulate organic carbon from stream ecosystems. Limnology and Oceanography: Methods 11, 247—261.

Richardson, D.C., Kaplan, L.A., Newbold, J.D., Aufdenkampe, A.K., 2009. Temporal dynamics of seston: a recurring nighttime peak and seasonal shifts in composition in a stream ecosystem. Limnology and Oceanography 54, 344—354.

Romito, A.M., Eggert, S.L., Diez, J.M., Wallace, J.B., 2010. Effects of seasonality and resource limitation on organic matter turnover by Chironomidae (Diptera) in southern Appalachian headwater streams. Limnology and Oceanography 55, 1083—1092.

Sedell, J.R., Naiman, R.J., Cummins, K.W., Minshall, G.W., Vannote, R.L., 1978. Transport of particulate organic matter in streams as a function of physical processes. Verhandlungen der Internationalen Vereinigung für Theoretische und Angewandte Limnologie 20, 1366—1375.

Schindler Wildhaber, Y., Liechti, R., Alewell, C., 2012. Organic matter dynamics and stable isotope signature as tracers of the sources of suspended sediment. Biogeosciences Discussions 9, 453—483.

Shafique, N.A., Autrey, B.C., Fulk, F., Cormier, S.M., 2001. Hydrospectral narrow wavebands selection for optimizing water quality monitoring on the Great Miami River, Ohio. Journal of Spatial Hydrology 1, 1–22.

Smock, L.A., Metzler, G.M., Gladden, J.E., 1989. Role of debris dams in the structure and functioning of low-gradient headwater streams. Ecology 70, 764–775.

Spaulding, S.A., Jewson, D.H., Bixby, R.J., Nelson, H., McKnight, D.M., 2012. Automated measurements of diatom size. Limnology and Oceanography: Methods 10, 882–890.

Tank, J.L., Rosi-Marshall, E.J., Griffiths, N.A., Entrekin, S.A., Stephen, M.L., 2010. A review of allochthonous organic matter dynamics and metabolism in streams. Journal of the North American Benthological Society 29, 118–146.

Thomas, S.A., Newbold, J.D., Monaghan, M.T., Minshall, G.W., Georgian, T., Cushing, C.E., 2001. The influence of particle size on seston deposition in streams. Limnology and Oceanography 46, 1415–1424.

Vannote, R.L., Minshall, G.W., Cummins, K.W., Sedell, J.R., Cushing, C.E., 1980. The river continuum concept. Canadian Journal of Fisheries and Aquatic Sciences 37, 130–137.

Voshell, J.R., Parker, C.R., 1985. Quantity and quality of seston in an impounded and a free-flowing river in Virginia, USA. Hydrobiologia 122, 271–288.

Wallace, J.B., Benke, A.C., Lingle, A.H., Parsons, K., 1987. Trophic pathways of macroinvertebrate primary consumers in subtropical blackwater streams. Archiv für Hydrobiologie Supplement 74, 423–451.

Wallace, J.B., Cuffney, T.F., Webster, J.R., Lugthart, G.J., Chung, K., Goldowitz, B.S., 1991. Export of fine organic particles from headwater streams: effects of season, extreme discharges, and invertebrate manipulations. Limnology and Oceanography 36, 670–682.

Wallace, J.B., Webster, J.R., 1996. The role of macroinvertebrates in stream ecosystem function. Annual Review of Entomology 41, 115–139.

Wallace, J.B., Merritt, R.W., 1980. Filter-feeding ecology of aquatic insects. Annual Review of Entomology 25, 103–132.

Wallace, J.B., Ross, D.H., Meyer, J.L., 1982. Seston and dissolved organic carbon dynamics in a southern Appalachian stream. Ecology 63, 824–838.

Wanner, S.C., Pusch, M., 2000. Use of fluorescently labeled *Lycopodium* spores as a tracer for suspended particles in a lowland river. Journal of the North American Benthological Society 19, 648–658.

Weber, C.I. (Ed.), 1973. Biological Field and Laboratory Methods for Measuring the Quality of Surface Water and Effluents. United States Environmental Protection Agency, Cincinnati, OH. EPA 640/4-73-001.

Webster, J.R., 2007. Spiraling down the river continuum: stream ecology and the U-shaped curve. Journal of the North American Benthological Society 26, 375–389.

Webster, J.R., Meyer, J.L., 1997. Organic matter budgets for streams: a synthesis. Journal of the North American Benthological Society 16, 141–161.

Webster, J.R., Golladay, S.W., 1984. Seston transport in streams at Coweeta Hydrologic Laboratory, North Carolina, USA. Verhandlungen der Internationalen Vereinigung für Theoretische und Angewandte Limnologie 22, 1911–1919.

Webster, J.R., Benfield, E.F., Ehrman, T.P., Schaeffer, M.A., Tank, J.L., Hutchens, J.J., D'Angelo, D.J., 1999. What happens to allochthonous material that falls into streams? A synthesis of new and published information from Coweeta. Freshwater Biology 41, 687–705.

Webster, J.R., Golladay, S.W., Benfield, E.F., D'Angelo, D.J., Peters, G.T., 1990. Effects of forest disturbance on particulate organic matter budgets of small streams. Journal of the North American Benthological Society 9, 120–140.

Welch, P.S., 1948. Limnological Methods. Blakiston Co., Philadelphia, PA.

Wotton, R.S., 1978. The feeding-rate of *Metacnephia tredecimatum* larvae in a Swedish lake-outlet. Oikos 30, 121–125.

Wotton, R.S., 1980. Coprophagy as an economic feeding tactic in blackfly larvae. Oikos 34, 282–286.

Wotton, R.S., 1984. The importance of identifying the origin of microfine particles in aquatic systems. Oikos 43, 217–221.

Wotton, R.S. (Ed.), 1990. Biology of Particles in Aquatic Systems. CRC Press, Boca Raton, FL.

Wotton, R.S., Malmqvist, B., Muotka, T., Larsson, K., 1998. Fecal pellets from a dense aggregation of suspension feeders in a stream: an example of ecosystem engineering. Limnology and Oceanography 43, 719–725.

Yoshimura, C., Gessner, M.O., Tockner, K., Furumai, H., 2008. Chemical properties, microbial respiration, and decomposition of coarse and fine particulate organic matter. Journal of the North American Benthological Society 27, 664–673.

Zetsche, E.-M., Mallahi, A.E., Dubois, F., Yourassowsky, C., Kromkamp, J.C., Meysman, F.J.R., 2014. Imaging-in-Flow: digital holographic microscopy as a novel tool to detect and classify nanoplanktonic organisms. Limnology and Oceanography: Methods 12, 757–775.

Chapter 26

Coarse Particulate Organic Matter: Storage, Transport, and Retention

Gary A. Lamberti[1], Sally A. Entrekin[2], Natalie A. Griffiths[3] and Scott D. Tiegs[4]

[1]Department of Biological Sciences, University of Notre Dame; [2]Department of Biology, University of Central Arkansas; [3]Environmental Sciences Division, Oak Ridge National Laboratory; [4]Department of Biological Sciences, Oakland University

26.1 INTRODUCTION

Coarse particulate organic matter, or CPOM, in streams is functionally defined as any organic particle larger than 1 mm in size (Cummins, 1974). CPOM can be further divided into nonwoody and wood material (Cummins and Klug, 1979), the former of which will be considered in this chapter. The nonwoody fraction includes allochthonous materials produced and donated by riparian organisms (e.g., leaves, needles, fruits, flowers, seeds, insects, frass) and autochthonous materials produced within the stream (e.g., algae, aquatic plants, dead aquatic animals). Smaller materials, including dissolved organic matter (DOM < 0.45 µm) and fine particulate organic matter (1 mm > FPOM > 0.45 µm), are primarily considered in Chapters 24 and 25, respectively, but are also discussed in the context of organic carbon spiraling as an advanced method in this chapter. Wood includes a broad range of size classes from branches to entire trees that fall into stream channels and perform important ecological functions (Bilby and Likens, 1980) that are considered in Chapter 29.

Allochthonous CPOM is a major energetic resource for most stream ecosystems, providing a large proportion of carbon in small streams of both deciduous and coniferous forests and a significant input to larger streams and rivers (Vannote et al., 1980; Cummins et al., 1983; Tank et al., 2010). Autochthonous CPOM can be an important resource in open-canopy streams or in forested streams prior to leaf-out in the spring (Minshall, 1978; Roberts et al., 2007). Regardless of source, this CPOM is broken down by stream biota during an activity known as organic matter processing. CPOM is transported downstream by the unidirectional flow of water, with very few mechanisms for upstream movement. The process of deposition and trapping of this material is termed "retention," which provides the critical link between input and the long-term storage and processing of CPOM. Retained CPOM can be measured as the areal amount in a particular habitat or an entire stream, often referred to as the *standing crop*. Standing crop suggests the amount available to the "stock" of micro- and macrodetritivores. Retention is therefore essential for subsequent microbial colonization and hydrolysis that precedes storage and consumption of CPOM (and associated microbes) by detritivores (Cummins and Klug, 1979; Graça, 2001).

CPOM retentive capacity of streams is a function of hydrologic, substrate-related, and riparian features (Speaker et al., 1984; Cordova et al., 2008). High roughness levels of the channel (*sensu* Chow, 1959) (e.g., large substrate size, streambed heterogeneity, abundant wood), combined with certain hydraulic conditions (e.g., presence of backwaters, interstitial flow), tend to increase channel CPOM retention efficiency. Large wood, especially in accumulations, provides particularly important retention structures (Bilby, 1981; Smock et al., 1989; Jones and Smock, 1991; Entrekin et al., 2008). Young et al. (1978) noted that the probability that a particle in transport will be retained is a function of the "active" entrainment efficiency of that particle size by a channel obstacle (e.g., rock, log, root, etc.) and the density of those obstacles within the channel. A particle also will be retained "passively" when current velocity is insufficient to transport it (Jones and Smock, 1991), and thus the particle "settles." Retention (R) can thus be expressed as a probability function:

$$P(R) = f(E, N, V) \tag{26.1}$$

where E = entrainment efficiency by channel obstacles, N = obstacle density in the channel, and V = critical velocity required to transport a particle. If an organic particle is retained, it subsequently can be respired by microorganisms, consumed by detritivores, or, if flow conditions change, be dislodged and transported further downstream (Speaker et al., 1984). This combined process of organic carbon transport, retention, and processing in streams is termed "organic carbon spiraling" (Newbold et al., 1982).

In this chapter, we describe basic methods to quantify CPOM (except for wood; see Chapter 29) stored in the stream channel. We also present a simple, quantitative field method to assess the CPOM retention efficiency of a specific stream reach. These methods are most easily used in small streams (orders 1—4), but can be adapted for larger streams and rivers. These approaches are also intended to relate storage and retention to hydraulics, streambed roughness, channel geomorphology, and riparian zone structure. We then present an advanced method to experimentally augment the retention capacity of a stream reach so that researchers can evaluate the effects that increased CPOM quantity may have on stream ecosystem processes (e.g., nutrient uptake, secondary production, invertebrate abundance). In a second advanced method, we demonstrate how to measure the transport and processing of organic carbon in stream ecosystems using a carbon spiraling approach. Our specific objectives are to (1) introduce the concepts of organic matter storage, retention, and dynamics; (2) demonstrate how to measure CPOM standing crop and retention, analyze data, and calculate indices of retention; (3) illustrate the utility of storage and retention measurements for assessing stream channel condition; and (4) describe field techniques to experimentally manipulate the retention capacity of a stream and to measure the longitudinal dynamics of organic carbon.

26.2 GENERAL DESIGN

Organic matter in streams provides short- and long-term resources for benthic biota including detritivorous micro- and macroinvertebrates, and even some vertebrates. Therefore, the measurement of CPOM benthic standing crop and its size fractions is important for assessing resource availability and potential energy flow among trophic levels of aquatic food webs (Wallace et al., 1997; Rosi-Marshall and Wallace, 2002). In fact, CPOM standing crops can predict aquatic community secondary production (Chadwick and Huryn, 2007; Entrekin et al., 2009). Standing crop is typically quantified as organic matter size fractions >1 mm in a particular habitat type (e.g., pools or riffles) or across all habitat types, as sampled with a benthic corer and expressed as dry mass (DM) or ash-free dry mass (AFDM) per unit area (see Chapter 12). Small, high-gradient streams in coniferous, deciduous, and boreal forests tend to store the most CPOM per unit area (cf. mid- to high-order streams and streams in arid/semiarid biomes), although standing crops vary with season, precipitation, channel gradient, and large wood storage (Jones, 1997). At the local scale, CPOM storage also varies among and within habitats; therefore, habitat-specific sampling may be informative. For example, in low-gradient, sandy-bottom streams, CPOM storage will be greatest along stream margins, in pools, and around wood accumulations (Jones and Smock, 1991). Habitat-specific CPOM sampling will thus provide information on how differences in channel form affect CPOM storage and availability.

Lotic retention can be quantified as the difference between the number of particles that enter a length of stream and the number transported through that same length (Speaker et al., 1984). Retention is most easily measured by releasing known numbers of readily distinguishable but representative particles into the channel. To compare different streams or stream reaches within a study, the experimental approach must be standardized for type and number of particles released, length of experimental reach, and duration of the retention measurement. Many types of CPOM have been released into streams, including leaves (Speaker et al., 1984), paper shapes (Webster et al., 1994), plastic strips (Bilby and Likens, 1980; Speaker et al., 1988), wood dowels (Ehrman and Lamberti, 1992), and even fish carcasses (Cederholm et al., 1989). In general, we believe that it is preferable to release natural (decomposable) materials into streams because analogs such as plastic items or paper strips (1) may not behave the same as natural materials and (2) will generally not be 100% retrievable, thereby leaving trash in the stream. In this chapter, we will demonstrate retention of leaves, but other materials significant to the specific stream can be substituted. For example, fruits and seeds are significant CPOM inputs in many tropical streams (Larned et al., 2001).

26.2.1 Site Selection

The selection of a study stream in which to conduct these exercises may be influenced by logistical considerations. Storage and carbon spiraling measurements can be made in any stream where the equipment can be safely operated. For the retention experiments, wadeable second- to fourth-order streams are ideal. Very small streams at low flow will have low transport capacity, and these methods can be challenging (and also dangerous) to conduct in large rivers. In general, however, these methods can be scaled to a wide variety of stream sizes. For CPOM release experiments, at least two reaches in the same stream (or two different streams) with contrasting channel features should be selected by the research coordinator. Ideally, one reach would have a relatively simple channel (straight, low roughness, limited hydraulic diversity, sparse wood), whereas the other reach should have a complex channel (sinuous, high roughness, diverse hydraulic conditions, abundant wood). Alternatively, the same stream reach can be studied at different discharges or before and after some event, such as a treefall or a channel modification associated with an experiment or stream restoration.

The length of the experimental reach should be scaled to stream size, with length increasing with stream order. As a rule of thumb, start with a stream length that is ∼ 10 times the wetted channel width. For example, 50 m may be an appropriate length for a 5-m wide, second-order stream, 100 m for a third-order stream, and 200 m for a fourth-order stream. If possible, use a pilot study to adjust reach length such that retention is not less than 10% nor greater than 90% of released particles.

26.2.2 Basic Methods

We describe below two basic exercises that quantify CPOM storage and retention that can be conducted by small groups of students in the field or that can be used as components of more sophisticated research or monitoring programs. CPOM storage typically increases with greater riparian inputs and under the same conditions as retention; however, storm flow, detritivore consumption, and microbial respiration result in depletion.

Within a stream, CPOM storage can be measured at the microsite (e.g., sediment type), habitat (e.g., riffle, pool), or stream-reach scale. Multiple, haphazard samples or stratified random sampling may be required to adequately characterize this often-patchy resource. In many cases, benthic corers (see Chapter 8) can be used to delineate a known area of streambed from which CPOM is collected and weighed to estimate standing crop. Studies that aim to quantify the importance of a given CPOM resource in a stream should at least sample weekly or biweekly during the period of greatest CPOM input (e.g., during autumnal leaf fall in deciduous forests, during algal senescence in open-canopy streams, or after major storms in intermittent streams). Monthly or biweekly sampling throughout the year may be required in other biomes to provide the most complete quantification of CPOM storage in regions without distinct seasonality in CPOM inputs.

Leaves are the major form of nonwoody CPOM input to most streams, and their retention is an important ecological process (Webster et al., 1999). In retention experiments, released leaves must be distinguishable from leaves found naturally in the channel and should be easy to obtain and manipulate. We have found that, for North American streams, abscised leaves of the exotic Asian ginkgo tree (*Ginkgo biloba*) meet these requirements. The leaves are tough even when wet, their size approximates that of many leaf types of riparian vegetation, and the bright yellow leaves are easily spotted in the channel. *Ginkgo* trees have been planted worldwide as ornamentals (and very often on college campuses). Other species of leaves can be substituted depending on their availability, visibility during experiments, and the composition of local riparian vegetation.

Physical measurements of the stream channel should be taken according to the level of effort possible and available equipment. Useful parameters include discharge, slope, sinuosity, cross-sectional area, and planar wetted area (see Chapters 2–5), along with volume of large wood (see Chapter 29). Retention data for leaves (using batch releases) should be fit to a negative exponential decay model, from which indices of retention (e.g., the retention coefficient, $-k$; average particle travel distance, $1/k$) can be calculated. Metrics for individual particle releases can be generated using simple statistics. If desired, CPOM releases can be conducted over longer periods of time, or at different seasons and discharges, to develop relationships between retention and stream temporal dynamics (e.g., Jones and Smock, 1991; Webster et al., 1999).

26.2.3 Advanced Methods

Two advanced exercises involving additional sophistication, time, and facilities are also presented in this chapter. These are "research-level" approaches suitable for incorporation into studies intended for theses and published papers. First, we describe an experiment to enhance stream retention via the installation of litter retention devices in the stream channel, after which other whole-reach responses such as metabolism (see Chapter 34) can be measured (Tiegs et al., 2008). We then describe an advanced method to assess organic carbon spiraling, which is an integrative measure of the transport, retention, and processing of organic carbon within a stream channel (Newbold et al., 1982). Carbon spiraling involves the measurement of organic carbon pools in transport, standing crops on the benthos, processing of organic carbon (via a measure or estimate of heterotrophic respiration), and physical measurements of stream depth, width, velocity, and discharge. From these measurements, three carbon spiraling metrics can be calculated: (1) V_{OC}, which is the downstream transport velocity of organic carbon (m/day), (2) K_{OC}, which is the processing rate of organic carbon (day^{-1}), and (3) S_{OC}, which is the organic carbon spiraling length and a measure of the distance organic carbon travels downstream before being respired (m).

Carbon spiraling metrics estimate all fates of organic carbon (transport, retention, and processing) and thus can be used to address hypotheses pertaining to whole-stream carbon dynamics. For instance, carbon spiraling metrics have been used to examine seasonal patterns in whole-stream carbon dynamics (Thomas et al., 2005) to compare the retentiveness and processing of carbon across a variety of streams (Webster and Meyer, 1997) and land use types (Griffiths et al., 2012), and to examine the effect of stream consumers on carbon processing (Taylor et al., 2006).

26.3 SPECIFIC METHODS

26.3.1 Basic Method 1: Coarse Particulate Organic Matter Storage and Measurement

26.3.1.1 Field Measurements

1. Select two stream reaches (e.g., 50–100 m long depending on stream width) with differing riparian canopy cover, slope, or other contrasting geomorphological features. Flag 5- to 10-m subreaches depending on chosen reach length, and haphazardly[1] select sampling locations within each subreach. If habitat-specific samples will be taken, identify the dominant habitat types and then haphazardly sample replicate habitats within each reach.
2. Using a coring device of known area (e.g., bucket with bottom removed, PVC pipe section), form a tight seal with the streambed (Fig. 26.1). By hand, remove as much coarse material as possible and place in a labeled paper bag. Then, stir the remaining material by hand, skim smaller pieces with a 1-mm mesh sieve or hand net, and place in a labeled paper bag.
3. Carefully fold the paper bag, and place in a labeled plastic bag for transport to the laboratory.

26.3.1.2 Laboratory Processing

1. Place all labeled paper bags in a labeled paper box and dry at 60°C for at least 24 h. Drying times will vary with amount and the type of litter. Once dried, litter types can be sorted (e.g., small wood, moss, leaves, other) and weighed and compared individually or together, depending on the research question.

FIGURE 26.1 Photograph of benthic corer (open-bottom bucket) used for coarse particulate organic matter storage measurements in an Ozark Mountain, USA, stream. *Photo: A. Bates.*

---

1. It will be challenging to truly sample randomly within the stream, and so "haphazard" sampling that attempts to minimize bias in sampling site selection is generally acceptable.

2. Remove litter from the drying oven, allow to cool to room temperature in a desiccator, and take a total weight. Then, with gloved hands, crush and homogenize all litter, place a subsample in a labeled, ashed, and weighed aluminum pan, and weigh the subsample. A grinder will be useful for tough material (e.g., seeds, wood fragments).

3. Place subsamples in a muffle furnace at 500°C for 2 h.[2] Remove litter, allow to cool at 60°C for 12 h, place in a desiccator until room temperature is reached, and then reweigh (see also Chapters 12 and 27).

26.3.1.3 Data Analysis

1. CPOM standing crop can be expressed as grams of DM or AFDM (dry mass minus ash mass after burning at 500°C) (Table 26.1; see also Chapter 12). Once DM or AFDM are estimated, convert to grams per unit area by dividing mass by corer area. To calculate proportion AFDM, subtract ash mass from DM and then divide by DM.

2. Various statistical analyses can be conducted depending on study objective. To compare two stream reaches, average the cores from the same habitats (or all benthic cores for a reach) and compare mean standing crop with a t-test. For a more advanced analysis, sample multiple times within a season or throughout the year and use one-way repeated measures analysis of variance (rmANOVA) to identify differences between stream reaches, or habitats, over time.

26.3.2 Basic Method 2: Coarse Particulate Organic Matter Transport and Retention

26.3.2.1 Laboratory Preparation

1. In the autumn, collect several thousand abscised (fallen) leaves of an exotic tree, such as *G. biloba*, or other readily identifiable species. Air-dry the leaves by spreading them over screens, netting (seines work well), or even on newspaper on the floor. Leaves can be stored dry in black garbage bags for a considerable length of time. Alternatively, you can use fresh-fallen leaves if the releases will be performed soon after collection (within days).

2. Count out two batches of 1000 leaves, used to conduct two releases in a third-order stream. Smaller or larger streams may require fewer or more leaves, respectively. The actual number of leaves is less important than knowing exactly how many are released.

3. The day before the release, soak leaves overnight in buckets of water to impart neutral buoyancy during transport. A soil sieve placed gently over dried leaves will help to keep them submersed. Drain most of the water before departing into the field.

26.3.2.2 Field Physical Measurements

1. Measure and flag an appropriate length (e.g., 100 m for a third-order stream) of at least two stream reaches differing in channel complexity, large wood abundance, or some other geomorphological feature. Stretch a meter tape along the bank over the length of the reach, with 100 m at the downstream end.

2. Measure major channel features at a level of intensity appropriate to the research objectives. We recommend working in a research team of three people (two making measurements and one recording data). Useful measurements include slope, average width, depth, sinuosity, and substrate composition (see Chapters 2 and 5). Determine discharge using the cross-sectional approach (see Chapter 3). Repeat for each reach.

3. Measure the abundance of large wood in the channel (see Chapter 29), noting if the wood is part of a dam (i.e., wood accumulation blocking some portion of stream flow).

4. If this method is being used for a class exercise, prior to the releases briefly discuss channel and riparian features. Have students predict retention for each reach (e.g., percentage of leaves that will be retained).

26.3.2.3 Field Coarse Particulate Organic Matter Releases

1. Position several researchers at the downstream end of the study reach. Release the leaf batch (e.g., 1000 leaves) at the upstream end of the reach (i.e., 0 m mark) by dispersing leaves over the entire width of the stream channel over a span of about 1 min (Fig. 26.2A).

2. Collect nonretained leaves at the downstream end of the reach (e.g., 100 m). Either of two approaches can be used to collect leaves. A beach seine can be stretched across the width of the channel (Fig. 26.2B), with the bottom lead line anchored, without gaps, to the streambed with rocks (in sandy-bottom streams, tent stakes can be substituted for rocks).

---

2. Wear appropriate personal protective equipment (PPE) including lab coat, oven mitts, and eyewear.

TABLE 26.1 Example laboratory processing spreadsheet for coarse particulate organic matter (CPOM) standing crop and size fractions (see also Online Worksheet 26.1 for detailed calculations).

Date:			Corer Area:		
Stream:			Team:		
Location:			Notes:		
Reach:					
Length:					
Location	Unit	CPOM	Sample	Subsample	
Meter Mark	Riffle or Pool	Fraction Type (e.g., Leaf, Moss, Small Wood)	Total Dry Weight (g)	Dry Weight (g)	Ash Weight (g)
100–95					
95–90					
90–85					
.					
.					
.					
5–0					

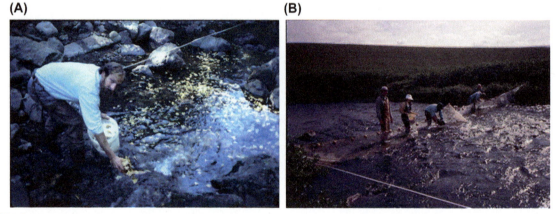

(A)   **(B)**

FIGURE 26.2 Photographs of coarse particulate organic matter releases. (A) *Ginkgo biloba* leaf release into a Oregon, USA, stream; (B) beach seine stretched across the channel of a northern Alaska, USA, stream (anchored to the bottom with rocks) for capturing unretained leaves. *Photos: G. Lamberti.*

The top of the seine should be held out of the water by attaching it to a taut rope tied to trees on both banks. In strong flows, it may be further necessary to support the top and rear of the seine with wood pieces driven into the substrate. Alternatively, researchers can line up across the channel and collect leaves in transport with hand-held dip nets (e.g., D-frame or delta nets) if safe to do so. The seine method is more efficient, especially if the number of researchers is low. The individual netting approach results in greater involvement of researchers in actual leaf collection, but some leaves may be missed.

3. Continue collecting leaves for a period of time specified by the coordinator, usually at least 15 min and up to 1 h, or when leaf transport ceases. This period should be consistent for all reaches. Count all collected (i.e., nonretained) leaves.

4. Move upstream to the next reach (or contrasting stream) and repeat the procedure.

26.3.2.4 Data Analysis

1. Calculate reach physical parameters, such as slope, planar surface, cross-sectional area, mean depth, current velocity, hydraulic radius, sinuosity, and discharge (see Chapters 2–5). These fundamental physical parameters can be related empirically or theoretically to observed retention values.
2. Determine the density (pieces per reach) and total volume (m^3) of wood in each reach, according to methods presented in Chapter 29.
3. Fit the leaf retention data to a negative exponential decay model of the form:

$$P_d = P_0 e^{-kd} \qquad (26.2)$$

where P_0 = number of particles released into the reach and P_d = number of particles still in transport (i.e., collected) at a known downstream distance d from the release point (m). Calculate the slope $-k$ (the instantaneous retention rate; $1/m$) and its reciprocal $1/k$ (the average distance traveled by a particle before it is retained; m). If particles are not inventoried after the release, then the model will be based on two data points, P_0 and P_d. See *Advanced Method 1* for data analysis if particles are reinventoried in the channel.

26.3.2.5 Option to Basic Method 2: Single-Particle Releases

1. As an alternative to batch releases of particles described above, the single-particle release method can be used (Webster et al., 1994). Single particles (e.g., leaves, fruits) or artificial analogs (e.g., "Rite in the Rain" field paper, cut into standard shapes) are released into the channel and individual travel distances are recorded.
2. Release a known number (e.g., 25–50) of visible particles one-by-one into the channel and record the distance traveled and retention structure for each particle. Repeat this procedure in as many stream reaches, or subreaches, as desired. Mean travel distances can be compared statistically among reaches using ANOVA, or relationships with stream characteristics such as discharge can be explored with regression (e.g., Webster et al., 1994, 1999).
3. This approach is especially useful in highly retentive streams where few or no particles may travel the entire stream reach, thereby invalidating the exponential decay model. However, this method requires relatively high water clarity and shallow depths to follow individual particles for their entire travel path.

26.3.3 Advanced Method 1: Enhancement of Stream Retentive Capacity

Retention of CPOM by the stream channel stems from two very different processes: deposition of CPOM due to insufficient transport capacity, and the forcing of CPOM by moving water against roughness elements (e.g., substrata, wood, tree roots, etc.) such that it is held in place (hereafter termed "active retention"). Below we describe an experimental procedure for enhancing the active retention of organic matter in a stream reach. Although experimental reduction of the retentive capacity of stream channels has been performed (e.g., Díez et al., 2000), enhancing retention through the installation of retention devices is more commonly done (e.g., Tiegs et al., 2008) and more readily standardized.

Retention devices can be customized to retain different types of CPOM (e.g., salmon carcasses, Tiegs et al., 2011; Fig. 26.3A and B), but in most instances researchers have used them to retain leaf litter (e.g., Dobson et al., 1995). In some instances, large wood has been introduced into streams to replicate past (prelogging) conditions with the explicit goal of retaining in-channel CPOM (Fig. 26.3C; see Entrekin et al., 2008). By installing litter retention devices, the standing crop of litter in the stream channel can be significantly increased above background levels, and the response of the stream ecosystem to this enhanced CPOM evaluated. Below we present a method for installing leaf litter retention devices made from metal or wooden stakes and pieces of extruded plastic mesh (Fig. 26.3D). The method as described is for small, shallow streams, with a study reach of 50 m in length, and will yield a retention device density of $\sim 1/m^2$ (after Dobson, 2005), but it can be adjusted to accommodate larger or smaller streams.

26.3.3.1 Laboratory Preparation

1. Assemble materials for field deployment of retention devices (see Materials and Supplies).

26.3.3.2 Field Measurements—Deployment of Retention Devices

1. Identify the stream reach in which retention is to be enhanced and a paired upstream reference reach that will not be enhanced. Retention will be most readily enhanced in reaches that lack abundant retention structures (e.g., pools, wood accumulations, or coarse substrate) and have significant CPOM inputs.

FIGURE 26.3 Photographs of coarse particulate organic matter (CPOM) retention devices. (A) Salmon carcass retention devices in a southeast Alaska, USA, stream; (B) close-up of single carcass retention device (1.5 m wide) with upright rebar stakes, capped for safety, and attached conduit pipe cross-member; (C) large wood (logs with cut ends) introduced into a Michigan, USA, stream; (D) CPOM retention devices (upright rebar stakes and green mesh fences) in a small stream in the Black Forest, Germany; each fence is ~20 cm × 20 cm. *(A) Photo: S. Tiegs. (B) Photo: G. Lamberti. (C) Photo: S. Entrekin. (D) Photo: F. Peter.*

2. Walk the length of the experimental reach, and distribute the materials for each retention device (two of the wood stakes, and one of the mesh squares) along the shoreline for every square meter of stream channel in the reach. For example, for a stream that is 1 m wide on average, distribute the materials for one retention device for every meter of stream length.[3]

3. Near the location where the materials for the retention devices were placed, begin installation by hammering one of the stakes into the stream substratum (Fig. 26.3C). Hammer until half the stake is in the stream substratum.

4. Attach one edge of the mesh square to the stake with binder clips or cable ties (three clips or ties along each edge, one at the top, one at the bottom, and the other near the middle should suffice). Do not tighten yet—wait until the other stake has been secured to the streambed.

5. Hammer the other stake into the substratum approximately 20 cm (i.e., width of mesh square) away from the first stake and aligned perpendicular to stream flow. Attach the clips or ties to the second stake, and tighten all the attachments securely.

6. Repeat this procedure for each of the retention devices. The CPOM retention of each trap will be maximized by locating them in areas of relatively fast water and by "staggering" the locations such that each trap is not located immediately downstream from the one above it.

7. After installation and following a suitable period for CPOM trapping (typically on the order of weeks to months), randomly select a subset of the retention devices, and collect all the CPOM trapped on each device. When removing CPOM from a retention device, make sure to position a 1-mm sieve or screen immediately downstream to capture detached particles.

8. Procedures described in *Basic Method 1* can then be used to estimate the DM and AFDM (see Chapter 12) of the retained material. To estimate the total mass of CPOM experimentally captured, extrapolate the subset to the total number of devices installed in the reach.

---

3. Install the devices during low flow conditions for ease and safety, such as before autumn rains and the onset of leaf fall.

26.3.3.3 Field Measurements—Related Ecosystem Measurements

Many ecosystem attributes can be measured and evaluated after the experimental augmentation of CPOM, such as nutrient cycling, leaf decomposition, invertebrate responses, and ecosystem metabolism in the paired reference and experimental reaches (see relevant chapters in this book). The timing of measurements should be adjusted to the length of the experiment and the period required for the response. Below, as an example, we describe an enhanced method for measuring leaf retention in the context of this experiment.

1. Prior to the installation of the retention devices, perform a leaf release in both reaches (i.e., an upstream reference reach, and the reach where the retention devices will be installed) as in *Basic Method 2*. After device installation, and under similar flow conditions, repeat the releases to estimate the enhanced retentiveness provided by the retention devices.
2. Inventory the location, number, and retention structure (including installed devices) for retained leaves. This is best accomplished by dividing the reach into longitudinal increments of 5 m using a bankside meter tape.
3. Researchers should move up the channel as a single line of observers, perpendicular to flow. Locate and count released leaves within each increment, noting also the retention structure (e.g., rock, wood, bank, retention device, etc.; see Table 26.2).
4. These inventory data can be used to refine the exponential model described in *Basic Method 2* and produce a more accurate estimate of $-k$, or to fit retention data to an alternate regression model (e.g., linear, power) more appropriate for the specific reach. The inventory most likely will not turn up all of the retained leaves; therefore, it is necessary to normalize inventory data to a percent of total leaves found. Graph the particle transport data for each release, using distance downstream from the release point as the x-axis and percent of particles still in transport as the y-axis (see Ehrman and Lamberti, 1992 for examples). Using a bar diagram, plot the percentage of leaves retained by specific channel structures in each reach. Describe the longitudinal pattern of retention, and identify important retention structures within the channel.
5. Data from the paired reference and experimental reaches can be analyzed by various means depending on study objectives and the frequency of sampling. If sufficient samples are taken in both reaches before and after the construction of retention devices, then a before-after-control-intervention (BACI) analytical approach can be used (Stewart-Oaten et al., 1986).

TABLE 26.2 Sample data sheet for inventory of retained coarse particulate organic matter (CPOM) particles in a 100-m reach; add a column for retention devices (*Advanced Method 1*) if employed (see also Online Worksheet 26.2).

Date:				CPOM Type:				
Stream:				Duration:				
Location:				Total Released:				
Reach:								
Length:				Total Captured:				
Team:								
Notes:				Total Retained:				

Location	Unit			Number of Particles Retained on Structure				
Meter Mark	Riffle or Pool	Rock	Root	Backwater	Bank	Wood	Debris Dam	Other
100–95								
95–90								
90–85								
.								
.								
.								
5–0								

26.3.4 Advanced Method 2: Measurement of Organic Carbon Spiraling

Organic carbon spiraling is a measure of the transport and turnover of organic carbon within a stream (Newbold et al., 1982). To estimate carbon spiraling in a given stream, carbon pools (in transport and as standing crops), carbon turnover rates (i.e., heterotrophic respiration), and physical characteristics (stream depth, width, velocity, and discharge) are measured. From these measurements, three carbon spiraling metrics are calculated: (1) the downstream transport velocity of organic carbon, (2) the processing rate of organic carbon, and (3) organic carbon spiraling length (Newbold et al., 1982; Thomas et al., 2005; Griffiths et al., 2012). All size fractions of organic carbon (coarse, fine, ultrafine, dissolved) are measured and described in the method below, but this method can also be modified to focus on only the CPOM fraction.

26.3.4.1 Site Selection

1. Select a study reach that is at least 100−200 m in length for a first- to second-order stream, or 500−1000 m in length for a wadeable mid-order stream. Ensure that there are minimal lateral inputs of water along the reach (i.e., tributaries).

26.3.4.2 Field Measurements—Transported Organic Carbon

Transported organic carbon (TOC) is the sum of CPOM (>1 mm), FPOM (52 μm−1 mm), ultrafine particulate organic matter (UPOM, 0.45−52 μm), and DOM (<0.45 μm).

1. To measure CPOM in transport, position a CPOM net (1-mm mesh size; drift net or similar, see Chapter 21) in an area with representative flow, with the bottom of the net slightly off the streambed. Allow CPOM to collect in the net, and remove the net before it begins to clog (usually 1−2 h). Record the length of time the net was in the stream collecting CPOM. Measure water velocity (see Chapter 3) and depth of the net at multiple (≥3) locations across the width of the net after the sample has been collected but before the net is removed from the water (Table 26.3). The net width and depth and water velocity are used to calculate discharge at the net, and the total volume of water that passed through the net during the CPOM collection is calculated based on the length of time the net was in the water. Rinse all organic matter caught in the net into a labeled sample cup or plastic bag, and keep the sample on ice until returning to the laboratory. If using a drift net, the net will likely not be wide enough to span the entire width of the stream. Therefore, sample multiple times at various locations along the reach to account for spatial and temporal variation in CPOM transport. Make sure to start downstream and work upstream to collect additional samples.

TABLE 26.3 Sample data sheet for total organic carbon collection in the field as part of the carbon spiraling analysis (see also Online Worksheet 26.3 for detailed calculations).

Date:

Stream:

Team:

Notes:

Location	Net Type	Sample ID	Time Net In	Time Net Out	Net Depth (m)	Water Velocity (m/s)	Net Width (m)
Meter Mark	CPOM, FPOM[a]	##	(hh:ss)	(hh:ss)	(3 Locations)	(3 Locations)	

[a]Make sure to collect an ultrafine particulate organic matter and dissolved organic matter sample with every coarse particulate organic matter (CPOM) and fine particulate organic matter (FPOM) sample.

2. To measure FPOM in transport, repeat the procedure for CPOM with an FPOM net (52 μm mesh size). Note that the FPOM net will clog more quickly (e.g., <5 min, depending on concentration and water velocity). A nested set of CPOM and FPOM nets could also be used (i.e., FPOM net downstream of CPOM net).
3. To measure UPOM, collect an unfiltered water sample (2 L or greater) along with each CPOM and FPOM sample (see also Chapter 25).
4. To measure DOM, collect a filtered (0.45 μm) 50-mL water sample into an amber glass vial, and acidify the sample with two drops of 6N hydrochloric acid[4] (see also Chapter 24).

26.3.4.3 Field Measurements—Benthic Organic Carbon

All organic matter fractions except for DOM are also measured on the streambed. Collect benthic organic carbon (BOC) at five locations along the study reach, and select more locations to sample if BOC in the reach is heterogeneous. Collect BOC samples 1–2 days after the TOC and heterotrophic respiration measurements as BOC sampling disrupts the streambed. The method to measure BOC described below does not distinguish among different organic matter substrata (e.g., leaves, algae). If this is desired, a habitat-weighted transect approach can be used (see Hoellein et al., 2007; Griffiths et al., 2012).

1. Measure CPOM on the benthos by placing a corer on the streambed and removing all CPOM contained within the core (see *Basic Method 1* above).
2. After all CPOM is removed from the core, use your hand to swirl the top 3–5 cm of sediments, causing the particles to be suspended in water. Collect all other particulate fractions by quickly sampling the suspension with a plastic cup, and then place the sample on ice.

26.3.4.4 Field Measurements—Organic Carbon Turnover

Multiple methods can be used to estimate heterotrophic respiration, which is a measure of organic carbon processing in stream ecosystems. Below, we describe a method for estimating heterotrophic respiration from whole-stream metabolism measurements. Heterotrophic respiration can also be measured using benthic chambers (see Chapter 27; Bott et al., 1978) or can be measured on different types of benthic organic matter (e.g., leaves, fine benthic organic matter) using small closed chambers (Hoellein et al., 2009; Griffiths et al., 2012).

1. Use the diel oxygen change method described in Chapter 34 to estimate gross primary production (GPP) and ecosystem respiration (ER). Measure GPP and ER on a day when no other measurements are occurring in the stream. If possible, use the two-station metabolism method (see Chapter 34) to measure GPP and ER in the study reach only.

26.3.4.5 Field Measurements—Physical Characteristics of the Stream

Measurements of stream discharge (Q), width (w), velocity (v), and depth (z) are needed to calculate metrics of organic carbon spiraling.

1. Measure stream discharge using the velocity–area protocol at the bottom and top of the study reach and calculate mean discharge for the reach from these measurements (see Chapter 3).
2. Measure stream width every ∼5 m along the study reach using a tape measure.
3. Measure average reach water velocity using the salt (NaCl) release method (see Chapter 30).
4. Calculate mean stream depth z from Q, w, and v as:

$$z = Q/v \times w \tag{26.3}$$

26.3.4.6 Laboratory Processing

1. Each fraction of TOC should be processed separately in the laboratory. For the CPOM samples, rinse the sample over a 1-mm sieve and place all detritus caught on the sieve into a precombusted and preweighed aluminum pan. For the FPOM samples, first pass the sample through a 1-mm sieve to remove the CPOM fraction (if the nets were not nested

---

4. Wear appropriate PPE including lab coat, eyewear, and gloves when handling HCl.

in the field), and then collect the filtrate on a precombusted and preweighed 0.7-μm glass fiber filter. For the UPOM samples, first pass the sample through the 52-μm mesh net to remove any CPOM and FPOM fractions, and then collect the filtrate on a precombusted, preweighed 0.45-μm filter. All pans and filters should then be dried at 60°C for 48 h, weighed, and then ashed in a muffle furnace at 500°C for 2 h. Calculate AFDM as described in *Basic Method 1* (see also Chapter 12). Convert mass from g AFDM to g organic carbon assuming that 48.4% of AFDM is organic carbon (Thomas et al., 2005). Measure DOC on the filtered (0.45 μm) water sample using the method described in Chapter 24.

2. The same processing steps are used to analyze the BOC samples. For the CPOM samples, pass the detritus through a 1-mm sieve and place the material captured by the sieve in a precombusted and preweighed aluminum pan. For the FPOM/UPOM sample, pass the sample through nested 1-mm and 52-μm sieves, collect the material on the 52-μm sieve for FPOM, and collect the filtrate on a 0.45-μm filter for UPOM. Measure g AFDM and convert to g C as described in step 1.

26.3.4 7 Data Analysis

The following describes the steps to calculate the three metrics of organic carbon spiraling: V_{OC}, the downstream transport velocity of organic carbon; K_{OC}, the processing rate of organic carbon, and S_{OC}, the organic carbon spiraling length:

1. Downstream transport velocity of organic carbon:

$$V_{OC} = TOC \times Q/BOC \times w \qquad (26.4)$$

TOC is the total organic carbon concentration in transport (g C/m^3) and is calculated as the sum of CPOC + FPOC + UPOC + DOC in transport. CPOC and FPOC in transport are calculated as the g organic carbon in the sample divided by the volume of water passing through the nets when the sample was being collected. UPOC in transport is calculated as the g organic carbon in the sample divided by the total volume of water in the sample. Q is stream discharge measured in the field (converted from L/s to m^3/day). BOC is the total benthic organic carbon standing stock (g C/m^2) and is calculated as the sum of CPOC + FPOC + UPOC on the streambed. CPOC, FPOC, and UPOC on the streambed are calculated as the g organic carbon in the sample divided by the surface area of the benthic corer/sampler. Mean stream width, w (m) is measured as described previously.

2. Processing rate of organic carbon:

$$K_{OC} = R_{het}/BOC + (TOCxz) \qquad (26.5)$$

BOC and TOC are benthic organic carbon (g C/m^2) and transported organic carbon (g C/m^3), as described above, and z is mean stream depth (m). Heterotrophic respiration (R_{het}) is estimated from measurements of GPP and ER using the following equation:

$$R_{het} = ER - \alpha GPP \qquad (26.6)$$

where α is the fraction of GPP that is respired by autotrophs; α has been estimated at 0.2 (Young and Huryn, 1999) to 0.5 (Webster and Meyer, 1997). The autotrophic respiration fraction may also be estimated from the relationship between GPP and ER, but only if continuous metabolism data are available (Hall and Beaulieu, 2013). Rates of heterotrophic respiration are then converted to g C m^{-2} d^{-1} by multiplying R_{het} (g O$_2$ m^{-2} d^{-1}) by a respiratory quotient of 0.85 and the molar ratio of atomic C to O$_2$ (12/32) (see Chapter 34).

3. Organic carbon spiraling length:

$$S_{OC} = V_{OC}/K_{OC} \qquad (26.7)$$

The organic carbon spiraling length, S_{OC} (m), is defined as the distance organic carbon travels downstream before being respired and is calculated as the downstream velocity of organic carbon, V_{OC} (m/day), divided by the processing rate of organic carbon, K_{OC} (day^{-1}).

26.4 QUESTIONS

1. How did CPOM storage differ between the stream reaches? Describe the conditions under which there was greater or lesser carbon storage.

2. Compare CPOM storage among stream habitats. What might explain the differences among habitats? Consider hydraulics, geomorphology, sediments, and other features.

3. In the leaf release experiment, to what features do you attribute any differences in retention between the two study reaches? Did retention exceed your expectations?

4. If leaves were reinventoried, what were the most important retention structures in the two reaches? Were more leaves retained in pools or in riffles? What were the mechanisms responsible for retention in these two habitat types?

5. Did the exponential model adequately describe the leaf retention patterns? What exactly do the parameters of this model describe? Under what conditions would alternate models be appropriate?

6. In the retention enhancement experiment, how would you expect detritivorous macroinvertebrates to respond in experimentally enhanced CPOM retention in the short term (i.e., scales of days to weeks) and long term (scales of months to years)? How about predators that may respond to invertebrate prey?

7. Describe how higher CPOM standing crops might influence carbon and nutrient (N, P) spiraling lengths.

8. How does carbon spiraling change seasonally in a temperature deciduous forested stream? In what season is spiraling length shortest? Longest? How would this differ in a tropical deciduous forest versus a prairie grassland stream?

9. In light of your findings, discuss the implications of stream and riparian management practices that tend to reduce the amount of wood in streams, to simplify stream channels, or to modify the hydrograph. What restoration approaches would you suggest to increase CPOM storage, retention, and consumption by stream biota?

26.5 MATERIALS AND SUPPLIES

Materials for CPOM storage measurement
  Benthic corer (e.g., bottomless bucket, PVC pipe section, or stovepipe corer)
  Meter stick to measure water depth
  1-mm sieve or hand net
  Labeled brown paper lunch bags to collect and dry leaves
  Labeled plastic bags to transport paper bags back to the laboratory
  Data sheets
  Small aluminum pans
  Top-loading balance
  Desiccator
  Drying oven
  Muffle furnace
Materials for CPOM release
  Dried or fresh-fallen leaves (e.g., 2000 abscised *G. biloba* leaves)
  Garbage bags (to store leaf batches until released)
  Buckets [two 20-L (5-gallon), to soak leaves] and brass sieves (if available)
  Current velocity meter (optional)
  Field notebook with data sheets
  Flagging tape
  Meter tape (100 m, 50 m)
  Dip (D-frame) nets (1 per investigator)
  Seine (1 cm mesh) with lead line (at least as long as channel width)
Materials for retention enhancement (50-m long, 1-m wide reach with a trap density of $1/m^2$)
  Drilling hammer
  60 cm L × 5 cm W × 5 cm D wooden stakes (100)
  Plastic mesh (~1 cm pore size)
  Cable ties or binder clips (300)
Materials for carbon spiraling estimate
  1-mm and 52-μm mesh nets
  Sample collection cups or plastic bags
  2-L and 60-mL HDPE bottles
  6N HCl (for acidifying DOC sample) and appropriate PPE
  Current velocity meter
  NaCl solution, conductivity meter

Meter stick and meter tape
Benthic corer (e.g., bottomless bucket or stovepipe core)
Logging dissolved oxygen sensor
Data sheets
Small aluminum pans
0.7- and 0.45-μm glass fiber filters
Filtration system
Top-loading balance
Drying oven
Muffle furnace
Total organic carbon analyzer

ACKNOWLEDGMENTS

We thank all those individuals who have inspired and instructed us in the wonders of organic matter dynamics, especially Mike Dobson, Mark Gessner, Stephen Golladay, Stanley Gregory, Patrick Mulholland, J. Denis Newbold, Jennifer Tank, J. Bruce Wallace, and Jack Webster.

REFERENCES

Bilby, R.E., 1981. Role of organic debris dams in regulating the export of dissolved and particulate matter from a forested watershed. Ecology 62, 1234—1243.

Bilby, R.E., Likens, G.E., 1980. Importance of organic debris dams in the structure and function of stream ecosystems. Ecology 61, 1107—1126.

Bott, T.L., Brock, J.T., Cushing, C.E., Gregory, S.V., King, D., Petersen, R.C., 1978. Comparison of methods for measuring primary productivity and community respiration in streams. Hydrobiologia 60, 3—12.

Cederholm, C.J., Houston, D.B., Cole, D.L., Scarlett, W.J., 1989. Fate of coho salmon (*Oncorhynchus kisutch*) carcasses in spawning streams. Canadian Journal of Fisheries and Aquatic Sciences 46, 2647—2655.

Chadwick, M.A., Huryn, A.D., 2007. Role of habitat in determining macroinvertebrate production in an intermittent-stream system. Freshwater Biology 52, 240—251.

Chow, V.T., 1959. Open-Channel Hydraulics. McGraw- Hill, New York, NY, 680 pp.

Cordova, J.M., Rosi-Marshall, E.J., Tank, J.L., Lamberti, G.A., 2008. Coarse particulate organic matter transport in low-gradient streams of the Upper Peninsula of Michigan. Journal of the North American Benthological Society 27, 760—771.

Cummins, K.W., 1974. Structure and function of stream ecosystems. BioScience 24, 631—641.

Cummins, K.W., Klug, M.J., 1979. Feeding ecology of stream invertebrates. Annual Review of Ecology and Systematics 10, 147—172.

Cummins, K.W., Sedell, J.R., Swanson, F.J., Minshall, G.W., Fisher, S.G., Cushing, C.E., Petersen, R.C., Vannote, R.L., 1983. Organic matter budgets for stream ecosystems. In: Barnes, J.R., Minshall, G.W. (Eds.), Stream Ecology: Application and Testing of General Ecological Theory. Plenum Press, New York, NY, pp. 299—353.

Díez, J.R., Larrañaga, S., Elosegi, A., Pozo, J., 2000. Effect of removal of wood on streambed stability and retention of organic matter. Journal of the North American Benthological Society 19, 621—632.

Dobson, M., Hildrew, A.G., Orton, S., Ormerod, S.J., 1995. Increasing litter retention in moorland streams: ecological and management aspects of a field experiment. Freshwater Biology 33, 325—337.

Dobson, M., 2005. Manipulation of stream retentiveness. In: Graça, M.A.S., Bärlocher, F., Gessner, M.O. (Eds.), Methods to Study Litter Decomposition: A Practical Guide. Springer, Dordrecht, The Netherlands, pp. 19—24.

Ehrman, T.P., Lamberti, G.A., 1992. Hydraulic and particulate matter retention in a 3rd-order Indiana stream. Journal of the North American Benthological Society 11, 341—349.

Entrekin, S.A., Tank, J.L., Rosi-Marshall, E.J., Hoellein, T.J., Lamberti, G.A., 2008. Responses in organic matter accumulation and processing to an experimental wood addition in three headwater streams. Freshwater Biology 53, 1642—1657.

Entrekin, S.A., Tank, J.L., Rosi-Marshall, E.J., Hoellein, T.J., Lamberti, G.A., 2009. Response of secondary production by macroinvertebrates to large wood addition in three Michigan streams. Freshwater Biology 54, 1741—1758.

Graça, M.A.S., 2001. The role of invertebrates on leaf litter decomposition in streams — a review. International Review of Hydrobiology 86, 383—393.

Griffiths, N.A., Tank, J.L., Royer, T.V., Warrner, T.J., Frauendorf, T.C., Rosi-Marshall, E.J., Whiles, M.R., 2012. Temporal variation in organic carbon spiraling in Midwestern agricultural streams. Biogeochemistry 108, 149—169.

Hall, R.O., Beaulieu, J.J., 2013. Estimating autotrophic respiration in streams using daily metabolism data. Journal of the North American Benthological Society 32, 507—516.

Hoellein, T.J., Tank, J.L., Rosi-Marshall, E.J., Entrekin, S.A., Lamberti, G.A., 2007. Controls on spatial and temporal variation of nutrient uptake in three Michigan headwater streams. Limnology and Oceanography 52, 1964—1977.

Hoellein, T.J., Tank, J.L., Rosi-Marshall, E.J., Entrekin, S.A., 2009. Temporal variation in substratum-specific rates of N uptake and metabolism and their contribution at the stream-reach scale. Journal of the North American Benthological Society 28, 305—318.

Jones, J.B., Smock, L.A., 1991. Transport and retention of particulate organic matter in two low-gradient headwater streams. Journal of the North American Benthological Society 10, 115—126.

Jones, J.B., 1997. Benthic organic matter storage in streams: influence of detrital import and export, retention mechanisms, and climate. Journal of the North American Benthological Society 16, 109—119.

Larned, S.T., Chong, C.T., Punewai, N., 2001. Detrital fruit processing in a Hawaiian stream ecosystem. Biotropica 33, 241—248.

Minshall, G.W., 1978. Autotrophy in stream ecosystems. BioScience 28, 767—770.

Newbold, J.D., Mulholland, P.J., Elwood, J.W., O'Neill, R.V., 1982. Organic carbon spiralling in stream ecosystems. Oikos 38, 266—272.

Roberts, B.J., Mulholland, P.J., Hill, W.R., 2007. Multiple scales of temporal variability in ecosystem metabolism rates: results from 2 years of continuous monitoring in a forested headwater stream. Ecosystems 10, 588—606.

Rosi-Marshall, E.J., Wallace, J.B., 2002. Invertebrate food webs along a stream resource gradient. Freshwater Biology 47, 129—141.

Smock, L.A., Metzler, G.M., Gladden, J.E., 1989. Role of debris dams in the structure and function of low-gradient headwater streams. Ecology 70, 764—775.

Speaker, R.W., Luchessa, K.J., Franklin, J.F., Gregory, S.V., 1988. The use of plastic strips to measure leaf retention by riparian vegetation in a coastal Oregon stream. American Midland Naturalist 120, 22—31.

Speaker, R.W., Moore, K., Gregory, S.V., 1984. Analysis of the process of retention of organic matter in stream ecosystems. Verhandlungen der Internationalen Vereinigung für Theoretische und Angewandte Limnologie 22, 1835—1841.

Stewart-Oaten, A., Murdoch, W.W., Parker, K.R., 1986. Environmental impact assessment: "Pseudoreplication" in time? Ecology 67, 929—940.

Tank, J.L., Rosi-Marshall, E.J., Griffiths, N.A., Entrekin, S.A., Stephen, M.L., 2010. A review of allochthonous organic matter dynamics and metabolism in streams. Journal of the North American Benthological Society 29, 118—146.

Taylor, B.W., Flecker, A.S., Hall, R.O., 2006. Loss of a harvested fish species disrupts carbon flow in a diverse tropical river. Science 313, 833—836.

Thomas, S.A., Royer, T.V., Snyder, E.B., Davis, J.C., 2005. Organic carbon spiraling in an Idaho river. Aquatic Sciences 67, 424—433.

Tiegs, S.D., Levi, P.S., Rüegg, J., Chaloner, D.T., Tank, J.L., Lamberti, G.A., 2011. Ecological effects of live salmon exceed those of carcasses during an annual spawning migration. Ecosystems 14, 598—614.

Tiegs, S.D., Peter, F., Robinson, C.T., Uelinger, U., Gessner, M.O., 2008. Leaf decomposition and invertebrate colonization responses to manipulated litter quantity in streams. Journal of the North American Benthological Society 27, 321—331.

Vannote, R.L., Minshall, G.W., Cummins, K.W., Sedell, J.R., Cushing, C.E., 1980. The river continuum concept. Canadian Journal of Fisheries and Aquatic Sciences 37, 260—267.

Wallace, J.B., Eggert, S.L., Meyer, J.L., Webster, J.R., 1997. Multiple trophic levels of a forest stream linked to terrestrial litter inputs. Science 277, 102—104.

Webster, J.R., Benfield, E.F., Ehrman, T.P., Schaeffer, M.A., Tank, J.L., Hutchens, J.J., D'Angelo, D.J., 1999. What happens to allochthonous material that falls into streams? Freshwater Biology 41, 687—705.

Webster, J.R., Covich, A.P., Tank, J.L., Crockett, T.V., 1994. Retention of coarse organic particles in streams in the southern Appalachian Mountains. Journal of the North American Benthological Society 26, 140—150.

Webster, J.R., Meyer, J.L., 1997. Organic matter budgets for streams: a synthesis. Journal of the North American Benthological Society 16, 141—161.

Young, S.A., Kovalak, W.P., Del Signore, K.A., 1978. Distances travelled by autumn-shed leaves introduced into a woodland stream. American Midland Naturalist 100, 217—222.

Young, R.G., Huryn, A.D., 1999. Effects of land use on stream metabolism and organic matter turnover. Ecological Applications 9, 1359—1376.

Chapter 27

Leaf-Litter Breakdown

E.F. Benfield[1], Ken M. Fritz[2] and Scott D. Tiegs[3]

[1]*Department of Biological Sciences, Virginia Polytechnic Institute and State University;* [2]*National Exposure Research Laboratory, U.S. Environmental Protection Agency;* [3]*Department of Biological Sciences, Oakland University*

27.1 INTRODUCTION

Streams have two main sources of energy: (1) in-stream photosynthesis by algae, mosses, and higher aquatic plants; and (2) imported organic matter from streamside vegetation (e.g., leaves and other parts of terrestrial vegetation). In small, heavily shaded streams, there is normally insufficient light (see Chapter 7) to support substantial in-stream photosynthesis (see Chapter 34), and thus energy pathways are supported largely by imported (i.e., allochthonous) organic matter. In such streams, the bulk of readily processed imported organic matter enters as leaf litter, wood, and other plant parts that fall into the stream, collectively termed "coarse particulate organic matter" (CPOM), which contributes to the annual organic matter budget. Additional dead leaves may slide or blow into streams from riparian zones during the rest of the year (Benfield, 1997). Dead woody material such as sticks, tree limbs, and even tree boles may enter streams due to tree death, animal activities, and wind. Leaves falling into streams may be transported short distances but are usually retained by structures in the streambed forming "leaf packs" (Fig. 27.1A) (Petersen and Cummins, 1974; see also Chapter 26) or accumulate around wood in debris dams (Fig. 27.1B) (Smock et al., 1989). Wood is an important contributor to the energetics of streams but on a much longer timescale than leaves (McTammany et al., 2008; see also Chapter 29). Leaves either in packs or debris dams are "processed" in place by components of the stream community in a series of well-documented steps.

Dead leaves entering streams are often nutritionally poor because trees resorb most of the soluble nutrients (e.g., sugars, amino acids, fatty acids) present prior to abscission (Suberkropp et al., 1976; Paul et al., 1978) or the leaves have high C:N ratios (Melillo et al., 1983). Within one or two days of entering a stream, soluble nutrients leach out of the leaf cellular matrix into the water, although there is evidence that some soluble materials remain long after dead leaves are immersed (Paul et al., 1978; France et al., 1997). After leaching, dead leaves are composed mostly of structural materials like cellulose and lignin, neither of which is very digestible by most animals. Within a few days of entering the water, fungi and bacteria begin to colonize the leaves leading to a process known as "microbial conditioning" (Bärlocher and Kendrick, 1975). Microbes produce suites of enzymes that can digest the remaining leaf constituents and begin the conversion of leaves to smaller particles (Suberkropp and Klug, 1976). After about 2 weeks of microbial conditioning, the leaves begin to soften, and some species may begin to fragment. Laboratory studies have shown that, given sufficient time, some species of aquatic hyphomycete fungi (see Chapter 10) can reduce whole leaves to small particles (Suberkropp and Klug, 1980). However, reduction in particle size from whole leaves as CPOM to fine particulate organic matter (FPOM) is generally thought to occur through the feeding activities of a variety of aquatic invertebrates collectively known as "shredders" (Cummins, 1974; Klug and Cummins, 1979; see also Chapter 20). Shredders help reduce the particle size of leaves through the production of "orts" (i.e., fragments shredded from leaves but not ingested) and fecal pellets. The particles can then be consumed by a variety of micro- and macroconsumers. FPOM may also come from a variety of other sources, both within and outside of streams (e.g., Klug and Cummins, 1979; Sollins et al., 1985; Ward and Aumen, 1986; see also Chapter 25). Leaves may also be fragmented by a combination of microbial activity and physical factors such as current and abrasion (Benfield et al., 1977; Paul et al., 1978; Ward and Aumen, 1986).

Leaves from various tree and shrub species break down at different rates. Thus, a "leaf processing continuum" exists in most forested streams (Petersen and Cummins, 1974) in which leaf mass from some species is lost rapidly ("fast processors"), some is lost moderately rapidly ("medium processors"), and others very slowly ("slow processors").

Methods in Stream Ecology. http://dx.doi.org/10.1016/B978-0-12-813047-6.00005-X

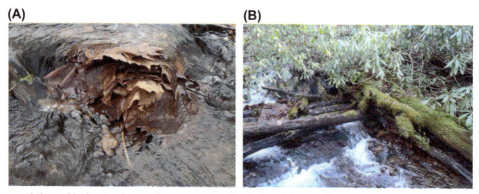

FIGURE 27.1 Accumulations of leaf matter in natural streams: (A) typical leaf pack in a small stream; (B) debris dam in a small stream where leaf matter is retained.

The consequence of this leaf processing continuum is that the stream community is supplied with leaf material as a food resource for much of the annual cycle (Petersen and Cummins, 1974). Differences in the rates at which "fast", "medium", and "slow" species break down in a particular stream appear to be largely a function of initial physical and chemical properties of leaves (Melillo et al., 1982; Webster and Benfield, 1986). Species-specific breakdown rates vary with stream, location in the stream, time of year, activity of microbes, presence of shredders, and other stream-specific factors (Webster and Benfield, 1986; Langhans et al., 2008; Lecerf and Chauvet, 2008; Woodward et al., 2012).

Leaf-litter breakdown is an integrative, ecosystem level process because it links various elements of stream systems (i.e., leaf species, microbial activity, invertebrates, and physical and chemical features of the stream). A major result of these linkages is that whole leaves are converted into fine particles, which are then distributed downstream (see Chapter 25) and used as an energy source by multiple components of stream food webs (see Chapter 20). Clearly, leaf-litter breakdown measured by the techniques described here is not equivalent to organic matter decomposition, as pointed out by Boulton and Boon (1991), Gessner et al. (1999), and others. Decomposition is probably best defined as the conversion of organic matter into its inorganic constituents (i.e., mineralization). However, decomposition is certainly a part of the breakdown process and understanding leaf-litter breakdown helps illuminate how energy flows through stream ecosystems (see Chapter 23).

Leaf-litter breakdown has been used to investigate long-term responses of streams to disturbance by logging (Benfield et al., 2001; Kreutzweiser et al., 2008), responses of streams to acid mine drainage (Niyogi et al., 2001) and acid rain (Simon et al., 2009), influences of varied land-use on stream structure and function (Sponseller and Benfield, 2001; Roy et al., 2016) responses of streams to a gradient of agricultural development (Niyogi et al., 2003; Hagen et al., 2006; Hladyz et al., 2010), and responses of streams to temperature and nutrient gradients (Boyero et al., 2011; Woodward et al., 2012; Griffiths and Tiegs, 2016). Indeed, leaf-litter breakdown and organic matter decomposition have been proposed as a useful method for assessing the functional integrity of streams (Gessner and Chauvet, 2002; Young et al., 2008; Feio et al., 2010; Tiegs et al., 2013). The objective of this chapter is to provide a basic protocol for performing leaf-litter breakdown experiments in streams, for measuring microbial activity through closed-chamber respiration, and to suggest some ways that leaf breakdown can be used to evaluate stream structure and function.

27.2 GENERAL DESIGN

The overall process of measuring leaf-litter breakdown rates involves placing a group of preweighed "leaf packs" in a stream, periodically sampling from the group over time, and estimating the rate at which the packs lose mass in the stream. Specifically, a large number of leaf packs are constructed and placed in a stream on Day 1 of the study. Three to five packs are retrieved regularly over the course of the study (perhaps 3–7 months), cleaned of debris and invertebrates, dried to constant mass, and weighed. Breakdown rates ($-k$) are computed using an exponential decay model that assumes the rate of mass loss from the packs is a constant fraction of the amount of material remaining. Operationally, $-k$ is the negative slope of the line produced by a linear regression of the natural log of percent leaf mass remaining plotted against time or cumulative temperature (i.e., degree-days).

For measuring microbial respiration, a known quantity of organic matter is placed in a small airtight chamber (e.g., centrifuge tubes, mason jars) along with stream water. The concentration of dissolved oxygen in the stream water is measured at the start and end of an incubation period. With a few calculations involving the volume of water in the chamber, the duration of the incubation, and the quantity of organic matter, the activity of microbes (as oxygen consumption) can be determined.

27.2.1 Site Selection

Leaf breakdown studies can be performed in virtually any size stream but small, shallow streams present fewer problems than do large ones, especially for a class experiment. Streams with gravel or cobble substrates are preferable to those with sandy bottoms because of the difficulty of anchorage and the likelihood of burial of packs in sandy-bottomed streams. Remote sites are preferable to sites that receive regular human traffic because leaf packs are attractive to the curious. Avoid spots that are likely to be significantly deeper during higher flows, areas of excessive erosion (e.g., next to cut banks) or deposition (e.g., point bars), and areas that may be unstable under higher flows (e.g., debris dams). Best results are usually obtained when leaf packs are placed in shallow riffles closer to the bank than the middle to avoid increased stream power during high flows.

27.3 SPECIFIC METHODS

27.3.1 General Protocol for Leaf-Litter Breakdown Experiments

1. Collect leaves from trees just before they are ready to fall (i.e., at abscission) or shortly after they fall but before they are exposed to rain.
2. Air dry the leaves in large cardboard boxes with many holes (about 3 cm in diameter) covered by plastic window screen. Place the boxes in a dry location. Invert the boxes daily and gently "fluff" the leaves to promote drying. Continue for 5–8 days until leaves reach a relatively constant dry mass (DM). Alternatively, leaves can be spread out on the floor, on tables, or in a suspended beach seine to dry. Thick piles of leaves should be turned over frequently to promote drying.
3. Weigh out 3–10 g (±0.1 g) portions on a top-loading analytical balance, and fashion them into leaf packs by one of several techniques described further. It is advisable to use approximately the same mass (weight) of leaves for each pack. Record the initial DM on a data sheet (see Table 27.1).
4. Construct mesh bags from bridal netting or similar material such as poultry fencing, hardware cloth, or large-mesh plastic screen. Commercial mesh bags used to package produce (e.g., oranges, grapes, etc.) also work well. Regardless of the material used, mesh openings should be large enough to allow access to consumers, yet small enough to retain the leaf material (Webster and Benfield, 1986).
5. Prepare enough packs for the entire exposure period. Variability in the amount of material lost from leaf packs can be relatively high, especially in the later stages of decomposition. Thus, a minimum of three packs per species per site should be retrieved on each date to calculate a mean and standard error. In experiments using more than one species, select a method to differentiate the leaf packs by species because it can be difficult to distinguish between species when the leaves are in the middle to later stages of decay. Bags of different colors are probably the best choice (Fig. 27.2A) but strips of colored flagging tape tied to the bags also work well.
6. By their very nature, dried leaves are easily broken in handling. Therefore, it is necessary to account for losses encountered in fashioning, transporting, and placing the packs in the stream. This can be accomplished by preparing an extra set of leaf packs that goes through the entire process but are not left in the stream to incubate. The extra packs are processed and used to correct for "handling losses" (described below).

TABLE 27.1 Example data sheet for leaf breakdown study (values in g)[1]. See also Online Worksheet 27.1.

	Stream: _____		Site: _____	Date: _____		Days of Incubation: _____			
Sample	LPDM	P	P + DM	P + AM	DM	AM	% Ash	% Org.	AFDM
Maple 1	8.27	1.000	1.2500	1.0250	0.2500	0.0250	10	90	7.443
Maple 2									
Maple 3									
Oak 1									
Oak 2									
Oak 3									

[1]AFDM, LPDM × % organic; AM, ash mass of milled sample; % Ash, AM/DM × 100; DM, dry mass of milled sample; LPDM, dry mass of leaf pack; % Organic, 100%−% ash; P, pan mass; P + AM, pan mass + postashed milled sample; P + DM, pan mass + preashed mass of milled sample.

(A) **(B)**

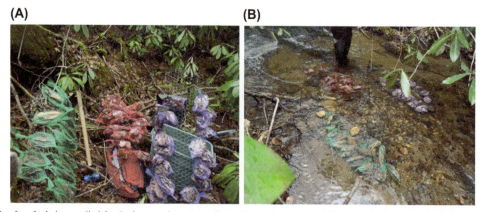

FIGURE 27.2 Leaf packs being readied for deployment in a stream for leaf breakdown study: (A) green (rhododendron, *Rhododendron maximum*), red (red maple, *Acer rubrum*), and purple (white oak, *Quercus alba*) leaf packs zip-tied to 2′ × 4′ sheets of poultry fencing, awaiting incubation in the stream; (B) fencing sheets are then staked to the streambed with rebar and tied to trees or woody shrubs on the bank for security against high flow.

7. Set up a retrieval schedule according to the leaf type used. Complete mass loss of "fast" leaf species may occur in 1–3 months and packs should be collected weekly or every 2 weeks. "Medium" and "slow" leaves may require 4–12 months for complete mass loss and may be collected at monthly intervals (Webster and Benfield, 1986). Note, however, that local stream conditions (e.g., flow, temperature, shredders) can strongly influence these durations.
8. Transport all packs to the stream site in a sturdy container (e.g., cooler), handling it carefully to avoid breakage.
9. Packs must be secured in the streambed. Depending on size and flow rate of the stream, various restraint systems are recommended. In small, shallow streams, leaf packs in mesh bags can be tied with polypropylene twine singly or in groups to "gutter" nails (9′ nails used to attach guttering to houses) pushed or driven into the streambed (Webster and Waide, 1982). In larger or faster-flowing streams, steel rods or metal fence posts driven into the streambed may be necessary to anchor the leaf packs. Alternatively, attach the packs with "zip-ties" to heavy wire tied to a tree along one bank (Benfield et al., 2000). If none of these techniques seem appropriate, employ stronger devices such as those described in Benfield and Webster (1985). For example, leaf packs can be zip-tied to a sheet of poultry fencing which, in turn, is staked to the stream bottom with rebar (Fig. 27.2B).
10. Place all packs, including those designated for "handling loss" correction, in the stream, spreading them out as much as possible as governed by the restraint system chosen, available space, and other considerations. After the leaf packs are secured, retrieve the packs designated for handling loss correction, while leaving the other packs for the experiment. A photograph of the site and/or drawing a simple map showing the location of the leaf packs may be helpful in locating the packs when it is time to retrieve them. Flagging tape tied to streamside vegetation is also helpful in relocating the sites, but be sure to remove it after the study is over.
11. Following the retrieval schedule, remove the appropriate number of leaf packs from the stream, discard any leaves attached to the outside of the mesh bags, and place each pack into an individual resealable plastic bag. Include an internal label (pencil or permanent marker on waterproof paper) identifying the sample with all pertinent information (e.g., retrieval date, site, species, etc.). Write the same information on the outside of the plastic bag, using a permanent marker. Place the samples on ice in a cooler and return them to the laboratory. Keep the bags in the cooler or refrigerate until processed.
12. Processing in the laboratory may involve several options depending on whether you decide to measure only leaf pack breakdown or conduct additional work with macroinvertebrates (see Chapters 15, 22), bacteria (see Chapter 9), fungi (see Chapter 10), leaf chemistry (see Chapter 36), stable isotopes (see Chapter 23), or other variables (see microbial respiration in the following sections). In any case, remove the leaf material and gently rinse the leaves of silt and debris. Because the exponential breakdown model represents loss from the original leaves, ignore small leaf fragments that may have been lost from the original mass but retained by the bag or strained from flow. Place the cleaned leaves in paper bags. Keep the field labels with the individual packs as you transfer them into the paper bags. Label the outside of the paper bags with the same information that appears on the inside labels. Hang the bags on a line stretched across the laboratory and allow the leaf material to air dry to constant mass. Alternatively, dry in a hot air oven at 50°C or less for at least 24 h. After drying, weigh leaf material and record the DM on the data sheet (Table 27.1). If the project calls for saving macroinvertebrates, perform the rinsing over a 250-μm sieve and place the macroinvertebrates in 70% ethanol with appropriate labeling inside and outside the containers (see Chapter 15). If you plan to do microbial analyses, subsample the leaves before drying and keep the subsamples cold and moist or frozen depending on the protocol required for the particular analyses planned (e.g., see Chapters 9 and 10).

13. In many cases, mineral deposits are not readily washed off the leaves and may result in errors in final DM. This problem can be mostly overcome by converting DM to ash-free dry mass (AFDM; see also Chapter 12). Organic matter combusts at about 550°C and the remaining material is mineral ash. When the mass of mineral ash is subtracted from initial DM, the result is the DM of the "organic fraction" of the leaf material (or "AFDM"). For each species, process the leaves by milling (e.g., by Wiley Mill, coffee grinder) or grinding in some way all or significant portions of the "handling loss" leaves and the leaves from each retrieval (all packs combined for each date) to a fine powder. Determine AFDM for the leaf packs as described further in Step 14. The AFDM of the "handling loss" leaves serves as the initial AFDM (i.e., the AFDM of the leaves before they were put into the stream).

14. Mark the underside of aluminum weighing pans by inverting them over the bottom of a beaker and impressing a code using a metal probe. Record the identification codes on the data sheet. To obtain the tare weight of each pan, heat the coded pans at 550°C for 30 min in a muffle furnace. Then, while handling with oven gloves and tongs or forceps as appropriate *throughout the process*, place the pans in a desiccator to cool. After cooling, weigh and record the mass (weight) of each pan at the appropriate place on the data sheet (Table 27.1, or see Chapter 12). Weigh out at least two subsamples of about 250 mg DM of each milled sample into a tared pan, oven-dry over night at 50°C, and place in a desiccator to cool. Weigh the pans plus milled samples and record on the data sheet. Place pans plus milled samples in a muffle furnace at 550°C for 20 min, remove and stir with a dissecting needle, then return pans and sample to the furnace and heat for an additional 20 min (Gurtz et al., 1980). Remove pans from the furnace and cool in a desiccator. After cooling, wet the material with distilled water, then oven dry at 50°C for 24 h. Remove pans with samples from the drying oven and desiccate. After cooling in the desiccator, weigh and record pan plus ash on the data sheet. Subtract the pan weights from the preashed pan plus sample and postash pan plus sample.

Compute % organic matter of the milled samples as follows:

$$\% \text{ organic matter} = [(\text{sample dry mass} - \text{sample ash mass})/(\text{sample dry mass})] \times 100 \qquad (27.1)$$

15. Convert leaf pack dry mass (LPDM) to AFDM:

$$\text{AFDM} = (\text{LPDM}) \times (\% \text{ organic matter}) \qquad (27.2)$$

16. Convert AFDM for each leaf pack to %AFDM remaining:

$$\%\text{AFDM remaining} = (\text{Final AFDM}/\text{Initial AFDM}) \times 100 \qquad (27.3)$$

17. Regress the natural log (ln) of mean %AFDM remaining (y-axis) on days of exposure (x-axis) using the AFDM of the "handling loss" leaf packs as 100% remaining for Day 0. The negative slope of the regression line is equal to the processing coefficient ($-k$).

27.3.2 Basic Method 1: Leaf Breakdown for One or More Leaf Species

Following the General Protocol aforementioned, perform a leaf breakdown study using one or more common leaf species incubated at one site in a stream. Install a water temperature monitoring device at the site (see Chapter 6). Compute the processing coefficient ($-k$) for AFDM loss over time and/or for cumulative temperature (degree-days) for each of the species tested.

The single-species protocol described earlier can easily be expanded to include more species. For example, one could contrast the breakdown rates of presumed "fast", "moderate", and "slow" species, as shown for in the linear plots for three species of leaves incubated in a stream over time (Fig. 27.3A). Clear differences are seen in the relative speed of breakdown (red maple > white oak > rhododendron) and the ranks of the three species with respect to longevity (red maple < white oak < rhododendron) in the stream. However, to quantitatively compare the speed of breakdown among the three species, it is necessary to convert the linear graphs to exponential functions (ln% AFDM remaining regressed against time; Fig. 27.3B). The slopes of the lines are equal to the breakdown rates ($-k$, the processing coefficient): red maple ($-k = 0.0225$), white oak ($-k = 0.0060$), and rhododendron ($-k = 0.0025$).

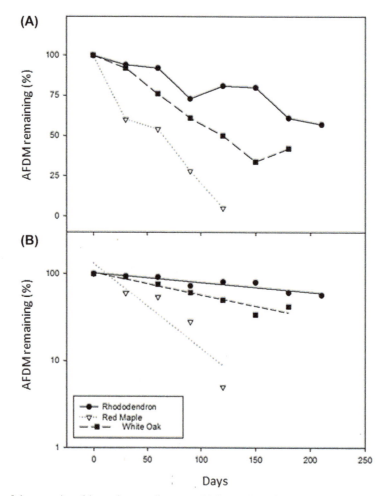

FIGURE 27.3 Breakdown of three species of leaves in a small stream: (A) linear plots of % ash-free dry mass (AFDM) remaining versus time; (B) log-normal plots of % AFDM versus time (E.F. Benfield, unpublished data).

Another experiment could be to contrast riparian shrub or herbaceous leaves with tree leaves, or perhaps deciduous and evergreen shrub and/or tree leaves (e.g., Maloney and Lamberti, 1995). Processing coefficients ($-k$) may be compared statistically using analysis of covariance (ANCOVA) or a Dummy Variable Regression to determine whether the $-k$ values are significantly different (see Sokal and Rohlf, 1995; Kleinbaum et al., 1988; or Zar, 2010).

To assess the effect of temperature on leaf breakdown, processing coefficients ($-k$) can be computed using cumulative temperature (degree-days) as values on the x-axis in place of days (Petersen and Cummins, 1974). Cumulative degree-days may be estimated by summing the average daily water temperature over each incubation period and entering the appropriate values (i.e., the degree-days accumulated from day 1 to the retrieval day) in place of days. For the three species used above, $-k$ values for ln% remaining can be computed based on days (Fig. 27.4A) and degree-days (Fig. 27.4B). Clearly, the ranks and patterns of breakdown among the three species are the same whether the $-k$ values are computed by days or degree-days since they were exposed to the same temperature in a single North Carolina stream. Other studies in which leaf packs were exposed to different temperature regimes show that temperature can be an important factor in regulating the rate of leaf breakdown (e.g., Robinson and Jolidon, 2005; Goncalves et al., 2006; Griffiths and Tiegs, 2016).

27.3.3 Basic Method 2: Effects of Spatially Varying Stream Features on Leaf Breakdown Rates

Many designs are possible depending on the question(s) of interest. For example, use one or more species of leaves at single or multiple sites in one or several streams. Site differences could include riffles versus pools, high elevation versus low elevation, cobble substrate versus bedrock or sand, or shaded versus unshaded reaches. Stream differences could be

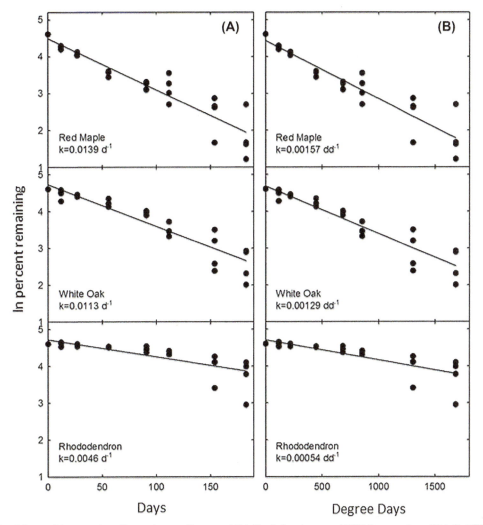

FIGURE 27.4 Breakdown of three species of leaves in a small stream: (A) ln% ash-free dry mass (AFDM) versus time; (B) ln% AFDM versus degree-days (E.F. Benfield, unpublished data).

based on stream order, gradient, geology, disturbance history, hardness, or nutrient level. In sand-bottom streams, leaves often become buried by transported bed material. The experiment may include comparing breakdown rates of buried versus unburied leaves (e.g., Webster and Waide, 1982; Tillman et al., 2003). Comparisons of site−species combinations or treatments can be analyzed in a manner similar to the statistical methods described in the General Protocol mentioned earlier.

27.3.4 Advanced Method 1: Effects of Anthropogenic Activities on Leaf Breakdown Rates

Investigate the impact of a municipal, industrial, or mining waste outfall on stream organic matter processes using leaf breakdown rate as an indicator (e.g., Paul et al., 1983; Niyogi et al., 2001; see also Chapter 40). The usual protocol for evaluating the impact of a waste outfall on streams is to compare some value(s) upstream and downstream of an outfall in similar habitats (for a general discussion see Plafkin et al., 1989). Establish an upstream "reference" site(s) that is totally removed from impacts of the outfall in question. In wider streams, you may also be able to use sites across the stream as reference sites if they are clearly removed from the outfall. Establish a site just downstream from the outfall where impact is likely to be maximal, and then several additional sites further downstream including one or more at which you judge the impact of the outfall to be abated. Proceed with the study as outlined earlier in the General Protocol and analyze for site differences as described in Basic Method l. For simplicity, it might be best to assess a single "index" leaf species with sufficient replicates to detect site differences.

27.3.5 Advanced Method 2: Assessing Relationships Among Leaf Breakdown Rates and Shredders

Exploring the relationships of shredders and microbes with leaf-breakdown rates in any of the protocols outlined earlier can be accomplished by planning ahead as described in General Protocol Step 12. Macroinvertebrates sorted from the leaf packs can be identified, enumerated, and inventoried among functional feeding groups or traits (as described in Chapter 20). Regression analysis can then be used to evaluate relationships between macroinvertebrate number per bag, density, or biomass and leaf-breakdown rates where there are sufficient data (e.g., Sponseller and Benfield, 2001; Niyogi et al., 2001). A word of caution: macroinvertebrate abundance in litter bags can be extremely variable, especially toward the end of the breakdown process.

27.3.6 Advanced Method 3: Assessing Microbial Activity During the Litter-Breakdown Process

Microbial activity and biomass on organic matter undergoing breakdown are important indicators of the decomposition process that is key to the reduction of CPOM to FPOM. Fungal biomass can be estimated by measuring ergosterol, the major sterol in the membranes of higher fungi (see also Chapter 10). Fungal production can be estimated by measuring the incorporation rate of radio-labeled acetate into ergosterol and glucosamine concentration (Kuehn et al., 2011). Bacterial and fungal activity can also be measured by microscopic and molecular techniques (Duarte et al., 2010; Blazewicz et al., 2013). See Chapters 10 and 11 for further discussion of emerging molecular methods for quantifying microbial activity.

Microbial respiration is a measure of microbial activity that integrates the effects of microbial biomass accumulation over time, community composition, and environmental factors such as temperature and nutrient availability. Microbial respiration tends to vary with the quality of organic matter (e.g., leaf species; Fig. 27.5). While microbial respiration is commonly expressed as the production of CO_2, it is measured more readily as the consumption of dissolved oxygen in either the laboratory or field using a dissolved oxygen probe (Niyogi et al., 2001; Carlisle and Clements, 2005; Griffiths and Tiegs, 2016). The following is a protocol used to quantify respiration (as oxygen consumption) by aquatic microbial communities that are growing on submerged leaf litter.

The general approach is to incubate samples of organic matter in airtight chambers that contain stream water and measure oxygen consumption over time. This technique can be used with organic matter that has been experimentally introduced into aquatic systems (Simon et al., 2009; Tiegs et al., 2013), or with naturally occurring material (Hoellein et al., 2009). Microbial respiration can be measured in conjunction with either the basic or advanced methods described previously.

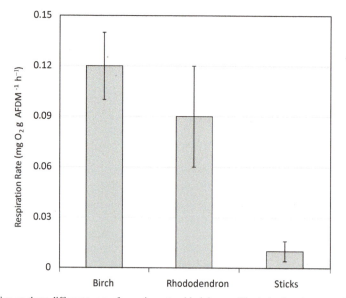

FIGURE 27.5 Microbial respiration on three different types of organic matter: birch leaves, *Rhododendron* leaves, and wooden sticks. Values shown are means (±1SD). *Data from Table 2 in Tank et al. (1993).*

Although the following protocol can be used in the field, such an approach may be neither practical nor desirable for a variety of reasons. Therefore, the protocol can be executed in the laboratory by subsampling organic matter such as leaf litter (General Protocol Step 11) and proceeding with the measurements. Water from the study stream should be returned to the lab and used as the incubation medium. The described protocol uses one control chamber (i.e., no introduced organic matter) and one treated chamber (with organic matter), but additional replicates of both conditions are needed to compute error (within and between treatment variation) for statistical analysis. For example, a minimum of 3–5 replicates per condition are advised for any incubation experiment.

27.3.7 Field Protocol

1. Place the organic matter (OM) sample of interest (e.g., discs cut using a cork borer or simply material cut from a leaf) into the respiration chamber (e.g., 60-mL plastic centrifuge tube). We will refer to this as the "OM" chamber.
2. Measure the dissolved oxygen concentration in the stream water with an oxygen meter to determine oxygen concentration at the start of the incubation period (see Chapter 34).
3. Slowly submerge the open OM chamber into the stream (or into stream water brought back to the laboratory), taking care to retain the leaf material, and to avoid introducing air bubbles. While the chamber is submerged, place the lid tightly on the chamber. Remove the chamber, turn it upside down, and tap the chamber with your index finger to make any air bubbles that may be present easier to detect. Note the time (to the nearest minute) at which the chamber was capped (i.e., time zero).
4. Fill a second OM chamber(s) with only stream water to serve as a control. As with Step 3, examine the chamber for air bubbles, and note the time at which the chamber was capped.
5. Wrap the chambers with aluminum foil to avoid potential oxygen production due to photosynthesis during the incubation period. In the laboratory, darkness can be achieved by incubating the chambers in a refrigerator or other darkened environment.
6. Place the chambers in the stream within some container (e.g., a mesh bag) that will allow for maintaining the chambers at stream temperature during the incubation. Anchor the container in place in the stream. In the laboratory, place the chambers in a rack within a controlled temperature environment.
7. After approximately 2 h, remove the chambers and measure the dissolved oxygen in both and record the data (Table 27.2). Note the time at which the incubation period is completed.
8. Transfer the organic matter to a prelabeled paper bag and transport to the lab for determining dry weight and AFDM as described earlier (General Protocol steps 12–16).

27.3.8 Data Analysis

Calculate the respiration rate of the microbes on the organic matter as follows (see also Table 27.2):

$$R_{OM} = [(DO_{OM\ start} - DO_{OM\ end}/t_{OM}) - (DO_{Control\ start} - DO_{Control\ end}/t_{Control})] \times Volume_{H_2O}Chamber/Mass_{OM} \quad (27.4)$$

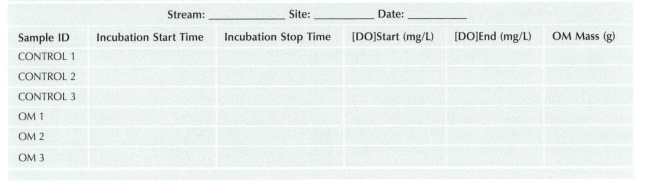

TABLE 27.2 Example data sheet for studying microbial respiration on organic matter. Each control chamber consists of stream water only while the organic matter chambers (OM) contain organic matter and stream water. Abbreviations correspond to those used in Eq. (27.2). See also Online Worksheet 27.2.

Stream: _____ Site: _____ Date: _____

Sample ID	Incubation Start Time	Incubation Stop Time	[DO]Start (mg/L)	[DO]End (mg/L)	OM Mass (g)
CONTROL 1					
CONTROL 2					
CONTROL 3					
OM 1					
OM 2					
OM 3					

where DO = dissolved oxygen (mg/L), R_{OM} = O_2 consumption in units of mg $O_2 \cdot$ g AFDM$^{-1} \cdot$ hr^{-1} by the microbes on the organic matter in the chamber, DO$_{OM \ start}$ and DO$_{Control \ start}$ = concentrations of dissolved oxygen in the respective chambers (in mg/L) at time zero, DO$_{OM \ end}$ and DO$_{Control \ end}$ = concentration of dissolved oxygen (in mg/L) in the respective chambers at the end of the incubation, Volume$_{H_2O}$ Chamber = volume of the chamber (in L), Mass$_{OM}$ = AFDM of the organic matter sample (in g), and t_{OM} and $t_{Control}$ = respective incubation period (in h).

27.4 QUESTIONS

1. Are leaf pack breakdown and organic material decay essentially the same process? Why or why not?
2. What might be the impact to energy flow in a woodland stream if streamside (riparian) vegetation composition were simplified by removing all but one or two species? Can you think of examples where this has been done?
3. By what mechanisms might a "pollution" source alter the process of leaf breakdown in streams?
4. How might you attempt to experimentally separate the importance of biological process (i.e., microbial conditioning and consumer feeding) from physical processes such as abrasion and fragmentation by currents in breaking down leaves in streams?
5. What are some of the variables that make some leaves more resistant or susceptible to leaf breakdown processes in streams?
6. What is the general pattern of microbial respiration over time? Does it differ with different leaf species?
7. Is microbial respiration a good integrator of microbial activity? Might other variables be more sensitive to changes in microbial activity over time during leaf breakdown?
8. What are some anthropogenic activities that could speed up, or slow down, the breakdown process?

27.5 MATERIALS AND SUPPLIES

Equipment
   Top-loading analytical balance accurate to 0.1 g
   Conventional analytical balance accurate to 0.1 mg
   Muffle furnace
   Forced air drying oven
   250-μm mesh sieve
   Grinding mill
   Portable dissolved oxygen probe and meter
   Air-tight respiration chambers (e.g., 60-mL centrifuge tubes)
   Cork borer or scissors
Supplies
   Mesh bags
   Gutter nails
   Steel rebar
   Resealable plastic bags
   Paper bags
   Labels
   Tape
   Waterproof marker and pencil
   Aluminum weighing pans
   Tongs
   Gloves
   Forceps

REFERENCES

Bärlocher, F., Kendrick, B., 1975. Leaf conditioning by microorganisms. Oecologia 20, 359–362.
Benfield, E.F., 1997. A comparison of litterfall input to streams. Journal of the North American Benthological Society 16, 104–108.
Benfield, E.F., Jones, D.R., Patterson, M.F., 1977. Leaf pack processing in a pastureland stream. Oikos 29, 99–103.
Benfield, E.F., Webster, J.R., 1985. Shredder abundance and leaf breakdown in an Appalachian mountain stream. Freshwater Biology 15, 113–120.

Benfield, E.F., Webster, J.R., Hutchens, J.J., Tank, J.L., Turner, P.A., 2000. Organic matter dynamics along a stream-order and elevational gradient in a southern Appalachian stream. Internationale Vereinigung für Theoretische und Angewandte Limnologie, Verhandlungen 27, 1341–1345.

Benfield, E.F., Webster, J.R., Tank, J.L., Hutchens, J.J., 2001. Long-term patterns in leaf breakdown in streams in response to watershed logging. International Revue of Hydrobiologie 86, 467–474.

Blazewicz, S.J., Barnard, R.L., Daly, R.A., Firestone, M.K., 2013. Evaluating rRNA as an indicator of microbial activity in environmental communities: limitations and uses. International Society for Microbial Ecology Journal 7, 2062–2068.

Boulton, A.J., Boon, P.I., 1991. A review of methodology used to measure leaf litter decomposition in lotic environments: time to turn over a new leaf? Australian Journal of Freshwater and Marine Research 42, 1–43.

Boyero, L., et al., 2011. A global experiment suggests climate warming will not accelerate litter decomposition in streams but might reduce carbon sequestration. Ecology Letters 14, 289–294.

Carlisle, D.M., Clements, W.H., 2005. Leaf litter breakdown, microbial respiration and shredder production in metal-polluted streams. Freshwater Biology 50, 380–390.

Cummins, K.W., 1974. Structure and function of stream ecosystems. BioScience 24, 631–641.

Duarte, S., Pascoal, C., Alves, A., Correia, A., Cassio, F., 2010. Assessing the dynamics of microbial communities during leaf decomposition in a low order stream by microscopic and molecular techniques. Microbiological Research 165, 351–362.

Feio, M.J., Alves, T., Boavida, M., Medeiros, A., Graca, M.A.S., 2010. Functional indicators of stream health: a river basin approach. Freshwater Biology 55, 1050–1065.

France, R., Culbert, H., Freeborough, C., Peters, R., 1997. Leaching and early mass loss of boreal leaves and wood in oligotrophic water. Hydrobiologia 345, 209–214.

Gessner, M.O., Chauvet, E., 2002. A case for using litter breakdown to assess functional stream integrity. Ecological Applications 12, 498–510.

Gessner, M.O., Chauvet, E., Dobson, M., 1999. A perspective on leaf breakdown in streams. Oikos 85, 377–384.

Goncalves, J.F., Graca, M.A.S., Callisto, M., 2006. Leaf-litter breakdown in 3 streams in temperate, Mediterranean, and tropical Cerrado climates. Journal of the North American Benthological Society 25, 344–355.

Griffiths, N.A., Tiegs, S.D., 2016. Organic matter decomposition along a temperature gradient in a forested headwater stream. Freshwater Science 35, 518–533.

Gurtz, M.E., Webster, J.R., Wallace, J.B., 1980. Seston dynamics in southern Appalachian streams: effects of clear-cutting. Canadian Journal of Fisheries and Aquatic Sciences 37, 624–631.

Hagen, E.M., Webster, J.R., Benfield, E.F., 2006. Are leaf breakdown rates a useful measure of stream integrity along an agricultural land-use gradient? Journal of the North American Benthological Society 25, 330–343.

Hoellein, T.J., Tank, J.L., Rosi-Marshall, E.J., Entrekin, S.A., Lamberti, G.A., 2009. Temporal variation in substratum-specific rates of N uptake and metabolism and their contribution at the stream-reach scale. Journal of the North American Benthological Society 28, 305–318.

Hladyz, S., Tiegs, S.D., Gessner, M.O., Giller, P.S., Risnoveanu, G., Preda, E., Nistorescu, M., Schindler, M., Woodward, G., 2010. Leaf-litter breakdown in pasture and deciduous woodland streams: a comparison among three European regions. Freshwater Biology 55, 1916–1929.

Kleinbaum, D.G., Kupper, L.L., Muller, K.E., 1988. Applied Regression Analyses and Other Multivariate Methods. Kent Publishing Company, Boston, MA.

Klug, M.J., Cummins, K.W., 1979. Feeding ecology of stream invertebrates. Annual Review of Ecology and Systematics 10, 147–172.

Kreutzweiser, D.P., Good, K.P., Capell, S.S., Holmes, S.B., 2008. Leaf-litter decomposition and macroinvertebrate communities in boreal forest streams linked to upland logging disturbance. Journal of the North American Benthological Society 27, 1–15.

Kuehn, K.A., Ohsowski, B.M., Francoeur, S.N., Neeley, R.K., 2011. Contributions of fungi to carbon flow and nutrient cycling from standing dead *Typha angustifolia* leaf litter in a temperate freshwater marsh. Limnology and Oceanography 56, 529–539.

Langhans, S.D., Tiegs, S.D., Gessner, M.O., Tockner, K., 2008. Leaf decomposition heterogeneity across a riverine floodplain mosaic. Aquatic Sciences 70, 337–346.

Lecerf, A., Chauvet, E., 2008. Intraspecific variability in leaf traits strongly affects alder leaf decomposition in a stream. Basic and Applied Ecology 9, 598–605.

Maloney, D.C., Lamberti, G.A., 1995. Rapid decomposition of summer-input leaves in a northern Michigan stream. American Midland Naturalist 133, 184–195.

McTammany, M., Benfield, E.F., Webster, J.R., 2008. Effects of agriculture on wood breakdown and microbial biofilm in southern Appalachian streams. Freshwater Biology 53, 842–854.

Melillo, J.M., Aber, J.D., Muratore, J.F., 1982. Nitrogen and lignin control of hardwood leaf litter decomposition dynamics. Ecology 63, 621–626.

Melillo, J.M., Naiman, R.J., Aber, J.D., Eshleman, K.N., 1983. The influence of substrate quality and stream size on wood decomposition dynamics. Oecologia 58, 281–285.

Niyogi, D.K., Lewis Jr., W.M., McKnight, D.M., 2001. Litter breakdown in mountain streams affected by mine drainage: biotic mediation of abiotic controls. Ecological Applications 11, 506–516.

Niyogi, D.K., Simon, K.S., Townsend, C.R., 2003. Breakdown of tussock grass in streams along a gradient of agricultural development in New Zealand. Freshwater Biology 48, 1698–1708.

Paul Jr., R.W., Benfield, E.F., Cairns Jr., J., 1978. Effects of thermal discharge on leaf decomposition in a river ecosystem. Internationale Vereinigung für Theoretische und Angewandte Limnologie, Verhandlungen 20, 1759–1766.

Paul Jr., R.W., Benfield, E.F., Cairns Jr., J., 1983. Dynamics of leaf processing in a medium-sized river. In: Fontaine, T.D., Bartell, S.M. (Eds.), Dynamics of Lotic Ecosystems. Ann Arbor Press, Ann Arbor, MI, pp. 403–423.

Petersen, R.C., Cummins, K.W., 1974. Leaf processing in a woodland stream. Freshwater Biology 4, 345–368.

Plafkin, J.L., Barbour, M.T., Porter, K.D., Gross, S.K., Hughes, R.M., 1989. Rapid Bioassessment Protocols for Use in Streams and Rivers: Benthic Macroinvertebrates and Fish. USEPA Document/444/4-89-001, Washington, DC.

Robinson, C.T., Jolidon, C., 2005. Leaf breakdown and the ecosystem functioning of alpine streams. Journal of the North American Benthological Society 24, 495–507.

Roy, A., Capps, K.A., El-Sabaawi, R.W., Jones, K.L., Parr, T.B., Ramirez, A., 2016. Urbanization and stream ecology: diverse mechanisms of change. Freshwater Science 35, 272–277.

Simon, K.S., Simon, M.A., Benfield, E.F., 2009. Variation in ecosystem function in Appalachian streams along an acidity gradient. Ecological Applications 19, 1147–1160.

Smock, L.A., Metzler, G.M., Gladden, J.E., 1989. Role of debris dams in the structure and functioning of low-gradient headwater streams. Ecology 70, 764–775.

Sokal, R.R., Rohlf, F.J., 1995. Biometry, third ed. W.H. Freeman and Co., San Francisco, CA.

Sollins, P., Glassman, C.A., Dahm, C.N., Clifford, N., 1985. Composition and possible origin of detrital material in streams. Ecology 66, 297–299.

Sponseller, R.A., Benfield, E.F., 2001. Influences of land use on leaf breakdown in southern Appalachian headwater streams: a multiple-scale analysis. Journal of the North American Benthological Society 20, 44–59.

Suberkropp, K., Godshalk, G., Klug, M., 1976. Changes in the chemical composition of leaves during processing in a woodland stream. Ecology 57, 720–727.

Suberkropp, K., Klug, M., 1976. Fungi and bacteria associated with leaves during processing in a woodland stream. Ecology 57, 707–719.

Suberkropp, K., Klug, M., 1980. The maceration of deciduous leaf litter by aquatic hyphomycetes. Canadian Journal of Botany 58, 1025–1031.

Tank, J.L., Webster, J.R., Benfield, E.F., 1993. Microbial respiration on decaying leaves and sticks in a southern Appalachian stream. Journal of the North American Benthological Society 12, 394–405.

Tiegs, S.D., Clapcot, J.E., Griffiths, N.A., Boulton, A.J., 2013. A standardized cotton strip assay for measuring organic-matter decomposition in streams. Ecological Indicators 32, 131–139.

Tillman, D.C., Moerke, A.H., Ziehl, C.L., Lamberti, G.A., 2003. Subsurface hydrology and degree of burial affect mass loss and invertebrate colonization of leaves in a woodland stream. Freshwater Biology 48, 98–107.

Ward, G.M., Aumen, N.G., 1986. Woody debris as a source of fine particulate organic matter in coniferous forest stream ecosystems. Canadian Journal of Fisheries and Aquatic Sciences 43, 1635–1642.

Webster, J.R., Benfield, E.F., 1986. Vascular plant breakdown in freshwater ecosystems. Annual Review of Ecology and Systematics 17, 567–594.

Webster, J.R., Waide, J.B., 1982. Effects of forest clear cutting on leaf breakdown in a southern Appalachian stream. Freshwater Biology 12, 331–344.

Woodward, G., et al., 2012. Continental-scale effects of nutrient pollution on stream ecosystem functioning. Science 336, 1438–1440.

Young, R.G., Matthaei, C.D., Townsend, C.R., 2008. Organic matter breakdown and ecosystems metabolism: functional indicators for assessing river ecosystem health. Journal of the North American Benthological Society 27, 605–625.

Zar, J.H., 2010. Biostatistical Analysis, fifth ed. Prentice-Hall, Inc., Upper Saddle River, NJ, USA.

Chapter 28

Riparian Processes and Interactions

Amanda T. Rugenski[1], G. Wayne Minshall[2] and F. Richard Hauer[3]

[1]Department of Ecology and Evolutionary Biology, Cornell University; [2]Stream Ecology Center, Department of Biological Sciences, Idaho State University; [3]Center for Integrated Research on the Environment and Flathead Lake Biological Station, University of Montana

28.1 INTRODUCTION

Derived from the Latin *riparius* (noun form; adjective *riparian*), meaning bank or shore, in stream ecology, the term relates to that environment that engages the open stream channel and is the principal interface between the terrestrial environment and streams. Because of the integral relationship that exists between the stream channel and riparian environments, the two often are regarded as constituting a single corridor that functions within a well-defined stream—riparian ecosystem (Minshall, 1988; Cummins et al., 1989; Gregory et al., 1991). The riparian zone often varies dramatically along a stream—river system from headwaters to mouth. It can be very narrow in confined stream reaches that are geomorphically characterized by canyons or narrow valley segments in mountain regions or by confining geologic formations outside of mountainous terrain (see also Chapter 2). In unconfined stream reaches, the riparian area can be expansive, where the main channel moves laterally across a floodplain over time (Gregory et al., 1991; Stanford et al., 2005) and encompasses streambank and floodplain vegetation, as well as vegetation lateral to stream channels likely to enter the stream by gravity (*recruitable debris*) (Minshall, 1994). Across landscapes throughout the world, riparian zones are disproportionately important in sustaining high diversity of habitats, supplying nutrients for primary and secondary production, and providing corridors of connectivity linking populations that would otherwise become isolated. Unconfined reaches, where streams and the riparian corridors form complex floodplains, are not only used by fish, but are essential to the life requirements of a wide variety of aquatic, avian, and terrestrial species (Hauer et al., 2016).

The riparian environment forms a transition zone between the open stream and the adjacent uplands and exerts major controls over the flux of materials and energy between the two systems (Ewel et al., 2001). The stream—riparian interface may be a sharp boundary with very recognizable edges or a gradual transition (i.e., ecotone) between the two. The size and distinctiveness of the riparian border and whether it is viewed as a distinct boundary or an ecotone depend on the gradient or rate of change in the environmental conditions encountered between the open channel and true uplands. The sharpness of the environmental gradient is a function of a number of factors including climate, geological controls, topography, surrounding land form, and fluvial processes (Gregory et al., 1991; Montgomery, 1999; Naiman et al., 2005). Shallow groundwaters, interconnecting with the riparius, supply dissolved nutrients to the open stream channel via the hyporheos through microbial transformations and transport, and can locally enhance stream productivity (Valett et al., 1994; Boulton et al., 2010). The hyporheos (see also Chapter 8) is a region of shallow groundwater (Fig. 28.1), which is intermediary between and interconnected to the surface water of the open stream channel and the deeper groundwater. Laterally, the hyporheos extends into the riparian region (Boulton et al., 1998) where it influences and is influenced by the riparian vegetation. Hyporheic groundwater may also contribute to higher plant species richness (Mouw and Alaback, 2003; Kuglerova et al., 2014) by providing nutrients, thereby having indirect effects on in-stream processes, such as changes in litter composition or amount of shading. Temporally, the interchanges are most active during periods of high hydraulic head, usually associated with flooding or extensive surface water runoff from the land, but are mediated by temperature (e.g., runoff over frozen soil). In rivers with extensive floodplains, annual floods result in lateral interactions between the open channel and the riparius. These interactions may be surficial, as explained by the flood pulse concept (Junk et al., 1989; Bayley, 1995) and/or may have a strong subsurface component associated with an expansive hyporheic zone and the alluvial aquifer of floodplain reaches (Stanford and Ward, 1993; Brunke and Gonser, 1997; also see Chapters 5 and 8).

Methods in Stream Ecology. http://dx.doi.org/10.1016/B978-0-12-813047-6.00006-1

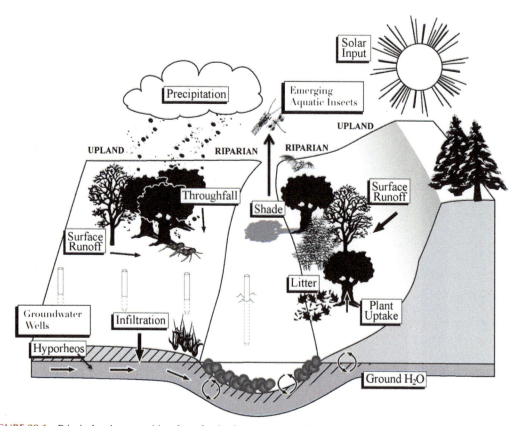

FIGURE 28.1 Principal pathways and interfaces for riparian processes and interactions. *Modified from Laird Duncan IRTC, ISU.*

Riparian habitats are especially important as refugia (Sedell et al., 1990) during periods of environmental stress—such as annual drought or rapid shifts in climate—because they provide ameliorated microclimate conditions along river valleys (Gregory et al., 1991; Palmer and Bennett, 2006). In addition, riparian habitats may harbor different species than upland habitats, leading to increased regional richness (Sabo et al., 2005). Plant community composition influences many processes along the stream–riparian corridor (Tabacchi et al., 2000; Richardson et al., 2010). The riparian environment strongly influences biophysical structure and the functions of microclimate, physical structure, and food resources driven by the distribution, abundance, and species composition of streamside and/or floodplain vegetation affecting alluvial water tables, soils and biogeochemical cycling, solar radiation and stream temperature regimes, and light conditions (see Chapters 6–8). For example, terrestrial leaf litter constitutes an important food resource for lotic consumers, while terrestrially derived woody debris provides physical habitat, modifies streamflow and channel conditions, retains organic matter of smaller sizes, and may provide additional food resources for benthic organisms (Richardson et al., 2010; Scholl et al., 2015). In undisturbed systems, the influence of riparian vegetation, the annual amount of terrestrial leaf litter in the channel, the availability of dissolved organic matter, and the modal size of particulate organic matter all vary with the distance from the headwaters of a stream system (Vannote et al., 1980). However, alteration of riparian plant communities, such as through land use or climate change, influence the dominant species (e.g., increase drought-tolerant species) and diversity of plant functional traits, which are likely to impact the quality and quantity of these resources in stream food webs (Kominoski et al., 2013).

Dissolved (DOM), fine (FPOM), and coarse particulate organic matter (CPOM) enter streams from the adjacent land (see Chapters 24–26). In addition, terrestrial invertebrates falling into the stream may fuel aquatic food webs through contributions to fish diets (Wipfli, 1997; Eberle and Stanford, 2010; Richardson et al., 2010), and large terrestrial mammals have recently been shown to have impacts on transporting substantial quantities of nutrients to streams (Subalusky et al., 2015) and influencing lotic trophic interactions (Nakano et al., 1999). Recent research has shown reciprocal flow of energy in stream–riparian environments and that these trophic linkages are fundamental to ecosystem processes (Baxter et al., 2004; Marczak et al., 2007; Bartels et al., 2012; Soininen et al., 2015). Flow of energy and transport of nutrients from aquatic to riparian ecosystems occur through predation by terrestrial animals including bats, birds, amphibians, or invertebrates (Collier et al., 2002; Baxter et al., 2005; Regester et al., 2006; Malison and Baxter, 2010; Subalusky et al., 2015). Awareness of the close relationship and importance of the association between lotic and terrestrial ecosystems has recently gained greater/renewed recognition among terrestrial and aquatic community and ecosystem ecologists (e.g., Soininen et al., 2015; Hauer et al., 2016).

Riparian vegetation also exerts significant control over the extent to which the surface of a stream is shaded and the dominance of the food type (allochthonous vs. autochthonous) driving stream organism trophic relationships. The availability of light regulates the occurrence and growth of algae and higher aquatic plants in streams (Hill and Dimick, 2002). Shading reduces light and moderates the thermal regime of stream communities by providing cooler temperatures, which benefit most aquatic life (Swanson et al., 1982; Kiffney et al., 2004). Removal of riparian vegetation can result in increases of water temperature and thus directly affect levels of dissolved oxygen, invertebrate species composition, and fish, which in turn generates changes in stream food webs (Tabacchi et al., 1998; England and Rosemond, 2004; Recalde et al., 2016). Invasion of nonnative riparian vegetation into a riparian habitat may alter shade and organic matter inputs and nutrient dynamics (Mineau et al., 2011, 2012). Riparian areas also perform a number of other important functions including (1) physical filtration of water, such as sediment and heavy metal removal (Cooper et al., 1987), (2) bank stabilization, (3) water storage and recharge of subsurface aquifers, (4) nutrient retention, transformation, and release (Lowrance et al., 1984; Triska et al., 1989; Brookshire et al., 2011), and (5) corridors for the dispersal of plants and animals (Gregory et al., 1991).

Below we present a suite of methods for measuring riparian structure and processes across spatial scales of stream reaches and segments. These methods may be scaled up to extend across multiple locations using a combination of on-site and remote-sensing techniques. At large spatial scales of watersheds and regions, additional GIS layers such as topography, vegetative cover, and road density may be included (see also Chapter 1) as part of an overall evaluation of riparian zone quantity and quality at watershed scales.

28.2 GENERAL DESIGN

A number of structural and functional attributes of riparian habitats can be measured and evaluated. Riparian zones, especially those that are within floodplain geomorphic settings, are spatially and temporally dynamic (Stanford et al., 2005). The riparian zone can be distinguished on the basis of hydrology and geomorphology that will express itself through highly characteristic responses in biogeochemical cycling, organic matter processing, soil development, vegetation, and aquatic, avian, and mammalian species habitat support (Swanson et al., 1982; Lowrance et al., 1984; Tabacchi et al., 1998; Naiman et al., 2005; Jansson et al., 2007; Kominoski et al., 2013; Hauer et al., 2016). From a functional perspective, the riparian zone and associated alluvial aquifer (i.e., hyporheic zone) is an area of direct interaction between terrestrial and aquatic systems involving exchanges of energy and matter (Baxter et al., 2004; Doucett et al., 2007; Allen et al., 2012; Valett et al., 2014). The focus of this chapter is on both the stream and the riparian zone as a functional corridor unit. Herein we present five methods of measurement of the riparius: (1) quantification and characterization of the vegetation; (2) the attenuation of solar radiation reaching the stream due to vegetation shading; (3) the input, transfer, and processing of coarse organic matter; (4) the fluxes of energy and nutrients involving reciprocal interactions affecting the stream channel specifically and the riparian corridor in general; and (5) the use of remote-sensing tools to provide insights into the character of the stream and riparian corridor.

28.2.1 Quantifying Riparian Vegetation Communities

As discussed in general terms above, most streams and rivers of the world alternate between confined and unconfined reaches (see also Chapters 1 and 2). In confined reaches the stream channel is constrained laterally, while in unconfined reaches the stream channel is unconstrained laterally, characterized by overbank flooding, cut-and-fill alluviation processes, and channel movement (see Chapter 5). The valley and reach type, along with climatic and biome setting, greatly affects the size, scope, and character of the riparian zone and the nature of the riparian corridor. In confined stream reaches, the riparian zone subject to flooding may be very narrow as erosional processes are down-cut into the surrounding landscape and the stream channel is well defined by stream banks and a rapidly rising slope lateral from the stream. In unconfined stream reaches, also generally associated with a geomorphic setting of a floodplain, the riparian zone is expansive with stream banks less well defined and off channel habitats, both aquatic and terrestrial, that have been formed largely by fluvial processes (Montgomery, 2002; Stanford et al., 2005).

The species composition, abundance, and diversity of vegetation are primary characteristics that define the riparian zone. In confined reaches, this vegetation may be only a narrow band found along the streamside. In unconfined reaches, characteristic riparian vegetation may extend hundreds of meters, and among large rivers kilometers, from the extant stream or river channel with vegetation characteristic of specific hydrogeomorphic settings created by past fluvial

FIGURE 28.2 Old growth riparian forest on the Nyack floodplain of the Middle Fork Flathead River, a western North America gravel-bed river, Montana, USA. The stream in the foreground is a floodplain springbrook. The mature cottonwood and later seral stages of conifer trees compose the dominant riparian species growing on the floodplain surface.

processes. For example, in the coastal plain region of southeastern United States, streams and rivers overflow their banks and flood across broad floodplains. This results in a highly characteristic vegetation, often referred to as bottomland hardwoods (Cronk and Fennessy, 2016), composed of red maple (*Acer rubrum*), bald cypress (*Taxodium distichum*), water tupelo (*Nyssa aquatic*), and a variety of other species all affected by the duration of floodplain inundation (Clark and Benforado, 2012). In another example, throughout the western United States and Canada, riparian zones of gravel-bed rivers are initially dominated by various pioneer species including cottonwood (*Populus* spp.), willow (*Salix* spp.), and alder (*Alnus* spp.) with spruce (*Picea* spp.), fir (*Abies* spp.), larch (*Larix* spp.), and cedar (*Thuja* spp.) dominating as surfaces mature in their succession processes (Fig. 28.2). Streamside forests provide an array of physical and biogeo-chemical attributes that protect stream water quality, habitat, and stream organisms (Sweeney and Newbold, 2014), yet they are greatly affected by stream and river regulation (Rood and Mahoney, 1990), invasive species (Glenn and Nagler, 2005), and declines in streamside and floodplain water tables (Scott et al., 1999).

Quantitative methods are typically used to evaluate vegetation composition and collect data suitable for statistical analysis. Presence/absence techniques can be used to rapidly collect large quantities of data. Two quantitative techniques utilizing this approach are the point intercept method and the line intercept method. These methods are widely used in upland plant ecology and are equally useful in riparian plant ecology (Greig-Smith, 1983). Both the point intercept method and the line intercept method are generally used to quantify herbaceous or shrub vegetation, but can also be readily adapted to large plot assessments utilizing remote-sensing and geographic information system (GIS) technologies.

The objective of the point intercept and line intercept techniques is to make measurements at regularly spaced, preselected, or defined locations and to avoid subjectively selecting locations in the field. Data can then be used in various analyses. Preselecting or defining points and finding them in the field are critical parts of this approach. Most GIS or mapping packages can generate grid points at a user-specified spacing. Selected points can then be entered into the navigation software of most GPS systems to rapidly find the points in the field. Alternatively, preselected point positions can be printed onto GIS maps and then navigated from one point to another in the field. The point intercept sampling technique is particularly good for herbaceous vegetation cover (i.e., grasses and forbs). The line intercept method has been widely used for gallery forest cover and large wood (see also Chapter 29 for an in-stream application). Generally, in the line intercept method, several transects are deployed in the study area, or around the riparian zone, to determine the percent occurrence and diversity of tree species in the plant community (Kimball et al., 2015). Line transects can be used for many purposes including directly measuring percent cover of species or marking set line intervals in which species presence is recorded. Species present should be recorded for each line transect, as is done for the point intercept method. Species are considered present in a given line transect if they intersect the plane of the line segment at some preselected distance laterally from the transect line.

The point quarter method (PQ), also called the point-centered quarter method, is commonly used to measure tree density and cover in forests (Silva et al., 2014). This is a "plotless" method that is particularly useful in riparian habitats when some species of specific interest are rare. For each target species, the PQ method uses the distance to the nearest individual in each of four quadrants. Each quadrant can be separated by the cardinal directions or by direction perpendicular to the stream bank or downslope energy line of a floodplain. This method measures relative and absolute density, percent cover, and frequency at which the species of interest occurs. Cover estimates derived from the PQ method assume a random distribution of the target species. The PQ method, and adaptations of the method, is particularly useful to calibrate remote-sensing imagery data, as well as validate the classifications following data analysis (Mouw et al., 2013; also see Advanced Method 2, below).

28.2.2 Attenuation of Solar Radiation—Shading

Solar radiation affects primary production and stream temperatures. A useful table comparing various methods for determining shading is provided by Davies-Colley and Payne (1998). The extent of stream shading by riparian vegetation (i.e., canopy cover) can be evaluated as (1) the degree (absolute or relative percent) to which solar radiation is diminished relative to that received on unobstructed bare ground and (2) the spatial extent and duration of shadows cast by overhead vegetation.

In the first instance, incoming solar radiation is measured at the water or soil surface with a suitable pyranometer or quantum sensor (see Chapter 7) at multiple locations along a stream segment and compared with values obtained from an unshaded sensor. Light values also can be integrated into a single value by using photodegrading organic dyes (e.g., rhodamine WT and fluorescein) in both aquatic (Bechtold et al., 2012) and terrestrial systems (Roales et al., 2013). Solar radiation changes seasonally with the angle of the sun, as well as with cloud cover, leaf development, and dominant plant species and age. Thus, these factors need to be considered in a comprehensive investigation.

In the second instance, measurement of the spatial extent and duration of shadows cast by overhead vegetation may be measured with a simple spherical densiometer, a Solar Pathfinder (Platts et al., 1987), or with fish eye lens photographs (Davies-Colley and Payne, 1998; Englund et al., 2000). The densiometer is a small, convex, spherical mirror with an engraved grid that reflects the canopy over the head of the observer. Canopy cover is measured by counting the grid intersections covered by vegetation. The densiometer reflects vegetation to the sides as well as overhead. Multiple measurements should be taken in different directions (e.g., facing downstream, left bank, right bank) and averaged.

The Solar Pathfinder instrument (www.solarpathfinder.com) is a relatively inexpensive and simple-to-use instrument that estimates energy input directly in energy units ($kW\ m^{-2}\ d^{-1}$) as a portion of total available energy reaching a site corrected for shading (Platts et al., 1987; Davis et al., 2001). To evaluate stream shading, the instrument is deployed in the middle of the stream. Obstructions that would block solar input are reflected on its domed surface. The reflection is then outlined on a solar chart or determined from a digital photograph using the manufacturer's software. Values representing the percent of total daily input relative to unshaded conditions are used to measure total annual solar energy input and percent shading. With modifications, this instrument can also provide daily values. Results are most accurate when only a few trees or shrubs are present. This instrument is of limited use with a dense riparian canopy.

Hemispherical photography (fisheye photography) provides a wide-angle measurement to estimate potential total solar radiation, canopy cover, and effective shade for a specified location (Kelley and Krueger, 2005; Julian et al., 2008). Photographs are taken with a standard digital camera fitted with a hemispherical lens and secured to a leveled tripod. Hemispherical canopy photographs are overlaid by an image of the sun path to calculate solar radiation transmitted through openings in the canopy (Fig. 28.3). To perform the analysis, HemiView software converts the hemispherical digital photo to a black-and-white image, where black pixels represent shading by vegetation or topography and white pixels represent sky pixels. For best results, photos should be taken under evenly overcast skies or in the early morning. Photos taken on clear sunny days may need image correction to enhance contrasts between riparian vegetation and the sky before being imported into the digitizing software. Though expensive to purchase, the canopy analyzer manufactured by LiCor (LAI-2200C) is the most versatile and informative instrument currently available for assessing shade and canopy change along stream reaches at multiple points (Davies-Colley and Payne, 1998).

Quantifying solar energy input to streams allows subsequent examination of interactions between light and algal production (Kiffney et al., 2004; Finlay et al., 2011) and the subsequent effects on invertebrate grazers. For example, as riparian plants gain and lose leaves in the spring and autumn and as the plants populating forests change, streambed light regimes change. This variation can affect periphyton photosynthetic characteristics and primary production in streams, as well as alter the importance of algae relative to leaf litter for macroinvertebrate consumers (Hill and Dimick, 2002; Schiller et al., 2007). Measurements of riparian canopy cover and light energy (also see Chapter 7) provide a better understanding of the effects that disturbances, such as agriculture (Hladyz et al., 2011), logging (Hauer et al., 2007), wildfire (Kleindl et al., 2015), or invasive species (Mineau et al., 2012), may have on stream systems.

FIGURE 28.3   Examples of hemispherical photos from (A) an open canopy stream and (B) a closed canopy stream, and photos of the streams where the hemispherical photos were taken for the (C) open and (D) closed stream. *Photos by C.L. Atkinson.*

28.2.3 Input and Decomposition of Coarse Organic Matter

Much of the organic matter in a stream originates from the surrounding terrestrial environment and be transported to the stream by wind, water, gravity, or direct deposition. Because of its origin outside of the stream boundaries, this material is referred to as *allochthonous* and is primarily of plant origin. Since much of this allochthonous plant litter, in the form of leaves, twigs, and other parts, is dead by the time it reaches the stream, it also is often referred to as *detritus* or *allochthonous detritus*. In many cases, particularly in shallow (wadeable) streams in forest or shrub lands, the composition of riparian plant communities and organic matter of terrestrial origin play a major role in establishing stream ecosystem structure and function and influencing aquatic food webs (e.g., Fisher and Likens, 1973; Minshall et al., 1983; Kominoski et al., 2011; Tank et al., 2010; Kiffney and Richardson, 2010; also see Chapters 20, 23, and 27). Valuable insights into the dynamics of flowing water ecosystems and terrestrial–aquatic linkages are based on the changing terrestrial dependence of these systems with respect to different biogeographical area; increasing channel size; varying types and amounts of streamside vegetation; different land use practices, and the input storage, processing, and output of organic matter (e.g., Cummins et al., 1989; Minshall et al., 1983; Duncan et al., 1989; Meyer, 1990; Colón-Gaud et al., 2008; Lagrue et al., 2011; Kominoski et al., 2013).

Accurate determination of leaf input and other forms of organic matter contributions to streams and soils by riparian plants is a daunting task. The most common approach has been to place collectors, such as plastic laundry baskets or large plastic storage containers, over open water or along the stream banks in an attempt to estimate these litter inputs. But such methods have suffered from a number of problems, including nonrandom placement and/or inadequate sample size. In addition, they only allow estimation of the portion that falls directly into the stream or on the ground and miss any lateral overland input by gravity, wind, or other means, which although often not quantified, can make up a large portion of total overland input to streams. We suggest, in conjunction with placing litter collectors in the stream corridor, that separate traps with mesh bags be secured along the stream banks at bankfull height facing away from the stream and with the bottom of the opening flush to the ground to collect *lateral litter inputs* (Wallace et al., 1997; Colón-Gaud et al., 2008; Kanasashi and Hattori, 2011; Hart et al., 2013). A more holistic approach is to collect all of the litter deposited on netting of known area spread over the stream at bank level and/or on the ground. Some researchers have attempted to estimate litter inputs by measuring litter occurrence in the benthic organic matter component of streams by using drift nets and benthos samples (see Chapters 15 and 26), but the values obtained are affected by (1) stream transport into and out of the location from which the collection is made and (2) the typically clumped distribution of leaves in streams.

Very few studies have attempted to measure or separate the individual components of litter input (e.g., direct fall in, lateral transport; but see Kanasashi and Hattori, 2011; Hart et al., 2013), and we know of no study that has measured litter production for an entire riparian zone or determined its relative contribution to the ground surface and adjacent stream. We regard both of these as worthwhile endeavors for future research. One approach, though labor intensive, would be to estimate litter production on a plant volume basis for each of the main forms of riparian plants, scale that up to the total volume of each of those plants within the area of the riparian zone of interest, and determine direct fall-in and lateral inputs to the stream as indicated above (Cummins et al., 1989). A simplification of this method, though less comprehensive and informative, is to use the line intercept method to determine riparian plant species composition and relative abundance, and relate this to the amount of leaf and needle litter that becomes trapped in the adjacent stream reach. Another approach is to determine the litter production for the vegetation within the riparian zone and then, by use of a suitable tag, such as a stable isotope, measure the portion that ends up in the stream. In this chapter, we describe a modification of the most commonly used procedure employing measurement of direct and lateral input of litter to streams. The litter fall procedure also is applicable to the riparius if suitable adjustments are made for losses to the stream and any export or import due to wind or other factors.

Litter decomposition and linkages to communities of stream invertebrates are strongly related to plant type (Cummins et al., 1989; Lagrue et al., 2011). Thus, it is important to characterize the detritus production according to the particular processing category to which various species belong. Leaves are generally recognized as belonging to one of three litter processing categories in terms of rates of decay: **fast** (>0.15% dry weight loss per day, normalized for temperature), **medium** (0.10%–0.15%), and **slow** (<0.10%). Common representatives of each category: fast—*Alnus, Cornus, Fraxinus, Liriodendron, Prunus*; medium—*Acer, Populus, Salix, Ulmus*; slow—*Pinus, Platanus, Quercus, Rhododendron, Tsuga* (Petersen and Cummins, 1974; Webster and Benfield, 1986; Cummins et al., 1989; see also Chapter 27). Leaf quality in terms of tannins (−), N (+), C:N (−), and lignin (−) is significantly correlated with processing rates (Ostrofsky, 1997).

Decomposition processes represent a major flux of both fixed carbon and nutrients in most riparian systems, and quantifying rates of litter mass loss or respiration and the concomitant changes in nutrients bound in the litter are important aspects of evaluating riparian ecosystem function. We refer the reader to Chapter 27 for a thorough discussion of the measurement and analysis of in-stream leaf litter breakdown and decomposition. Litter decomposition on riparian soil is controlled to varying degrees by abiotic and biotic conditions (Wagener et al., 1998). The decomposition process transforms senescent plant material into labile and refractory organic matter, both above and below ground (Harmon et al., 1999). The dynamics of riparian decomposition can act as either a nutrient sink (i.e., nutrients retained in the riparian area) or a source (i.e., nutrients transported to the stream). Standing stocks of litter are important carbon and nutrient reservoirs of the riparian zone. The sizes of these reservoirs are influenced by rates of litter production and decomposition.

Different kinds of leaves have variable decay rates, and a mix of leaf species may affect leaf pack decomposition (McArthur et al., 1994; Kaneko and Salamanca, 1999; Swan and Palmer, 2004; Leroy and Marks, 2006; Schindler and Gessner, 2009). The use of cotton strip assays has been proposed as a standardized method for measuring decomposition (Tiegs et al., 2007, 2013). On riparian soils, litter provides a direct source of food to invertebrate consumers, serves as a substrate for microbes, and through leaching and decomposition releases nutrients needed by riparian plants. In streams, allochthonous detritus serves a similar function and is especially tightly linked to characteristic fungi (see Chapters 10 and 20 for fungi and shredders, respectively). Therefore, in addition to measuring the rate of leaf decay, the strength of these linkages on riparian soils and in streams can be determined through assessment of fungal and detritivore standing crops and litter nutrients. On riparian soils, additional indicators include (1) terrestrial plant growth rates and size and (2) soil nutrient concentrations.

28.2.4 Transfer of Dissolved Organic Matter and Nutrients

Riparian zones play a dominant role in energy and material flow to streams and can serve both as sources or sinks for energy and matter (Naiman and Décamps, 1997; Brookshire et al., 2011). DOM and nutrients are transported into, through, and out of riparian habitats, primarily by precipitation, surface runoff, and groundwater carrying leachates from organic horizons of soils (Fig. 28.1). Water chemistry is altered as it passes through the riparian zone along hydrologic pathways from uplands to streams. Thus, riparian zones function as control points for fluxes of carbon, nitrogen, and other nutrients from terrestrial to aquatic systems (Hedin et al., 1998; Ranalli and Macalady, 2010; Vidon et al., 2010; Grabs et al., 2012). For example, Mulholland (1992) found higher concentrations of soluble reactive phosphorus and inorganic nitrogen (N) in riparian groundwater and springs than in upslope soil solution or stream water. DOM and nutrient inputs to riparian zones can be measured through collection and analysis of precipitation, surface runoff from uplands, and groundwater to investigate transformations occurring as water moves through the riparian zone. These values may differ depending on time of collection relative to hydrologic conditions, whether during baseflow, rainstorms, or snowmelt and spring runoff. In this section, we examine the processes and interactions in the riparian zone involving exchanges of energy and matter, resulting in the regulation of the movement of materials in the soil and groundwater, and the effects of these processes on food web structure in streams.

Precipitation and canopy leaching (via throughfall) can potentially be large sources of important nutrients, such as N and phosphorus (P), to the stream, whereas soils can be a major sink (Mulholland, 1992; McGlynn and McDonnell, 2003; Hoffmann et al., 2009). Riparian vegetation removes and retains particulates from uplands, which favors soil microbiological processes and increases soil nutrient cycling, thus reducing nutrient inputs to streams. Furthermore, groundwater also plays an important role in structuring plant community composition, which in turn affects nutrient dynamics (Kuglerova et al., 2014). In most forested watersheds, biological and geochemical processes in upper soil horizons effectively retain N and P, thus reducing inputs to streams (Wood et al., 1984; Mayer et al., 2007; Hoffmann et al., 2009; Brookshire et al., 2011). Soil texture also plays an important role in determining relative proportions of surface water and groundwater inputs. Processes that occur in the soil are influenced by redox (reduction or oxidation) conditions. Reduction processes require that soils be anaerobic or of low redox potential (Eh), and oxidation processes require the opposite conditions. Low Eh is a result of belowground processes consisting of biogeochemical reactions that transfer electrons from organic matter released from plants to various terminal electron acceptors (Tabacchi et al., 1998). For example, increasing oxidation of soils leads to increased nitrate concentrations due to nitrification, whereas reduction leads to increased ammonium concentrations through denitrification (see Chapter 32). These changes in redox conditions can be measured through the analysis of water samples taken from the stream, groundwater, and upland areas (Dwire et al., 2006; Brookshire et al., 2005; Groffman et al., 2009).

The main control on the interaction of groundwater within stream riparian zones is the hydrogeologic setting, which encompasses surface topography, soils, and the composition, stratigraphy, and hydraulic characteristics of the underlying geological deposits (Frank et al., 1994; Kuglerova et al., 2014). To examine the role of groundwater in riparian—stream interactions, we refer the reader to Chapter 8. Individual wells or clusters of wells may be used depending on the scale of study and the questions being addressed. Water samples collected from the wells can be analyzed for nitrate, nitrite, total organic N, ammonium N, total P, orthophosphate P, and organic matter concentrations (see Chapters 24, 31−33, 36).

Stable isotopes are also a powerful tool to understand stream—riparian interactions and processes and address complex questions dealing with trophic interactions (see also Chapter 23). Stable isotope analysis (SIA) can be conducted for C, N, and H in each of the main stages that may be encountered as a nutrient makes its way across the riparius and into the stream (e.g., abscised leaves, leaves conditioned in the stream and on the forest floor, periphyton, soil, groundwater, surface water, and invertebrates). This information can be used to determine pathways into, within, and out of the riparius, compare riparian zone processes in different ecoregions, and determine the effect of alterations, such as deforestation and agricultural practices on riparian—stream processes. Stable isotopes also can serve as tracers of energy flow and nutrient subsidies within food webs (Peterson and Fry, 1987; Allen et al., 2012; Reisinger et al., 2013; Atkinson et al., 2014) and can be useful in establishing the relative importance of terrestrial versus aquatic energy sources (Finlay, 2001; Doucett et al., 2007; Finlay and Kendall, 2007). Stable isotopes of C and N can be used to discriminate between allochthonous and autochthonous pathways in food webs at specific sites (Thorp and Delong, 2002; Lau et al., 2009), but is dependent on the degree of isotopic differentiation between these two food resources (Fig. 28.4; also see Chapter 23).

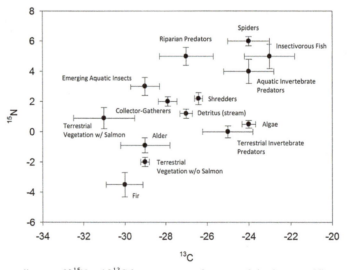

FIGURE 28.4 Sample stable isotope diagram of $\delta^{15}N$ and $\delta^{13}C$ for components of stream and riparian areas. Nitrogen isotopes are indicators of trophic level (2−3‰ fractionation for each level), while carbon isotopes indicate which plants (terrestrial and aquatic) are potential sources for consumers.

Current research is also using hydrogen isotopes (δD) to determine allochthonous and autochthonous energy and water sources, and may be a powerful complement to δ^{13}C because environmental conditions known to affect δ^{13}C (e.g., water discharge, seasonality) do not greatly influence δD (Doucett et al., 2007; Solomon et al., 2009; Jardine et al., 2009; Finlay et al., 2010); however, δD should not be used as a substitute for δ^{13}C (Jardine et al., 2009). The ratio of C isotopes changes little as carbon moves through food webs, and therefore δ^{13}C can typically be used to evaluate the ultimate sources of carbon for an organism when the isotopic signature of the sources is different (Collier et al., 2002). In aggregate, δ^{13}C, δ^{15}N, and δD signatures for stream algae, terrestrial leaves, and stream invertebrates are useful for distinguishing aquatic versus terrestrial energy sources to consumers (England and Rosemond, 2004; Finlay, 2001; Jardine et al., 2009). Furthermore, δD may be used as an environmental marker in fish to provide insight into their environmental history (Whitledge et al., 2006). The extent to which nitrate decreases in riparian groundwater due to denitrification and/or plant uptake can be determined through measurements of nitrogen isotopes in both groundwater nitrate and riparian plant tissues (Clément et al., 2003).

Other advantages in using stable isotopes are (1) distinguishing between marine and terrestrial sources of nitrogen; (2) tracing the transfer of carbon and nitrogen to riparian predators; and (3) tracking N derived from consumers into food webs. Past research has mainly focused on examining N derived from marine sources (MDN) to examine the effects on nutrient cycling and riparian and stream productivity (Cederholm et al., 1999; Wipfli et al., 1998; Thomas et al., 2003). Isotopic ratios of ^{15}N:^{14}N are generally higher in marine systems, and elevated ratios in terrestrial systems are indicative of marine enrichment. Riparian plants adjacent to spawning streams may derive up to 25% of foliar N from salmon (Bilby et al., 1996; Merz and Moyle, 2006; Koshino et al., 2013), thereby enhancing vegetation growth rates (Helfield and Naiman, 2002; Morris and Stanford, 2011). Recent research focuses on the interaction between salmon and bears in stimulating riparian processes such as soil and plant nutrient cycling (Helfield and Naiman, 2006; Holtgrieve et al., 2009; Koshino et al., 2013). In the field, ^{15}N tracer experiments combined with δ^{15}N and δ^{13}C natural abundances and measures of diversity are proving to be powerful tools in elucidating the transfer of energy, nutrients, and biomass between aquatic and terrestrial systems.

Stable isotopes are also being used as tracers to track nutrients from enriched consumers through food webs. Recent research focused on resource subsidies shows the movement and transfer of energy and nutrients by mobile organisms, which may spend only brief periods in or adjacent to streams, such as bears and moose (Helfield and Naiman, 2006; Bump et al., 2009; Holtgrieve et al., 2009), amphibians (Regester et al., 2006; Capps et al., 2015), and invertebrates (Nakano et al., 1999; Sanzone et al., 2003). Allen et al. (2012) documented a trophic cascade from aquatic to terrestrial ecosystems through increased aquatic insect emergence when mussel biodiversity was high, which led to an increase in terrestrial spider abundance tracking higher abundances of emerging insects. Trophic cascades have also been documented from riparian to aquatic habitats where terrestrial arthropods controlled algal biomass through changes in fish predation (Nakano et al., 1999). Moreover, spiders in the riparian area may obtain 65%−100% of their carbon from in-stream sources in desert streams (Sanzone et al., 2003). These studies show that these transfers are significant for community structure and function in recipient systems.

28.2.5 Remote Sensing of the Riparian Zone

Remote sensing is the acquisition of information about an object or phenomenon without making physical contact with the object that is being measured. This contrasts to on-site observation in which researchers or instruments make contact with, or are in close proximity of, that which is being measured. In stream ecology, remote sensing often refers to geospatially explicit electromagnetic radiation data collected from drones, aircraft, or satellites (Fig. 28.5). These data are often converted into imagery. The integration of remote sensing and GIS has been a prominent feature in vegetation mapping and data analysis going back several decades (Goodchild, 1994). Land cover classification over large geographic areas using remotely sensed data is increasingly common as a result of national inventory and monitoring programs, scientific modeling, and international environmental treaties (Wulder et al., 2006). However, remote-sensing systems of even a few years ago produced data that integrate information over tens or even hundreds of square meters (e.g., Landsat) and thus data are too coarse to detect fine-scale attributes such as variation in vegetation or plant diversity on a river floodplain or riparian zone along a stream.

Advances in the spatial and spectral resolutions of sensors now available make direct remote sensing of species-specific variation and density feasible for stream ecologists interested in large-scale patterns and processes. For example, distinguishing species assemblages or identifying species composition of patches of gallery forest or of individual trees is now possible with high-resolution, multispectral imagers deployed from aircraft. A growing number of studies have focused on evaluating vegetation parameters and canopy reflectance to produce vegetation maps that evaluate river riparian zones and floodplains (Carbonneau and Piegay, 2012; Mouw et al., 2013). Advances in the spatial and spectral resolutions

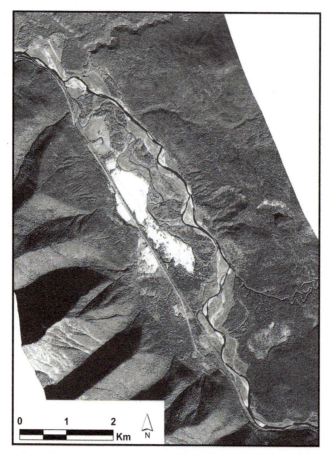

FIGURE 28.5 Quickbird Satellite panchromatic band image at 0.6-m spatial resolution available from Digital Globe of the Nyack floodplain of the Middle Fork Flathead River, a western North America gravel-bed river, Montana, USA.

of sensors now available to ecologists are making remote sensing of attributes requiring high resolution of vegetation cover to be increasingly feasible (Turner et al., 2003; Mouw et al., 2013). Pixel resolution <10 cm^2 obtained using multispectral (RGB) imagers, including high-resolution digital cameras deployed from aircraft, are now routine. Classification of spectral reflectance from such high-resolution imagery allows spatial analysis of riparian vegetation at the submeter, pixel-level of resolution.

Airborne LiDAR (light detection and ranging) is a surveying method that measures distance to a target by illuminating that target with a laser light. When deployed from an aircraft or helicopter, LiDAR is popularly used to make high-resolution "bare earth" digital elevation models (DEMs) or 3D maps. LiDAR maps can be highly detailed and data can be displayed to determine the height of forest canopies (Lefsky et al., 2002) and provide spatially explicit, three-dimensional distribution analysis of plants. This allows discrimination of vegetation species and analysis of species distribution and abundance.

Functional aspects of forests, such as productivity, carbon storage, and litter production, are related to forest canopy structure (Hardiman et al., 2013). Riparian ecologists have long understood that specific organisms, or overall richness of wildlife communities, can be highly dependent on the three-dimensional spatial pattern of vegetation, from flying vertebrates (Rotenberry, 1985; Müller et al., 2012) and boreal mammals (Melin et al., 2016) to arthropods (Normann et al., 2016). Linking data from passive optical systems to high-resolution LiDAR data enables researchers to analyze the spatial pattern of classification data.

28.3 SPECIFIC METHODS

28.3.1 Site Selection

The procedures described here lend themselves to virtually all stream or river riparian zone conditions, including large floodplains. In constrained valleys, the width of the riparian zone is commonly regarded as extending to a distance equivalent to the height of the tallest trees growing on the hillslope above and beyond the stream bank. In unconstrained valleys, the riparian zone extends across the floodplain from stream channel and, similar to the constrained valley, to the height of the tallest trees or shrubs growing on the hillslope (see also Chapters 1, 2, and 5).

Depending on project objectives, a common approach to site selection is to choose a 100- to 250-m reach for small-scale exercises or monitoring or a 1-km-long segment or segments of stream length for large-scale projects. The area of the riparian zone(s) is then determined as the product of the reach or segment length times the mean width of the riparian zone. The width of the riparian zone is determined from 5 m or more (e.g., at 25- to 50-m intervals) transects oriented perpendicular to the channel and extending from the edge of the stream bank to the hillslope on both sides of the stream or river. In confined reaches this distance may be on the order of a few meters, but in unconfined reaches this distance may extend hundreds or thousands of meters (see Chapter 5). Record the stream widths separately for these same cross-sectional locations. Generally, the segment(s) will contain a reasonable degree of environmental heterogeneity to provide a range of conditions. However, the investigator may prefer to examine a gradient of conditions. For example, a variety of habitat types within one segment or segments of different stream orders or streams within the same ecoregion can be compared across a set of contrasting conditions (e.g., constrained vs. unconstrained, grassland vs. forest, logged vs. unlogged, roaded vs. unroaded, grazed vs. ungrazed, burned vs. unburned). A single, paired comparison may be instructive for learning purposes, but adequate replication is needed for research projects.

28.3.2 Basic Method 1: Quantifying Riparian Vegetation Communities

The point intercept and line intercept methods work well for understory, shrub, or herbaceous plant coverage; however, the PQ method described here is preferred for riparian gallery forests. The PQ method can also be used, in conjunction with detailed GPS measures of the forest canopy, to provide the data needed for calibrating remotely sensed imagery data (see Advanced Method 2 below).

28.3.2.1 Point-Centered Quarter Method

1. Determine the general area to be sampled. For certain research studies, aerial or satellite photos may be needed. The number of sampling sites and the number of plots (n) sampled will depend on the questions being asked in the research and the variation in the riparian area being investigated.
2. In an unconfined reach of stream or river length, the width of the riparian area or the size of the floodplain to be sampled may extend several hundred meters from the stream channel. Locations of plots for the PQ method should be chosen based on the density of the trees and/or shrubs in the riparian area and any variation in riparian forest cover. For example, if the patches of vegetation are highly varied, then the frequency of plots and plot location must reflect that variation. If the riparian cover is more homogeneous, then the frequency of plots may be less numerous.
3. Based on the variation in the riparian vegetation to be sampled, using the aerial or satellite imagery, select points with a 50-m radius of uniform stand of forest around each point. Using GIS, register the points and enter the geospatial coordinates into a GPS field instrument for locating field position. These plots will constitute the samples replicated within each plant community.
4. In the field, locate the preselected point center for each plot. After locating the point center of a plot, lay out the plot either on a cardinal NS-EW quartering or a quartering of the plot based on perpendicular orientation with the stream-channel energy slope. If the riparian area to be sampled is on a large floodplain, use the cardinal orientation. If the area being sampled is near a hillslope, use the stream-channel energy slope orientation for the quartering.
5. Determine the number of trees within each of the four quarters of the plot to a distance of 50 m or $n = 20$, whichever occurs first. Identify, measure, and record on the plot data sheets (see Online Worksheet 28.1) the species, angle from the northerly direction (using the cardinal direction protocol), distance from the point center to the center of the tree bole, diameter of the tree bole at breast height (DBH), and the distance from the bole to the outer edge of the tree canopy for each tree in the plot population. Note: these data will be used for calibrating (ground truth) the tree canopy data for Advanced Method 2: Remote Sensing (see below).
6. Determine the gallery forest tree composition. Estimate the percentage of cover in the gallery canopy that is attributable to each of the trees in the plot sample. (Use of the protocol for determination of shading, given below in Basic Method 2, is highly useful in determining percent canopy coverages; see also Fig. 28.3.)
7. Use the point intercept method or line intercept method for determining shrub or herbaceous cover within one of each of the quarters of the gallery forest plots.

28.3.3 Basic Method 2: Attenuation of Solar Radiation—Shading

Select a series of riparian habitat conditions ranging from sunlit to heavily shaded, or if conducting a specific research project, use a randomized sampling design to compare reference and treatment segments.

28.3.3.1 Pyranometer or Quantum Sensor

1. Measure midday light levels, with a pyranometer or quantum sensor, in midstream and in the center of the riparian zone at multiple points along each stream—riparian segment and in an open (reference) area in the general vicinity.
2. Also measure air and water or soil temperatures at depths of 0 and 5 cm at each of these locations. Improvements to this basic approach would be to integrate values over a day, season(s), or year by obtaining representative sets of multiple measurements (e.g., hourly throughout a day) or through use of a data logger (see Chapter 6). In addition, the temperature measurements could be linked to microbial activity or (in terms of cumulative *degree days*) to leaf decay rates (see below).

28.3.3.2 Spherical Densiometer Measurements

1. Determine the absolute and relative degree of shading and its effect on the thermal environment in each segment and compare the results between/among segments. Take a densiometer reading every 5—25 m over a 250-m reach. The densiometer should be held ~10—20 cm in front of your body and at elbow height.
2. Take four measurements from the center of the stream facing different directions (e.g., N, E, S, W or upstream, downstream, right bank, left bank), and one measurement on each stream bank, for a total of six readings per transect. Note that the reflective mirror is engraved with a grid of 24 0.64-cm (¼″) squares; each square is visually divided into 4 smaller squares or imagine 4 dots in each corner for a total of 96.
3. Count the number of dots in each individual square covered by canopy, and record this number ($n = 24$ readings/transect). For each transect, every measurement represents one-fourth of the total density, and the sum of the individual measurements is multiplied by 1.04 (Platts et al., 1987) to determine the canopy closure or density for that transect (Platts and Nelson, 1989).
4. By expressing the results relative to readings from a totally unshaded area, measurements may be obtained in terms of "% shaded."

28.3.3.3 Hemispherical Photo Measurements

1. The camera is mounted on a tripod or placed in a shallow tray near the water or soil surface and oriented to the north. Photos are best taken on cloudy days, as opposed to midday sunny conditions.
2. Photos can be analyzed using HemiView software or free online programs, such as Gap Light Analyzer. Percent effective shade is calculated by using the Global Site Factor (GSF) in the HemiView software.

$$\text{Effective Shade } \% = (1 - \text{GSF}) \times 100 \qquad (28.1)$$

28.3.4 Basic Method 3: Input and Decomposition of Coarse Particulate Organic Matter

The approach generally employed for quantifying CPOM input from the riparian environment to the stream is to measure the mean amount of riparian plant leaf and needle litter deposited in the stream during leaf abscission per unit area (e.g., square meter) (see also Chapter 26). This is commonly done using a suitable number of strategically placed litter traps fashioned from plastic containers with small drainage holes in the bottom. For easier retrieval and efficient handling of the captured material, these containers may be lined with netting or with plastic trash bags that have small holes for drainage. For many purposes, a stream segment of 250 m long will be adequate for study, although appropriate adjustments should be made depending on the specific objectives. For illustrative purposes, it will be assumed that the mean width of the riparian corridor is 20 m, evenly divided on each side of the 5-m wide stream segment.

The preferred approach is to measure the mean amount of CPOM deposited in litter traps (=direct inputs) by each of the major plant species, expressed on a per area basis [e.g., g dry mass (DM)/m^2 or g ash-free dry mass (AFDM)/m^2] or scaled up for a stream reach or segment. An added refinement would be to stratify the sampling across the moisture gradient, extending from stream channel to outer edge of the riparian zone, to obtain mean values for multiple locations along the stream reach or segment. In addition, the estimates of the amounts falling directly into the stream can be further improved by adding in the amounts contributed by lateral transport (g/m² for each of the stream edges).

28.3.4.1 Leaf Litter Traps

1. Construct vertical litter-fall traps from laundry baskets or large plastic storage containers lined with plastic bags or netting. Small (<1-mm) holes placed in the bottom of the bags and traps will allow rainwater to drain out. Lateral input litter traps can be made by constructing 20-cm high × 50-cm wide rectangular wood or metal frames and attaching fine mesh bags to them. Orient the opening of the frame perpendicular to and facing away from the stream, with the frame staked upright and the bottom flush with the ground.

2. Place sets of the vertical and two lateral traps along the stream transect. On the order of 5−10, transects along a 250-m reach should be adequate for many purposes. Decide whether to place these sets randomly or in representative locations. Leaf litter traps can be suspended over the stream as close to the water surface as possible by tying them to nearby trees with ropes or securing them just above the water to logs or other material. Traps can also be placed on the banks. It is important to make sure that they will not become inundated with stream water, especially during high flow events. Lateral input traps can be placed on the bank at the high flow mark or ∼0.5−1 m away from the edge of the bank.

3. Empty the vertical (direct input) litter traps every 1−2 weeks until leaf fall has ceased. Continue sampling the lateral input traps until the organic material on the ground is stabilized, usually by rain or snow. Keep the results for each individual trap and date separate at this point. If the results are to be separated by species and/or decay rate (fast, medium, slow), do this before further processing. Dry material at 60°C until all moisture is removed—i.e., no further weight loss occurs—and record the final weight.

4. Sum the results from all sampling dates for each trap, convert the values to a square meter or linear meter equivalent based on the area of the vertical trap or width of the lateral trap, and calculate the mean and standard deviation for all traps of a similar type (vertical or lateral) to quantify total inputs of dry mass (or AFDM; see Chapter 27) as direct fall (per m²) or as lateral input (per m of stream length) can be calculated as the mean total combined litter input (MTCLI) per meter of stream segment, either composited or by species:

$$\text{MTCLI/m} = \left(\overline{V} \times \overline{W}\right) + \overline{L} + \overline{R} \tag{28.2}$$

where $\overline{V}$ = grand mean of direct input/m² for the vertical traps multiplied by the mean width of the stream $\left(\overline{W}\right)$, plus grand means per meter for the left $\left(\overline{L}\right)$ and right bank $\left(\overline{R}\right)$ lateral inputs. This value, multiplied by the length of the stream segment, equals the mean total litter input per segment (e.g., for use when comparing two or more segments having different treatments).

28.3.4.2 Measurement of Coarse Particulate Organic Matter Decomposition Rates

1. Leaf decomposition methods are described in detail in Chapter 27, so here we describe the general protocol. The common/recommended procedure is to place packets of weighed leaves of a given species in the stream or on the riparian soil surface for a series of different exposure times. For the purposes of this protocol, it will be assumed that measurements will be made both in the stream and in its adjacent riparian zone.

2. 5- to 10-g packs generally are used, and the leaves are either fastened loosely together with monofilament or placed in mesh bags. Aquaculture cage netting and potato/onion/orange bags with 5-mm openings have proven satisfactory. In streams, the packs usually are held in place by attaching the netting/bags to a stationary object. Bricks are often used for this purpose because they are relatively inert and stable, easy to locate for retrieval, and convenient to obtain in quantity. If mesh bags or other enclosures are used, care must to be taken to use sufficiently large mesh openings to ensure aerobic conditions and allow access by shredding invertebrates.

3. Make up a set of leaf packs for each tree species of interest. Determine the number of leaf packs of each species by multiplying the desired number of replicates (usually 3−5) by the habitats to be treated. For example, if you are interested in decomposition rates of leaves from 4 different species, in a mix of 6 different habitats (example: 3 stream and 3 riparian), then 4 species × 5 replicates × 6 habitats = 120 packs. For teaching and some research purposes, leaves may be stockpiled in advance, air-dried, and stored in labeled 113.5-L (30-Gal) plastic trash bags until needed. Note: for short-term studies, such as a class laboratory exercise, *fast* leaves will prove the most satisfactory.

4. At the start of the study, disperse the entire set of packs throughout the study area and expose them to the stream or riparian conditions until they are retrieved. Allow 24–48 h for leaching of water-soluble substances before collecting the first subset of packs, then collect additional subsets every 150–300 degree days depending on the leaf-processing category of the leaf (the faster the expected processing rate, the shorter the thermal interval).

5. Degree days are calculated by summing the mean diel (24-h) temperature for each day of exposure; a miniature data logger, in a waterproof container, is ideal for obtaining precise mean diel temperature values, but a maximum–minimum recording thermometer will suffice. To determine the approximate sampling frequency: take the mean of the maximum and minimum stream/soil (or air) temperatures, measured over several days at the time of the study,

$$[(T_{max} + T_{min})/2] \qquad (28.3)$$

and divide it into the desired degree-day value. For example, for a maximum temperature of 14°C, a minimum of 6°C, and a desired 150 degree-day exposure period, the sampling interval would be 150°C days/10°C = 15 days. The intervals should be selected to yield packs that have lost approximately 25%, 50%, and 75% of their initial dry weight after leaching.

6. At the time of removal, place a 250-μm mesh dip net under the leaf pack, transfer the pack and net contents to a labeled plastic bag, and temporarily refrigerate until drying and weighing (60°C until a constant weight is attained). If the leaf packs and associated material cannot be processed within a few days after arrival in the laboratory, they should be frozen until processing can be done. Prior to drying, rinse leaves over a sieve to remove sediment and invertebrates. Invertebrates can then be removed from the sieve and frozen or saved in labeled vials or small plastic bags containing a preservative, for use in the next section.

7. Calculate the decay rate (k) from the slope of the best-fit line in a semilogarithmic plot of percent dry mass remaining versus exposure time (on x-axis), or as the least squares fit of the data to an exponential the function:

$$W_f = W_i e^{-kt} \qquad (28.4)$$

where W_i and W_f = initial and final weights and t = amount of time (days) that leaves were in the field (riparian or stream habitats).

$$-k = \log_e(\%R/100)/t \qquad (28.5)$$

where $\%R$ = percent remaining at any time (t).

$$\%R = W(t_f)/W(t_i) \times 100 \qquad (28.6)$$

The rate coefficient can be converted into mean % daily loss by:

$$\%R/day = (1 - e^{-k}) \times 100 \qquad (28.7)$$

Leaf decomposition is strongly controlled by the thermal environment, and temperature may explain much of the difference between riparian and stream decay rates (see Section 28.2.2). For comparative purposes, decay rates may be standardized for temperature by substituting degree days of exposure in place of time in the preceding plot or by using the relationship in Eq. (28.4), where t = cumulative degree days.

28.3.4.3 Assessment of Detritivore Standing Crops

Standing crops of stream macroinvertebrate shredders may be quantified using quadrat sampling techniques and devices such as a Hess or Surber sampler or, if in cobble-bed rivers, the Hauer-Stanford net (see Merritt et al., 2008 and Chapter 15). Cummins et al. (1989) suggested that shredders will maximize their biomass at the time of greatest availability of litter in a given processing class that can support maximal growth (approximately the 50% weight loss point).

1. Collect a minimum of five samples from each stream segment of interest, remove all of the shredders from each (see Chapter 20), determine the shredder biomass (DM at 60°C) in each sample, and calculate the mean biomass and SD.

2. Express the results per square meter, based on the area of the sampler. In general, there should be a direct relationship between the amount of CPOM in a stream and the biomass of shredders, which can be tested with the data collected thus far.

3. By initiating the measurement of CPOM decomposition rates (see above) at the time of maximum leaf drop, and by sampling shredder biomass at regular intervals until at least 50% of the mass of the leaf pack of the target species above has disappeared, the hypothesis above can be tested.

4. Ideally, sampling intervals will be selected to yield leaf packs and shredder biomass when approximately 25%, 50%, and 75% of the initial leaf pack biomass after leaching has been lost. However, if the timing is not known from previous study, it can be approximated from published values. For example, for fast-decaying plant species, sampling every 10−14 days over a 2-month period would be adequate to encompass the entire decay sequence. In the case of medium or slow decay leaf species, attaining the 50% loss point indicated above, possibly using a logarithmic scheme of increasing sampling intervals, should provide results satisfactory for illustrative purposes.

5. Plot the weight of shredder biomass per mass of remaining leaves (y-axis) against the percentage of leaf mass remaining (x-axis). If the hypothesis is supported, the peak shredder biomass will occur at approximately the 50% weight loss point.

6. Similar procedures and rationales can be used for investigating the riparian soil fauna. One approach is to collect soil samples (leaf litter can be collected separately) using a bulb planter, plastic cylinder, or special soil corer and then physically remove all of the shredder-type organisms (e.g., earthworms) from each core by sieving samples to remove fauna either in the laboratory or through heat and/or light extraction methods (see Chapter 14). After fauna are collected, measure their biomass per unit area (based on the cross-sectional area of the sampling device), and relate it to riparian litter standing crops or decay rates as described above.

28.3.5 Advanced Method 1: Transfer of Dissolved Organic Matter and Nutrients from the Riparius to the Stream

As described above (also see Chapters 24 and 30), water is the principal mechanism for transporting DOM and nutrients from the riparius to the stream, and hence their transfer is closely linked to the hydrologic cycle (Fig. 28.1). In addition, riparian zones can be potential sources or sinks for DOM and nutrients, depending on redox conditions (Mulholland, 1992; Costello and Lamberti, 2008). The approach taken here is to isolate each of the important compartments along the hydrologic cycle in which DOM and nutrients occur, and measure their concentrations. After collection, all water samples can be analyzed according to APHA standard methods (Rice et al., 2012) for nitrite, nitrate, total Kjeldahl N, ammonium N, total P, orthophosphate P, and DOM, depending on the research objectives (see also Chapters 24, 31−33, 36). The study of nitrogen transfers in riparian environments is especially valuable, both because of the critical role that nitrogen plays as a limiting nutrient and because its various forms are diagnostic of particular states (aerobic vs. anaerobic) and biological transformations occurring there.

28.3.5.1 General Measures of Dissolved Organic Matter and Nutrient Transfer

Precipitation can be measured with a rain gauge, and samples for chemical analysis may be collected in clean polyethylene wide-mouth containers that are placed level on the ground and covered with mesh to keep out insects and large debris. Samples should be obtained soon after a rain event. Surface runoff measurements are made with surface water collectors, which consist of lidded polyethylene bottles placed into holes in the soil in an inverted position, with a rectangular slot cut into the uphill side at ground level. A small plastic sheet is attached to the bottom of the slot and spread uphill to direct flow into the bottle (Peterjohn and Correll, 1984). For maximum information, each runoff event should be sampled separately. Groundwater measurements can be made through collection of water samples from slotted PVC wells placed in the riparian zone and in the stream (see Chapter 8). Soil measurements can be made from samples collected from riparian and upland areas with a soil corer, dried at 60°C, and analyzed for nitrate and ammonium (Page, 1982). Lysimeters can be placed at various depths to collect water in moist soils, while capacitance rods (e.g., Trutrack) can be used to record groundwater table positions.

1. In confined-reach riparian zones or on floodplains, wells can be installed to access the water table for sampling. Samples can be collected from riparian wells using a bailer, and then filtered and stabilized in the field, before being returned to the laboratory for analysis employing APHA standard methods (Rice et al., 2012).

2. After filtration, samples can be analyzed for nitrite, nitrate, ammonium-N, total-P, orthophosphate-P, and organic matter concentrations (e.g., TOC, DOC, DON). Analyses of the various forms of nitrogen can be used to detect patterns in soil biogeochemistry from different environments to measure the effects of N inputs. Further understanding of stream/riparian zone biogeochemistry and microbial activity can be gained through measurement of redox conditions (Hedin et al., 1998).

3. The results from riparian wells can be compared to stream concentrations to determine if correlations are present and whether the riparian soils are a potential nutrient source or sink. This will vary with the geology, soil chemistry, local geomorphology, and species composition of the riparian vegetation or the watershed (e.g., steep slopes, floodplains, land use, wetlands, small moist riparian, desert, species dominated by N-fixing plants).

4. For a more detailed analysis of soil, microbial biomass can be estimated through several methods that include staining and counting of microbial cells; physiological parameters, such as ATP, respiration, and heat output; or fumigation techniques (Page, 1982).

28.3.5.2 Nitrogen Mineralization Potential

A soil's capacity to transform organic nitrogen in soil organic matter to inorganic nitrogen (i.e., *nitrogen mineralization potential*) is often used as an index of nitrogen availability (Robertson et al., 1999). Nitrogen (N) mineralization releases large amounts of ammonium and measures the net increase in both ammonium (NH_4) and nitrate (NO_3) in soil, since any NO_3 formed must first have been NH_4.

1. The relative availability of N can be measured with ion-exchange resin bags (Binkley and Matson, 1983; Binkley et al., 1986). The resin bags mimic nutrient uptake by plant roots by the adsorption and accumulation of nitrate and ammonium to the resin beads, forming an ionic bond with positively ($+$) or negatively ($-$) charged particles on the beads.
2. Resin bags can be placed along transects that run perpendicular to the stream or they may be placed in random locations in the riparian zone, as well as in upland areas, depending on the question being addressed. Resin bags placed in the riparian zone can be compared in terms of N pools with those placed in the uplands; this will provide insights into the processes taking place in the riparius (Frank et al., 1994).
3. Resin bags are prepared by placing mixed-bed ion-exchange resin ($\sim$5 g) in nylon stockings or swimming suit liner material ($\sim$10 cm $\times$ 10 cm square/bag) (see Binkley et al., 1986) and pretreated in 0.5 M HCl for around 30 min, and then rinsed with deionized water until the pH of the rinse water is neutral and stable.
4. After the bags are prepared, number them, place them in a plastic bag, label with the name/number of the batch (each bottle of resin constitutes a batch and should be placed in separate plastic bags), and store in a refrigerator. Record the number of the resin bag before placement in the soil.
5. For placement in the soil, cut out an area about 10 cm deep (slightly wider than the resin bag) and then, at a 90-degree angle, cut into the wall of the hole (about 5 cm from the bottom), making a shelf on which the resin bag may sit without disturbing the soil above (this is very important).
6. After resin bags are in place, fill in the hole and mark with an adjacent surveyor's stake. Time of incubation will vary depending on environment of placement (wet vs. dry). Incubation periods can vary from weeks to months. Take care not to damage the resin bag when removing it. Resin bags should be placed in individual Ziploc bags for transport to the laboratory, and then the resin should be extracted with 1N KCl (1 g resin/15 mL 1N KCl). Filtration of extract must be done 24 h after resin is mixed with KCl and shaken. Dry the resin bag at 60°C and then cut open and weigh the dry resin and record the resin mass.
7. If the soil is not disturbed, another resin bag can be placed on the same shelf in the original hole to assess long-term or seasonal processes. The extracts are analyzed for ammonium and nitrate by standard soil chemistry methods (Page, 1982).
8. Results are reported as resin ammonium and nitrate accumulation in milligram per day. Competition with plants and soil microbes may strongly reduce available N. Increases in water flow to the resin bags may increase ammonium capture more than that of nitrate (Binkley, 1984).

28.3.5.3 Stable Isotope Analysis

SIA can be used to examine trophic status and energy flow pathways (see also Chapter 23). Representative qualitative samples of material generally will suffice for these analyses. Herein we describe methods used to determine linkages between riparian vegetation, soil, groundwater, stream water, invertebrates, and reciprocal flow effects to better understand riparian stream processes.

1. Using a combination of C, N, and D isotopes for all samples collected will help determine organic matter transfers and give insight into trophic structures. Numerous laboratories throughout the United States and Europe can run samples for SIA at reasonable prices (about \$12−30/sample). Results obtained are expressed as $\delta^{13}C$, δD, or $\delta^{15}N$ using the following equation:

$$[(R_{sample} - R_{standard})/(R_{standard}) \times 1000] \tag{28.8}$$

where R = ratio of ^{15}N to ^{14}N or ^{13}C to ^{12}C and the standards are Pee Dee Belemnite (PDB) carbonate for $\delta^{13}C$ and atmospheric N for $\delta^{15}N$.

2. Algae, leaf, and soil samples should be dried at 60°C, ground, and placed in labeled covered glass vials until analysis (samples should be stored in a desiccator and/or dried again before analysis). Qualitative samples of algae can be collected using methods described in Chapter 12.

3. The substrates where algae are collected should be similar for each replicate. Leaves can be collected from vegetation during the senescent period or collected from the ground or the stream. Extra leaf packs can be placed in the riparian zone and in the stream for separate SIA analysis. Soil samples can be collected from upland areas and compared to riparian soil samples.

4. Invertebrates should be allowed to clear their guts before they are dried and ground. Gut clearance can be done by first separating the invertebrates (to avoid predation or cannibalism) and then leaving them overnight in aerated flasks of stream water. Cleared guts will eliminate measurement of unassimilated food particles in the gut.

5. Water samples should either be placed on ice immediately and kept cold until analysis, or frozen if analysis cannot be completed within a few days. Water samples also can be analyzed for δ^{15}N-NO_3 and δ^{15}N-NH_4. Carbon and nitrogen isotopes are useful indicators of trophic status and terrestrial versus aquatic sources (Fig. 28.3). Comparison of δ^{15}N values among vegetation, soil, and water samples can provide information on the processes taking place.

6. Separation in δ^{13}C for consumers eating different sources (periphyton vs. detritus) should be easily detected. Finlay (2001) found that, in forested headwater streams, δ^{13}C in algae was distinct from terrestrial detritus δ^{13}C. Theory predicts a 3–5‰ enrichment in δ^{15}N of consumers relative to their food (Peterson and Fry, 1987). Analysis also can be conducted, where a marine signature is expected, to investigate the effects that anadromous fishes (e.g., salmon) may have on riparian—stream processes.

7. Collect samples from all compartments, as previously described (see Chapter 23), and compare the results with those collected from areas where no marine signature is present. For more advanced techniques, these methods can also be used for other organisms, such as mussels (Allen et al., 2012; Atkinson et al., 2014), emerging aquatic insects, and spiders (Sanzone et al., 2003; Davis et al., 2011) to quantify the amount of stream-derived nitrogen or carbon from terrestrial food webs.

8. Both natural and enriched ^{15}N tracer experiments (see Chapter 31) allow for tracking of animal-derived nutrients from enhanced food resources that are incorporated into animal tissue through consumption, and then tracked into both aquatic and terrestrial food webs. A mixing model can then be used to determine the proportion of ^{15}N from the aquatic system into the terrestrial (see Sanzone et al., 2003; Allen et al., 2012 for detailed equations).

28.3.6 Advanced Method 2: Remote Sensing of the Riparian Zone

The advanced methods presented here serve as an example of how various remote-sensing technologies can be used, either individually or in combination, to provide landscape to submeter scale analysis of the riparian zone along streams and rivers. There are several types and sources of data, from satellite-based passive reflectance imaging data to aircraft-based active laser pulsed data, the various technologies and types of data described below can be employed to pursue research questions at spatial scales that would otherwise be nearly impossible to pursue. These technologies, whether data are acquired through government or commercial sources, allow spatially explicit linkages of both riparian structure and function. However, we advise that research questions be carefully considered and approaches derived prior to the pursuit of any specific techniques or technologies.

28.3.6.1 Satellite Imagery

Several sources of satellite passive reflectance imaging exist. Satellite imagery (e.g., Quickbird) consisting of four multispectral bands (blue: 450–520 nm, green: 520–600 nm, red: 630–690 nm, NIR: 760–900 nm) at a 2.4-m spatial resolution and a panchromatic band at a 0.6-m spatial resolution is available from Digital Globe (Fig. 28.5).

1. The image (Fig. 28.5) was orthorectified using a combination of digital orthoquadrangle data (DOQs) and ground control points (GCPs) located throughout the floodplain.

2. Satellite imaging makes an excellent base image from which other remote-sensing data can be geospatially located.

28.3.6.2 Airborne-Based Imagery

Airborne multispectral data (blue: 450–520 nm, green: 520–600 nm, red: 630–690 nm) can be obtained from vendors or by using high-resolution digital photographic cameras (Fig. 28.6) deployed by the researcher. Aircraft that have been fitted

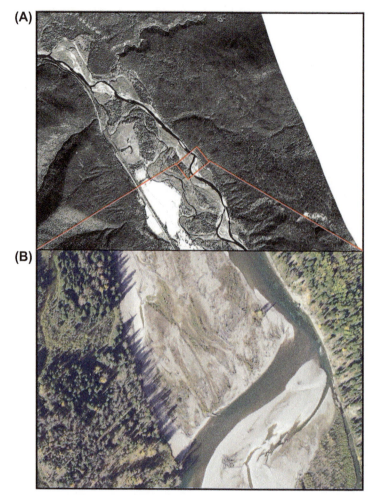

FIGURE 28.6 (A) The lower river segments of the Nyack floodplain as in Fig. 28.5; the *red box* indicates the location of the image below. (B) Airborne digital image of the visible spectral data (blue: 450−520 nm, green: 520−600 nm, red: 630−690 nm) obtained using a high-resolution (<10 cm²) digital photographic camera.

with either wing or fuselage photographic mounts are readily available with a commercial pilot from many general aviation fixed-base operators that provide aeronautical services.

1. All data capture is synchronized and streamed to an onboard laptop computer. A fully programmable image capture system can be configured to capture data in coordination with the flight of the aircraft.
2. With the camera sensor deployed with a 50 mm "normal" lens, the aircraft needs to be flown at 1000 m above ground level, a ground speed of 87 kn (161 km/h), and a frequency of frame-by-frame capture with a capture every 3−4 s to have 50% overlap of images. Ultrahigh-resolution imagery using 20 megapixel imagers acquire image data of <10 cm pixel resolution.
3. Data are collected along predetermined flight lines oriented along the long axis of the riparian zone. Distance between flight lines should be flown to produce a 40%−50% overlap among all neighboring flight lines.
4. Image data are georectified using Caligeo software from Spectral Imaging. Final georectification and mosaiking are completed in ERDAS Imagine from Hexagon Geospatial. A combination of DOQs, GCPs, and satellite imaging is used to complete the orthorectification of airborne data. Minor color-balancing between image frames can be applied during the mosaiking process.

28.3.6.3 Light Detection and Ranging

LiDAR data can be acquired from a variety of sources. For example, the data illustrated in Fig. 28.7 were acquired in cooperation with the National Center for Airborne Laser Mapping (NCALM). Data were collected using an Optech ALTM

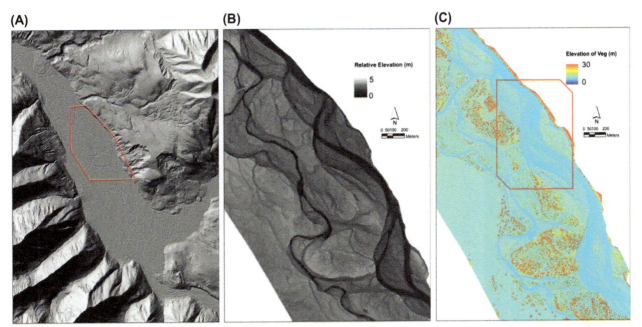

FIGURE 28.7 (A) Light detection and ranging data creating a bare earth digital elevation model (DEM) of the Nyack floodplain of the Middle Fork Flathead River, a western North America gravel-bed river, Montana, USA. The *red polygon* corresponds to the floodplain area shown in Panels B and C. (B) A DEM from Panel A corrected to the mean elevation of the floodplain energy slope. This process allows for relative elevations to be determined at any point on the floodplain for determination of floodplain surface elevations and vegetation. (C) A DEM from Panel B with first return data of the riparian vegetation growing on the floodplain. The *red polygon* on Panel C corresponds to the area classified and illustrated in Fig. 28.9.

(Airborne Laser Terrain Mapper) 1233. This system uses a laser beam pulsing at 33 kHz and is directed by a scanning mirror to record the time and amplitude of a laser pulse reflected off target surfaces. The following steps outline the specific methods a researcher might use to derive the spatially explicit and elevation-specific data illustrated in Fig. 28.7.

1. Terrasolids TerraScan Lidar (http://terrasolid.fi) processing software was used by NCALM to process the raw LiDAR data.
2. Data are unfiltered (i.e., first return) as DEM gridded to 1 m, as well as 1 m filtered (i.e., bare earth) DEM (see Fig. 28.7A).
3. Based on the energy slope of the stream, create a relative elevation grid of the riparian zone (as in Fig. 28.7B of the floodplain surface) derived from the bare earth DEM to estimate elevations above the water surface.
4. To create the relative elevation grid, the water surface and its associated elevation must be extracted from the DEM. These elevations are then interpolated across the entire riparian zone or floodplain to create a water surface elevation grid.
5. The floodplain water surface elevation grid is then subtracted from the bare earth DEM to produce the relative elevation grid.
6. A vegetation canopy DEM (as in Fig. 28.7C) is calculated as the difference between the bare earth DEM and the first-return LiDAR data.

28.3.6.4 Classification of Vegetation and Riparian Patches

Floodplain vegetation cover types can be classified using a combination of the satellite imagery and airborne spectral imagery data to produce detailed vegetation cover maps of riparian zones (Fig. 28.8). Here, we present a hierarchical approach used to generate the vegetation cover types that are illustrated in Figs. 28.8 and 28.9.

1. An unsupervised classification (ERDAS, 2015) is generated from the satellite image using the four multispectral bands and the NDVI (Normalized Differential Vegetation Index) to identify six general cover types—water, cobble, coniferous trees, deciduous trees, grasslands, and agricultural pasture.
2. The airborne image data are used to differentiate reflectance characteristics within the deciduous vegetation classification. Using ENVI software, spectral signatures are generated from known ground truth plots among the dominant deciduous community types. Use geospatially explicit vegetation data that are collected employing the

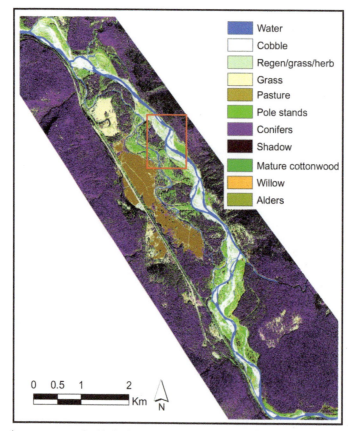

FIGURE 28.8 Floodplain vegetation cover types classified using a combination of the satellite imagery illustrated in Fig. 28.5 and airborne spectral imagery data in Fig. 28.6 to produce a detailed vegetation cover map of the riparian zone of the Nyack floodplain of the Middle Fork Flathead River, a western North America gravel-bed river, Montana, USA. The vegetation classification is generated from an unsupervised classification (ERDAS, 2015). The *red rectangle* corresponds to the area classified and illustrated in Fig. 28.9.

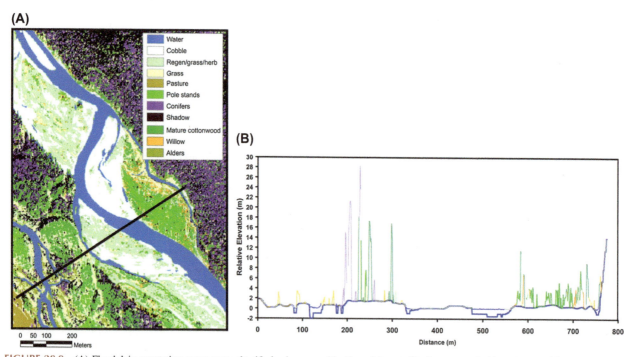

FIGURE 28.9 (A) Floodplain vegetation cover types classified using a combination of the satellite imagery and airborne spectral imagery data corresponding to the area identified by the *red rectangle* in Fig. 28.8. The *black line* indicates an 800-m long "virtual transect" across the floodplain and the classified vegetation and surface cover types. (B) The relative elevation (m) of the floodplain surface and vegetation along the transect indicated in Panel A by the *black line*. Color of the vegetation line corresponds with the classification legend in Panel A. Data are derived from the vegetation classification in Fig. 28.8 and the relative elevation bare earth digital elevation model (DEM) and vegetation DEM in Figs. 28.7B and C.

protocols above in Basic Method 1. Using the mean spectral signature for the deciduous cover types, employ a Mixed Tuned Matching Filter (MTMF) to classify the deciduous vegetation into its major constituents of cottonwood, willow, alder, and others.

3. Tree species succession in the riparian area of our example floodplain (western United States), begins as regeneration of cottonwood, willow, and alder. As the surface of riparian patches matures through natural succession, cottonwood dominates the gallery forest canopy excluding willow and alder. The old cottonwood trees are eventually replaced by spruce, larch, or fir (see Fig. 28.2).

4. LiDAR data are used to further differentiate age categories of trees in the riparian area. In our example of riparian floodplain, pixels classified as cottonwood, conifer, etc., can be further evaluated to determine canopy height as the difference between the bare earth DEM (Figs. 28.7A and B) and the first-return DEM of vegetation heights (Fig. 28.7C).

5. Thus, in our example, pixels classified as cottonwood with height of ≥ 10 m can be further classified as mature cottonwoods, first as pixels and then can be aggregated as needed into patches and areas. Pixels with DEM heights of between 1.5 and 10 m were identified as pole stand cottonwood, and patches less than 1.5 m in height were classified as regeneration cottonwood patches (Fig. 28.9). We also advise the researcher or instructor to visit the online site for this book and download the Chapter 28 movie (Online Supplement 28.2) to visualize a "digital fly-through" of the data based on the digital elevations of the vegetation and floodplain surfaces.

28.4 QUESTIONS

1. The riparian zones of streams vary tremendously across biomes, stream order, land use, and many other variables. Using the point-centered quarter method, did you see a difference in riparian vegetation between stream corridors that you are investigating? What is the range of variation? What species dominate the riparian? What is the sequence of vegetation from early seral stages to mature vegetation?

2. Following your quantification of solar input and canopy cover as a stream increases in size, did you observe any differences among sites? In what ways did they differ? If you compare sites across stream order (e.g., first to fifth order), how did they differ between or among stream segments? If you compare streams with different riparian vegetation (e.g., adjacent to meadow, intact forest, shrubland, agricultural field, clear-cut or selectively logged forest, etc.) or hydrology (e.g., springbrook, runoff-dominated/flashy, semiarid, desert, north- vs. south-facing drainage system, etc.), what differences did you observe?

3. Did you find a relationship between solar input and stream temperature, dissolved oxygen, or periphyton abundance? How did these relationships change quantitatively on an hourly, daily, or seasonal basis? How did these relationships change between reference and treatment sites of your choosing?

4. What differences did you find in riparian litter input and decomposition rates among sites (such as those listed in Question 1), and how does this relate to soil, groundwater, or other properties? Compare the in-stream decomposition rates with those in the adjacent riparius and account for any differences you observed.

5. Calculate the contributions of direct and lateral litter inputs from the riparian zone into the stream to total litter biomass and by processing categories. How do these measurements compare with those of CPOM collected in the stream? Compare the proportion of the input from direct versus lateral sources. What relationships and patterns are present, and how do these patterns vary with stream order, slope, composition of riparian vegetation, or land use?

6. Was there a quantitative relationship between macroinvertebrate shredder biomass and time of greatest leaf litter availability? Suggest reasons if this relationship does not exist.

7. Describe any differences in gut contents of stream or riparian consumers that feed on leaf litter (shredders) that you observed with increases in sunlight or between/among stream—riparian conditions, such as those described in Question 1.

8. What differences did you observe in water quality (DO, temperature, pH, alkalinity, hardness, conductivity, and specific nutrient concentrations) between stream water and groundwater? What may account for these differences? Consider differences in topography, geology, soil type, and land use.

9. Did you observe any differences in nitrate and ammonium among wells located in the lateral transect across the riparian zone. How do you account for these differences? What implications do these findings have for the stream?

10. What differences occur seasonally in riparian soil N, as reflected in resin bags, and how does N content change at increasing distance away from the stream to upland areas?

11. Alder plants and their leaves are in the *fast* decay category and have special physiological properties. What differences did you find in soil and groundwater properties in riparian areas dominated by alder, mixed alder, or no alder?

12. Based on your data, describe important ways that changes in land use (e.g., deforestation, agriculture, and urban) affect stream–riparian interactions. What management recommendations could you make based on your findings?

13. What changes occur in $\delta^{15}N$ values for leaves collected from vegetation during senescence, those decomposing on the forest floor, and those in the stream? How do these values relate to soil properties and in-stream water chemistry?

14. In systems receiving anadromous fish migrations, what differences did you find in $\delta^{15}N$ and $\delta^{13}C$ among riparian vegetation, periphyton, and macroinvertebrates in streams and riparian areas with or without marine-derived nutrients? How do you explain these differences?

15. From isotopic analysis of $\delta^{15}N$, $\delta^{13}C$ and δD, what do you conclude regarding the dominant energy source to invertebrate consumers in stream and riparian habitats? How could the timing of sampling during the year influence your results?

16. What riparian processes are you able to distinguish from satellite or airborne remote-sensing data? What is the range of variation in vegetation percent cover within the riparian zone of the stream(s) you are considering? What is the variation in species composition? Are these differences supported by the variation in the imagery? How have these procedures enhanced you understanding of riparian structure and/or function?

28.5 MATERIALS AND SUPPLIES

The investigator is encouraged to exercise a bit of ingenuity in acquiring the materials and equipment used in the procedures described in this chapter. Considerable savings can be made and a broader array of solutions obtained by using sources other than specialized scientific equipment suppliers. Local sources such as building supply, farm, general merchandise, hardware, and lawn and garden stores can supply many of the needed items either directly or for subsequent fabrication by the investigator. Additional options are to buy in bulk from the manufacturer or primary supplier, or to purchase used or surplus materials from a secondhand or salvage store (e.g., plastic insect netting from an army/navy surplus store). However, for those with limited time or abilities or with unlimited bankrolls, scientific products suppliers will be the obvious source of choice.

Field Materials and Equipment
Establishment of sampling locations
    Surveyor's stakes (~2 dozen per treatment type, such as a forest or pasture)
    Meter tapes—25, 50, and 100 m
    Meter stick
    Field notebook
    Compass
    Inclinometer
Quantifying riparian vegetation communities
    DBH tape
Attenuation of solar radiation—shading
    Pyranometer or quantum sensor
    Spherical densiometer, Solar pathfinder, or digital camera with a 180-degree fish eye lens
    Thermometers for air, water, and soil point measurements
    Thermistor with data logger or maximum–minimum recording thermometer (at least one each for land and water)
Input and decomposition of coarse organic matter
    Plastic laundry baskets or similar for leaf litter collectors
    Rope, nylon (e.g., parachute cord) for suspending leaf input containers
    Metal or wood frames 20 cm high × 50 cm wide for lateral input collectors (10–20 per 250-m reach)
    Netting, 5-mm mesh (in bulk quantity) for fabrication of lateral input collection bags
    Surveyor's stakes (46 cm), 2 per lateral input frame
    Labeling materials
    Bricks or other means for holding leaf packs in place (~20–50 per treatment)
    Thermometer, recording type (same as above)
    Collecting nets (250-μm mesh recommended)
    Dip net
    Surber, Hess, or Hauer-Stanford sampler (see Chapter 15)
    Bulb planter, plastic cylinder, or special soil corer for collecting soil shredder-type invertebrates

Sieve or Berlese—Tullgren apparatus for extracting invertebrates from soil

Plastic bags

    Large (garbage bag) size for bulk leaves, lining for litter collectors, etc.

    Ziplocs (sandwich bag size or larger) for leaf packs, invertebrates, etc.

Vials, small with tight-fitting lids and preservative (or refrigeration) for invertebrate storage

Transfer of dissolved organic matter and nutrients

Polyethylene bottles

Whirl Paks

Meter tape

Rain gauge

Fine mesh (small enough openings to keep out debris from polyethylene precipitation bottles)

Scintillation vials

Plastic lining

Surveyors flagging

Sharpie

Whatman filters or Millipore HA membrane filters (0.45 μm) Ziploc bags

PVC wells 5—7 cm (2—3″)

Well screens

PVC plugs

Well bailer

Auger or fence post driver (for placement of wells)

YSI meter or other portable field probe

Soil corer

Periphyton sampling equipment (see Chapter 12)

Surber or Hess sampler (see Chapter 15)

Mortar and pestle or Wiley Mill

Laboratory Materials and Equipment

Input and decomposition of coarse organic matter

Leaves, large trash bags full of individual species or a representative composite

Aquaculture cage netting or potato/onion/orange bags with 5-mm mesh or nylon monofilament line for constructing leaf packs (36—100 per species)

Drying oven (60°C, large capacity)

Muffle furnace, crucibles, and desiccators (for AFDM determinations)

Laboratory balance, 15.00-g capacity

Refrigerator for short-term storage of leaf packs and other samples until they can be processed or frozen for longer-term storage

Laboratory balances

Drying oven (60°C, large capacity)

Transfer of dissolved organic matter and nutrients

Drying oven

1 M KCl

Resin beads (Supelco Corp., MTO-Dowex)

Nylon stockings

Flask (large enough to hold resin extract)

Parafilm

Vacuum filter

Whatman filters #1, 70 mm

REFERENCES

Allen, D.C., Vaughn, C.C., Kelly, J.F., Cooper, J.T., Engel, M.H., 2012. Bottom-up biodiversity effects increase resource subsidy flux between ecosystems. Ecology 93, 2165—2174.

Atkinson, C.L., Kelly, J.F., Vaughn, C.C., 2014. Tracing consumer-derived nitrogen in riverine food webs. Ecosystems 17, 485—496.

Bartels, P., Cucherousset, J., Steger, K., Eklöv, P., Tranvik, L.J., Hillebrand, H., 2012. Reciprocal subsidies between freshwater and terrestrial ecosystems structure consumer resource dynamics. Ecology 93, 1173—1182.

Baxter, C.V., Fausch, K.D., Murakami, M., Chapman, P.L., 2004. Fish invasion restructures stream and forest food webs by interrupting reciprocal prey subsidies. Ecology 85, 2656—2663.

Baxter, C.V., Fausch, K.D., Carl Saunders, W., 2005. Tangled webs: reciprocal flows of invertebrate prey link streams and riparian zones. Freshwater Biology 50, 201—220.

Bayley, P.B., 1995. Understanding large river: floodplain ecosystems. BioScience 45, 153—158.

Bechtold, H.A., Rosi-Marshall, E.J., Warren, D.R., Cole, J.J., 2012. A practical method for measuring integrated solar radiation reaching streambeds using photodegrading dyes. Freshwater Science 31, 1070—1077.

Bilby, R.E., Fransen, B.R., Bisson, P.A., 1996. Incorporation of nitrogen and carbon from spawning coho salmon into the trophic system of small streams: evidence from stable isotopes. Canadian Journal of Fisheries and Aquatic Sciences 53, 164—173.

Binkley, D., 1984. Ion exchange resin bags: factors affecting estimates of nitrogen availability. Soil Science Society of America Journal 48, 1181—1184.

Binkley, D., Aber, J., Pastor, J., Nadelhoffer, K., 1986. Nitrogen availability in some Wisconsin forests: comparisons of resin bags and on-site incubations. Biology and Fertility of Soils 2, 77—82.

Binkley, D., Matson, P., 1983. Ion exchange resin bag method for assessing forest soil nitrogen availability. Soil Science Society of America Journal 47, 1050—1052.

Boulton, A.J., Datry, T., Kasahara, T., Mutz, M., Stanford, J.A., 2010. Ecology and management of the hyporheic zone: stream-groundwater interactions of running waters and their floodplains. Journal of the North American Benthological Society 29, 26—40.

Boulton, A.J., Findlay, S., Marmonier, P., Stanley, E.H., Valett, H.M., 1998. The functional significance of the hyporheic zone in streams and rivers. Annual Review of Ecology and Systematics 29, 59—81.

Brookshire, E.N.J., Gerber, S., Webster, J.R., Vose, J.M., Swank, W.T., 2011. Direct effects of temperature on forest nitrogen cycling revealed through analysis of long-term watershed records. Global Change Biology 17, 297—308.

Brookshire, E.N.J., Valett, H.M., Thomas, S.A., Webster, J.R., 2005. Coupled cycling of dissolved organic nitrogen and carbon in a forest stream. Ecology 86, 2487—2496.

Brunke, M., Gonser, T.O.M., 1997. The ecological significance of exchange processes between rivers and groundwater. Freshwater Biology 37, 1—33.

Bump, J.K., Tischler, K.B., Schrank, A.J., Peterson, R.O., Vucetich, J.A., 2009. Large herbivores and aquatic—terrestrial links in southern boreal forests. Journal of Animal Ecology 78, 338—345.

Capps, K.A., Berven, K.A., Tiegs, S.D., 2015. Modelling nutrient transport and transformation by pool-breeding amphibians in forested landscapes using a 21-year dataset. Freshwater Biology 60, 500—511.

Carbonneau, P.E., Piegay, H., 2012. Remote Sensing and River Management. John Wiley & Sons, Chichester, UK.

Cederholm, C.J., Kunze, M.D., Murota, T., Sibatani, A., 1999. Pacific salmon carcasses: essential contributions of nutrients and energy for aquatic and terrestrial ecosystems. Fisheries 24, 6—15.

Clark, J.R., Benforado, J. (Eds.), 2012. Wetlands of Bottomland Hardwood Forests, vol. 11. Elsevier, Amsterdam, The Netherlands.

Clément, J.C., Holmes, R.M., Peterson, B.J., Pinay, G., 2003. Isotopic investigation of denitrification in a riparian ecosystem in western France. Journal of Applied Ecology 40, 1035—1048.

Collier, K.J., Bury, S., Gibbs, M., 2002. A stable isotope study of linkages between stream and terrestrial food webs through spider predation. Freshwater Biology 47, 1651—1659.

Colón-Gaud, J.C., Peterson, S., Whiles, M.R., Kilham, S.S., Lips, K.R., Pringle, C.M., 2008. Allochthonous litter inputs, organic matter standing stocks, and organic seston dynamics in upland Panamanian streams: potential effects of larval amphibians on organic matter dynamics. Hydrobiologia 603, 301—312.

Cooper, J.R., Gilliam, J.W., Daniels, R.B., Robarge, W.P., 1987. Riparian areas as filters for agricultural sediment. Soil Science Society of America Journal 51, 416—420.

Costello, D.M., Lamberti, G.A., 2008. Non-native earthworms in riparian soils increase nitrogen flux into adjacent aquatic ecosystems. Oecologia 158, 499—510.

Cronk, J.K., Fennessy, M.S., 2016. Wetland Plants: Biology and Ecology. CRC Press, Boca Raton, FL.

Cummins, K.W., Wilzbach, M.A., Gates, D.M., Perry, J.B., Taliaferro, W.B., 1989. Shredders and riparian vegetation. BioScience 39, 24—30.

Davies-Colley, R.J., Payne, G.W., 1998. Measuring stream shade. Journal of the North American Benthological Society 17, 250—260.

Davis, J.C., Minshall, G.W., Robinson, C.T., Landres, P., 2001. Monitoring Wilderness Stream Ecosystems. U.S.D.A. Forest Service General Technical Report RMRS-GTR-70, Washington, DC.

Davis, J.M., Rosemond, A.D., Small, G.E., 2011. Increasing donor ecosystem productivity decreases terrestrial consumer reliance on a stream resource subsidy. Oecologia 167, 821—834.

Doucett, R.R., Marks, J.C., Blinn, D.W., Caron, M., Hungate, B.A., 2007. Measuring terrestrial subsidies to aquatic food webs using stable isotopes of hydrogen. Ecology 88, 1587—1592.

Duncan, W.F.A., Brusven, M.A., Bjornn, T.C., 1989. Energy-flow response models for evaluation of altered riparian vegetation in three southeast Alaskan streams. Water Research 23, 965—974.

Dwire, K.A., Kauffman, J.B., Baham, J.E., 2006. Plant species distribution in relation to water-table depth and soil redox potential in montane riparian meadows. Wetlands 26, 131—146.

Eberle, L.C., Stanford, J.A., 2010. Importance and seasonal availability of terrestrial invertebrates as prey for juvenile salmonids in floodplain spring brooks of the Kol River (Kamchatka, Russian Federation). River Research and Applications 26, 682–694.

England, L.E., Rosemond, A.D., 2004. Small reductions in forest cover weaken terrestrial-aquatic linkages in headwater streams. Freshwater Biology 49, 721–734.

Englund, S.R., O'Brien, J.J., Clark, D.B., 2000. Evaluation of digital and film hemispherical photography and spherical densiometry for measuring forest light environments. Canadian Journal of Forest Research 30, 1999–2005.

ERDAS, 2015. IMAGINE, Hexagon Geospatial Software. Norcross, Georgia.

Ewel, K.C., Cressa, C., Kneib, R.T., Lake, P.S., Levin, L.A., Palmer, M.A., Snelgrove, P., Wall, D.H., 2001. Managing critical transition zones. Ecosystems 4, 452–460.

Finlay, J.C., 2001. Stable-carbon-isotope ratios of river biota: implications for energy flow in lotic food webs. Ecology 82, 1052–1064.

Finlay, J.C., Doucett, R.R., McNeely, C., 2010. Tracing energy flow in stream food webs using stable isotopes of hydrogen. Freshwater Biology 55, 941–951.

Finlay, J.C., Hood, J.M., Limm, M.P., Power, M.E., Schade, J.D., Welter, J.R., 2011. Light-mediated thresholds in stream-water nutrient composition in a river network. Ecology 92, 140–150.

Finlay, J.C., Kendall, C., 2007. Stable isotope tracing of temporal and spatial variability in organic matter sources to freshwater ecosystems. In: Michener, R., Lajtha, K. (Eds.), Stable Isotopes in Ecology and Environmental Science. Blackwell, Oxford, UK, pp. 283–333.

Fisher, S.G., Likens, G.E., 1973. Energy flow in Bear Brook, New Hampshire: an integrative approach to stream ecosystem metabolism. Ecological Monographs 43, 421–439.

Frank, D.A., Inouye, R.S., Huntly, N., Minshall, G.W., Anderson, J.E., 1994. The biogeochemistry of a north-temperate grassland with native ungulates: nitrogen dynamics in Yellowstone National Park. Biogeochemistry 26, 163–188.

Glenn, E.P., Nagler, P.L., 2005. Comparative ecophysiology of *Tamarix ramosissima* and native trees in western US riparian zones. Journal of Arid Environments 61, 419–446.

Goodchild, M.F., 1994. Integrating GIS and remote sensing for vegetation analysis and modeling: methodological issues. Journal of Vegetation Science 5, 615–626.

Grabs, T., Bishop, K., Laudon, H., Lyon, S.W., Seibert, J., 2012. Riparian zone hydrology and soil water total organic carbon (TOC): implications for spatial variability and upscaling of lateral riparian TOC exports. Biogeosciences 9, 3901–3916.

Gregory, S.V., Swanson, F.J., McKee, W.A., Cummins, K.W., 1991. An ecosystem perspective of riparian zones. BioScience 41, 540–551.

Greig-Smith, P., 1983. Quantitative Plant Ecology, vol. 9. University of California Press.

Groffman, P.M., Butterbach-Bahl, K., Fulweiler, R.W., Gold, A.J., Morse, J.L., Stander, E.K., Tague, C., Tonitto, C., Vidon, P., 2009. Challenges to incorporating spatially and temporally explicit phenomena (hotspots and hot moments) in denitrification models. Biogeochemistry 93, 49–77.

Hardiman, B.S., Gough, C.M., Halperin, A., Hofmeister, K.L., Nave, L.E., Bohrer, G., Curtis, P.S., 2013. Maintaining high rates of carbon storage in old forests: a mechanism linking canopy structure to forest function. Forest Ecology and Management 298, 111–119.

Hauer, F.R., Locke, H., Dreitz, V.J., Hebblewhite, M., Lowe, W.H., Muhlfeld, C.C., Nelson, C.R., Proctor, M.F., Rood, S.B., 2016. Gravel-bed river floodplains are the ecological nexus of glaciated mountain landscapes. Science Advances 2, e1600026.

Hauer, F.R., Stanford, J.A., Lorang, M.S., 2007. Pattern and process in northern Rocky Mountain headwaters: ecological linkages in the headwaters of the Crown of the Continent. Journal of the American Water Resources Association 43, 104–117.

Harmon, M.E., Nadelhoffer, K.J., Blair, J.M., 1999. Measuring decomposition, nutrient turnover, and stores in plant litter. In: Robertson, G.P., Coleman, D.C., Bledsoe, C.S., Sollins, P. (Eds.), Standard Soil Methods for Long-Term Ecological Research. Oxford University Press, Oxford, UK, pp. 202–240.

Hart, S.K., Hibbs, D.E., Perakis, S.S., 2013. Riparian litter inputs to streams in the central Oregon Coast Range. Freshwater Science 32, 343–358.

Hedin, L.O., von Fischer, J.C., Ostrom, N.E., Kennedy, B.P., Brown, M.G., Robertson, G.P., 1998. Thermodynamic constraints on nitrogen transformations and other biogeochemical processes at soil-stream interfaces. Ecology 79, 684–703.

Helfield, J.M., Naiman, R.J., 2002. Salmon and alder as nitrogen sources to riparian forests in a boreal Alaskan watershed. Oecologia 133, 573–582.

Helfield, J.M., Naiman, R.J., 2006. Keystone interactions: salmon and bear in riparian forests of Alaska. Ecosystems 9, 167–180.

Hill, W.R., Dimick, S.M., 2002. Effects of riparian leaf dynamics on periphyton photosynthesis and light utilisation efficiency. Freshwater Biology 47, 1245–1256.

Hladyz, S., Åbjörnsson, K., Chauvet, E., Dobson, M., Elosegi, A., Ferreira, V., Fleituch, T., Gessner, M.O., Giller, P.S., Gulis, V., 2011. Stream ecosystem functioning in an agricultural landscape: the importance of terrestrial-aquatic linkages. Advances in Ecological Research 44, 211–276.

Hoffmann, C.C., Kjaergaard, C., Uusi-Kämppä, J., Hansen, H.C.B., Kronvang, B., 2009. Phosphorus retention in riparian buffers: review of their efficiency. Journal of Environmental Quality 38, 1942–1955.

Holtgrieve, G.W., Schindler, D.E., Jewett, P.K., 2009. Large predators and biogeochemical hotspots: brown bear (*Ursus arctos*) predation on salmon alters nitrogen cycling in riparian soils. Ecological Research 24, 1125–1135.

Jansson, R., Laudon, H., Johansson, E., Augspurger, C., 2007. The importance of groundwater discharge for plant species number in riparian zones. Ecology 88, 131–139.

Jardine, T.D., Kidd, K.A., Cunjak, R.A., 2009. An evaluation of deuterium as a food source tracer in temperate streams of eastern Canada. Journal of the North American Benthological Society 28, 885–893.

Julian, J.P., Doyle, M.W., Stanley, E.H., 2008. Empirical modeling of light availability in rivers. Journal of Geophysical Research – Biogeosciences 113, G03022.

Junk, W.J., Bayley, P.B., Sparks, R.E., 1989. The flood pulse concept in river-floodplain systems. Canadian Special Publication of Fisheries and Aquatic Sciences 106, 110−127.

Kanasashi, T., Hattori, S., 2011. Seasonal variation in leaf-litter input and leaf dispersal distances to streams: the effect of converting broadleaf riparian zones to conifer plantations in central Japan. Hydrobiologia 661, 145−161.

Kaneko, N., Salamanca, E., 1999. Mixed leaf litter effects on decomposition rates and soil microarthropod communities in an oak−pine stand in Japan. Ecological Research 14, 131−138.

Kelley, C.E., Krueger, W.C., 2005. Canopy cover and shade determinations in riparian zones. Journal of the American Water Resources Association 41, 37−46.

Kiffney, P.M., Richardson, J.S., 2010. Organic matter inputs into headwater streams of southwestern British Columbia as a function of riparian reserves and time since harvesting. Forest Ecology and Management 260, 1931−1942.

Kiffney, P.M., Richardson, J.S., Bull, J.P., 2004. Establishing light as a causal mechanism structuring stream communities in response to experimental manipulation of riparian buffer width. Journal of the North American Benthological Society 23, 542−555.

Kimball, S., Lulow, M., Sorenson, Q., Balazs, K., Fang, Y.C., Davis, S.J., O'Connell, M., Huxman, T.E., 2015. Cost-effective ecological restoration. Restoration Ecology 23, 800−810.

Kleindl, W.J., Rains, M.C., Marshall, L.A., Hauer, F.R., 2015. Fire and flood expand the floodplain shifting habitat mosaic concept. Freshwater Science 34, 1366−1382.

Kominoski, J.S., Marczak, L.B., Richardson, J.S., 2011. Riparian forest composition affects stream litter decomposition despite similar microbial and invertebrate communities. Ecology 92, 151−159.

Kominoski, J.S., Shah, J.J.F., Canhoto, C., Fischer, D.G., Giling, D.P., Gonzalez, E., Griffiths, N.A., Larranaga, A., LeRoy, C.J., Mineau, M.M., 2013. Forecasting functional implications of global changes in riparian plant communities. Frontiers in Ecology and the Environment 11, 423−432.

Koshino, Y., Kudo, H., Kaeriyama, M., 2013. Stable isotope evidence indicates the incorporation into Japanese catchments of marine-derived nutrients transported by spawning Pacific Salmon. Freshwater Biology 58, 1864−1877.

Kuglerova, L., Jansson, R., Ågren, A., Laudon, H., Malm-Renöfält, B., 2014. Groundwater discharge creates hotspots of riparian plant species richness in a boreal forest stream network. Ecology 95, 715−725.

Lagrue, C., Kominoski, J.S., Danger, M., Baudoin, J., Lamothe, S., Lambrigot, D., Lecerf, A., 2011. Experimental shading alters leaf litter breakdown in streams of contrasting riparian canopy cover. Freshwater Biology 56, 2059−2069.

Lau, D.C.P., Leung, K.M.Y., Dudgeon, D., 2009. What does stable isotope analysis reveal about trophic relationships and the relative importance of allochthonous and autochthonous resources in tropical streams? A synthetic study from Hong Kong. Freshwater Biology 54, 127−141.

Lefsky, M.A., Cohen, W.B., Parker, G.G., Harding, D.J., 2002. Lidar remote sensing for ecosystem studies: Lidar, an emerging remote sensing technology that directly measures the three-dimensional distribution of plant canopies, can accurately estimate vegetation structural attributes and should be of particular interest to forest, landscape, and global ecologists. BioScience 52, 19−30.

Leroy, C.J., Marks, J.C., 2006. Litter quality, stream characteristics and litter diversity influence decomposition rates and macroinvertebrates. Freshwater Biology 51, 605−617.

Lowrance, R., Todd, R., Fail, J., Hendrickson, O., Leonard, R., Asmussen, L., 1984. Riparian forests as nutrient filters in agricultural watersheds. BioScience 34, 374−377.

Malison, R.L., Baxter, C.V., 2010. The fire pulse: wildfire stimulates flux of aquatic prey to terrestrial habitats driving increases in riparian consumers. Canadian Journal of Fisheries and Aquatic Sciences 67, 570−579.

Marczak, L.B., Thompson, R.M., Richardson, J.S., 2007. Meta-analysis: trophic level, habitat, and productivity shape the food web effects of resource subsidies. Ecology 88, 140−148.

Mayer, P.M., Reynolds, S.K., McCutchen, M.D., Canfield, T.J., 2007. Meta-analysis of nitrogen removal in riparian buffers. Journal of Environmental Quality 36, 1172−1180.

McArthur, J.V., Aho, J.M., Rader, R.B., Mills, G.L., 1994. Interspecific leaf interactions during decomposition in aquatic and floodplain ecosystems. Journal of the North American Benthological Society 13, 57−67.

McGlynn, B.L., McDonnell, J.J., 2003. Role of discrete landscape units in controlling catchment dissolved organic carbon dynamics. Water Resources Research 39, 1090.

Melin, M., Matala, J., Mehtätalo, L., Pusenius, J., Packalen, P., 2016. Ecological dimensions of airborne laser scanning—analyzing the role of forest structure in moose habitat use within a year. Remote Sensing of Environment 173, 238−247.

Merritt, R.W., Cummins, K.W., Berg, M.B. (Eds.), 2008. An Introduction to the Aquatic Insects of North America, fourth ed. Kendall Hunt Publishing, Dubuque, IA.

Merz, J.E., Moyle, P.B., 2006. Salmon, wildlife, and wine: marine-derived nutrients in human-dominated ecosystems of central California. Ecological Applications 16, 999−1009.

Meyer, J.L., 1990. A blackwater perspective on riverine ecosystems. BioScience 40, 643−651.

Mineau, M.M., Baxter, C.V., Marcarelli, A.M., 2011. A non-native riparian tree (*Elaeagnus angustifolia*) changes nutrient dynamics in streams. Ecosystems 14, 353−365.

Mineau, M.M., Baxter, C.V., Marcarelli, A.M., Minshall, G.W., 2012. An invasive riparian tree reduces stream ecosystem efficiency via a recalcitrant organic matter subsidy. Ecology 93, 1501−1508.

Minshall, G.W., 1988. Stream ecosystem theory: a global perspective. Journal of the North American Benthological Society 7, 263−288.

Minshall, G.W., 1994. Stream-riparian ecosystems: rationale and methods for basin-level assessments of management effects. In: Jensen, M.E., Bourgeron, P.S. (Eds.), Eastside Forest Ecosystem Health Assessment. Volume II: Ecosystem Management: Principles and Applications. Pacific Northwest Research Station, Portland, OR, pp. 153–177. General Technical Report PNW-GTR-318. U.S. Forest Service.

Minshall, G.W., Petersen, R.C., Cummins, K.W., Bott, T.L., Sedell, J.R., Cushing, C.E., Vannote, R.L., 1983. Interbiome comparison of stream ecosystem dynamics. Ecological Monographs 53, 2–25.

Montgomery, D.R., 1999. Process domains and the river continuum. Journal of the American Water Resources Association 35, 397–410.

Montgomery, D.R., 2002. Valley formation by fluvial and glacial erosion. Geology 30, 1047–1050.

Morris, M.R., Stanford, J.A., 2011. Floodplain succession and soil nitrogen accumulation on a salmon river in southwestern Kamchatka. Ecological Monographs 81, 43–61.

Mouw, J.E.B., Alaback, P.B., 2003. Putting floodplain hyperdiversity in a regional context: an assessment of terrestrial–floodplain connectivity in a montane environment. Journal of Biogeography 30, 87–103.

Mouw, J.E.B., Chaffin, J.L., Whited, D.C., Hauer, F.R., Matson, P.L., Stanford, J.A., 2013. Recruitment and successional dynamics diversify the shifting habitat mosaic of an Alaskan floodplain. River Research and Applications 29, 671–685.

Mulholland, P.J., 1992. Regulation of nutrient concentrations in a temperate forest stream: roles of upland, riparian, and instream processes. Limnology and Oceanography 37, 1512–1526.

Müller, J., Mehr, M., Bässler, C., Fenton, M.B., Hothorn, T., Pretzsch, H., Klemmt, H.J., Brandl, R., 2012. Aggregative response in bats: prey abundance versus habitat. Oecologia 169, 673–684.

Naiman, R.J., Bechtold, J.S., Drake, D.C., Latterell, J.J., O'Keefe, T.C., Balian, E.V., 2005. Origins, patterns, and importance of heterogeneity in riparian systems. In: Lovett, G.M., Jones, C., Turner, M.G., Weathers, K.C. (Eds.), Ecosystem Function in Heterogeneous Landscapes. Springer, New York, NY, pp. 279–309.

Naiman, R.J., Décamps, H., 1997. The ecology of interfaces: riparian zones. Annual Review of Ecology and Systematics 28, 621–658.

Nakano, S., Miyasaka, H., Kuhara, N., 1999. Terrestrial-aquatic linkages: riparian arthropod inputs alter trophic cascades in a stream food web. Ecology 80, 2435–2441.

Normann, C., Tscharntke, T., Scherber, C., 2016. Interacting effects of forest stratum, edge and tree diversity on beetles. Forest Ecology and Management 361, 421–431.

Ostrofsky, M.L., 1997. Relationship between chemical characteristics of autumn-shed leaves and aquatic processing rates. Journal of the North American Benthological Society 16, 750–759.

Page, A.L., 1982. Methods of Soil Analysis. Part 2. Chemical and Microbiological Properties, second ed. American Society of Agronomy, Madison, WI.

Palmer, G.C., Bennett, A.F., 2006. Riparian zones provide for distinct bird assemblages in forest mosaics of south-east Australia. Biological Conservation 130, 447–457.

Peterjohn, W.T., Correll, D.L., 1984. Nutrient dynamics in an agricultural watershed: observations on the role of a riparian forest. Ecology 65, 1466–1475.

Petersen, R.C., Cummins, K.W., 1974. Leaf processing in a woodland stream. Freshwater Biology 4, 343–368.

Peterson, B.J., Fry, B., 1987. Stable isotopes in ecosystem studies. Annual Review of Ecology and Systematics 18, 293–320.

Platts, W.S., et al., 1987. Methods for Evaluating Riparian Habitats with Applications to Management. U.S.D.A. Forest Service General Technical Report INT-221, Washington, DC.

Platts, W.S., Nelson, R.L., 1989. Stream canopy and its relationship to salmonid biomass in the intermountain west. North American Journal of Fisheries Management 9, 446–457.

Ranalli, A.J., Macalady, D.L., 2010. The importance of the riparian zone and in-stream processes in nitrate attenuation in undisturbed and agricultural watersheds—a review of the scientific literature. Journal of Hydrology 389, 406–415.

Recalde, F.C., Postali, T.C., Romero, G.Q., 2016. Unravelling the role of allochthonous aquatic resources to food web structure in a tropical riparian forest. Journal of Animal Ecology 85, 525–536.

Regester, K.J., Lips, K.R., Whiles, M.R., 2006. Energy flow and subsidies associated with the complex life cycle of ambystomatid salamanders in ponds and adjacent forest in southern Illinois. Oecologia 147, 303–314.

Reisinger, A.J., Chaloner, D.T., Rüegg, J., Tiegs, S.D., Lamberti, G.A., 2013. Effects of Pacific salmon spawners on the isotopic composition of biota differ across Southeast Alaska streams. Freshwater Biology 58, 938–950.

Rice, E.W., Baird, R.B., Eaton, A.D., Clesceri, L.S. (Eds.), 2012. Standard Methods for the Examination of Water and Wastewater, twenty-second ed. American Public Health Association, American Water Works Association, and Water Environment Federation, Washington, DC.

Richardson, J.S., Zhang, Y., Marczak, L.B., 2010. Resource subsidies across the land – freshwater interface and responses in recipient communities. River Research and Applications 26, 55–66.

Roales, J., Durán, J., Bechtold, H.A., Groffman, P.M., Rosi-Marshall, E.J., 2013. High resolution measurement of light in terrestrial ecosystems using photodegrading dyes. PLoS One 8, e75715.

Robertson, G.P., Coleman, D.C., Bledsoe, C.S., Sollins, P., 1999. Standard Soil Methods for Long-Term Ecological Research. Oxford University Press, Oxford, UK.

Rood, S.B., Mahoney, J.M., 1990. Collapse of riparian poplar forests downstream from dams in western prairies: probable causes and prospects for mitigation. Environmental Management 14, 451–464.

Rotenberry, J., 1985. The role of habitat in avian community composition: physiognomy or floristics? Oecologia 67, 213–217.

Sabo, J.L., Sponseller, R., Dixon, M., Gade, K., Harms, T., Heffernan, J., Jani, A., Katz, G., Soykan, C., Watts, J., 2005. Riparian zones increase regional species richness by harboring different, not more, species. Ecology 86, 56−62.

Sanzone, D.M., Meyer, J.L., Martí, E., Gardiner, E.P., Tank, J.L., Grimm, N.B., 2003. Carbon and nitrogen transfer from a desert stream to riparian predators. Oecologia 134, 238−250.

Schiller, D.V., Martí, E., Riera, J.L., Sabater, F., 2007. Effects of nutrients and light on periphyton biomass and nitrogen uptake in Mediterranean streams with contrasting land uses. Freshwater Biology 52, 891−906.

Schindler, M.H., Gessner, M.O., 2009. Functional leaf traits and biodiversity effects on litter decomposition in a stream. Ecology 90, 1641−1649.

Scholl, E.A., Rantala, H.M., Whiles, M.R., Wilkerson, G.V., 2015. Influence of flow on community structure and production of snag-dwelling macroinvertebrates in an impaired low-gradient river. River Research and Applications 32, 677−688.

Scott, M.L., Shafroth, P.B., Auble, G.T., 1999. Responses of riparian cottonwoods to alluvial water table declines. Environmental Management 23, 347−358.

Sedell, J.R., Reeves, G.H., Hauer, F.R., Stanford, J.A., Hawkins, C.P., 1990. Role of refugia in recovery from disturbances: modern fragmented and disconnected river systems. Environmental Management 14, 711−724.

Silva, H.C.H., Caraciolo, R.L.F., Marangon, L.C., Ramos, M.A., Santos, L.L., Albuquerque, U.P., 2014. Evaluating different methods used in ethnobotanical and ecological studies to record plant biodiversity. Journal of Ethnobiology and Ethnomedicine 10, 1−48.

Soininen, J., Bartels, P., Heino, J., Luoto, M., Hillebrand, H., 2015. Toward more integrated ecosystem research in aquatic and terrestrial environments. BioScience 65, 174−182.

Solomon, C.T., Cole, J.J., Doucett, R.R., Pace, M.L., Preston, N.D., Smith, L.E., Weidel, B.C., 2009. The influence of environmental water on the hydrogen stable isotope ratio in aquatic consumers. Oecologia 161, 313−324.

Stanford, J.A., Ward, J.V., 1993. An ecosystem perspective of alluvial rivers: connectivity and the hyporheic corridor. Journal of the North American Benthological Society 12, 48−60.

Stanford, J.A., Lorang, M.S., Hauer, F.R., 2005. The shifting habitat mosaic of river ecosystems. Internationale Vereinigung für Theoretische und Angewandte Limnologie Verhandlungen 29, 123−136.

Subalusky, A.L., Dutton, C.L., Rosi-Marshall, E.J., Post, D.M., 2015. The hippopotamus conveyor belt: vectors of carbon and nutrients from terrestrial grasslands to aquatic systems in Sub-Saharan Africa. Freshwater Biology 60, 512−525.

Swan, C.M., Palmer, M.A., 2004. Leaf diversity alters litter breakdown in a Piedmont stream. Journal of the North American Benthological Society 23, 15−28.

Swanson, F.J., Gregory, S.V., Sedell, J.R., Campbell, A.G., 1982. Land-water interactions: the riparian zone. In: Edmonds, R.L. (Ed.), Analysis of Coniferous Forest Ecosystems in the Western United States, US/IBP Synthesis Series, vol. 14. Hutchinson Ross Publishing Company, Stroudsburg, PA, pp. 267−291.

Sweeney, B.W., Newbold, J.D., 2014. Streamside forest buffer width needed to protect stream water quality, habitat, and organisms: a literature review. Journal of the American Water Resources Association 50, 560−584.

Tabacchi, E., Correll, D.L., Hauer, F.R., Pinay, G., Planty-Tabacchi, A.M., Wissmar, R.C., 1998. Development, maintenance and role of riparian vegetation in the river landscape. Freshwater Biology 40, 497−516.

Tabacchi, E., Lambs, L., Guilloy, H., Planty-Tabacchi, A.-M., Muller, E., Decamps, H., 2000. Impacts of riparian vegetation on hydrological processes. Hydrological Processes 14, 2959−2976.

Tank, J.L., Rosi-Marshall, E.J., Griffiths, N.A., Entrekin, S.A., Stephen, M.L., 2010. A review of allochthonous organic matter dynamics and metabolism in streams. Journal of the North American Benthological Society 29, 118−146.

Thomas, S.A., Royer, T.V., Minshall, G.W., Snyder, E., 2003. Assessing the historic contribution of marine-derived nutrients to Idaho streams. In: American Fisheries Society Symposium, vol. 34, pp. 41−58.

Thorp, J.H., Delong, M.D., 2002. Dominance of autochthonous autotrophic carbon in food webs of heterotrophic rivers. Oikos 96, 543−550.

Tiegs, S.D., Clapcott, J.E., Griffiths, N.A., Boulton, A.J., 2013. A standardized cotton-strip assay for measuring organic-matter decomposition in streams. Ecological Indicators 32, 131−139.

Tiegs, S.D., Langhans, S.D., Tockner, K., Gessner, M.O., 2007. Cotton strips as a leaf surrogate to measure decomposition in river floodplain habitats. Journal of the North American Benthological Society 26, 70−77.

Triska, F.J., Kennedy, V.C., Avanzino, R.J., Zellweger, G.W., Bencala, K.E., 1989. Retention and transport of nutrients in a third-order stream in northwestern California: hyporheic processes. Ecology 70, 1893−1905.

Turner, W., Spector, S., Gardiner, N., Fladeland, M., Sterling, E., Steininger, M., 2003. Remote sensing for biodiversity science and conservation. Trends in Ecology and Evolution 18, 306−314.

Valett, H.M., Fisher, S.G., Grimm, N.B., Camill, P., 1994. Vertical hydrologic exchange and ecological stability of a desert stream ecosystem. Ecology 75, 548−560.

Valett, H.M., Hauer, F.R., Stanford, J.A., 2014. Landscape influences on ecosystem function: local and routing control of oxygen dynamics in a floodplain aquifer. Ecosystems 17, 195−211.

Vannote, R.L., Minshall, G.W., Cummins, K.W., Sedell, J.R., Cushing, C.E., 1980. The river continuum concept. Canadian Journal of Fisheries and Aquatic Sciences 37, 130−137.

Vidon, P., Allan, C., Burns, D., Duval, T.P., Gurwick, N., Inamdar, S., Lowrance, R., Okay, J., Scott, D., Sebestyen, S., 2010. Hot spots and hot moments in riparian zones: potential for improved water quality management. Journal of the American Water Resources Association 46, 278−298.

Wagener, S.M., Oswood, M.W., Schimel, J.P., 1998. Rivers and soils: parallels in carbon and nutrient processing. BioScience 48, 104−108.

Wallace, J.B., Eggert, S.L., Meyer, J.L., Webster, J.R., 1997. Multiple trophic levels of a forest stream linked to terrestrial litter inputs. Science 277, 102–104.

Webster, J.R., Benfield, E.F., 1986. Vascular plant breakdown in freshwater ecosystems. Annual Review of Ecology and Systematics 17, 567–594.

Whitledge, G.W., Johnson, B.M., Martinez, P.J., 2006. Stable hydrogen isotopic composition of fishes reflects that of their environment. Canadian Journal of Fisheries and Aquatic Sciences 63, 1746–1751.

Wipfli, M.S., 1997. Terrestrial invertebrates as salmonid prey and nitrogen sources in streams: contrasting old-growth and young-growth riparian forests in southeastern Alaska, USA. Canadian Journal of Fisheries and Aquatic Sciences 54, 1259–1269.

Wipfli, M.S., Hudson, J., Caouette, J., 1998. Influence of salmon carcasses on stream productivity: response of biofilm and benthic macroinvertebrates in southeastern Alaska, USA. Canadian Journal of Fisheries and Aquatic Sciences 55, 1503–1511.

Wood, T., Bormann, F.H., Voigt, G.K., 1984. Phosphorus cycling in a northern hardwood forest: biological and chemical control. Science 223, 391–393.

Wulder, M.A., Franklin, S.E., White, J.C., Linke, J., Magnussen, S., 2006. An accuracy assessment framework for large-area land cover classification products derived from medium-resolution satellite data. International Journal of Remote Sensing 27, 663–683.

Chapter 29

Dynamics of Wood

Stanley V. Gregory[1], Angela Gurnell[2], Hervé Piégay[3] and Kathryn Boyer[1]

[1]*Department of Fisheries and Wildlife, Oregon State University;* [2]*School of Geography, Queen Mary University of London;* [3]*National Center for Scientific Research, University of Lyon*

29.1 INTRODUCTION

In the mid-1970s, fluvial geomorphologists and stream ecologists began to recognize the potential role of wood in stream ecosystems (Heede, 1972; Swanson et al., 1976; Gregory and Walling, 1973). Since the early pioneering research on wood in streams, aquatic scientists throughout the world have documented the terrestrial, riparian, and floodplain processes that contribute wood to river networks and the influences of wood on the physical and biological structure and function of streams and rivers (Gregory et al., 2003a). Wood is a major roughness element in streams that influences channel morphology, decreases the average velocity within a reach, physically traps material in transport, creates complex habitat, and provides an abundant but relatively low quality food resource for aquatic organisms (Benda and Sias, 2003; Gregory et al., 2003a; Montgomery et al., 2003; Mutz, 2003).

Several major reviews provide an excellent foundation for designing studies of wood in aquatic ecosystems. The first global perspective on wood in streams and rivers was "The Ecology and Management of Wood in World Rivers" (Gregory et al., 2003a), which explored the role of wood in physical and ecological processes in river networks throughout the world rather than a local region. Recent syntheses (Hering et al., 2000; Gurnell et al., 2005; LeLay et al., 2013; Gurnell et al., 2016; Wohl et al., 2016) have strengthened and expanded the conceptual frameworks and analytical methods for studying wood in aquatic ecosystems and provide more thorough reviews of the literature on wood dynamics. In addition, large wood and riparian and floodplain vegetation interact to shape river channels, which in turn modify the sources, storage, and transport of wood, resulting in self-organization of hydrogeomorphological properties of streams and rivers (Gurnell et al., 2016). As knowledge of wood in rivers has grown, models of wood dynamics have become more powerful and allow investigators to integrate complex physical and biological processes across different spatial scales and timeframes (Benda and Sias, 2003; Meleason et al., 2003; Gregory et al., 2003b; Ruiz-Villanueva et al., 2014).

A confusing array of terms has been used to describe wood in streams, including large woody debris, coarse woody debris, and large organic debris. This confusion increases even more as different researchers and agencies adopt different acronyms (e.g., LWD, CWD, LOD) and codify their use through field datasheets, agency standards, and management policies. In particular, the term "debris" became associated with wood when timber harvest practices left large amounts of logging debris after harvest (see preface in Gregory et al., 2003a; Wohl et al., 2016). The term wood debris soon was applied to all wood, both natural and woody material left after logging. At the same time, forest managers and fish biologists referred to accumulations of wood as "debris dams." Unfortunately, public audiences often associate the term "debris" with more negative connotations, such as garbage or human litter. The term "debris" has no scientific or public outreach advantage, and we encourage scientists, educators, and managers to use the term wood, a simple term that accurately describes this important component of streams and avoids confusion in communicating to diverse audiences.

Dimensions of wood pieces are an important determinant of the roles of wood in both geomorphic and ecological processes in stream ecosystems. All wood, which may range in size from microscopic particles to pieces 10s of meters in length, is potentially important in ecological and geomorphic processes. Large pieces of wood can form stable accumulations (often referred to as log jams or debris dams) and strongly influence channel scour and sediment deposition, while smaller wood can filter particulate organic matter from transport and provide complex cover for aquatic organisms. As a

Methods in Stream Ecology. http://dx.doi.org/10.1016/B978-0-12-813047-6.00007-3

113

result, most investigations distinguish large wood (i.e., logs, whole trees, root wads, major branches) from smaller sizes of wood (i.e., twigs, branches, fine particulate wood). One of the original and most common criteria for distinguishing large wood from small wood is a minimum length of 1 m and minimum diameter of 10 cm (Swanson et al., 1976), but researchers and agencies may use different dimensions to classify wood size, depending on the objectives of the research. The relationship between the size of wood and the size of the river channel is useful in assessing the likely geomorphological importance (Gurnell et al., 2002), a factor that shapes the upstream to downstream transition in wood accumulation types (Abbe and Montgomery, 2003). Use of consistent size classes facilitates comparisons with other studies, so we advise investigators to consider relevant studies in their research area and use similar criteria.

29.1.1 Dynamics of Wood in Streams

The amount of wood (W) in a stream channel is a function of the lateral input from the riparian forest (F_{in}), transport into a reach from upstream (T_{in}), biological decomposition (D), physical abrasion (A), and transport out of the reach to downstream areas (T_{out}):

$$W = f(F_{in}, T_{in}, D, A, T_{out})$$ (29.1)

This relationship is a fundamental context for all studies of wood in streams and must be considered, even when only studying a single aspect of the physical or ecological properties and processes of wood. Differences in wood dynamics between streams or regions often are outcomes of differences in these fundamental processes.

29.1.1.1 Storage

One of the most commonly measured characteristics of wood in streams is storage or the amount of wood in a given reach or area. The amount of wood can be reported as number of pieces, volume, or mass of wood, and methods for these measurements will be described in Section 29.3. The two major approaches for measuring the amount of wood are the direct count method and the line intersect method (LIM). Both methods are valid for quantifying the abundance of large wood, but the LIM is more feasible for small wood where sticks, twigs, and small branches are abundant. A comparison of the two methods in northwestern United States concluded the LIM slightly overestimated wood volume and underestimated wood frequency (Warren et al., 2008), and similar differences have been observed in other studies (Wallace et al., 2001; Miesbauer, 2004). Differences between the two methods depend on the size variation, wood abundance, and frequency of transects. Wood accumulations or dams can be extremely large with numerous pieces of wood, and alternative techniques based on geometry of the accumulation and subsampling the number and size class distributions may be required for extremely large wood dams. Measuring their spatial distribution and abundance over long longitudinal distances may require simpler and faster techniques based on statistical relationships of the overall shape of the accumulation to wood mass estimated by weighing pieces cut from the accumulation (Piégay, 2003; Boivin et al., 2015).

In addition to the descriptions of quantities of wood, factors that are determinants of storage, stability, or retention can be described at a site as well. Channel dimensions (depth, width, sinuosity), channel bedforms (pool, riffle), and substrates (boulder, cobble, gravel, sand) provide information relevant to the probability of wood being removed from transport or entrained into transport. Channel roughness elements, such as boulders, trees, and log accumulations are important factors that influence movement or retention. Regenerating wood (i.e., wood from species that can sprout once deposited, such as willows and poplars) further enhances retention and development of new wood accumulations (Gurnell et al., 2005). Exchange between river channels and floodplains is an important aspect of wood dynamics, and measurements of wood storage should account for storage in both the active channel and its floodplain (Piégay et al., 1999; Wohl, 2013). The spatial extent and geomorphic complexity of floodplains in river channels often requires alternative methods for assessing the abundance and distribution of wood (Thévenet et al., 1998; Piégay, 2003). Because much of the stored wood is delivered or removed during floods, discharge records are valuable for assessing changes in storage and processes responsible for change. While measurement of storage in a single year is informative, repeated measures of storage provide a more dynamic perspective of interannual variation in amounts and distribution of wood and their relationship with flow events. For long-term studies of wood, annual measurements should be conducted during the low flow season of the water year if possible both to facilitate measurement in the channel and to reflect the input and transport of the previous high flow season. Analysis of distributions and abundances of wood at the scale of catchments requires alternative field methods and landscape analysis approaches (Benda and Sias, 2003; Seo et al., 2008).

The ecological and geomorphic influences of a piece of wood are related to its position in the channel, and a single piece of wood can occupy multiple geomorphic zones. To account for numbers and volumes of wood in different functional areas of a channel, researchers have developed a classification of geomorphic zones for wood surveys (Robison and Beschta, 1990). Zone 1 is the area within the low-flow wetted channel, which may have greater importance as aquatic habitat for invertebrates and fish (Fig. 29.1). Zone 2 is the area above the low-flow wetted channel but below the water surface at bankfull flow, which may be important refuge for aquatic organisms at high flow. Zone 3 is the area directly above the water surface at bankfull flow and is biologically important during major floods. Zone 4 is the area lateral to the bankfull channel and encompasses the floodplain and lateral hillslope up to the height of maximum historical flooding. In most cases, Zone 4 is equivalent to the floodplain. The portion of each log in any of the four zones can be estimated and recorded (maximum of 100% for each log). Numbers, volume, and mass of wood in a stream reach can be summarized by geomorphic zone to provide additional functional interpretation of wood surveys.

A final aspect of wood dynamics that is important in some river systems is the role of deposited wood that is capable of sprouting. This reproductive trait is possessed by many riparian poplar and willow species, and it has a fundamental effect on retention and stability of wood in river systems where the river margins are dominated by such species (Gurnell et al., 2001, 2005; Gurnell, 2014). While the same methods can be used to quantify wood storage in all river systems, those that are dominated by wood that can sprout display different patterns of wood storage and turnover. First, wood pieces, including entire uprooted trees deposited on suitable moist exposed river bed sediments, become rapidly anchored by developing root systems, which enhance wood stability. Second, development of shoots and foliage from deposited wood provides flow resistance and thus an enhanced potential to retain transported sediments. Third, the combined effect of enhanced anchorage and sediment retention leads to the rapid disappearance of these pieces from measured wood storage and their replacement by a stand of young trees. Nevertheless, the "pioneer" landforms (Gurnell, 2014) formed during this sequence of root anchorage, sprouting, and sediment retention provide important wood retention structures in their own right. Finally, species that propagate freely from deposited wood also tend to produce wood that decays rapidly in fluvial systems. Therefore, in systems dominated by regenerating wood, the long-term retention of dead wood is replaced by a more dynamic turnover of retained wood either through sprouting or decay.

29.1.1.2 Breakdown and Decomposition

Once a piece of wood is retained in a stream reach, its dimensions and mass can decrease through biological decomposition and physical abrasion (Bilby, 2003). These two processes are influenced by hydrology, geomorphology of the channel, water chemistry and temperature, location in the channel or floodplain, and exposure to drying or wetting (Ruiz-Villanueva et al., 2016a). The chemistry and structure of the wood are fundamental determinants of the decomposition of wood, and wood is a relatively refractory source of organic matter and food for aquatic communities. With carbon:nitrogen ratios ranging from 50 to 400 and lignin content ranging from 15% to 36%, decomposition rates of wood are slow, contributing to their persistence in streams and long-term consequences for channel geomorphology and aquatic habitat (Bilby, 2003). Nutrient concentrations in water also affect the rate of decomposition and microbial activity (Aumen et al., 1985; Mellilo et al., 1984). Mechanical abrasion and breakage also increase the breakdown rates of wood (Harmon et al., 1986; Merten et al., 2013). The state of decomposition can be quantified by measuring wood density. As an example, green riparian wood in the Rhône, France, had densities of 800 ± 170 kg/m^3, whereas instream wood had average densities of 660 ± 200 kg/m^3 with minimum of 400 kg/m^3 (Ruiz-Villanueva et al., 2016a). Difference in density is a reasonable proxy

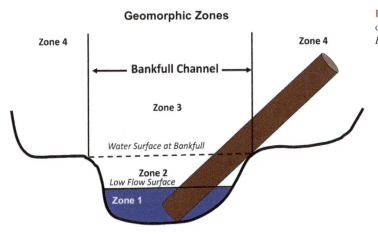

Geomorphic Zones

Zone 4

Zone 4

Bankfull Channel

Zone 3

Water Surface at Bankfull

Zone 2
Low Flow Surface

Zone 1

FIGURE 29.1 Geomorphic zones for classifying the location of wood or portions of individual pieces. *Based on Robison and Beschta (1990). Illustration by S. Gregory.*

for decomposition but both are not always equivalent or related to time since entering the stream depending on local conditions, location in the channel, or burial (MacVicar et al., 2009). During high flows, wood literally is exposed to sandblasting by sediment in suspension, and the forces of water and impacts with material in transport can break and splinter pieces stored in the channel.

29.1.1.3 Input

Wood is delivered laterally from adjacent riparian areas, floodplains, and hillsides, longitudinally from upstream fluvial reaches, and even from downstream sources in estuaries and large floodplains. The most commonly considered sources of lateral inputs are tree fall and blowdown and amounts of wood in channels are related to adjacent stand ages and volumes (Warren et al., 2009). Mass failures are major sources in many steep landscapes (Reeves et al., 2003) and bank erosion is a major factor along laterally dynamic streams in less confined landscapes (Beechie et al., 2006). Transport from upstream reaches can be a major source to a specific section of stream. Though volumes of wood delivered to channels can be predicted from the numbers and sizes of riparian trees, sizes of wood in streams are determined by both delivery of trees from the riparian zones or upslope sources and breakage of wood during delivery and transport (Van Sickle and Gregory, 1990). Input from floodplain erosion or avulsion can be estimated by quantifying eroded floodplain area (based on comparison of GIS maps of channel position at two dates) and standing wood volume on those floodplains based on field measurements (assuming forest characteristics along the eroded bank are similar to the forest previously eroded). Wood abundance and distribution in stream networks may be significantly related to characteristics of adjacent riparian forests, bank erosion, mass failures, and floodplain dynamics within a catchment, but differences in hydrology, geomorphology, climate, and forest ecology between geographic regions may limit application to other areas (Warren et al., 2009). Riparian forest management and adjacent land use practices are determinants of the amounts and types of wood inputs along stream and river networks, and riparian conservation and silviculture are major tools in restoration of wood inputs to stream ecosystems (Boyer et al., 2003).

29.1.1.4 Transport

One of the most dynamic aspects of wood in streams and rivers is fluvial transport during high flows. The dimensions of the piece of wood and the dimensions of the channel strongly influence the probability of a piece of wood being entrained during high flows and the distance it is likely to travel before being retained (Braudrick and Grant, 2001; Gurnell, 2003). A rough rule of thumb is that the probability of transport increases as (1) the ratio of the length of the piece of wood to the width of the active channel decreases and (2) the ratio of the diameter of the piece of wood to the depth of the active channel decreases. Once in transport, many other factors influence the average travel distance before being retained. For example, channel morphology is very important, as illustrated by Bertoldi et al. (2013) on a large braided river, where a sizeable proportion of the wood input by bank erosion was trapped on the first major bar downstream. Measurement is complicated by the dangerous conditions during wood movement. New remote camera systems offer options for quantifying wood movement continuously in numerous locations in a river system (MacVicar et al., 2009; MacVicar and Piégay, 2012; Kramer and Wohl, 2014). RFID (Radio-Frequency Identification) Transponders or GPS devices also can be used to track wood and assess their transport distance after or even during floods (MacVicar et al., 2009; Schenk et al., 2014; Ravazzolo et al., 2015).

Standing stocks of wood, spatial extents of river channels, slow breakdown rates of wood, wood regeneration to produce new trees, and seasonal and interannual variation in environmental drivers create challenges for quantifying wood dynamics in stream and river networks. Specific methods of wood measurement and its spatial distribution and timing are determined by the research questions and the physical and ecological characteristics of the river networks and their catchments.

29.1.2 Models of Wood Dynamics

Causal factors for amounts of wood measured locally in a stream reach cannot be determined from simple inventories. Long-term measurements of input rates, sources, breakdown rates, regeneration rates, and transport can provide the information necessary to interpret local wood abundance, but such extensive studies are costly and time consuming. Simulation modeling is an alternative to measuring all physical and biological processes that affect storage of wood. Many models of wood dynamics have been developed for different regions (Gregory et al., 2003b). Some models address specific processes, such as input (Van Sickle and Gregory, 1990; Benda and Sias, 2003) or transport (Ruiz-Villanueva et al., 2014).

Other models provide quantitative integration of riparian forest growth and mortality, input processes, disturbance processes, and in-channel processes that shape longitudinal patterns of wood storage (Benda and Sias, 2003; Meleason et al., 2003). Studies of large wood and its influence on coarse particulate organic matter (CPOM) retention may require use of a regionally relevant model of stream wood dynamics (e.g., see example from the US Pacific Northwest later in this chapter). Several models offer the ability to alter critical parameters in the model so the user can adapt the model to different tree species, flow regimes, and channel structure.

29.1.3 Humans and the Dynamics of Wood

Most of this chapter describes methods for investigating wood in streams and rivers and the physical and ecological processes that shape the distributions of wood throughout river networks. Throughout most of the world, the patterns observed differ substantially from those expected on the basis of hydrogeomorphology and the adjacent terrestrial vegetation. Human practices that remove wood from stream channels create substantially lower abundances of wood than would be predicted from adjacent forests in many regions of the world (Harmon et al., 1986; Wohl, 2014). The most daunting challenge for managing wood in world rivers is managing people and their modifications of rivers and floodplains (Petts and Welcomme, 2003). In the face of changing patterns of human development and climatic and environmental uncertainty, providing space within the landscape for dynamic rivers and less constraining infrastructure (e.g., bridges, roads, levees) can be both ecologically beneficial and economically sound (Buijse et al., 2005; Buffin-Bélanger et al., 2015).

Large wood in rivers and streams originates from both riparian and upland forests. Sound riparian management practices that maximize the potential for recruitment of large wood to stream channels are mandated for some land uses in some regions of the world, especially in North America and Western Europe. These practices include (1) conservation of intact functional riparian and floodplain forests, (2) riparian buffers and reserves, (3) restoration of degraded riparian forests, and (4) implementation of silvicultural prescriptions that assure an adequate supply of large wood for recruitment directly to stream channels (Boyer et al., 2003). Silvicultural practices and riparian management policies can restore the composition of riparian forests and allow them to mature and deliver wood to adjacent streams and rivers through mortality and mass failures. Riparian management guidelines differ greatly by land use type, region, and country, but these policies are essential to sustain recruitment of wood to streams and rivers.

We commonly assume that science is neutral and unbiased, but society and social experience potentially shape our observations and interpretations. Wood in rivers is an excellent example of such biases and scientific methods that can be used to explore the social context of our science (Gregory and Davis, 1993). Researchers showed images of streams and large rivers to both university students (Piégay et al., 2005) and resource managers (Chin et al., 2014) in nine countries without mentioning the presence or absence of wood in the images. These groups were asked to rate the streams and rivers for esthetic value, ecological value, perception of danger, and need for improvement. Groups in many countries rated the scenes with wood as more dangerous, less esthetically pleasing, and in greater need for improvement. A few regions [Oregon (USA), Sweden, and Germany] rated the scenes with wood as more esthetically and ecologically positive. However, Oregon differed significantly in perception compared to seven other US states surveyed—Texas, Colorado, Connecticut, Georgia, Illinois, Iowa, Missouri—almost as great as differences between countries (Chin et al., 2008). Interestingly, land managers in the United States generally perceived rivers with wood as significantly more esthetically pleasing, less dangerous, and requiring less active management than rivers without wood (Chin et al., 2014). These views were consistent among different types of managers (resource conservation, fisheries, forestry, water, and recreation), suggesting that graduate education, on the job training, and field experience resulted in managers having more positive attitudes about wood in river networks. Social sciences provide important tools for exploring cultural contexts for our understanding of the ecology and management of wood in streams and rivers and highlight the importance of public education and policy.

Transport of wood in rivers inherently creates risks to adjacent human property and human safety. Wood can lodge against infrastructures, such as bridge pilings, and create major accumulations (Lassettre and Kondolf, 2012). Such phenomena can have adverse effects such as undermining abutments with subsequent bridge collapse as well as significant reduction in channel flow conveyance, increases in upstream water level, and flooding of adjacent areas. Changes in flow patterns in a channel reach can increase near-bank shear stress and cause bank erosion. Wood movement also can adversely affect river users, such as hydroelectric power facilities, spill-ways and locks, diversions, and fish ladders and screens (Moulin and Piégay, 2004).

Development of management strategies to reduce adverse effects of wood movement in rivers requires risk assessment based on a holistic perspective of both hydro-ecological processes and social systems. Assessment of natural hazards of wood in rivers can incorporate spatially explicit descriptions of processes that deliver wood into the river, transport characteristics of the reach, sites of more likely deposition and delivery, and vulnerability of human communities and their infrastructure to identify where wood can have substantial negative consequences. Such a framework can provide a balanced context for determining river reaches where we can conserve wood for ecological purposes and others where we can design management alternatives to prevent problems related to wood (Piégay and Landon, 1997). Practices that reduce risk in sections of rivers where potential wood-related damage to human infrastructure is high include clearing riparian vegetation and removing wood, building structures to trap the wood upstream of sensitive areas, and adapting existing structures to reduce their retention capacity (Diehl, 1997). River managers have redesigned existing bridges, removed unsafe structures, and rebuilt bridges that have collapsed or been damaged (Ruiz-Villanueva et al., 2016b). More importantly, future development and building along rivers can be located in areas farther from the active channel and floodplain where risk of damage from floods and wood transport are inherently lower.

In this chapter, we describe quantitative field methods to assess the input, storage, residence, and transport of wood in a specific stream reach. We discuss approaches to quantify rates of input of wood into stream channels. We then present two methods to quantify the abundance of large wood in both small stream reaches and large rivers. We describe methods for measuring residence time and rates of decomposition of wood. We then introduce approaches for measuring transport of wood in stream channels. Finally, we illustrate the application of two publicly available models to explore the dynamics of large wood under differing scenarios of catchment management. Our specific objectives are to (1) introduce the concept of a mass balance of wood in rivers, (2) describe the measurement of wood input into channels, (3) describe the direct count and line-intersect methods for estimating large wood abundance in stream channels and geomorphic information required to understand the dynamics of wood, (4) discuss alternatives for estimating transport of wood, and (5) demonstrate the use of a simulation model for wood input and dynamics. Note that this chapter addresses methods of measuring processes related to wood because they differ substantially from measurements appropriate for other types of CPOM in streams (see Chapter 26).

29.2 GENERAL DESIGN

29.2.1 Site Selection

The selection of a study stream in which to conduct this exercise may be influenced by logistical considerations. In general, wadeable 3rd to 4th-order streams are ideal for field classes or workshops. Very small streams at low flow have low transport, and the method described in this chapter is difficult (and can be dangerous) to conduct in large rivers. In general, however, this method can be scaled to a wide variety of stream sizes. For field classes, comparison of channels with contrasting channel features or adjacent land uses may illustrate fundamental relationships related to input, retention, and management of wood. For example, one reach could have a relatively simple channel or riparian forest (straight channel, low roughness, limited hydraulic diversity, sparse riparian forest, and low wood abundance) whereas the other reach could have a complex channel and mature riparian forest (sinuous channel, high roughness, diverse hydraulic conditions, older riparian forest, and abundant wood). This exercise can be combined effectively with the study of retention of CPOM described in Chapter 26.

Length of the experimental reach should be scaled to the size of the stream channel and size of wood pieces and accumulations, with length increasing with stream order. As a rule of thumb, start with a stream length that is at least 20 times the wetted channel width or more. For example, 100 m may be an appropriate length for a 2nd-order stream, 200 m for a 3rd-order stream, and 400 m for a 4th-order stream. Streams of the same size in different riparian settings and with different planforms will differ in the distributions and sources of wood. If possible, use a pilot study to adjust reach length such that multiple aggregations of wood are contained within the reach and the sizes and types of wood reflect the characteristics of the adjacent forests (e.g., diameter, length, and species).

Physical data from the stream channel should be analyzed according to the geomorphic processes to be represented and the types of measurements taken (see Chapters 1–5). At a minimum, the following parameters should be measured for each study reach: width, depth, length, cross-sectional area, planar wetted area, and volume of large wood (using direct counts or the line-intersect method). Sinuosity and slope may also be useful geomorphic measurements. Geomorphic classification systems provide a useful context for comparison of reaches with different channel morphology, because different types have different lateral mobility (bank erosion) and susceptibility to inputs from hillslope failure (see review of channel classification systems by Buffington and Montgomery, 2013). River typology systems have been developed in North America (Montgomery and Buffington, 1998) and Europe (Rinaldi et al., 2016) based on valley confinement, river planform, and bed material size and have been designed for application by river managers in rivers ranging from steep mountain torrents to sluggish anastomosing lowland systems. Discharge measurements also can be useful; but discharge is highly variable temporally and longer discharge records will be more informative. Discharge estimates can be related to available hydrologic analyses of return intervals (e.g., median, 2-years, or 5-years annual flood), which could be interpreted as the bankfull or "channel forming" event. Discharge for ungaged sites could be estimated from a regional analysis based on catchment area.

29.2.2 Marking Techniques for Repeated Surveys

Repeated surveys are necessary to quantify interannual variation in storage, breakdown rates, input processes, transport phenomena, and other episodic or temporally dynamic properties of wood in streams. Pieces of wood often must be marked individually and reinventoried at intervals or identified after being transported downstream. The most common method of wood tagging is nailing numbered metal or plastic tags to the log surface. Surface tags can be lost if the log is floated downstream, and some researchers countersink the tags by drilling holes in the logs and placing the tags beneath the log surface to minimize abrasion and tag loss. Because portions of a log may be broken or buried, placing multiple tags on each log (e.g., on both sides of the ends and middle of the log) may increase the chances of relocating a log. Electronic tags, such as Passive Integrated Transponder tags, radios, and GPS recorders, can be detected more easily or over greater distances, but such techniques are more costly and better suited for tracking a subset of the population of wood pieces.

In the following section, we will illustrate measurement of several of these aspects of wood dynamics in a series of field studies and a computer modeling exercise. First, we compare two approaches for quantifying the abundance and characteristics of wood in two stream reaches and explain how to measure basic geomorphic and hydraulic features of a channel that are related to wood dynamics. Second, a method for measuring transport and retention of large wood will be described and related to the measurement of CPOM retention in Chapter 26. Third, the dynamics of large wood will be modeled with simulations using a publicly available computer model. Physical characteristics of channels can be measured at various levels of detail and longitudinal extent (see Chapters 2−5) and will only be described briefly in this chapter.

29.3 SPECIFIC METHODS

29.3.1 Basic Method 1: Estimation of Standing Stocks of Wood

1. Measure major channel features at a level of intensity appropriate for the research objectives. We recommend working in a research team of three people (two making measurements and one recording data). Minimally, measurements should include average width, average depth, length, cross-sectional area, planar wetted area, channel slope, sinuosity, and substrate composition. Discharge can be measured using the cross-sectional approach (see Chapter 3). Repeat for each reach in the study or exercise.

2. Characteristics of adjacent riparian forests (see Chapter 28) can be useful in assessment of wood sources for streams and are required for models of wood input (see below Advanced Method 1: Modeling Wood Accumulation). Density, diameter at breast height, height, and distance from the active channel are required by most models of wood input. Such silvicultural information may be available for local forests, though riparian stands may differ substantially.

3. Measure the length (L) and average diameter (D) of all wood contacting the channel and larger than a minimum size (e.g., 1 m L × 10 cm D; Fig. 29.2). Estimate and record the portion of each piece of wood in the four zones described in Fig. 29.1. Note if the wood is part of a dam (i.e., wood accumulation blocking some portion of stream flow) and if it is stabilized by a large roughness element (i.e., boulder, tree, large log) so that it would be less likely to be transported during a flood. Option 2 to the Basic Method (further) describes an alternate estimation approach if wood is extremely abundant.

4. If this method is being used for a class demonstration, prior to the survey briefly discuss channel and riparian features. Have students predict amounts of wood for each reach, proportions in each of the four geomorphic zones (Fig. 29.1), and the factors responsible for differences.

29.3.1.1 Data Analysis

1. Calculate reach physical parameters, such as average width, average depth, planar surface area, cross-sectional area, average velocity, hydraulic radius, slope, sinuosity, and discharge (see Chapters 2−4). These fundamental physical parameters can be related empirically or theoretically to observed wood retention values.

FIGURE 29.2 Methods for measuring large wood in stream channels: top: direct count method with formula for volume estimation; bottom: line-intersect method with formula for volume estimation. *Illustrations by J. Miesbauer.*

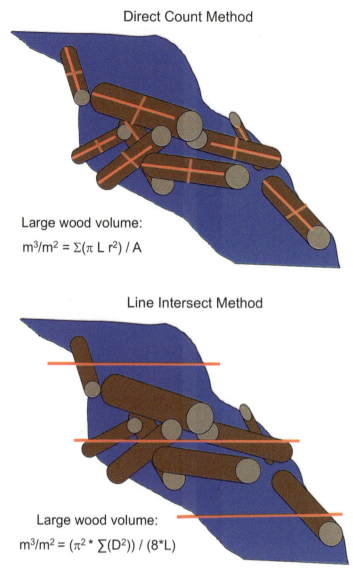

Direct Count Method

Large wood volume:

m³/m² = Σ(π L r²) / A

Line Intersect Method

Large wood volume:

m³/m² = (π² * Σ(D²)) / (8*L)

2. Determine the density (pieces per reach) and total volume (in m³) of wood in each reach, assuming that a cylinder approximates the geometry of a log such that:

$$\text{volume} = \pi L r^2 \tag{29.2}$$

where L = length of the piece (m) and r = radius (m), and then summing for all wood pieces in the reach. Alternatively, you can calculate the volume of wood (m³) per unit area (A, in m²) of stream channel:

$$\text{volume per unit area} = \frac{\Sigma(\pi L r^2)}{A} \tag{29.3}$$

3. Calculate the volume and number of pieces in the four geomorphic zones.
4. Mass of wood can also be calculated based on either empirical estimates or literature values of wood density (mass/volume), but densities of wood differ greatly between species, size of wood, stage of decomposition, and recent history of wetting or drying. Estimates of dry mass or ash-free dry mass may be more relevant for organic matter budgets, but wet mass may be more relevant for investigations of wood transport (Ruiz-Villanueva et al., 2014).

29.3.2 Basic Method 2: Line-Intersect Estimation of Large Wood

1. The LIM can be used in place of the direct count method of large wood in streams having high volumes of wood. For LIM, diameters are measured for all pieces of wood intersecting multiple line transects placed perpendicular to the longitudinal axis of flow (Fig. 29.2). LIM was designed to estimate wood on the forest floor (Van Wagner, 1968; DeVries, 1974), but has been adapted for both small streams (Wallace et al., 2001; Warren et al., 2008) and large rivers (Wallace and Benke, 1984; Benke and Wallace, 1990).

2. In each study reach, use a tape measure to establish a transect every 5 or 10 m along the reach perpendicular to streamflow. Measure the diameter of all large wood pieces intersecting the transect line, using log calipers if available.

3. Compute the wood volume per unit area (m^3/m^2) for each transect using the following equation:

$$\text{volume} = \frac{\pi^2 \Sigma d^2}{8L} \tag{29.4}$$

where d = diameter of a wood piece (m) and L = length of the transect line (m) across the stream (Van Wagner, 1968). To estimate the average large wood volume (m^3/m^2) for a reach, sum wood volumes for each transect and then divide by the total number of transects.

4. In large rivers or in streams with large amounts of wood, LIM may reduce the effort required to estimate large wood volume. In particular, LIM can provide estimates of the abundance of small wood, but the direct count method may not be possible if twigs, sticks, and branches are numerous. However, LIM may overestimate or underestimate the actual large wood volume, determined by direct counts, depending on stream characteristics and large wood distribution (Wallace et al., 2001; Miesbauer, 2004; Warren et al., 2008).

5. As for the direct count method aforementioned, observations and estimates for the LIM can be differentiated between the four geomorphic zones.

29.3.3 Basic Method 3: Long-Term Wood Retention and Transport

1. Release tagged pieces of wood (e.g., logs, sticks, dowels depending on the size of wood in the stream reach) over a time span of several weeks or months, depending on the stream and research objectives. Spraying the ends of wood pieces with a fluorescent paint is an effective and inexpensive tagging method.

2. Inventory the location of tagged wood or dowels in the channel, but leave them in place and reinventory after varying periods of time. Different sizes of wood also can be released. Year-classes of wood can be marked differently (e.g., paint color), permitting year-to-year evaluation of transport. Additional releases of leaves and wood can be conducted in different seasons or at different discharges to describe more precisely the temporal dynamics of retention.

3. Note that it is also possible to conduct short-term wood releases to estimate channel retention, similar to procedures used for leaf releases described in Chapter 26. Such releases are practical for student exercises and involve reinventory of released wood during the same field session followed by retention modeling with an exponential decay function.

29.3.4 Advanced Method 1: Modeling Wood Accumulation

Processes of input, retention, breakdown, and movement of large wood can be represented by simulation models (see review of 14 models in Gregory et al., 2003b). To illustrate the application of simulation modeling for evaluating ecological and physical processes that influence the storage of wood, we will use a publicly available version of OSU Streamwood (Meleason et al., 2003), an integrated model of riparian stand dynamics and wood dynamics in streams of the US Pacific Northwest.

1. Download OSU Streamwood from the H.J. Andrews LTER website http://andrewsforest.oregonstate.edu/lter/data/tools/models/streamwood.cfm?topnav=148 as a compressed file (ZIP file). Also download the User's Guide to assist in running the model for future applications.

2. Create a folder on your computer's directory named OSU_Streamwood. Unzip the model and place the files in your OSU_Streamwood folder.

3. Double left-click (PC users) to open the StreamWood application file (StreamWood.exe or "StreamWood MFC Application"). Mac users will need to "select" files in all instances.

4. Double left-click on "Environment" under streamwood.bsn in the left window. In this window you can specify the wood dimensions, key wood processes, operation of the riparian stand model, and flow regime.

5. Click the box next to "Use Forest Model" and then click OK.

6. Double left-click on "Sections." OSU Streamwood allows the user to set up a network of stream sections with different riparian conditions or different geomorphic characteristics. The default version has three sections composed of four reaches. For example, S1R1 is "Section 1 Reach 1" and is the downstream reach of the network. S2R1 is immediately above S1R1. S2 indicates that it is the second section and R1 indicates that it flows into Section 1. S2R2 is still in Section 2 and flows into S2R1 (i.e., longitudinal series of reaches). S3R1 is the third section that flows into Section 1. That means that it is a tributary to the mainstem with its confluence at the boundary between Sections 1 and 2 of the mainstem (S1R1 and S2R1).

7. Double left-click on each of the reaches. An "environment" tab will appear for each of the four reaches.

8. Double left-click on the "Environment" tab for S1R1. The reach characteristics are described and can be modified. Click on "Same Forest Model Conditions for Both Riparian Zones." Click box next to "Grow a Riparian Forest from 76−100 m from Stream Bank." Note that riparian forest automatically grows from 0 to 75 m, but this can be modified (unselect) for different forest management regimes. Then click on "Define Riparian Forest Management Regime." This allows you to define the management of both the riparian management zone and the upslope forest. For this phase of the exercise, use the default values and click OK. Note that this will mean that there is no forest harvest in this model run.

9. Repeat this step for the other three reaches.

10. Left click on the "Results" tab on the upper toolbar and then click on "Set Results." This allows the user to set the interval at which the model records the results. The default is a 10-years interval. Click OK.

11. Left click on the "Run" tab on the upper toolbar and then click on "Model." Type in the name of the simulation. This window allows you to change the time extent for the model run. Change the time from 400 to 600 years. The model is a probabilistic model and the Monte Carlo simulation can be used to explore the variance in model output. For this exercise, do not click "use Monte Carlo" and we will generate a single run of the model.

12. Click "Run."

13. When the hourglass disappears the model run is complete. You can display the outcomes for each reach. If you click on the down arrow to the right of "Source," you can select the reach to be displayed. If you click on "Variables," you can select the variable to be displayed. The choices include "NuChLog"—number of logs in the active channel, "NuToLog"—number of logs in the active channel, floodplain, and hillslope that touch the channel, "ChanVol"—volume of logs in the active channel, and "Tot_Vol"—volume of logs in the active channel, floodplain, and hillslope that touch the channel. Multiple graphs can be displayed simultaneously by clicking on "Multiline."

14. Click on S2R2. Then click on "ChanVol." Then click on "Multiline." Then click on S2R1 (the downstream reach). Then click on S1R1 (the most downstream reach). Note that the wood storage in the channel tends to reach an inflection at approximately 300 years and the downstream reach continues to accumulate wood because of transport from upstream.

15. You can obtain the numerical values for this model run in the files for Forest and Stream under the Results folder in the OSU_Streamwood folder that you initially created prior to running the model.

29.3.5 Advanced Method 2: Modeling Effects of Timber Harvest on Wood Accumulation

This is a modification of the previous exercise and illustrates the effects of timber harvest on the accumulation of wood in stream channels.

1. Follow steps 1−7 in Advanced Method 1 aforementioned.

2. Under the "Environment" tab for S1R1, click on "Same Forest Model Conditions for Both Riparian Zones." Click box next to "Grow a Riparian Forest from 76−100 m from Stream Bank." Then click on "Define Riparian Forest Management Regime." Under the Riparian Management Area box on the left side of the window, set "Years Between Cut" to 50. In this box, "Total RMA Width," "No-Cut Width," "Min Basal Area for Cut," "Min Num of Leave Trees for Cut," and "Min DBH of Leave Trees" should be automatically set at zero. Under the Riparian Forest Outside of RMA box on the right side of the window, set "Years Between Cut" to 50. Then click OK. This simulates a 50-year harvest rotation with no riparian buffer or management area. Repeat this step for the other three reaches.

3. Left click on the "Run" tab on the upper toolbar and then click on "Model." Type in the name of the simulation. This window allows you to change the time extent for the model run. Change the time from 400 to 600 years. Again, for this exercise, do not click "use Monte Carlo," and we will generate a single run of the model.

4. Click "Run."

5. When the hourglass disappears the model run is complete.

6. In the "Source" box, click on S2R2. Then click on "ChanVol." Then click on "Multiline." Then click on S2R1 (the downstream reach). Then click on S1R1 (the most downstream reach). Note that the wood storage in the channel tends to reach an inflection at approximately 200 years instead of 300 years as in the previous nonharvest exercise. Also note the sequence of peaks and declines that reflect the impact of harvest on recruitment of wood to the channel. Lastly, the storage of wood in the channel under a 50-years harvest cycle was less than 10% of the volume that would accumulate without harvest (or other forms of forest disturbance).

29.3.6 Advanced Method 3: Modeling Wood Abundance in a River Network

The model in Advanced Methods 1 and 2 operates at the spatial scale of stream reaches (i.e., sections of designated channel lengths typically over 100–10,000 m or between major tributary junctions) and at annual time increments. Other models are available for modeling wood dynamics over river networks and large catchments—for example NetMap, a catchment-level model of wood dynamics based on the conceptual framework of Benda and Sias (2003). This model is available at http://www.terrainworks.com/ and requires GIS skills and annual software licenses. Models of specific processes, such as wood transport (Ruiz-Villanueva et al., 2014), are also available from individual researchers.

29.4 QUESTIONS

1. What physical, ecological, or social features of the study reaches might account for any differences in amounts of wood?
2. What are the mechanisms responsible for retention of wood in the two reaches?
3. Compare the wood volumes estimated by direct counts and LIM. Were they similar? How did they differ? What would account for any observed differences?
4. How did the volumes of wood differ between the four geomorphic zones in the two reaches? Describe potential ecological and geomorphic consequences of the differences?
5. How would you model the dynamics of wood in these two stream reaches? Which physical characteristics of the channel would have the greatest influence on the abundance of wood in the stream reaches?
6. In light of your findings, discuss the implications of stream and riparian management practices that (1) reduce the amount of wood loading to streams, (2) simplify morphology of stream channels, or (3) modify the hydrograph.
7. How would you expect the decay rates of wood (e.g., different tree species, different temperature) to influence the accumulation of wood in a stream reach? How could you use the model to explore this question?
8. How would stream discharge influence the storage of wood in stream reaches? How could you use the model to examine the potential consequences of altered hydrologic patterns on wood dynamics?
9. The earliest pioneering research on the geomorphic and ecological roles of wood in streams were first published in the early 1970s. Scientists prior to the 1970s were intelligent and perceptive, but the importance of wood in streams somehow was overlooked. What process or critical component of stream ecosystems are we overlooking today? What can you do to open your eyes and minds to see streams in a new light and possibly recognize phenomena we currently do not?
10. How would you rate the esthetic quality, ecological value, perception of risk, and need for improvement in your study streams? How have your experience and cultural background shaped your perception? Do these influence your design and interpretation of scientific studies and experiments? How could you use social sciences to complement geomorphic and ecological investigations of the role of wood in streams and rivers?

29.5 MATERIALS AND SUPPLIES

Materials
   Field notebook with waterproof data sheets
   Log calipers (if available)
   Meter sticks
   Metric tapes (10, 50, 100 m)
   Stadia rod and clinometer or hand level (for measuring slope)
   Surveying Total Station (if available)
   Log tagging equipment (as appropriate)
   Stopwatch

REFERENCES

Abbe, T.B., Montgomery, D.R., 2003. Patterns and processes of wood debris accumulation in the Queets river basin, Washington. Geomorphology 51, 81–107.

Aumen, N.G., Bottomley, P.J., Gregory, S.V., 1985. Nitrogen dynamics in stream wood samples incubated with ^{14}C lignocellulose and potassium ^{15}N nitrate. Applied and Environmental Microbiology 49, 1119–1123.

Beechie, T.J., Liermann, M., Pollock, M.M., Baker, S., Davies, J., 2006. Channel pattern and river-floodplain dynamics in forested mountain river systems. Geomorphology 78, 124–141.

Benda, L.E., Sias, J.C., 2003. A quantitative framework for evaluating the mass balance of in-stream organic debris. Forest Ecology and Management 172, 1–16.

Benke, A.C., Wallace, J.B., 1990. Woody dynamics in coastal plain blackwater streams. Canadian Journal of Fisheries and Aquatic Sciences 47, 92–99.

Bertoldi, W., Gurnell, A.M., Welber, M., 2013. Wood recruitment and retention: the fate of eroded trees on a braided river explored using a combination of field and remotely-sensed data sources. Geomorphology 180–181, 146–155.

Bilby, R.E., 2003. Decomposition and nutrient dynamics of wood in streams and rivers. In: Gregory, S.V., Boyer, K.L., Gurnell, A.M. (Eds.), The Ecology and Management of Wood in World Rivers, vol. 37. American Fisheries Society, Symposium, Bethesda, Maryland, USA, pp. 135–148.

Boivin, M., Buffin-Bélanger, T., Piégay, H., 2015. The raft of the Saint-Jean River, Gaspé (Québec, Canada): a dynamic feature trapping most of the wood transported from the catchment. Geomorphology 231, 270–280.

Boyer, K.L., Berg, D.R., Gregory, S.V., 2003. Riparian management for wood in rivers. In: Gregory, S.V., Boyer, K.L., Gurnell, A.M. (Eds.), The Ecology and Management of Wood in World Rivers, vol. 37. American Fisheries Society, Symposium, Bethesda, Maryland, USA, pp. 405–420.

Braudrick, C.A., Grant, G.E., 2001. Transport and deposition of large woody debris in streams: a flume experiment. Geomorphology 41, 263–283.

Buffin-Bélanger, T., Biron, P., Larocque, M., Demers, S., Olsen, T., Choné, G., Ouellet, M.A., Cloutier, C.A., Desjarlais, C., Eyquem, J., 2015. Freedom space for rivers: an economically viable river management concept in a changing climate. Geomorphology 251, 137–148.

Buffington, J.M., Montgomery, D.R., 2013. Geomorphic classification of rivers. In: Shroder, J., Wohl, E. (Eds.), Treatise on Geomorphology, vol. 9. Academic Press, San Diego, CA, pp. 730–767.

Buijse, A.D., Klijn, F., Leuven, R.S.E.W., Middelkoop, H., Schiemer, F., Thorp, J.H., Wolfert, H.P., 2005. Rehabilitating large regulated rivers. Archiv für Hydrobiologie Supplement 155 (Large Rivers 15). 738 pp.

Chin, A., Daniels, M.D., Urban, M.A., Piégay, H., Gregory, K.J., Bigler, W., Butt, A.Z., Grable, J.L., Gregory, S.V., Lafrenz, M., Laurencio, L.R., Wohl, E., 2008. Perceptions of wood in rivers and challenges for stream restoration in the United States. Environmental Management 41, 893–903.

Chin, A., Laurencio, L.R., Daniels, M.D., Woho, E., Urban, M.A., Boyer, K.L., Butt, A., Piegay, H., Gregory, K., 2014. The significance of perceptions and feedbacks for effectively managing wood in rivers. River Research and Applications 30, 98–111.

DeVries, D.G., 1974. Multi-stage line intersect sampling. Forest Science 20, 129–133.

Diehl, T.H., 1997. Potential Drift Accumulation at Bridges. U.S. Department of Transportation, Federal Highway Administration Research and Development, McLean, Virginia, USA.

Gregory, K.J., Walling, D.E., 1973. Drainage Basin Form and Process. Arnold, London, UK.

Gregory, K.J., Davis, R.J., 1993. The perception of riverscape aesthetics: an example from two Hampshire rivers. Journal of Environment Management 39, 71–185.

Gregory, S.V., Boyer, K.L., Gurnell, A.M. (Eds.), 2003a. The Ecology and Management of Wood in World Rivers, vol. 37. American Fisheries Society, Symposium, Bethesda, Maryland, USA.

Gregory, S.V., Meleason, M., Sobota, D.J., 2003b. Modeling the dynamics of wood in streams and rivers. In: Gregory, S.V., Boyer, K.L., Gurnell, A.M. (Eds.), The Ecology and Management of Wood in World Rivers, vol. 37. American Fisheries Society, Symposium, Bethesda, Maryland, USA, pp. 315–336.

Gurnell, A.M., Petts, G.E., Hannah, D.M., Smith, B.P.G., Edwards, P.J., Kollmann, J., Ward, J.V., Tockner, K., 2001. Riparian vegetation and island formation along the gravel-bed Fiume Tagliamento, Italy. Earth Surface Processes and Landforms 26, 31–62.

Gurnell, A.M., Piégay, H., Swanson, F., Gregory, S.V., 2002. Large wood and fluvial processes. Freshwater Biology 74, 601–619.

Gurnell, A.M., 2003. Wood storage and mobility. In: Gregory, S.V., Boyer, K.L., Gurnell, A.M. (Eds.), The Ecology and Management of Wood in World Rivers, vol. 37. American Fisheries Society, Symposium, Bethesda, Maryland, USA, pp. 75–91.

Gurnell, A.M., Tockner, K., Edwards, P., Petts, G., 2005. Effects of deposited wood on biocomplexity of river corridors. Frontiers in Ecology and the Environment 3, 377–382.

Gurnell, A.M., 2014. Plants as river system engineers. Earth Surface Processes and Landforms 39, 4–25.

Gurnell, A.M., Corenblit, D., García de Jalón, D., González del Tánago, M., Grabowski, R.C., O'Hare, M.T., Szewczk, M., 2016. A conceptual model of vegetation-hydrogeomorphology interactions within river corridors. River Research and Applications 32, 142–163.

Harmon, M.E., Franklin, J.F., Swanson, F.J., Sollins, P., Gregory, S.V., Lattin, J.D., Anderson, N.H., Cline, S.P., Aumen, N.G., Sedell, J.R., Lienkaemper, G.W., Cromack Jr., K., Cummins, K.W., 1986. Ecology of coarse woody debris in temperate ecosystems. Advances in Ecological Research 15, 133–302.

Heede, B.H., 1972. Influences of a forest on the hydraulic geometry of two mountain streams. Water Resources Bulletin 8, 523–530.

Hering, D., Kail, J., Eckert, S., Gerhard, M., Meyers, E., Mutz, M., Reich, M., Weiss, I., 2000. Coarse woody debris quantity and distribution in Central European streams. International Review of Hydrobiologie 85, 5–23.

Kramer, N., Wohl, E., 2014. Estimating fluvial wood discharge using time-lapse photography with varying sampling intervals. Earth Surface Processes and Landforms 39, 844–852.

Lassettre, N.S., Kondolf, G.M., 2012. Large woody debris in urban stream channels: redefining the problem. River Research and Applications 28, 1477−1487.

LeLay, Y.F., Piégay, H., Moulin, B., 2013. Wood entrance, deposition, transfer and effects on fluvial forms and processes: problem statements and challenging issues. Treatise on Geomorphology 12, 20−36.

MacVicar, B.J., Henderson, A., Comiti, F., Oberlin, C., Pecorari, E., 2009. Quantifying the temporal dynamics of wood in large rivers: field trials of wood surveying, dating, tracking, and monitoring techniques. Earth Surface Processes and Landforms 2031−2046.

MacVicar, B., Piégay, H., 2012. Implementation and validation of video monitoring for wood budgeting in a wandering piedmont river, the Ain River (France). Earth Surface Processes and Landforms 37, 1272−1289.

Meleason, M.A., Gregory, S.V., Bolte, J., 2003. Implications of selected riparian management strategies on wood in streams of the Pacific Northwest. Ecological Applications 13, 1212−1221.

Mellilo, J.M., Naiman, R.J., Aber, J.D., Linkens, A.E., 1984. Factors controlling mass loss and nitrogen dynamics of plant litter decaying in northern streams. Bulletin of Marine Science 35, 341−356.

Merten, E.C., Vaz, P.G., Decker, F.J., Fritz, J., Finlay, J.C., Stefan, H.G., 2013. Relative importance of breakage and decay as processes depleting large wood from streams. Geomorphology 190, 40−47.

Miesbauer, J.M., 2004. An Assessment of Large Woody Debris, Fish Populations, and Organic Matter Retention in Upper Midwestern Streams (M.S. thesis). University of Notre Dame, Notre Dame, Indiana.

Montgomery, D.R., Buffington, J.M., 1998. Channel processes, classification, and response. In: Naiman, R.J., Bilby, R.E. (Eds.), River Ecology and Management. Springer-Verlag, New York, NY, pp. 13−42.

Montgomery, D.R., Collins, B.D., Buffington, J.M., Abbe, T.B., 2003. Geomorphic effects of wood in rivers. In: Gregory, S.V., Boyer, K.L., Gurnell, A.M. (Eds.), The Ecology and Management of Wood in World Rivers, vol. 37. American Fisheries Society, Symposium, Bethesda, Maryland, USA, pp. 21−48.

Moulin, B., Piégay, H., 2004. Characteristics and temporal variability of large woody debris trapped in a reservoir on the river Rhône (Rhône), implications for river basin management. River Research and Applications 20, 79−97.

Mutz, M., 2003. Hydraulic effects of wood in streams and rivers. In: Gregory, S.V., Boyer, K.L., Gurnell, A.M. (Eds.), The Ecology and Management of Wood in World Rivers, vol. 37. American Fisheries Society, Symposium, Bethesda, Maryland, USA, pp. 93−108.

Petts, G., Welcomme, R., 2003. River and wood: a human perspective. In: Gregory, S.V., Boyer, K.L., Gurnell, A.M. (Eds.), The Ecology and Management of Wood in World Rivers, vol. 37. American Fisheries Society, Symposium, Bethesda, Maryland, USA, pp. 421−431.

Piégay, H., Landon, N., 1997. Promoting ecological management of riparian forests on the Drôme River, France. Aquatic Conservation: Marine and Freshwater Ecosystems 7, 287−304.

Piégay, H., Thévenet, A., Citterio, A., 1999. Input, storage and distribution of large woody debris along a mountain river continuum, the Drôme River, France. Catena 35, 19−39.

Piégay, H., 2003. Dynamics of wood in large rivers. In: Gregory, S.V., Boyer, K.L., Gurnell, A.M. (Eds.), The Ecology and Management of Wood in World Rivers, vol. 37. American Fisheries Society, Symposium, Bethesda, Maryland, USA, pp. 109−134.

Piégay, H., Gregory, K.J., Bondarev, V., Chin, A., Dahlstrom, N., Elosegi, A., Gregory, S.V., Joshi, V., Mutz, M., Rinaldi, M., Wyzga, B., Zawiejska, J., 2005. Public perception as a barrier to introducing wood in rivers for restoration purposes. Environmental Management 36, 665−674.

Ravazzolo, D., Mao, L., Picco, L., Lenzi, M.A., 2015. Tracking log displacement during floods in the Tagliamento River using RFID and GPS tracker devices. Geomorphology 228, 226−233.

Reeves, G.H., Burnett, K.M., McGarry, E.V., 2003. Sources of large wood in the main stem of a fourth-order watershed in Coastal Oregon. Canadian Journal of Forest Research 33, 1363−1370.

Rinaldi, M., Gurnell, A.M., González del Tánago, M., Bussettini, M., Hendriks, D., 2016. Classification of river morphology and hydrology to support management and restoration. Aquatic Sciences 78, 17−33.

Robison, E.G., Beschta, R.L., 1990. Characteristics of coarse woody debris for several coastal streams of southeast Alaska, U.S.A. Canadian Journal of Fisheries and Aquatic Sciences 47, 1684−1693.

Ruiz-Villanueva, V., Bladé Castellet, E., Díez-Herrero, A., Bodoque, J.M., Sánchez-Juny, M., 2014. Two-dimensional modelling of large wood transport during flash floods. Earth Surface Processes and Landforms 39, 438−449.

Ruiz-Villanueva, V., Piégay, H., Gaertner, V., Perret, F., Stoffel, M., 2016a. Wood density and moisture sorption and its influence on large wood mobility in rivers. Catena 140, 182−194.

Ruiz-Villanueva, V., Piégay, H., Gurnell, A.M., Marston, R.A., Stoffel, M., 2016b. Recent advances quantifying the large wood dynamics in river basins: new methods and remaining challenges. Reviews of Geophysics 54, 611−652, 1−92.

Schenk, E.R., Moulin, B., Hupp, C.R., Richter, J.M., 2014. Large wood budget and transport dynamics on a large river using radio telemetry. Earth Surface Processes and Landforms 39, 487−498.

Seo, J., Nakamura, F., Nakano, D., Ichiyanagi, H., Chun, K.W., 2008. Factors controlling the fluvial export of large woody debris, and its contribution to organic carbon budgets at watershed scales. Water Resources Research 44, W04428.

Swanson, F.J., Lienkaemper, G.W., Sedell, J.R., 1976. History, Physical Effects, and Management Implications of Large Organic Debris in Western Oregon Streams. USDA Forest Service, pp. 1−15. General Technical Report.

Thévenet, A., Citterio, A., Piégay, H., 1998. A new methodology for the assessment of large woody debris accumulations on highly modified river (example of two French Piedmont rivers). River Research and Applications 14, 467−483.

Van Sickle, J., Gregory, S.V., 1990. A model of woody debris input into streams. Canadian Journal of Forest Research 20, 1593−1601.

Van Wagner, C.E., 1968. The line intersect method in forest fuel sampling. Forest Science 14, 20–26.

Wallace, J.B., Benke, A.C., 1984. Quantification of woody habitat in subtropical coastal plain streams. Canadian Journal of Fisheries and Aquatic Sciences 41, 1643–1652.

Wallace, J.B., Webster, J.R., Eggert, S.L., Meyer, J.L., Siler, E.R., 2001. Large woody debris in a headwater stream: long-term legacies of forest disturbance. International Review of Hydrobiology 86, 501–513.

Warren, D.R., Keeton, W.S., Kraft, C.E., 2008. A comparison of line-intercept and census techniques for assessing large wood volume in streams. Hydrobiologia 598, 123–130.

Warren, D.R., Kraft, C.E., Keeton, W.S., Nunery, J.J., Likens, G.E., 2009. Dynamics of wood recruitment in streams of the northeastern US. Forest Ecology and Management 258, 804–813.

Wohl, E., 2013. Floodplains and wood. Earth-Science Reviews 123, 194–212.

Wohl, E., 2014. A legacy of absence: wood removal in U.S. rivers. Progress in Physical Geography 38, 637–663.

Wohl, E., Bledsoe, B.P., Fausch, K.D., Kramer, N., Bestgen, K.R., Gooseff, M.N., 2016. Management of large wood in streams: an overview and proposed framework for hazard evaluation. Journal of the American Water Resources Association 1–21.

Section E

Ecosystem Processes

Gary A. Lamberti and F. Richard Hauer

Stream ecosystem research has progressed from largely focused on biophysical structure to the comprehensive consideration of functional properties of these highly reactive systems. This evolution was initiated by the recognition that streams operate "differently" from many other ecosystems because of the downstream movement of water and the elements carried with that flow. The concept of a downstream "nutrient spiral" for important elements, such as carbon (C), nitrogen (N), and phosphorus (P), has stimulated much research on how nutrients cycle through microbial and macrobiotic compartments in streams. The seven chapters in this section address the dynamics of important functional processes in streams with a focus on the cycling of important elements and their ultimate fate in lotic biota, which may culminate in primary and secondary production. Chapter 30 presents the fundamental concepts and methods of investigation of nutrient spiraling in streams, with a focus on the theoretical basis for measuring the downstream advection and dispersion of bioreactive and nonreactive elements carried with water. Chapters 31–33 build on this foundation by presenting approaches for measuring the dynamics and transformations of specific reactive solutes in streams. Chapter 31 describes the use of nutrient diffusing substrates for determining the nutrient status of streams, as well as using constant releases of stable isotopes for measuring rates of nutrient uptake along stream courses. In Chapter 32, methods to measure specific transformations of N, a critical and highly reactive element in streams, are described in coupled field and laboratory protocols. Chapter 33 presents methods for determining the dynamics of P, a major nutrient playing a critical role in the biochemical apparatus of all organisms, and includes exercises using both stable and radioactive forms of P to interpret uptake and cycling. This section then shifts to methods to measure the productivity of streams and the incorporation of these critical elements into the biota. In Chapter 34, major advances in methods to determine the "metabolism" of streams from measures of dissolved oxygen dynamics are described, including modern modeling techniques for assessing production and respiration. Classical methods for measuring the secondary production of streams, with a focus on benthic macroinvertebrates, are described with detailed exercises in Chapter 35. Finally, Chapter 36 explains the emerging field of "ecological stoichiometry" whereby useful insights into stream function are provided by measuring the elemental content of stream biota, especially C, N, and P. Overall, this section presents a sound foundation for measuring the cycling and fate of important elements in streams and rivers and highlights how much is yet to be discovered about the functioning of these unique ecosystems.

Chapter 30

Conservative and Reactive Solute Dynamics

Michelle A. Baker[1] and Jackson R. Webster[2]

[1]Department of Biology and the Ecology Center, Utah State University; [2]Department of Biology, Virginia Polytechnic Institute and State University

30.1 INTRODUCTION

Solutes are materials that are chemically dissolved in water. These include cations (positively charged ions) such as calcium, magnesium, sodium, and potassium; anions (negatively charged ions) including chloride, sulfate, silicate, and bicarbonate; and organic molecules. In comparison to these common solutes, which are found in relatively large concentrations in many natural waters, more biologically important solutes such as phosphate and ammonium are normally present at very low concentrations. Solutes enter streams from three natural sources. First, the atmosphere (e.g., rainwater) is often the major source of chloride, sodium, and sulfate. Second, other solutes come from soil and rock weathering, including calcium, phosphate, silica, and magnesium. Third, biological processes may be important. For example, while nitrate may enter from the atmosphere or from weathering, it also may be generated from nitrogen that was biologically fixed by cyanobacteria. In addition, inorganic carbon (i.e., carbon dioxide, bicarbonate, or carbonate) comes from the atmosphere and weathering, but it also comes from respiration by soil and stream organisms. Point sources (such as pipes) and nonpoint sources (e.g., agricultural runoff) are often major inputs of solutes to streams.

Solutes in water can be classified according to their biological and chemical reactivity. *Conservative* solutes are those that do not react chemically or biologically, and thus their concentration is not changed by in-stream processes other than dilution from groundwater or tributaries. As such, conservative solutes mimic downstream transport of water. Examples of conservative solutes include lithium and bromide (e.g., Bencala et al., 1991). On the other hand, solutes whose concentration is changed by chemical and/or biological transformations are referred to as nonconservative or *reactive* solutes. Nutrients such as nitrate and phosphate are examples of reactive solutes. Some reactive solutes may be so abundant that biological and chemical transformations do not measurably influence stream concentration and may in fact be treated as conservative solutes. Chloride is an example of such a solute that, although is essential to organisms, exists in most streams in concentrations that far exceed biological needs. As such, chloride is often used as a conservative solute in stream studies (e.g., Triska et al., 1989).

Solute dynamics describe the coupled physical, chemical, and biological processes that govern transformations of materials dissolved in water. As such, the term describes the spatial and temporal patterns of solute transport and transformation (Stream Solute Workshop, 1990). Solute dynamics are tightly coupled to the physical movement of water in all ecosystems, but in streams this coupling between transport and transformation is particularly important. Material cycling in a conventional, ecological, sense describes the continued recycling of solutes between inorganic and organic (living or dead) forms (Fig. 30.1). When considering a single point in space and time, the unidirectional flow of stream water prohibits material cycling in this conventional view (Webster and Patten, 1979). At the reach scale, however, cycling becomes apparent as inorganic materials are mineralized and transported downstream before being reacted upon biologically or chemically. Stream ecologists term this combination of cycling and longitudinal transport a *spiral* (Fig. 30.1; Webster, 1975). The balance between how much and for how long solutes are retained versus being transported longitudinally is central to understanding stream ecosystem functioning.

Methods in Stream Ecology. http://dx.doi.org/10.1016/B978-0-12-813047-6.00008-5

129

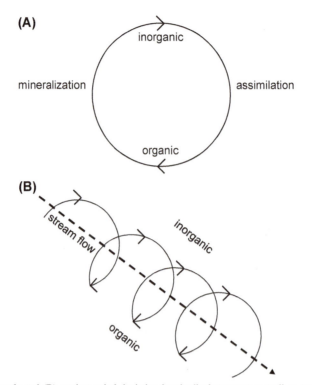

FIGURE 30.1 (A) Simple nutrient cycle and (B) nutrient spiral depicting longitudinal transport as well as cycling between inorganic and inorganic forms. *After Newbold (1992).*

While the dynamics of many solutes are determined primarily by biogeochemical and hydrologic interactions occurring in the whole watershed (Webb and Walling, 1992), important in-stream dynamics also occur (e.g., Peterson et al., 2001; Bernhardt et al., 2003). Studies of solute dynamics in streams provide two types of information. First, dynamics of conservative solutes can be used to quantify various hydrodynamic properties of a stream reach or segment. Second, comparing conservative and reactive solute dynamics can provide information on rates of transport and transformation of the solutes themselves, which is important to the understanding of their availability and importance. In this chapter, we describe investigations of solute dynamics from both perspectives.

30.1.1 Conservative Solute Dynamics

The dynamics of conservative solutes in streams are driven largely by two processes—advection and dispersion. *Advection* is downstream transport of the solute with the water itself, occurring at average water velocity. *Dispersion* is the spreading of the solute in the medium and can occur by molecular diffusion, but in streams diffusion is primarily caused by turbulence. Dispersion allows some solute molecules to move more rapidly or more slowly than the bulk transport of materials due to advection. Mathematical models derived from the advection—dispersion equation are used to quantify solute dynamics in streams (Bencala and Walters, 1983; Stream Solute Workshop, 1990). These models can take on varying degrees of complexity depending on what aspects of solute dynamics are being considered. The following description was adapted from the more complete derivations presented in the Stream Solute Workshop (1990).

For a uniform channel with constant discharge, advection and dispersion can be expressed in the partial differential equation

$$\frac{\partial C}{\partial t} = -u\frac{\partial C}{\partial x} + D\frac{\partial^2 C}{\partial x^2}$$ (30.1)

where C = solute concentration; t is time; x = distance in the downstream direction; u = water velocity; and D = dispersion coefficient. This equation means that the rate of change in concentration with time is a function of advection plus dispersion in the downstream (x) direction. Other terms can be added to this equation to include variable stream morphology, groundwater inputs, and transient storage. *Transient storage* refers to the temporary storage of water that is traveling much more slowly than the main body of water (Bencala and Walters, 1983), such as water in hyporheic flow paths, surface pools or backwaters, and macrophyte beds (e.g., Bencala et al., 1984; Harvey et al., 1996; Ward, 2015). Adding these factors, the mathematical model becomes a pair of equations:

$$\frac{\partial C}{\partial t} = -\frac{Q}{A}\frac{\partial C}{\partial x} + \frac{1}{A}\frac{\partial}{\partial x}\left[\frac{AD\partial C}{\partial x}\right] + \frac{Q_L}{A}(C_L - C) + \alpha(C_S - C)$$

and

(30.2)

$$\frac{\partial C_S}{\partial t} = -\alpha\frac{A}{A_S}(C_S - C)$$

where Q = discharge; A = stream cross-sectional area; Q_L = lateral inflow from groundwater; C_L = solute concentration of the lateral inflow; α = exchange coefficient between the main channel and transient storage zones; A_S = size (cross-sectional area) of the transient storage zone; and C_S = solute concentration in the transient storage zone. The first equation is the advection–dispersion equation (as in Eq. (30.1), but allowing for changes in stream discharge and cross-sectional area) with lateral inflow and transient storage, while the second equation describes the rate of concentration change in the transient storage zone as a function of exchange rate and the size of the storage zone relative to the main channel (both as cross-sectional area). Metrics to compare transient storage among streams can be derived from these parameters and include A_S/A (from Eq. 30.2) and F_{med} (the fraction of median transport time in transient storage) along with others (e.g., Harvey and Wagner, 2000; Runkel, 2002).

30.1.2 Reactive Solute Dynamics

Dynamics of reactive solutes are more complicated because of the production and consumption of these solutes by in-stream processes. In streams, the majority of these processes occur on the stream bottom, although processes in the water column also can play a role (Reisinger et al., 2015). Such processes may be abiotic, such as adsorption, desorption, precipitation, and dissolution. Many biotic processes, such as assimilation by microbes and plants, as well as mineralization, are important to the dynamics of reactive solutes in streams. Generally, abiotic and biotic processes that remove solutes from the water column are termed *immobilization*. In streams the most important immobilization processes for nutrients are adsorption (especially for phosphate) and microbial uptake (by heterotrophs and algae). Revising the advection–dispersion equation for uniform channel and discharge to account for immobilization, the expression becomes:

$$\frac{\partial C}{\partial t} = -u\frac{\partial C}{\partial x} + D\frac{\partial^2 C}{\partial x^2} - \lambda_C C$$

(30.3)

where C = reactive solute concentration, λ_C = dynamic uptake rate (units of time^{-1}), and the other terms are as described for Eq. (30.1). Nutrients that are immobilized will eventually be mineralized and returned to the water column. This can be most simply expressed by adding another term to Eq. (30.3) and adding another equation for the immobilized nutrient:

$$\frac{\partial C}{\partial t} = -u\frac{\partial C}{\partial x} + D\frac{\partial^2 C}{\partial x^2} - \lambda_C C + \frac{1}{z}\lambda_B C_B$$

and

(30.4)

$$\frac{\partial C_B}{\partial t} = z\lambda_C C - \lambda_B C_B$$

where C_B = immobilized nutrient standing crop (usually considered to be benthic, as mass per unit area), z = stream depth, and λ_B = rate of mineralization.

As a nutrient atom cycles between inorganic and organic forms (Fig. 30.1), the spiraling length (S) is the distance it travels while completing this cycle (Newbold et al., 1981; Elwood et al., 1983). Over the length of a spiral, the nutrient changes from abiotic to biotic and back to abiotic form. Thus, the spiraling length has two components: (1) the *uptake length* (S_W) is the distance traveled in dissolved inorganic form before the nutrient is removed from solution, and (2) the *turnover length* (S_B) is the distance traveled before being mineralized and returned to the water column:

$$S = S_W + S_B$$

(30.5)

Because much of the organic material in streams resides in the benthic sediments (e.g., Fisher and Likens, 1973) and movement of these particles is far slower than movement of dissolved constituents (Newbold et al., 1983; Minshall et al., 2000), the uptake length dominates total spiraling length (Newbold et al., 1983; Mulholland et al., 1985). Accordingly, we focus on dynamics of dissolved inorganic nutrients as addressed by S_W and related measures.

Uptake length can be mathematically related to Eqs. (30.3) and (30.4) as the inverse of the longitudinal uptake rate:

$$S_W = \frac{1}{k_W} \tag{30.6}$$

where the longitudinal uptake rate (k_W) is the dynamic uptake rate (λ_C) divided by water velocity:

$$k_W = \frac{\lambda_C}{u} \tag{30.7}$$

Because S_W is a displacement distance, it is strongly influenced by stream discharge and velocity. To correct for this influence of stream size, S_W is often standardized to allow comparison of solute dynamics across systems (Davis and Minshall, 1999). This standardization converts S_W to a mass transfer coefficient, which describes a theoretical velocity at which a nutrient moves toward the location of immobilization (Stream Solute Workshop, 1990). This mass transfer coefficient is referred to in the literature as the uptake velocity (v_f, Davis and Minshall, 1999). Uptake velocity is related to the uptake rate coefficient (k_C) through stream depth (z) as $v_f = zk_C$. As such, v_f is related to S_W and can be calculated as:

$$v_f = \frac{uz}{S_W} \tag{30.8}$$

In some instances, it is useful to describe nutrient uptake per unit area of stream bottom (mass area^{-2} time^{-1}), as is done in other ecosystems. Areal uptake (U) is calculated as:

$$U = v_f C \tag{30.9}$$

where C = ambient nutrient concentration. *Areal uptake* refers to the mass of solute immobilized by an area of streambed per unit time and reflects the magnitude of the flux of inorganic solute from the water column to the biota. In the literature, U is sometimes referred to as "uptake rate"; however, this is inaccurate—the units of U represent a transport flux—in this case the mass of solute flow across a unit area. Stream ecologists use the term "uptake" synonymously with immobilization. Certain uptake processes may be assimilatory while others may be dissimilatory (e.g., denitrification) or abiotic (e.g., adsorption). It is important to note that U represents gross uptake and not net retention, as mineralization releases some portion of nutrients back to the water column (Stream Solute Workshop, 1990).

Together, these measures (S_W, v_f, and U) provide insight into the dynamics of nutrients in streams, and these metrics are mathematically related (Fig. 30.2; Webster and Valett, 2007). Uptake length is a reach- or segment-scale estimate of retention efficiency and gives explicit information (as distance) about the spatial extent over which nutrient uptake occurs. Areal uptake conveys important information on biological assimilation or other immobilization processes and is useful for comparison with other ecosystems but does not provide this spatial context. Uptake velocity standardizes uptake length for discharge (depth and velocity). Uptake velocity is also a measure of nutrient demand (areal uptake) relative to nutrient availability (Davis and Minshall, 1999), as seen by rewriting Eq. (30.9) as:

$$v_f = \frac{U}{C} \tag{30.10}$$

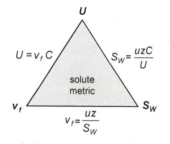

FIGURE 30.2 Metric triad for determining nutrient dynamics in stream ecosystems. Symbols are defined in the text. *After Webster and Valett (2007).*

Conceptually, S_W is useful as an overall index of nutrient retention by streams because it accounts for physical transport processes leading to export and biogeochemical processes that retain nutrients in the stream for some time. S_W increases with stream discharge (Peterson et al., 2001; Tank et al., 2008; Hall et al., 2013) and decreases with autotrophic processes (e.g., Hall and Tank, 2003) and heterotrophic processes (Mulholland et al., 1985) as microbes that mediate both types of processes require nutrients for their activities. Stream substrates with greater surface area and/or higher biofilm biomass may also decrease S_W (Davis and Minshall, 1999; Arp and Baker, 2007). Shorter uptake lengths are expected for nutrients that limit biological processes (Newbold et al., 1983; King et al., 2014).

Because of the influence of stream discharge on S_W, v_f and U are better metrics for assessing the influence of nutrient concentrations and biomass on nutrient uptake (Stream Solute Workshop, 1990; Davis and Minshall, 1999; Dodds et al., 2002). Three models describing the relationship between uptake and concentration have been proposed. The first is a linear relationship between nutrient concentration and U (Dodds et al., 2002; O'Brien et al., 2007). Under this scenario, U increases linearly with concentration, and mass transfer across a stream-averaged boundary layer limits uptake according to the equation:

$$U = kC \tag{30.11}$$

where k = constant and C = ambient concentration (Dodds et al., 2002).

At the other extreme, the relationship between uptake and concentration may be controlled by reaction kinetics and follow a Michaelis–Menten type model (Stream Solute Workshop, 1990; Bernot and Dodds, 2005). In this case, the relationship between U and concentration is hyperbolic and reaches a saturation point above which U no longer increases with concentration. Mathematically this model is of the form:

$$U = \frac{C_b U_{max}}{K_m + C_b} \tag{30.12}$$

where U = areal uptake, C_b = background concentration, U_{max} = maximum uptake rate, and K_m = Michaelis, or half-saturation, constant. By substitution, the Michaelis–Menten model predicts a positive linear relationship between S_W and concentration and a negative nonlinear relationship between v_f and concentration (Earl et al., 2006, 2007).

A third model, termed the Efficiency Loss model, is similar to the Michaelis–Menten model in that the relationship between uptake and concentration is nonlinear. It suggests that as concentrations increase, the efficiency of uptake decreases such that the relationship between U and concentration is a power law with exponent (b) less than one:

$$U = kC^b \tag{30.13}$$

In this scenario, S_W increases nonlinearly with concentration while v_f decreases nonlinearly with concentration (O'Brien et al., 2007).

These models have been tested at the reach scale with regard to P uptake and N uptake (as ammonium and nitrate; Dodds et al., 2002; Bernot et al., 2006; Earl et al., 2006; Newbold et al., 2006; O'Brien et al., 2007). For nitrate, uptake appears to follow Michaelis–Menten kinetics for many streams, but the Efficiency Loss model better explains data across Kansas streams that vary widely in N loads (O'Brien et al., 2007). In contrast, nitrate uptake attributable to denitrification has a first-order (linear) relationship with concentration, suggesting that if substrate is available, denitrification is limited by mass transfer (O'Brien et al., 2007; Mulholland et al., 2008). For oligotrophic streams, the Michaelis–Menten model tends to best fit observed data for ammonium or phosphate uptake (e.g., Payn et al., 2005; Bernot et al., 2006; Newbold et al., 2006).

The experiments described in this chapter allow exploration of dynamics of both conservative and reactive (nutrient) solutes in flowing waters. Because of differences in equipment that might be available and the highly variable nature of stream chemistry, we have provided a number of procedural options. At a minimum, you should be able to measure discharge, velocity, the importance of transient storage, and estimate nutrient uptake. The procedure for estimating nutrient uptake described here requires elevating nutrient concentration. Because uptake rate (k_C) is proportional to concentration, experiments that increase concentrations above ambient conditions tend to overestimate S_W (Mulholland et al., 2002; Dodds et al., 2002). Isotopic tracers such as ^{15}N and ^{32}P do not elevate nutrient concentrations and therefore provide the most accurate estimates of nutrient update. The stable isotope ^{15}N has been widely used to study N cycling in streams (e.g., Peterson et al., 2001; Mulholland et al., 2008)—its major limitation being economic. The radioisotope ^{32}P was used as a tracer in the pioneering work of Elwood et al. (1981), but this tracer is not feasible for use in most environments because of its radioactivity. See Chapters 31 and 32 for procedures involving isotopes.

Alternatives to nutrient enrichment procedures include sequential addition of nutrients at increasingly higher concentrations (Payn et al., 2005; Earl et al., 2007; Demars, 2008) and back-calculating nutrient uptake metrics at ambient concentration. While more effort is needed, this approach not overestimate S_W. A more recent alternative is the Tracer Additions for Spiraling Curve Characterization (TASCC) approach (Covino et al., 2010), which involves using an instantaneous pulse of concentrated solutes to take advantage of the dynamic concentration range observed at a single location as the solutes move through a stream reach. Each data point along the breakthrough curve (BTC) is used to estimate S_W and net areal uptake at each concentration of nutrient compared to a conservative solute. These data then can be used to estimate functional relationships across a range of concentrations in a manner similar to that initially proposed by Payn et al. (2005) for plateau tracer tests. It is beyond the scope of this chapter to describe and compare multiple methods for measuring nutrient uptake metrics in the field (but see Alvarez et al., 2010; Trentman et al., 2015 for a comparison of approaches) and for discussion of the various reactive transport models (e.g., Runkel, 2007; Payn et al., 2008; Claessens and Tague, 2009) that can be used to solve for nutrient uptake metrics. Several studies have also used nutrient declines downstream of point sources such as wastewater treatment plants or springs to estimate nutrient removal (e.g., Gibson and Meyer, 2007; Hensley et al., 2014). However, this technique measures net nutrient removal and the difference between gross uptake and mineralization and provides less insight into nutrient dynamics occurring in the stream. These measurements of net nutrient removal can be useful in determining the capacity of a stream to remove dissolved nutrients from downstream transport by processes such as denitrification, but net nutrient removal cannot be directly compared with measurements of uptake described in this chapter.

Here, we provide a general field-based approach to study conservative and reactive solute dynamics using pulse and steady-state (plateau) approaches. The general approaches described here could subsequently be used for more advanced analyses such as sequential nutrient or stable isotope releases, TASSC analysis, and/or solute transport modeling.

30.2 GENERAL DESIGN

The general design of these experiments involves releasing a known concentration of solute either as a pulse (instantaneous release) or at a constant rate (plateau release) and making measurements at a downstream location. Care should be taken in site selection, choice of solute(s), method of release, and data analysis as described below.

30.2.1 Site Selection

Nearly 1000 solute releases have been performed in first- to fourth-order streams that range in discharge from <1 to 2000 L/s (Tank et al., 2008; Hall et al., 2013). Streams of this size allow wadeable access for physical measurements and sampling, and most of these studies have used constant-rate releases. At larger stream flows, the pulse release method and sampling design are preferred for logistical reasons (e.g., Dodds et al., 2008; Tank et al., 2008).

Choice of a stream or section of stream will depend on the question posed (e.g., single reach or comparison of multiple reaches). Ideally, a stream or set of streams should be selected that provide a range of physical and biological conditions. A comparison of hydraulic properties between two reaches might include one simple reach (e.g., straight channel, homogenous substrate, low amount of wood) and one more complex reach (e.g., sinuous channel, heterogeneous substrate, high amount of wood). Avoid reaches with tributary inputs or water diversions. Experimental reach length will vary with flow (higher flows require longer reaches), but reach length must be long enough for complete mixing of the released solute(s). Complete mixing can be evaluated using a preliminary dye release (e.g., Rhodamine WT or fluorescein) and visual observation. Typical reach lengths range from 50 m for small headwater streams to several hundreds of meters for mid-order streams, or to several kilometers for rivers.

30.2.2 Tracer Selection

Selection of a conservative solute is a function of local geology, ambient solute concentration, research budget, and analytical capacity. It is essential to raise stream concentration of the solute sufficiently above ambient concentration to be analytically detectable, while avoiding potential harm to biota (e.g., Flury and Papritz, 1993; Stewart and Kzsos, 1996). A number of hydrological studies have used the dye, Rhodamine WT, as a conservative tracer, but recent research (Runkel, 2015) shows that Rhodamine WT can sorb to hyporheic sediments and thus exhibits nonconservative behavior and should not be used to characterize transient storage processes. Typical conservative solutes used are salts of chloride, sodium, lithium, potassium, and bromide. Of these, chloride is the most common. Chloride can easily be obtained as NaCl at a grocery store or feed store, but choose a nonionized form. Most commercial NaCl contains a small amount of cornstarch or other anticaking agent, which will cause a slightly cloudy solution but should not be a problem. Chloride measurements in the field can be made conveniently with a conductivity meter, and conductivity itself can be used as a conservative measure (e.g., Mulholland et al., 1994).

Conductivity is very sensitive to temperature, but most handheld meters and data sondes can be set to measure specific conductance, which corrects for temperature. Portable ion-specific electrodes also are available for chloride, bromide, sodium, and other ions. Bromide has the advantage of very low ambient concentrations in most waters (e.g., Flury and Papritz, 1993) and may be preferred in streams where ambient chloride concentrations are high. A disadvantage to using sodium as a conservative tracer is that it loses 5–10% by mass through sorption to stream bottom materials compared to almost no loss of chloride (Bencala, 1985). A disadvantage to using ion-specific electrodes is that they are sensitive to matrix effects; bromide electrodes are influenced by variations in ambient chloride, for example. Both conductivity sensors and ion-specific electrodes can be calibrated using solutions prepared with water from the study stream for better performance. If portable instruments are not available, samples can be collected and filtered in the field and then analyzed by ion chromatography or other means in the laboratory.

One or more reactive solutes (e.g., nutrients) can be added along with a conservative solute to estimate uptake metrics. Choice of a nonconservative solute for study will depend on your specific research objectives and knowledge of the streams of interest. If choosing a nutrient, phosphate and inorganic forms of nitrogen (nitrate or ammonium) are obvious candidates. Your choice may depend on the availability of analytical instrumentation for measuring concentrations of these nutrients. Be sure not to choose a nutrient that will interact with the conservative solute. For example, calcium and phosphate cannot be used together because they form a highly insoluble salt. Also take care to choose nonconservative solutes that do not interfere with analysis and interpretation of data. For example, nitrate and ammonium should not be used together if nitrification is an important process because nitrate uptake could be masked to an extent by nitrate produced by nitrification.

30.2.3 Release Techniques

In this chapter, we describe techniques for two types of solute releases. The first is the slug release, also termed a gulp or pulse release. In this method, a preweighed mass of salt is dissolved in stream water, and the entire volume is dumped nearly instantaneously at an upstream location. Concentration of solute at a site downstream is measured as a function of time until the concentration recedes to ambient levels (Fig. 30.3). This method can be used to estimate discharge (dilution gauging; see Chapter 3) or used for analyses of solute dynamics.

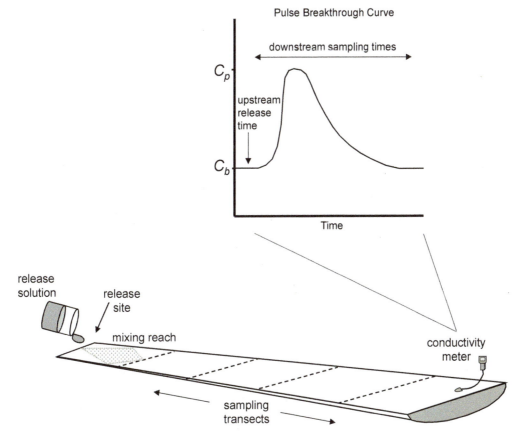

FIGURE 30.3 Diagram of setup for a pulse release and an idealized breakthrough curve for a pulse release of conservative solute. C_b = background concentration, C_p = peak concentration. See also Online Worksheet 30.1.

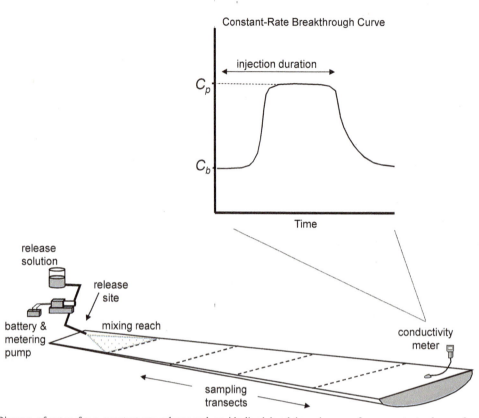

FIGURE 30.4 Diagram of setup for a constant-rate release and an idealized breakthrough curve for a constant release of conservative solute. C_b = background concentration, C_p = plateau (peak) concentration. See also Online Worksheet 30.1.

Constant-rate releases are useful for measuring solute dynamics in small streams and are advantageous over pulse releases in which fewer samples are required for laboratory analysis. In this technique, solutes are dissolved in stream water and then added at a constant rate to the stream for some time period until the added solute concentration reaches a plateau above ambient concentration at a downstream site (Fig. 30.4). A simple, inexpensive, and nonelectrical release apparatus is the Mariotte bottle (Webster and Ehrman, 1996); however, battery-powered metering pumps (e.g., Fluid Metering, Inc., Syosset, NY, USA) are generally more reliable, and pump rates can be easily adjusted to field conditions. At discharge above 2000 L/s it is challenging to pump sufficient solute(s) to the stream using this technique, and the pulse method would be preferred.

30.2.4 Data Analyses

The most important data to be collected and analyzed during a solute tracer test is a concentration-time profile at one or more downstream locations from the point of release. This profile is called a *breakthrough curve*. Beyond the single reach release, solute dynamics can be compared spatially among the reaches of one to several streams, before and after manipulation, and over time at different flows. Reactive solute (nutrient) releases can be performed simultaneously with conservative tracer, and the BTCs for both solutes simulated and compared.

Essential physical measurements to be made in studies of solute dynamics include ambient water chemistry (especially for the solutes that are to be used as tracers), water temperature, stream discharge, average water depth, and average wetted-channel width for the stream reach over which the release is being conducted (see Chapters 2—5). These data should be included in any publication so that the published data set is useful for future syntheses and metaanalyses. A number of other potentially interesting and important measurements that can be made to aid in interpretation are described elsewhere in Volumes 1 and 2 of this book, including, for example, thalweg velocity (Chapter 4), gradient (Chapter 2), sediment size distribution (Chapter 5), coarse wood volume or area (Chapter 29), benthic biomass (Chapter 12), and metabolism (Chapter 34).

One can calculate hydraulic characteristics such as discharge and velocity by graphical analysis of the BTC resulting from the experiment; however, it is necessary to have a reasonable estimate of discharge prior to the experiment to calculate expected solute release concentrations. If one has a measurement of discharge independent of the BTC data, tracer recovery can be calculated, which can give insight into losses of water from the main channel to the subsurface (e.g., Payn et al., 2009). Uptake length and rate can be calculated from nutrient data fit to a negative exponential model. For each solute release, a computer model can be used to simulate the observed BTC and calculate hydraulic parameters such as dispersion and transient storage size and exchange rate. These techniques are described later in this chapter.

30.3 SPECIFIC METHODS

30.3.1 Basic Method: Dynamics of Conservative Solutes

In this exercise, we use chloride as the conservative solute and derive concentration from data obtained with a temperature-correcting specific conductance meter. For brevity we call this conductivity. We describe laboratory and field preparations to be done prior to conducting conservative solute tracer tests in the field as well as two field techniques—a slug (pulse) release and a constant rate (plateau) release.

30.3.1.1 Laboratory Preparation

1. Prepare a stock solution of sodium chloride in deionized or distilled water. A stock solution of 238 g NaCl/L (144 g Cl/L) is two-thirds the saturation of NaCl in cold water and is fairly easily dissolved. The mixture can be heated in a water bath to aid dissolution. Mix vigorously and repeatedly to be certain that the salt is completely dissolved.
2. Make serial dilutions from the stock solution to prepare a series of chloride standards across the range of expected chloride concentrations (e.g., 0.25–20 mg/L). The standards can be made by diluting the stock solution in deionized water, but could also be made by diluting the stock solution in stream water. The latter is generally more accurate especially if there are matrix effects or if ionselective probes are used to collect data.
3. Large volumes of stock solution could be prepared to use in the field for releases. Alternatively, the salt can be preweighed in 500-g increments into labeled zipper top plastic bags and used to prepare tracer solutions stream side.

30.3.1.2 Field Preparation—Prerelease

1. In the field, walk the intended study reach, and identify the location of the release site above a good mixing reach. A good mixing reach is one that is not wide and slow-moving, rather above a small riffle or swift moving water would help mix the tracer laterally and vertically. Identify the location of the end of the study reach where the conductivity meter will be deployed.
2. Place the chloride standards in the stream to equilibrate with ambient stream temperature. Calibrate the conductivity meter with the standards per the manufacturer's instructions.
3. Use a tape measure to delimit the extent of the study reach between the release point and end of study reach. Mark every 5 m (for a 100-m reach) with labeled flagging tape. At each 5-m transect, measure wetted-channel width, depth (c.10 measurements at each cross section), and thalweg velocity (optional). Often, "effective depth" calculated from discharge, velocity, and width will be more useful than measured depth (see Chapter 34). Additional data such as benthic biomass, substrate size distribution, and gradient could also be collected at these locations.
4. Measure stream flow at the top and bottom of the stream reach using the velocity–area method or other means as described in Chapter 3.
5. For constant-rate releases, working in a downstream–upstream direction, measure stream temperature and ambient chemistry (either with the conductivity meter or by collecting a grab sample from the thalweg) every 10 m.
6. For both release types, place the conductivity meter in a well-mixed area at the downstream site, and assign a person to record conductivity during the release. Synchronize stopwatches among all team members (and the conductivity meter if it automatically records data). For pulse releases, collect several measurements of temperature and ambient chemistry (conductivity) before the release.

30.3.1.3 Field Procedure—Pulse Release

1. For a pulse release, prepare the tracer solution in a 5-gallon ($\sim$20-L) bucket at the top of the reach. Either use a preprepared solution or dissolve a known mass of salt using 2.5 gallons ($\sim$10 L) of stream water. For conductivity, it may be sufficient to increase stream concentration by 10 μS/cm if the conductivity meter reads to 0.1 μS. We have found that a mass of 500 g NaCl can increase the peak conductivity by about 50 μS/cm in a stream flowing at 100 L/s, but this depends on transport characteristics and dilution along the reach. Make sure to record the mass of salt and volume of water added to the bucket. A small (1 mL or less) aliquot of tracer solution should be collected to measure concentration of the tracer at a later time.
2. Add the dissolved tracer by quickly pouring the bucket's contents across the width of the stream, and quickly rinse the bucket and mixing stick in the stream. This will ensure that all the tracer mass enters the stream. Be sure to record the time of tracer addition.
3. At the downstream location, record conductivity every minute or two until the tracer pulse begins to arrive, and then increase recordings every 15 s as the conductivity increases rapidly. If using a data sonde or recording meter, set the instrument to record conductivity every 2 s. Collect data in this manner until the tracer completely passes the downstream location and conductivity returns to ambient conditions.

30.3.1.4 Field Procedure—Constant-Rate Release

1. Calculate the release (pumping) rate and solute concentration necessary to raise stream concentration measurably above background for a given flow rate. A target plateau increase of 10 μS/cm may be sufficient as stated above. Use the measured discharge to calculate the release rate (Q_R) as

$$Q_R = \frac{Q \times C_S}{C_I} \tag{30.14}$$

where Q = stream discharge, C_S = target stream concentration of added solute, and C_I = concentration of the solute in the release solution.
2. At the top of the reach, prepare the release solution at C_I by dissolving the preweighed salt in the appropriate volume of stream water, or use a preprepared solution. Use a sufficiently large volume so that the solution does not run out before plateau is reached at the downstream location. Release duration can be anywhere from 30 min to several hours depending on stream flow and reach length. Make sure to record mass of salt and volume of water used. Reserve a small aliquot (1 mL or less) of tracer solution to verify C_I at a later time.
3. Set up the pump as described by the manufacturer. Set the pump rate to Q_R as calculated above using a graduated cylinder and stopwatch. Make sure to keep a bucket under the release hose to avoid premature tracer addition to the stream. During the release, periodically check and record the release rate as the pump rate can drift over time, and empty the collected solute into the stream.
4. Begin the release by turning on the pump and recording the start time. The person at the downstream site should begin recording conductivity every 1–5 min until the tracer arrives and then every 15 s as conductivity increases rapidly.
5. At plateau, that is, when the conductivity is no longer changing, work in a downstream to upstream direction, and measure conductivity (or collect grab samples) in the stream at 10-m intervals. If you only have one conductivity meter, the break in the data at the downstream site won't be a problem. After taking the upstream measurements, return the meter to the downstream site. Then shut off the release. Record the total time of release (i.e., the duration of the solute addition).
6. Continue recording conductivity at the downstream site until conditions approach background or ambient levels. We have frequently found that conductivity readings never return to background levels, either because of real change in background concentration or because of instrument drift. To correct for either of these problems, it is useful to measure conductivity above the release site several times during the experiment.

30.3.1.5 After the Release

1. Summarize the physical parameters measured in the field [mean width, and depth at each cross section over the whole reach, mean velocity (optional), gradient (optional)].
2. Make a 1:10,000 dilution of the release solution, and measure the conductivity (or chloride concentration). Convert the measured conductivity values to chloride concentration using the known concentrations from the chloride standards.

3. Graph the conservative solute concentration versus time since tracer addition start at the downstream end of the reach—this plot is called a breakthrough curve.

30.3.1.6 Estimating Discharge

1. For a slug release, discharge at the downstream location can be estimated as (see Chapter 3):

$$Q = \frac{M}{\int_0^t C(t)} \tag{30.15}$$

where $M =$ mass of tracer added, $C =$ concentration, and $t =$ time. This equation makes two important assumptions: first, we assume that all of the added tracer was recovered at the downstream sampling location; second, we assume that the added solute was completely mixed across the stream channel where it passed the downstream sampling location.

2. For a constant-rate release, discharge can be calculated as:

$$Q = \frac{(C_R - C_b) \times Q_R}{C_p - C_b} \tag{30.16}$$

where $Q_R =$ release rate; $C_R =$ chloride (or conductivity) concentration of the release solution, $C_P =$ plateau chloride (or conductivity) concentration, and $C_b =$ background (i.e., ambient) chloride (or conductivity) concentration. As for Eq. (30.15) (above), this equation assumes perfect mixing and 100% tracer recovery. Use this equation to calculate discharge at each of the transects where plateau measurements were made. Make a graph of discharge versus distance to see if there is evidence of groundwater input. If there is evidence of a flow increase at a specific point, go back to the stream and see if you can identify landscape features associated with this subsurface input.

3. Compare your estimate of discharge with the direct measurements you made in the field.

30.3.1.7 Estimating Other Hydraulic Parameters

4. Nominal travel time (NTT) for a pulse release is the time required for tracer to reach peak concentration and is the time required for the tracer concentration to reach half of the plateau concentration in the case of constant-rate releases. Using your BTC, estimate NTT.

5. The median travel time (MTT) is another measure of hydraulic retention, defined as the time required for half of the tracer mass to pass out of the stream reach (Runkel, 2002). Tracer mass can be estimated by integration of the BTC:

$$M_{REC} = Q_D \int_0^t C_D(t) \tag{30.17}$$

where $M_{REC} =$ tracer mass recovered at the downstream site, $Q_D =$ measured discharge (this must be an independent measure and not estimated from the tracer test), $C_D =$ concentration at the downstream location, and $t =$ time. Such integration can be done with many graphics or spreadsheet programs, and MTT is the time for 50% of the mass to pass the downstream station.

6. Average solute velocity can be calculated by dividing the length of the reach by MTT. Compare this calculated value with the direct measurements of thalweg velocity made in the field.

7. Once tracer mass recovery has been estimated using Eq. (30.17) (above), it is useful to compare this mass to the mass that was added during the tracer test. Calculate percent recovery as:

$$\%rec = \frac{M_{REC}}{M} 100 \tag{30.18}$$

where $M_{REC} =$ mass recovered at the downstream sampling station and $M =$ mass added during the tracer test. Compare your recovered mass with the mass you added in this way. If less tracer is recovered than was added, this could represent tracer that entered a flow path that left the main channel and did not return—in other words, a loss of water. This sort of analysis can allow for estimation of water gains and losses along the study reach (Payn et al., 2009).

30.3.1.8 Transient Storage

A number of solute transport models can be used to simulate the BTC, which allow estimation of transient storage parameters (Eq. 30.2). One such model is a program called OTIS (One-dimensional Transport with Inflow and

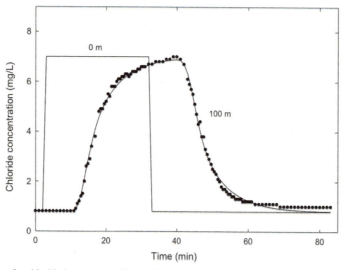

FIGURE 30.5 Breakthrough curve for chloride in a stream with considerable transient storage and no increase in flow over the reach. Square wave at 0 m shows tracer input. At 100 m, the dots are actual data and the solid line is a computer simulation of these data using a transient storage model.

Storage; Runkel, 1998) produced by the U.S. Geological Survey and free for download (https://water.usgs.gov/software/OTIS/). It is beyond the scope of this chapter to discuss model assumptions and limitations in simulating solute BTCs to estimate transient storage parameters. The reader is reminded that models that simulate multiple unknown parameters are subject to equifinality (multiple plausible parameter estimates).

Comparing MTT among different stream reaches can give an indication of the relative importance of transient storage. MTT should be longer if water and the conservative tracer it transports enter slower flow paths. The fraction of MTT due to transient storage, or F_{med}, can be calculated after simulating the BTC using a solute transport model with and without transient storage (Runkel, 2002).

The shape of the BTC also can give you some idea of the transient storage in the experimental reach. A reach with little transient storage will have a nearly rectangular BTC (Fig. 30.5, similar to 0 m). If transient storage is important, the leading edge to peak or plateau concentration will have a rounded shoulder, and the falling limb of the BTC will have a long tail (Fig. 30.5, similar to 100 m). In many streams with transient storage, the BTC tail is characterized by a power law. You can compare BTCs among streams by plotting the tail (concentration vs. time) on $\log_{10}$ scales and compare the slopes of the lines—a steeper slope would indicate transient storage as less important than lines with shallower slopes. It is important to note that such analysis assumes that the stream concentration returns to background. Such assumptions and other aspects of BTC tail analysis are given by Drummond et al. (2012).

30.3.2 Advanced Method: Reactive (Nonconservative) Solute Dynamics

Simultaneously with the conservative solute, a reactive solute may be released to determine nutrient uptake. Either a pulse or a constant rate approach can be used, as described previously (see also Chapter 31). Determine the needed level of nutrient addition with the constant-rate method (above). Make a stock solution of nutrient, calculate the necessary release solution concentration based on the release rate previously determined for chloride, and add the appropriate amount of stock nutrient solution to the release solution. As nutrient uptake is sensitive to concentration, care should be taken to not excessively elevate nutrient concentration. We have found that with good analytical chemistry, it is possible to raise concentration above ambient by about 20 µg/L and estimate uptake parameters. In the case of a pulse release, one could monitor the reactive tracer breakthrough at the lower-most site, collecting samples manually as conductivity increases, and then analyzed with a computer model to estimate uptake parameters (e.g., Runkel, 2007; Lin and Webster, 2012). Alternatively, multiple sampling stations can be monitored and data analyzed (this approach is presented in greater detail in Chapter 31) as described for constant-rate data. Make sure to collect several samples of ambient concentration before the pulse arrives. It is important to note that pulse releases can introduce nutrient concentrations that approach or exceed saturation, especially close to the point of release, in which case sensitivity to model assumptions (i.e., first-order vs. Monod kinetics) is important (see, for example, Covino et al., 2010; Lin and Webster, 2012).

For constant-rate releases, as with the conductivity measurements, samples for nutrient concentration should be taken from the stream before the release and at the plateau of the release. Collect at least three replicate samples at each site. These samples can be taken in any type of clean container. Many researchers use acid-washed high-density polyethylene (HDPE) containers (e.g., Nalgene), whereas others use disposable centrifuge tubes. The samples should be filtered either as they are collected or as soon as possible once the samples are taken to the lab. Methods of sample preservation vary depending on the nutrient you are using, and you should consult a manual such as *Standard Methods for the Examination of Water and Wastewater* (Clesceri et al., 1998). In most cases it is best to keep samples on ice or refrigerated in the dark and analyze samples within 24 h of collection.

Graph normalized nutrient concentration versus distance, and calculate the longitudinal uptake rate (k_W) and uptake length (S_W) (Fig. 30.6). Nutrient concentrations of the samples collected at plateau must be corrected for background levels (C_b) to get the added nutrient level. Then calculate normalized added nutrient concentrations (C_N) by dividing the nutrient concentrations at a specific site (C_x) by the conservative solute (C_c, corrected for background) concentrations at the site:

$$C_N = \frac{(C_x - C_b)}{C_c} \tag{30.19}$$

By doing this you are essentially correcting for decline in nutrient concentration that may result from lateral inflow over the reach. For steady conditions (e.g., at plateau) the solution of Eq. (30.19) is a negative exponential:

$$C_N = C_{N0}e^{-k_W x} \tag{30.20}$$

where C_{N0} = added nutrient concentration at the release site, and x = distance downstream from the release site. Taking the logarithm of both sides of Eq. (30.20) gives:

$$\ln(C_N) = \ln(C_{N0}) - k_W x \tag{30.21}$$

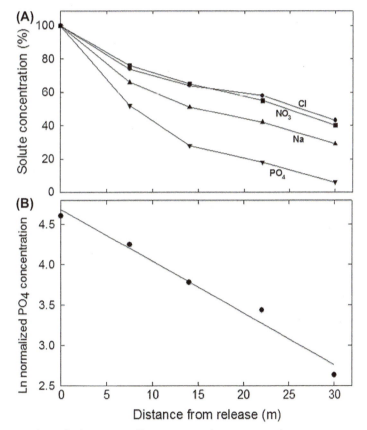

FIGURE 30.6 (A) Plateau concentrations of solutes versus distance expressed as a percent of upstream concentrations. In this stream, NO_3 was relatively abundant and acted like a conservative solute. PO_4 was rapidly removed from the stream water column. (B) Semilog plot of normalized PO_4 concentration versus distance. The slope of this line is the PO_4 longitudinal uptake rate (k_W).

This is the equation for a straight line with intercept of $\ln(C_{N0})$ and a slope of $-k_W$. So, if you use your data to determine a regression of $\ln(C_N)$ versus x, the k_W will be an estimate of the longitudinal uptake rate, and uptake length (S_W) is the inverse of this (Stream Solute Workshop, 1990). Uptake (U) and uptake velocity (v_f) can then be calculated using the metric triad (Fig. 30.2).

30.4 QUESTIONS

1. What are the causes of hydraulic retention (transient storage) in a stream? That is, what causes temporary retention of conservative solutes?
2. Which stream features affect solute retention?
3. If you recovered less mass of conservative tracer than you added, how might this influence your interpretation of the discharge estimate you calculated from your tracer test?
4. Which factors determine the usefulness of various conservative and reactive solute tracers?
5. How does stream size affect hydraulic parameters? Nutrient uptake?
6. What is the significance of wood in streams in terms of solute dynamics? How do you think historical wood removal from streams and rivers might affect solute dynamics?
7. Consider how other human modifications of streams and stream channels may affect solute dynamics. Think about such changes as nutrient enrichment from point and nonpoint sources, dams, channelization, and modification of riparian vegetation.
8. If you conducted conservative and reactive solute tracer tests at multiple plateau concentrations, what was the relationship between areal uptake and concentration?

30.5 MATERIALS AND SUPPLIES

Laboratory materials
    Conservative solute salts
    Reactive solute salts (nutrients)
    Carboys for stock solutions
    Containers for standard solutions (HDPE)
    Deionized or distilled water
    Volumetric flasks (acid washed)
Laboratory equipment
    Analytical instruments for measuring solute concentrations
    Computer with spreadsheet or graphing software
    Analytical balance (± 0.001 g or better)
    Filtering apparatus and glass fiber filters
Field materials
    Water-resistant paper or notebook, pencils
    Flagging tape
    Meter stick
    Tape measure (100 m)
    Stopwatches
    Buckets or carboy for solute mixing and delivery
    Graduated cylinder (1000 mL for solute mixing and 10−20 mL for pump calibration)
    Metering pump with tubing and charged batteries
    Sample bottles
    Conductivity meter (with temperature)
    Chloride standard solutions
    Velocity meter and top-setting wading rod or other means to measure stream discharge

ACKNOWLEDGMENTS

We thank the numerous collaborators who have worked with us in field studies of solute dynamics to conduct tracer tests around the globe. The accompanying spreadsheet was developed based on past work with Drs. Tim Covino, Mike Gooseff, and Brian McGlynn. Dr. Maury Valett and Fr. Terry Ehrman made significant contributions to this chapter in earlier editions of *Methods in Stream Ecology*.

REFERENCES

Alvarez, M., Proia, L., Ruggiero, A., Sabater, F., Butturini, A., 2010. A comparison between pulse and constant rate additions as methods for the estimation of nutrient efficiency in streams. Journal of Hydrology 388, 273–279.

Arp, C.D., Baker, M.A., 2007. Discontinuities in stream nutrient uptake below lakes in mountain drainage networks. Limnology and Oceanography 52, 1978–1990.

Bencala, K.E., 1985. Performance of Sodium as a Transport Tracer: Experimental and Simulation Analysis. Water Supply Paper 2270:83–89. U.S. Geological Survey, Reston, VA.

Bencala, K.E., Kennedy, V.C., Zellweger, G.W., Jackman, A.P., Avanzino, R.J., 1984. Interactions of solutes and streambed sediment 1. An experimental analysis of cation and anion transport in a mountain stream. Water Resources Research 20, 1797–1803.

Bencala, K.E., Kimball, B.A., McKnight, D.M., 1991. Use of variation in solute concentration to identify interactions of the substream zone with instream transport. In: Mallard, G.E., Aronson, D.A. (Eds.), U.S. Geological Survey Toxic Substances Hydrology Program, Water-Resources Investigations Report 91-4034. U.S. Geological Survey, Reston, VA, pp. 377–379.

Bencala, K.E., Walters, R.A., 1983. Simulation of solute transport in a mountain pool-and-riffle stream: a transient storage model. Water Resources Research 19, 718–724.

Bernhardt, E.S., Likens, G.E., Buso, D.C., Driscoll, C.T., 2003. In-stream uptake dampens effects of major forest disturbance on watershed nitrogen export. Proceedings of the National Academy of Sciences of the United States of America 100, 10304–10308.

Bernot, M.J., Dodds, W.K., 2005. Nitrogen retention, removal, and saturation in lotic ecosystems. Ecosystems 8, 442–453.

Bernot, M.J., Tank, J.L., Royer, T.V., David, M.B., 2006. Nutrient uptake in streams draining agricultural catchments of the midwestern United States. Freshwater Biology 51, 499–509.

Claessens, L., Tague, C.L., 2009. Transport-based method for estimating in-stream nitrogen uptake at ambient concentration from nutrient addition experiments. Limnology and Oceanography: Methods 7, 811–822.

Clesceri, L.S., Greenberg, A.E., Eaton, A.D., 1998. Standard Methods for the Examination of Water and Wastewater, twentieth ed. American Public Health Association, Washington, DC.

Covino, T.P., McGlynn, B.L., McNamara, R.A., 2010. Tracer additions for spiraling curve characterization: quantifying stream nutrient uptake kinetics from ambient to saturation. Limnology and Oceanography: Methods 8, 484–498.

Davis, J.C., Minshall, G.W., 1999. Nitrogen and phosphorus uptake in two Idaho (USA) headwater wilderness streams. Oecologia 119, 247–255.

Demars, B.O.L., 2008. Whole-stream phosphorus cycling: testing methods to assess the effect of saturation of sorption capacity on nutrient uptake length measurements. Water Research 42, 2507–2516.

Dodds, W.K., Lopez, A.J., Bowden, R.D., Gregory, S.V., Grimm, N.B., Hamilton, S.K., Hershey, A.E., Marti, E., McDowell, W.H., Meyer, J.L., Morrall, D.D., Mulholland, P.J., Peterson, B.J., Tank, J.L., Valett, H.M., Webster, J.R., Wollheim, W.M., 2002. N uptake as a function of concentration in streams. Journal of the North American Benthological Society 21, 206–220.

Dodds, W.K., Beaulieu, J.J., Eichmiller, J.J., Fischer, J.R., Franssen, N.R., Gudder, D.A., Makinster, A.S., McCarthy, M.J., Murdock, J.N., O'Brien, J.M., Tank, J.L., Sheibley, R.W., 2008. Nitrogen cycling and metabolism in the thalweg of a prairie river. Journal of Geophysical Research 113, G04029.

Drummond, J.D., Covino, T.P., Aubeneau, A.F., Leong, D., Patil, S., Schumer, R., Packman, A.I., 2012. Effects of solute breakthrough curve tail truncation on residence time estimates: a synthesis of solute tracer injection studies. Journal of Geophysical Research 117, G00N08.

Earl, S.R., Valett, H.M., Webster, J.R., 2006. Nitrogen saturation in stream ecosystems. Ecology 87, 3140–3151.

Earl, S.R., Valett, H.M., Webster, J.R., 2007. Nitrogen spiraling in streams: comparisons between stable isotope tracer and nutrient addition experiments. Limnology and Oceanography 52, 1718–1723.

Elwood, J.W., Newbold, J.D., Trimble, A.F., Stark, R.W., 1981. The limiting role of phosphorus in a woodland stream ecosystem — effects of P enrichment on leaf decomposition and primary producers. Ecology 62, 146–158.

Elwood, J.W., Newbold, J.D., O'Neill, R.V., VanWinkle, W., 1983. Resource spiraling: an operational paradigm for analyzing lotic ecosystems. In: Fontaine III, T.D., Bartell, S.M. (Eds.), Dynamics of Lotic Ecosystems. Ann Arbor Science, Ann Arbor, MI, pp. 3–27.

Fisher, S.G., Likens, G.E., 1973. Energy flow in Bear Brook, New Hampshire: an integrative approach to stream ecosystem metabolism. Ecological Monographs 43, 421–439.

Flury, M., Papritz, A., 1993. Bromide in the natural environment: occurrence and toxicity. Journal of Environmental Quality 22, 747–758.

Gibson, C.A., Meyer, J.L., 2007. Nutrient uptake in a large urban river. Journal of the American Water Resources Association 43, 576–587.

Hall, R.O., Tank, J.L., 2003. Ecosystem metabolism controls nitrogen uptake in streams in Grand Teton National Park, Wyoming. Limnology and Oceanography 48, 1120–1128.

Hall, R.O., Baker, M.A., Rosi-Marshall, E.J., Tank, J.L., Newbold, J.D., 2013. Solute specific scaling of inorganic nitrogen and phosphorus uptake in streams. Biogeosciences 10, 7323–7331.

Harvey, J.W., Wagner, B.J., 2000. Quantifying hydrologic interactions between streams and their subsurface hyporheic zones. In: Jones, J.B., Mulholland, P.J. (Eds.), Streams and Ground Waters. Academic Press, San Diego, CA, pp. 3–44.

Harvey, J.W., Wagner, B.J., Bencala, K.E., 1996. Evaluating the reliability of the stream tracer approach to characterize stream-subsurface water exchange. Water Resources Research 32, 2441–2451.

Hensley, R.T., Cohen, M.J., Korhnak, L.V., 2014. Inferring nitrogen removal in large rivers from high-resolution longitudinal profiling. Limnology and Oceanography 59, 1152–1170.

King, S.A., Heffernan, J.B., Cohen, M.J., 2014. Nutrient flux, uptake, and autotrophic limitation in streams and rivers. Freshwater Science 33, 85–98.

Lin, L., Webster, J.R., 2012. Sensitivity analysis of a pulse nutrient addition technique for estimating nutrient uptake in large streams. Limnology and Oceanography: Methods 10, 718–727.

Minshall, G.W., Thomas, S.A., Newbold, J.D., Monaghan, M.T., Cushing, C.E., 2000. Physical factors influencing fine organic particle transport and deposition in streams. Journal of the North American Benthological Society 19, 1–16.

Mulholland, P.J., Elwood, J.W., Newbold, J.D., Ferren, L.A., 1985. Effect of a leaf-shredding invertebrate on organic matter dynamics and phosphorus spiralling in heterotrophic laboratory streams. Oecologia 66, 199–206.

Mulholland, P.J., Steinman, A.D., Marzolf, E.R., Hart, D.R., DeAngelis, D.L., 1994. Effect of periphyton biomass on hydraulic characteristics and nutrient cycling in streams. Oecologia 98, 40–47.

Mulholland, P.J., Tank, J.L., Webster, J.R., Bowden, W.B., Dodds, W.K., Gregory, S.V., Grimm, N.B., Hamilton, S.K., Johnson, S.L., Marti, E., McDowell, W.H., Merriam, J.L., Meyer, J.L., Peterson, B.J., Valett, H.M., Wollheim, W.M., 2002. Can uptake length in streams be determined by nutrient enrichment experiments? Results from an interbiome comparison study. Journal of the North American Benthological Society 21, 544–560.

Mulholland, P.J., Helton, A.M., Poole, G.C., Hall, R.O., Hamilton, S.K., Peterson, B.J., Tank, J.L., Ashkenas, L.R., Cooper, L.W., Dahm, C.N., Dodds, W.K., Findlay, S.E.G., Gregory, S.V., Grimm, N.B., Johnson, S.L., McDowell, W.K., Meyer, J.L., Valett, H.M., Webster, J.R., Arango, C.P., Beaulieu, J.J., Bernot, M.J., Burgin, A.J., Crenshaw, C.L., Johnson, L.T., Niederlehner, B.R., O'Brien, J.M., Potter, J.D., Scheibley, R.W., Sobota, D.J., Thomas, S.M., 2008. Stream denitrification across biomes and its response to anthropogenic nitrate loading. Nature 452, 202–205.

Newbold, J.D., Elwood, J.W., O'Neill, R.V., VanWinkle, W., 1981. Measuring nutrient spiralling in streams. Canadian Journal of Fisheries and Aquatic Sciences 38, 860–863.

Newbold, J.D., 1992. Cycles and spirals of nutrients. In: Calow, P., Petts, G.E. (Eds.), The Rivers Handbook. Blackwell Scientific, Oxford, UK, pp. 370–408.

Newbold, J.D., Elwood, J.W., O'Neill, R.V., Sheldon, A.L., 1983. Phosphorus dynamics in a woodland stream ecosystem: a study of nutrient spiraling. Ecology 64, 1249–1265.

Newbold, J.D., Bott, T.L., Kaplan, L.A., Dow, C.L., Jackson, J.K., Aufdenkampe, A.K., Martin, L.A., Van Horn, D.J., de Long, A.A., 2006. Uptake of nutrients and organic C in streams in New York City drinking-water-supply watersheds. Journal of the North American Benthological Society 25, 998–1017.

O'Brien, J.M., Dodds, W.K., Wilson, K.C., Murdock, J.N., Eichmiller, J., 2007. The saturation of N cycling in Central Plains streams: [15]N experiments across a broad gradient of nitrate concentrations. Biogeochemistry 84, 31–49.

Payn, R.P., Webster, J.R., Mulholland, P.J., Valett, H.M., Dodds, W.K., 2005. Estimation of stream nutrient uptake from nutrient addition experiments. Limnology and Oceanography: Methods 3, 174–182.

Payn, R.A., Gooseff, M.N., Benson, D.A., Cirpka, O.A., Zarnetske, J.P., Bowden, W.B., McNamara, J.P., Bradford, J.H., 2008. Comparison of instantaneous and constant-rate stream tracer experiments through non-parametric analysis of residence time distributions. Water Resources Research 44, W06404.

Payn, R.A., Gooseff, M.N., McGlynn, B.L., Bencala, K.E., Wondzell, S.M., 2009. Channel water balance and exchange with subsurface flow along a mountain headwater streams in Montana, United States. Water Resources Research 45, W11427.

Peterson, B.J., Wollheim, W., Mulholland, P.J., Webster, J.R., Meyer, J.L., Tank, J.L., Martí, E., Bowden, W.B., Valett, H.M., Hershey, A.E., McDowell, W.H., Dodds, W.K., Hamilton, S.K., Gregory, S.V., Morrall, D.D., 2001. Control of nitrogen export from watersheds by headwater streams. Science 292, 86–90.

Reisinger, A.J., Tank, J.L., Rosi-Marshall, E.J., Hall, R.O., Baker, M.A., 2015. The varying role of water column nutrient removal along river continua in contrasting landscapes. Biogeochemistry 125, 115–131.

Runkel, R.L., 1998. One-dimensional Transport with Inflow and Storage (OTIS): A Solute Transport Model for Streams and Rivers. Water–Resources Investigations Report 98-4018. U.S. Geological Survey, Denver, CO.

Runkel, R.L., 2002. A new metric for determining the importance of transient storage. Journal of the North American Benthological Society 21, 529–543.

Runkel, R.L., 2007. Toward a transport-based analysis of nutrient spiraling and uptake in streams. Limnology and Oceanography: Methods 5, 50–62.

Runkel, R.L., 2015. On the use of rhodamine WT for the characterization of stream hydrodynamics and transient storage. Water Resources Research 51, 6125–6142.

Stewart, A.J., Kzsos, L.A., 1996. Caution on using lithium (Li) as a conservative tracer in hydrological studies. Limnology and Oceanography 41, 190–191.

Stream Solute Workshop, 1990. Concepts and methods for assessing solute dynamics in stream ecosystems. Journal of the North American Benthological Society 9, 95–119.

Tank, J.L., Rosi-Marshall, E.J., Baker, M.A., Hall, R.O., 2008. Are rivers just big streams? Using a pulse method to measure nitrogen demand in a large river. Ecology 89, 2935–2945.

Triska, F.J., Kennedy, V.C., Avanzio, R.J., Zellweger, G.W., Bencala, K.E., 1989. Retention and transport of nutrients in a third-order stream in northwestern California: hyporheic processes. Ecology 70, 1893–1905.

Trentman, M.T., Dodds, W.K., Fencl, J.S., Gerber, K., Guarneri, J., Hitchman, S.M., Peterson, Z., Ruegg, J., 2015. Quantifying ambient nitrogen uptake and functional relationships of uptake versus concentration in streams: a comparison of stable isotope, pulse, and plateau approaches. Biogeochemistry 125, 65–69.

Ward, A.S., 2015. The evolution and state of interdisciplinary hyporheic research. WIREs Water 3, 83–103.

Webb, B.W., Walling, D.E., 1992. Water quality. Chemical characteristics. In: Calow, P., Petts, G.E. (Eds.), The Rivers Handbook. Blackwell Scientific, Oxford, UK, pp. 73—100.

Webster, J.R., 1975. Analysis of Potassium and Calcium Dynamics in Stream Ecosystems of Three Southern Appalachian Watersheds of Contrasting Vegetation (Ph.D. dissertation). University of Georgia, Athens, GA.

Webster, J.R., Patten, B.C., 1979. Effects of watershed perturbation on stream potassium and calcium dynamics. Ecological Monographs 49, 51—72.

Webster, J.R., Ehrman, T.P., 1996. Solute dynamics. In: Hauer, F.R., Lamberti, G.A. (Eds.), Methods in Stream Ecology. Academic Press, San Diego, CA, pp. 145—160.

Webster, J.R., Valett, H.M., 2007. Solute dynamics. In: Hauer, F.R., Lamberti, G.A. (Eds.), Methods in Stream Ecology, second ed. Academic Press, San Diego, CA, pp. 169—185.

Chapter 31

Nutrient Limitation and Uptake

Jennifer L. Tank[1], Alexander J. Reisinger[2] and Emma J. Rosi[2]

[1]Department of Biological Sciences, University of Notre Dame; [2]Cary Institute of Ecosystem Studies

31.1 INTRODUCTION

All organisms need nutrients, such as nitrogen (N) and phosphorus (P), for growth and reproduction. Nitrogen circulates through the atmosphere and landscape in a complex cycle composed of biotic and abiotic transformations. The most abundant form, dinitrogen gas (N_2), composes nearly 78% of the atmosphere (Lutgens and Tarbuck, 1992; Schlesinger and Bernhardt, 2013). Prior to human alteration of the global N cycle, the two mechanisms by which N_2 entered the bioavailable N pool were via biologically mediated N-fixation and, secondarily, lightning. Because only a few specialized N-fixing organisms can directly use this gaseous pool, frequently N is a limiting nutrient in ecosystems (Vitousek and Howarth, 1991). In contrast to N, P does not have a major gaseous form, and natural sources of bioavailable P include decomposition of dead biomass and mineralization of mineral-bound P (Smil, 2000). Due to its relative scarcity, as well as the spatial heterogeneity of geological deposits, P is also frequently a limiting nutrient in ecosystems (Elser et al., 2007). While N or P may limit freshwater ecosystems separately, colimitation by N and P is also common (Elser et al., 2007; Harpole et al., 2011).

Dissolved N and P concentrations in stream ecosystems are determined in part by watershed geology and vegetation (e.g., Gregory et al., 1991; Dodds, 1997). Although stream flow continuously delivers dissolved inorganic nutrients to biofilms colonizing submerged surfaces, many studies have shown that N and P are often limiting to the algae, bacteria, and fungi that make up these biofilms (Pringle et al., 1986; Tank and Webster, 1998; Francoeur et al., 1999; Wold and Hershey, 1999; Tank and Dodds, 2003; Johnson et al., 2009; Reisinger et al., 2016). In addition to an adequate supply of N and P, organisms may be limited by the stoichiometric balance of available nutrients (e.g., N:P ratio). To address the almost ubiquitous problems associated with increased nutrient loading in streams around the world, it is critical to understand the factors controlling nutrient limitation and uptake in stream ecosystems.

31.1.1 Elevated Nutrient Loading to Streams

Anthropogenic activities, including burning fossil fuels, planting N-fixing crops, fertilizer production, and wastewater disposal, (David and Gentry, 2000; Schlesinger and Bernhardt, 2013) have more than doubled N inputs into the global cycle (Vitousek et al., 1997; Galloway et al., 2004; Gruber and Galloway, 2008), and bioavailable P consumption has increased by more than $10\times$ since the industrial revolution (Smil, 2000). Due to the relative scarcity of economically available P, world phosphate rock reserves have been estimated to be depleted sometime between 2050 and 2100 (Cordell and White, 2011 and references therein). While not all pristine ecosystems have low nutrient availability (e.g., Subalusky et al., 2015), in general, excess N and P availability has converted once nutrient-limited systems to ones exhibiting nutrient saturation (Fenn et al., 1998; Duff and Triska, 2000; Bernot and Dodds, 2005).

Stream ecosystems are particularly threatened by anthropogenic increases of inorganic P and N availability, which often enters as ammonium (NH_4^+) in wastewater effluent (Duff and Triska, 2000) and as nitrate (NO_3^-) via agricultural runoff (Howarth et al., 1996; David et al., 1997; David and Gentry, 2000; Kemp and Dodds, 2001). Increased N and P in stream ecosystems results in problems including eutrophication, hypoxia, and loss of species diversity, issues that are common worldwide (Vitousek et al., 1997; Carpenter et al., 1998; Rabalais et al., 2002; Diaz and Rosenberg, 2008). In the human water supply, excess NO_3^- can cause methemoglobinemia in infants (Fewtrell, 2004) and increased occurrence of non-Hodgkin's lymphoma (Ward et al., 1996). These problems represent losses of ecosystem and human health as well as

Methods in Stream Ecology. http://dx.doi.org/10.1016/B978-0-12-813047-6.00009-7

losses of economic goods and services (Dodds et al., 2009). The ultimate fate of anthropogenic NH_4^+, NO_3^-, and dissolved P varies with differences in biotic demand (Duff and Triska, 2000; Mulholland et al., 2008), abiotic sorption characteristics, substrata type (Kemp and Dodds, 2002), and the physical retention characteristics of stream channels (Bernot and Dodds, 2005). Continued research is therefore needed to identify how nutrients are assimilated, retained, and transformed within stream ecosystems.

31.1.2 Capacity for Nutrient Retention in the Landscape

Nutrient cycling in headwater streams has received significant attention due to the potential for small streams to process nutrients, especially N, and thereby influence downstream export (Alexander et al., 2000; Sabater et al., 2000; Peterson et al., 2001; Mulholland et al., 2008,2009; Hall et al., 2009). For example, mechanisms of dissolved N uptake and removal from the stream water column include assimilatory uptake, denitrification, adsorption, burial, and volatilization (Peterson et al., 2001; Bernot and Dodds, 2005; Mulholland et al., 2008, 2009). Other forms of dissimilatory N-transformations occur in freshwaters, including dissimilatory nitrate reduction to ammonium (Burgin and Hamilton, 2008) and anaerobic ammonium oxidation (Lansdown et al., 2016), but the ubiquity and importance of these novel pathways remain unclear.

The degree of nutrient limitation influences the nutrient uptake capacity of streams. In a cross-site comparison among streams in multiple biomes, Peterson et al. (2001) found that headwater streams are generally effective at removing and transforming dissolved inorganic nutrients (in this case N) from the water column because of their high biological activity combined with increased sediment/water contact time. As nutrient availability increases, biological demand and nutrient limitation decreases (Reisinger et al., 2016). For example, in another cross-site comparison of 72 streams representing reference, agricultural, and urban lands, Mulholland et al. (2008) found that both denitrification and whole-stream demand for NO_3^- decreased with increasing NO_3^- availability, while permanent NO_3^- removal via denitrification accounted for $\sim 10-20\%$ of reach-scale NO_3^- uptake. Thus, biotic assimilation represents the majority of NO_3^- removal from the water column (Hall et al., 2009), which enters autotrophic and heterotrophic pathways and ultimately moves via trophic transfer through the food web. Because uptake of dissolved inorganic nutrients is especially high in shallow streams, these systems may play a key role in reducing nutrient export to downstream ecosystems (Alexander et al., 2000, 2008), yet more recent work suggests the role of rivers in reducing nutrient export may be more important than previously thought and should not be overlooked (Tank et al., 2008; Hall et al., 2013). Stream biota can respond to increased nutrient supply with increased uptake and growth, but this response is limited by the processing capacity of the biota and physical factors such as temperature and light availability. Eventually, with increased nutrient loading, nutrient uptake in streams will decrease in efficiency, eventually becoming saturated with excess dissolved nutrients being exported downstream (Bernot and Dodds, 2005; O'Brien et al., 2007; Helton et al., 2011).

Although streams are capable of processing and retaining a large portion of the nutrients entering from the watershed, downstream nutrient export is inevitable, particularly in streams with high background concentrations (Mulholland et al., 2008). Despite potentially high nutrient uptake rates, the ultimate fate of nutrients taken up via biotic assimilation deserves further study as much of the assimilated nutrients are exported downstream as organic N or P, or mineralized back to dissolved forms (Hall et al., 2009). Numerous ecological and economic problems are associated with excess nutrient loading and subsequent eutrophication in downstream ecosystems, including the recurring hypoxic zone in the Gulf of Mexico (Rabalais et al., 2002) and harmful algal blooms in the Western Lake Erie Basin (Scavia et al., 2014), but the effects of nutrient pollution causing eutrophication are apparent at river outlets worldwide (Diaz and Rosenberg, 2008).

31.1.3 Overview of Chapter

Ecologists have described nutrient dynamics in streams using the concept of uptake length, which is the average distance traveled by a dissolved nutrient before biotic uptake (Webster and Patten, 1979; Newbold et al., 1981; Stream Solute Workshop, 1990). By quantifying nutrient uptake lengths, one can estimate reach-scale demand for dissolved nutrients including nutrient removal rates from the water column per area of stream bed (Newbold et al., 1981; Davis and Minshall, 1999). In combination with nutrient uptake length, quantifying N and P limitation of stream biofilms allows for the assessment of demand and uptake efficiency of nutrients relative to nutrient supply, with implications for downstream export.

In this chapter, we describe a simple but powerful approach for assessing whether stream biofilms are nutrient limited using nutrient diffusing substrata (NDS), followed by two advanced methods for quantifying whole-stream nutrient uptake. The first whole-stream approach we describe is a short-term nutrient release, where nutrient concentrations are slightly elevated above background in the water column, and the decline in concentration, relative to a conservative tracer, is measured downstream. Whole-stream nutrient uptake reflects the apparent downstream decrease in dissolved inorganic nutrients in the water column and represents the combination of nutrient uptake from stream water and remineralization back into stream water from biota. A more in-depth discussion of short-term solute dynamics can be found in Chapter 30 in this volume, which will complement methods described here.

The second whole-stream method for nutrient uptake we describe is for dissolved N in particular, using a short-term stable isotope (^{15}N) tracer release, which has the benefit of maintaining water column N concentrations at ambient levels. This tracer method allows one to use the ^{15}N labeling of a food web compartment, such as algae, to quantify whole-stream N uptake, while avoiding the challenges of quantifying dissolved ^{15}N in the water column, either as NH_4^+, NO_3^-, or dissolved organic N (Peterson et al., 2001; Mulholland et al., 2008, 2009 and references therein). Using the ^{15}N tracer approach, whole-stream uptake is represented by ^{15}N assimilated into biotic compartments, reflected in the changing δ^{15}N values of their tissue. Unfortunately, there is no stable isotope of P, necessitating nonisotopic methods to estimate P dynamics in streams. Both short-term nutrient additions as well as stable isotope tracer experiments have been used successfully in a wide variety of streams, although empirical data on larger streams and rivers are lacking (Tank et al., 2008; Hall et al., 2013). Both Tank et al. (2008) and Hall et al. (2013) reviewed previous studies of nutrient uptake and found that uptake length increases with system size, but solute-specific differences exist in the relationship between uptake metrics and discharge. All three methods described in this chapter are complementary and, in combination, can provide an assessment of nutrient dynamics in one stream through time or provide a comparison among streams. The overall objective of this chapter is to present methods to quantify nutrient limitation and uptake in streams to gain a better understanding of the potential for streams to retain nutrients in the landscape.

31.2 GENERAL DESIGN

31.2.1 Index of Limitation: Nutrient Diffusing Substrata

NDS provide a means for measuring whether the growth and/or activity of stream biofilms (autotrophic algae and heterotrophic bacteria and fungi) are nutrient limited. The Basic Method described below represents a modification from previously published work (Fairchild et al., 1985; Winterbourn, 1990; Corkum, 1996; Tank and Webster, 1998; Tank and Dodds, 2003; Johnson et al., 2009). NDS are constructed using small plastic cups filled with nutrient-amended agar and topped with either an inorganic (i.e., fritted glass disc) or organic (i.e., cellulose sponge disc) surface that selects for mainly autotrophic or heterotrophic colonization, respectively (Johnson et al., 2009). Cups are then incubated in the stream for up to 21 days given that the rate of nutrient diffusion from the plastic cups, filled with agar and nutrient, is nearly constant through this time. Once surfaces have been colonized, multiple metrics can be quantified to assess the structure and function of the biofilm community. For functional metrics, metabolism incubations are performed on NDS discs to quantify rates of gross primary production (GPP) and ecosystem respiration (ER) of biofilms, either in the field or in the laboratory. After quantifying biofilm metabolism, measurements of chlorophyll *a* standing stock and ash-free dry mass (AFDM) can be used to estimate the autotrophic and total biomass of the biofilm as a structural metric. In the method below, we recommend constructing five replicates of each substrate—nutrient treatment combination as well as five replicate control NDS (no nutrient added to agar). Variability between NDS within a treatment can sometimes be high, and five replicates allow for increased statistical power for determining treatment differences.

A two-factor analysis of variance (ANOVA), with presence of N and/or P as the main factors, is used to test whether biofilm structure and/or function are significantly influenced by nutrient enrichment (Dube et al., 1997; Tank and Dodds, 2003). Possible outcomes from the ANOVA on the bioassays are summarized in Table 31.1. Single nutrient limitation is indicated when just one of the nutrients (N or P) elicits a positive response, and the interaction term in the ANOVA is not significant. If neither N nor P alone significantly increases algal biomass or metabolism ($p > .05$), but N and P added together (N + P) do (i.e., the interaction term in the ANOVA is significant; $p < .05$), then the algal biofilm is considered to be colimited by both N and P. Similarly, there also could be colimitation by both N and P if, when added separately, they each stimulate algal biomass relative to controls, but the N and P responses are not different from each other. Secondary limitation is indicated if N or P alone significantly increases algal biomass, both N and P added together result in an even greater increase in biomass, and the interaction term for the ANOVA is significant.

In addition to characterizing the nutrient limitation status of a stream using two-way ANOVA, the magnitude of limitation can be assessed by calculating the nutrient response ratio (NRR) for each treatment disc (Johnson et al., 2009; Reisinger et al., 2016). For example, for GPP, the NRR is calculated as follows:

$$NRR_{j,i} = GPP_{j,i}/average(GPP_{cont}) \tag{31.1}$$

TABLE 31.1 Interpretation of nutrient diffusing substrata responses to N and P addition.

Interpretation[a]	N Effect	P Effect	N × P Interaction
N limited	◆		
P limited		◆	
N and P colimited			◆
	◆	◆	
	◆	◆	◆
1°N limited, 2°P limited	◆		◆
1°P limited, 2°N limited		◆	◆
Not limited by N or P			

[a]*Diamond in N or P effect indicates a significant N or P limitation in the two-factor analysis of variance ($p < .05$) and a diamond in the Interaction $N \times P$ indicates a significant interaction between the two treatments.*
From Tank and Dodds (2003).

where $NRR_{j,i}$ = nutrient response ratio of the jth disc in the ith treatment, $GPP_{j,i}$ = gross primary production of the jth disc in the ith treatment, and average(GPP_{cont}) denotes that the average GPP for all control discs is included in the denominator. Values >1 denote limitation by the specific treatment, whereas values <1 denote inhibition by the specific treatment.

Potential problems associated with the NDS method include loss of substrata due to floods or vandalism. We suggest using long, heavy stakes to secure the NDS bars to the stream bottom. Alternatively, two long metal fence posts can be placed perpendicular to flow in a stream (as in rails of a ladder), and then NDS can be cable tied to the posts parallel to the direction of flow (as the rungs of the ladder). If the ladder deployment is used, metal fence posts must then be secured to the streambed or bank in some way. Placing stream rocks on the rails of the ladder in combination with tethering to a nearby tree is an effective method (T.V. Royer, personal communication). If destruction of NDS by animals becomes a problem, cover NDS with a mesh cage for protection. In general, the NDS method provides a simple, cost-effective way to measure the nutrient limitation status of autotrophic and heterotrophic biofilms and is generally highly successful. However, NDS only assess nutrient limitation or use by a singular stream compartment (i.e., algae and microbes colonizing inorganic substrata), whereas whole-stream nutrient uptake can be estimated using short-term nutrient enrichments (Advanced Method 1) or short-term isotopic tracer additions (Advanced Method 2).

An additional problem frequently encountered with NDS deployment is the apparent biofilm inhibition in response to nutrient amendments (Francoeur, 2001; Tank and Dodds, 2003; Bernhardt and Likens, 2004; Hoellein et al., 2010; Reisinger et al., 2016). Multiple potential mechanisms for this apparent inhibition have been suggested in the literature, including the production of hydrogen peroxide during agar preparation (Tanaka et al., 2014), shifts in microbial community composition (Bechtold et al., 2012), excess nutrients leading to toxicity (Fairchild et al., 1985), inhibition of growth by the cation associated with the nutrient amendment (Lehman, 1976; Fairchild et al., 1985), and preferential grazing of biofilms on nutrient-amended NDS (Bernhardt and Likens, 2004; Hood et al., 2014). It is unclear which of these mechanisms are responsible for inhibition, and it is likely that multiple mechanisms may lead to inhibition under certain environmental conditions. An additional modification of this approach is to amend NDS with contaminants of emerging concern, such as pharmaceuticals and personal care products (Rosi-Marshall et al., 2013), or heavy metals (Costello et al., 2016). If this approach is used, the chemical structure, in particular the solubility of the compound, should be considered and diffusion kinetics, such as how much compound diffuses out of the agar, should be examined (Costello et al., 2016).

31.2.2 Short-Term Nutrient Addition

Uptake of nutrients can be measured using the short-term addition of a concentrated nutrient solution (with a conservative tracer) dripped continuously into the stream to elevate concentrations of both the nutrient of interest and the conservative tracer. When concentrations reach plateau (i.e., when in-stream concentrations reach a stable maximum), water samples are collected at stations downstream of the injection point, and nutrient uptake over the study reach can be calculated. To calculate a nutrient uptake length (S_w), an exponential decay model is used:

$$\ln N_x = \ln N_0 - kx \tag{31.2}$$

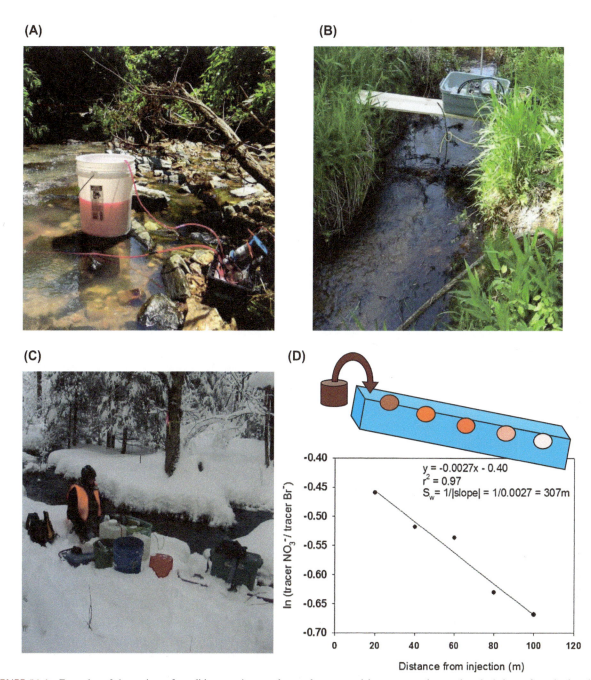

FIGURE 31.1 Examples of the variety of conditions nutrient uptake can be measured in streams and example calculations of uptake length. (A) Measuring NO_3^- uptake in an urban stream in Baltimore, MD, USA, using Br^- and Rhodamine-WT as conservative tracers; (B) measuring NH_4^+ uptake in an agricultural stream in Indiana; (C) measuring NH_4^+, NO_3^- and P uptake during winter in a forested stream in the Upper Peninsula of Michigan, USA. (D) Example data used to calculate NO_3^- uptake length in Shane Creek, MI, USA. Tracer NO_3^- and Br^- are plateau concentrations minus background concentrations. Uptake length, S_w, is the inverse of the slope of the regression line. Br^- is used as a conservative tracer to account for dilution along the stream reach. The diagram above panel depicts how nutrient concentrations should decline moving downstream. *Credits: L.T. Johnson provided the photo for panel B and the cartoon schematic in panel D.*

where N_x = background-corrected plateau nutrient concentration at x meters downstream of the injection point, N_0 = background-corrected nutrient concentration at the site of injection, designated as 0 m, and k = exponential decay rate (Newbold et al., 1981; Stream Solute Workshop, 1990; see also Fig. 31.1). A conservative tracer is added concurrently to account for dilution along the reach (see Chapter 30) and nutrient concentrations are divided by plateau background-corrected tracer concentrations at each sampling station (Fig. 31.1). A linear regression is then used to calculate the decay rate (k) and test for significance of the relationship. Uptake length (S_w) is calculated as the inverse of the decay rate:

$$S_w \text{ (m)} = k^{-1} \tag{31.3}$$

Because uptake length is strongly influenced by discharge (Q), and Q may vary among streams or within a stream over time, nutrient uptake velocity (V_f) should also be calculated. Uptake velocity can be thought of as the biotic demand for the nutrient relative to in-stream concentration, and it is functionally the velocity at which the nutrient is removed from the water column via uptake. Uptake velocity is calculated as:

$$V_f \text{ (m/min)} = Qk/w \tag{31.4}$$

where Q = discharge (m^3/min) and w = mean stream wetted width (m). Precise estimates of Q can be made using the conservative tracer (see Chapter 30). In addition to S_w and V_f, the areal uptake rate (U, mg N m^{-2} min^{-1}) provides an estimate of nutrient removal expressed per unit area of stream bed. For example, for N, the U of the nutrient (N) is calculated as follows:

$$U = V_f N_b \tag{31.5}$$

where N_b = background nutrient concentration prior to release. When these parameters of nutrient uptake (S_w, V_f, U) are used together, a greater understanding of the factors controlling uptake in stream ecosystems is possible, as uptake can then be understood relative to stream size, discharge, and nutrient availability.

Short-term nutrient additions to streams have been used frequently because of the ease of studying multiple streams within a short period of time and the low cost of materials. Because NH_4^+ is one of the most labile forms of N in streams, downstream declines in concentration occur more quickly compared to NO_3^- or P additions. Thus, we describe an NH_4^+ release in Advanced Method 1, although modifications to this approach include changes in the form of the nutrient released, which may result in the need to extend the length of the study reach and increase release duration to insure a decline in concentrations downstream (e.g., for NO_3^- or P release). It is also possible to increase the target enrichment level, depending on background concentrations. For example, in high NO_3^- streams (e.g., >1 mg/L), 100 μg NO_3^--N/L may be the target enrichment, rather than 10 μg NO_3^--N/L. The NH_4^+ release described in Advanced Method 1 is conducted in conjunction with a chloride (Cl^-) conservative tracer release to correct for downstream dilution (see Chapter 30 for more detail). Chloride is a low-cost choice as a conservative tracer, as increased chloride can easily be measured as the concurrent change in conductivity (using a field meter). However, if background water-column conductivity in the study reach is high (such as in urban or agricultural streams), bromide (Br^-) can be substituted as the conservative tracer and measured using ion chromatography in the laboratory, because sufficiently sensitive bromide-specific probes are not available for field use. An alternative tracer that can be measured in the field is rhodamine-WT, which can be added to the release solution to determine when the stream reaches plateau if you are using Br^- as your analytical conservative tracer in the lab. Despite previous concerns, rhodamine-WT does not interfere with laboratory analyses of NO_3^- via either ion chromatography or flow-injection colorimetric analysis (Reisinger, unpublished data). Although there are drawbacks to using an approach that elevates background nutrient concentration (see below), short-term nutrient additions are cost-effective and particularly useful for comparing uptake between streams or within one stream over time (Mulholland et al., 2002; Hoellein et al., 2007).

31.2.3 Short-Term ^{15}N Tracer Release

The goal of this method is to use the isotopic labeling of primary uptake compartments (i.e., epilithon, biofilm) to estimate whole-stream N uptake using the relative change in ^{15}N downstream. In Advanced Method 2, we describe a short-term ^{15}N tracer addition where the longitudinal decay of label in a rapid-turnover compartment (e.g., filamentous green algae, epilithon, or organic sediments) is used as a surrogate for water-column ^{15}N label, and it is sampled ~6 h after a continuous release of a ^{15}N-enriched solution. Stable isotope ^{15}N tracer additions eliminate the potential effects of N enrichment associated with short-term nutrient additions, including the potential for artificially long uptake lengths due to saturation of N demand and/or stimulation of uptake in the presence of ample nutrient (Mulholland et al., 2002). However, the use of ^{15}N isotopes is more costly both for the short-term addition itself and for subsequent required sample analyses via mass spectrometry (see Chapter 23). Using a rapid-turnover biomass (i.e., primary uptake) compartment avoids the difficult isotopic analysis of water samples. The biomass compartment will label with ^{15}N quickly, so isotope additions of shorter duration can be used, thereby reducing isotope costs. Because the overall removal of ^{15}N from the water column will be reflected in the ^{15}N label in the biomass compartment, nitrogen uptake metrics (S_w, V_f, U) are calculated in the same way as for short-term releases of N. Substitute background-corrected $\delta^{15}N$ values of the biomass compartment at each station for water column N concentrations used in Advanced Method 1.

If there are no fast-turnover compartments available within a study stream (e.g., a sandy-bottomed agricultural stream), an experimental compartment can be added to the stream prior to the short-term addition. These experimental compartments could include precolonized ($\sim$2 weeks) ceramic tiles or landscaping rocks of uniform size at the multiple sampling stations along the stream reach, thereby providing epilithon for sampling with the short-term ^{15}N tracer.

31.3 SPECIFIC METHODS

31.3.1 Basic Method: Nutrient Diffusing Substrata

31.3.1.1 Laboratory Procedures

General Laboratory Protocol

1. Make one agar type (control, +N, +P, +N + P) at a time.
2. See Table 31.2 for specific quantities of nutrient salts to add.
3. Polycon cups should be prelabeled (using a Sharpie) with agar type on both sides and bottom of cup to ensure that the label remains legible during deployment in the stream.
4. Make a copy of Fig. 31.2 for the NDS preparation checklist and gather items on the equipment checklist.

Agar Preparation (see Table 31.2)

1. Bring 1 L of water to boil and stir continuously with a 2.5-cm stir bar.
2. While stirring, slowly add agar powder (2−3% by weight depending on nutrient treatments; Table 31.2).
3. Continue to boil the solution while stirring to prevent burning.
4. To minimize the potential for H_2O_2 formation (Tanaka et al., 2014), we recommend adding nutrients after boiling. Once agar has dissolved, allow the solution to cool slightly. Once the solution is no longer boiling but has not solidified, add the appropriate nutrient salt. Stir vigorously until dissolved.
5. *Tips*:
   a. Solution is ready to pour when it becomes transparent.
   b. Bubbles will disappear as solution cools.
   c. Do not let the agar cool below 50°C or it will solidify.

Pouring Agar Solution

Safety Note: Always wear personal protective equipment when preparing and pouring agar solution, including lab coat, goggles, and closed-toe shoes. Additionally, when preparing agar solution, wear latex gloves to prevent cross-contamination of agar treatments, and add heat-resistant gloves when pouring agar.

1. Carefully pour hot agar solution into cups until they are almost overfilled (rounded meniscus forms).
2. Allow at least 15 min for agar to cool and solidify. The agar will settle slightly when cooled.
3. Place fritted glass or cellulose sponge discs on agar surfaces and snap caps shut.
4. *Tips*:
   a. If lids do not close completely, carve out slight depression in agar for discs.
   b. Be sure that discs are held firmly in place and not loose under the cap.

Attaching Cups to L-Bars and Storage

1. Attach the cups to the L-bars using small cable ties (Fig. 31.3).
2. Secure each cup bottom to the L-bar using a small bead of rubber silicon glue.
3. Cut ends of cable ties close to the fastener to avoid sharp tips.
4. Cover each bar with plastic wrap to keep moist and refrigerate until stream deployment.
5. *Tips*:
   a. Polycon cups should be secured close to each other along the L-bars.
   b. Agar types (control, +N, +P, +N + P) should be randomly arranged among the L-bars. Ideally a stratified random scheme is employed with each bar having a replicate of each treatment; therefore if an entire bar is lost or interacts differently with the environment, no individual treatment is compromised.
   c. Tighten cable ties with needle-nose pliers.

TABLE 31.2 Basic Method: Calculations for preparation of nutrient diffusing substrata.

Date _____
Stream _____

Project Investigators _____

(A) Target Nutrient	(B) Salt A	(C) Salt B	(D) Molecular Weight Salt A	(E) Molecular Weight Salt B	(F) Desired Molarity (M)	(H) g Salt A/L Agar Solution	(I) g Salt B/L Agar Solution	(J) g Agar/L Solution	(K) Total Number Cups	(L) mL/Cup	(M) Total Volume Solution Needed (L)
NH_4^+-N	NH_4Cl	–	53.5	–	0.5	26.7	–	20	5	30	0.150
NO_3^--N	KNO_3	–	101.1	–	0.5	50.6	–	20	5	30	0.150
NO_3^--N	$NaNO_3$	–	85.0	–	0.5	42.5	–	20	5	30	0.150
PO_4^{3-}-P	KH_2PO_4	–	136.1	–	0.5	68.0	–	20	5	30	0.150
NH_4^+-N and PO_4^{3-}-P	NH_4Cl	KH_2PO_4	53.5	136.1	0.5	26.7	68.0	30	5	30	0.150
NO_3^--N and PO_4^{3-}-P	KNO_3	KH_2PO_4	101.1	136.1	0.5	50.6	68.0	30	5	30	0.150
NO_3^--N and PO_4^{3-}-P	$NaNO_3$	KH_2PO_4	85.0	136.1	0.5	42.5	68.0	30	5	30	0.150
Calculation						D^aF	E^aF				$(K^aL)/1000$
Explanation					0.5 M unless specific change			When adding two nutrients, an additional 10 g/L of agar should be dissolved into solution	Generally 5 replicates per treatment per stream	Volume of agar solution to fill each polycon cup	Minimum volume of agar needed to fill the respective number of replicates[a]

[a]As some volume will evaporate during the agar preparation, we recommend preparing extra volume (e.g., if 0.15 L is a minimum, we would recommend preparing at least 0.2 L of agar).

Stream name _____ Date_____
Crew _____
Date NRS placed in field_____
Date NRS collected from field_____
NRS deployment locaton_____

NDS construction
- ❏ Purchase materials
- ❏ Drill openings in polycon cup lids
- ❏ Drill holes in L-bars
- ❏ Prepare agar with nutrient treatments
- ❏ Fill Polycon cups with solution and allow to harden
- ❏ Place fritted glass discs on top of dried agar and close
- ❏ Secure Polycons to L-bar tightly with cable ties
- ❏ Place silicon beads on bottom of polycons to secure in place
- ❏ Cover L-bars with plastic wrap to keep moist and refrigerate

NDS deployment and collection
- ❏ Secure L-bars to benthos with stakes at each end
- ❏ Incubate ~17 days
- ❏ Collect L-bars, remove discs, and place in individually labeled Ziplocs
- ❏ Extract and measure concentrations of chlorophyll *a* on discs
- ❏ Perform metabolism incubations if desired

FIGURE 31.2 Datasheet and checklist for nutrient diffusing substrata construction, deployment in the field, and laboratory processing (Basic Method).

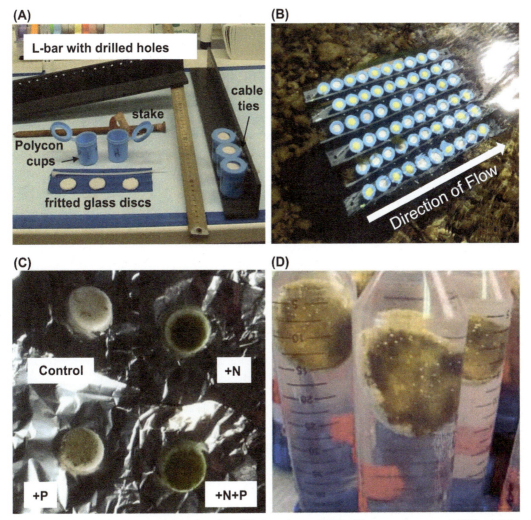

FIGURE 31.3 (A) Equipment needed for preparation of nutrient diffusing substrata (NDS), (B) stream placement of NDS parallel to stream flow, (C) fritted glass discs exemplifying extreme N limitation following a 15-day incubation in the Green River near Seedskadee Wildlife Refuge, WY, USA, and (D) depiction of incubations performed on fritted glass discs to estimate biofilm metabolism. *Photo credit: L.T. Johnson provided the photo for panel D.*

31.3.1.2 Field Procedures

Placement of Nutrient Diffusing Substrata in Stream (Fig. 31.3)

1. In the stream, place the L-bars next to each other in a riffle.
2. Position the bars parallel to flow to help prevent sedimentation. Bars can be elevated slightly above the stream bed by placing them on concrete blocks if sedimentation is a concern (e.g., in a sandy-bottomed river; Reisinger et al., 2016).
3. Pound stakes into holes at each end of the L-bar to secure into stream bottom.
4. If you can, check on NDS periodically over 18−20 days incubation period.
5. *Tip*: If some NDS are lost within the first 48 h, they can be replaced but make sure to note difference in incubation time.

Retrieval From Stream

1. After 18−20 days, gently remove one L-bar at a time from the stream.
2. Uncap each cup and immediately place each disc into a labeled 50-mL centrifuge tube. If metabolism incubations will be performed in the laboratory, fill centrifuge tubes with water to completely submerge discs, and store tubes on ice in the dark until incubations commence. If incubations will be performed in the field, proceed to "Metabolism Incubation of Discs."
3. *More Tips*:
   a. Centrifuge tubes need to be "flattened" slightly for fritted glass discs to fit inside the tubes.
   b. When filling centrifuge tubes with water, first collect a bucket of site water and then slowly fill tubes from the bucket to prevent turbulence from displacing biofilms colonizing discs.

31.3.1.3 Metabolism Incubation of Discs

1. Fill a 20-L bucket with site water (this can be filled directly from stream if incubations are performed in the field, or carboys can be filled in the field and returned to the lab).
2. Using a hand-held dissolved oxygen (DO) probe, record the time, temperature (°C), and DO (mg/L and % saturation) in the bucket of stream water used to fill tubes. Repeat this reading after all tubes have been filled. If DO is different between the two readings, a correction must be applied either by using the mean of the two readings, or calculating the change in DO over time throughout filling the tubes, and then applying this correction to each respective tube.
3. Discard old site water from centrifuge tubes containing NDS discs and then slowly fill centrifuge tubes from bucket. Cap the tubes underwater to prevent headspace and ensure that there are no air bubbles in the tubes. Record the time each tube was filled.
4. Incubate tubes in the light—this can either be in the stream or under a grow lamp in the laboratory. *Tip*: If incubations are performed in the stream, a wire-mesh "rack" can be used to hold centrifuge tubes in place and prevent losing them downstream.
5. After approximately 2 h, measure the final DO from each tube [record time, temperature, and DO (mg/L and % saturation)]. The difference between final DO and initial DO in the light is net ecosystem production.
6. Repeat steps 2 and 3 to begin dark incubations.
7. Instead of incubating tubes in the light, incubate tubes in the dark. *Tip*: If incubations are in situ, aluminum foil "sleeves" can be constructed by wrapping aluminum foil tightly around centrifuge tubes and then using duct tape to insure the sleeves retain their shape. If incubations are performed in the laboratory, darkness can be insured by wrapping NDS in opaque garbage bags or incubating in the dark inside of a closed cooler (or any other device that eliminates ambient lighting from reaching incubating tubes).
8. After approximately 2 h in the dark, measure the final DO [record time, temperature, and DO (mg/L and % saturation)]. The difference between final and initial DO in the dark is ER.
9. Calculate GPP as follows:

$$GPP = NEP + abs(ER) \qquad (31.6)$$

where abs(ER) represents the absolute value of ER.

10. Express GPP and ER per unit surface area per unit time.
11. Calculate nutrient limitation status using two-way ANOVA (Tank and Dodds, 2003). Calculate the NRR (Eq. 31.1) to provide a quantitative metric of nutrient limitation that can then be used in other statistical analyses (e.g., simple linear regression; Johnson et al., 2009; Reisinger et al., 2016)

12. *More Tips*:
    a. While filling tubes from the bucket, water should be gently stirred approximately every five samples to insure nothing settles on the bottom of the bucket.
    b. Centrifuge tube caps can shade NDS during light incubations. Therefore, light incubations should be performed with tubes inverted, allowing full light to reach to the disc inside the centrifuge tube.
    c. Incubation lengths can be varied depending on the expected activity of the biofilm. For example, if a thick, bright green biofilm is on the disc, incubations may only be 1 h, as 2 h in the light may lead to supersaturation of oxygen.

31.3.1.4 Chlorophyll a and Ash-Free Dry Mass Analysis on Discs

1. Extract chlorophyll *a* directly from discs by placing each disc in a labeled film canister and covering disc with acetone (normally 10 or 15 mL). See Chapter 12 in Volume 1 of this book for details on chlorophyll extraction and analysis.
2. Quantify AFDM on discs by drying each disc in a preweighed aluminum weighing pan for at least 48 h. Weigh pan and disc, and then combust discs in a muffle furnace at 500°C for at least 2 h. Following combustion, allow discs to cool, then rewet discs to replace water lost from inorganic compartments (e.g., water bound within clay), dry for an additional 48 h, and reweigh ashed pan and discs. The difference in weight between dried pan + disc and ashed pan + disc provides an estimate of AFDM. See Chapter 12 for details on AFDM processing and analysis, along with a data recording table.
3. Express chlorophyll *a* and AFDM per unit surface area and compare across treatments using the same approach as for metabolism (Tank and Dodds, 2003; Johnson et al., 2009; Reisinger et al., 2016).

Tip: If you want to quantify chlorophyll *a* and AFDM from the same disc, first quantify chlorophyll *a* via appropriate methods, and then place the entire extracted disc and any extractant (e.g., chlorophyll *a* extracted in acetone) in an aluminum pan. Let the acetone evaporate in a fume hood prior to placing pan + disc in a drying oven.

31.3.1.5 Modifications and Enhancements

As mentioned above, the NDS approach can be modified to assess the influence of other compounds on microbial biofilms. Pharmaceutical compounds and heavy metals have been successfully used in place of nutrient amendments (Costello et al., 2016). When using novel compounds, first ensure that the compound is soluble in water to allow for proper agar preparation and diffusion while deployed. Additionally, the method is designed to test nutrient limitation by supplying nutrients in excess. One can modify the amount and ratio of nutrients to modify the research question. Also, we advise checking the diffusion rates when using novel compounds or nutrient concentrations (Capps et al., 2011; Costello et al., 2016). Additional response metrics may also be measured for NDS. For example, microbial community composition has been assessed using molecular techniques (Hoellein et al., 2010; Rosi-Marshall et al., 2013; Costello et al., 2016; see also Chapter 9) as has extracellular enzyme activity (e.g., Hill et al., 2012) to isolate selected microbial metabolic pathways in response to nutrient addition.

31.3.2 Advanced Method 1: Short-Term Nutrient Release

31.3.2.1 General Preparations

Site Selection and Solute Decisions

1. Select an appropriate stream reach for the release. The stream reach should have minimal inputs from groundwater sources, tributaries, or point discharges. Groundwater inputs can be detected by identifying changes in water temperature or by conducting a conservative tracer release prior to the short-term nutrient release. The selected stream should be relatively homogenous (i.e., no large changes in structure or composition) over the entire experimental reach.
2. Before going into the field, determine the quantity of nutrient and conservative tracer salts needed for the injectate, keeping the following relationship in mind:

$$Q_I = Q \cdot C_s / C_I \tag{31.7}$$

where Q = stream discharge, C_s = background nutrient concentration, Q_I = injectate drip rate into the stream, and C_I = nutrient concentration in the carboy (see Chapter 30).

3. For precise measurements, use a recent estimate of background nutrient concentrations and stream discharge.

4. *Tips*:
   a. If you are unsure of the stream discharge (i.e., there is no recent estimate available), weigh salts in increments needed for 5 or 10 L/s of stream discharge. You can add salts accordingly once you arrive at the stream and measure stream discharge directly (see Chapters 3 and 30 for methods).
   b. If not enough salt is available, the drip rate can be altered. For example, you can either use *X* amount of salt and a drip rate of 0.1 L/min or you can use half of *X* amount of salt and a drip rate of 0.2 L/min. Make sure to check the total volume needed for the release to ensure you have enough injectate for the duration of the release.

Calculating the Amount of Salt to Be Added to the Carboy (see Table 31.3)

1. The quantity of salt added can be calculated as follows (using NH_4Cl as an example):

$$g \text{ as } NH_4Cl = \underbrace{\left(\frac{L}{sec}\right)}_{Q} * \underbrace{\left(\frac{\mu g\ N}{L}\right)}_{\substack{\text{Target} \\ \text{addition}}} * \underbrace{\left(\frac{min}{L}\right)}_{\substack{\frac{1}{\text{drip rate}}}} * \underbrace{\left(\frac{20L}{1}\right)}_{\substack{\text{L in} \\ \text{carboy}}} * \underbrace{\left(\frac{NH_4Cl \text{ mol. wt. } 53.49}{N \text{ mol. wt } 14}\right)}_{\substack{\text{N in} \\ NH_4Cl}} * \underbrace{\left(\frac{60 \text{ sec}}{min}\right) * \left(\frac{1g}{1000000\mu g}\right)}_{\text{conversions}} \tag{31.8}$$

2. *Tip*: Optimal target additions should be just high enough above background to be analytically detectable at the most downstream site, but not so high at the top site so that N demand is saturated (see Table 31.3 for examples).

Laboratory Preparations for the Field

1. Choose five sampling stations spaced approximately evenly downstream of the injection site (e.g., 20, 40, 60, 80, and 100 m from injection point).
2. For each station, label four clean 60-mL wide-mouth bottles, one for background (BKD) nutrient samples and three for plateau (PLT) nutrient samples.
3. Place the bottles for each station (1 BKD and 3 PLT) in separate ziploc bags.
4. Also, label one 60-mL bottle "caution: release solution" and place in a separate bag.
5. Make a copy of Fig. 31.4 for a field datasheet and gather items on checklist.

31.3.2.2 Field Procedures

Before Turning on the Dripper

1. Mark sampling stations with flagging tape, using a meter tape to measure distances, and set bottles at each station.
2. Collect BKD water samples and measure conductivity at each station.
3. *Tips*:
   a. Take care not to disturb the stream bed before and during release.
   b. Water samples can be filtered in the field using a syringe-mounted 25-mm filter or back in the laboratory prior to water chemistry analyses. Water samples should be frozen until nutrient analysis can be performed.
   c. If background conductivity is too high, you will need to use Br^- instead of Cl^- as a conservative tracer during the release. Bromide concentrations can be measured via ion chromatography in the lab. To ensure concentrations have reached plateau prior to collecting PLT samples, either calculate travel time and estimate the time to plateau or Rhodamine-WT can be used as an additional semiconservative tracer, quantified using a field fluorometer.

Preparing Injectate

1. Fill a 20-L carboy with 18 L of stream water, pour the preweighed salts into the carboy, and mix well to fully dissolve all the salts. Bring the volume to 20 L. Both N and conservative tracer salts should be added to the carboy.
2. Record the initial and final volume of the carboy before and after the N release so you can calculate the volume you injected over the release.
3. *Tip*: Use caution with injectate as this is a highly concentrated salt solution. Be careful not to splash injectate solution into any sample bottles or spill any into the stream.

TABLE 31.3 Advanced Method 1: Calculation of solute to add to carboy for short-term nutrient releases.

Date _____

Stream _____

Project Investigators _____

(A) Target Solute	(B) Background Concentration (µg/L)	(C) Q (L/s)	(D) Enrichment Concentration (µg/L)	(E) Drip Rate (L/min)	(F) Salt	(G)[a] Proportion Target Solute:Salt	(H) Carboy Volume (L)	(I) Salt to Add to Carboy (g)	(J)[b] Maximum Solubility at 10°C (g/L)
Br^-	<10	10	40	0.1	$NaBr$	0.776	20	6.19	459
Cl^-	10	10	1000	0.1	$NaCl$	0.603	20	199	358
NH_4^+-N	10	10	20	0.1	NH_4Cl	0.260	20	9.23	333
PO_4^{3-}-P	10	10	20	0.1	NaH_2PO_4	0.258	20	9.23	699
NO_3^--N	150	10	150	0.1	$NaNO_3$	0.165	20	109	209
Calculation								$\dfrac{(C \times D \times H \times 60)}{(G \times E \times 1{,}000{,}000)}$	

[a]Molecular weight of target solute divided by molecular weight of salt (e.g., molecular weight on N divided by molecular weight of NH_4Cl).

[b]If g salt/L exceeds maximum solubility, then salt will not dissolve. Try increasing drip rate and carboy volume.

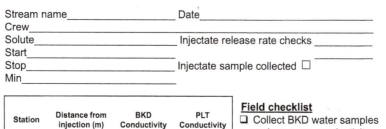

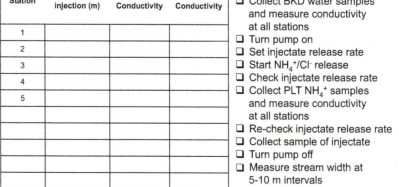

FIGURE 31.4 Datasheet and checklist for short-term nutrient releases (Advanced Method 1; *BKD*, background; *PLT*, plateau).

Adding Injectate to Stream and Sampling at Plateau

1. Set up the pump and a place for the tubing to drip the injectate into the stream. It may be useful to cable tie the tubing to a stake and secure into benthos, suspending the tubing slightly above water surface.

2. Place inlet and outlet of pump ends into the stream to fill tubing with stream water and rinse pump.

3. Test drip rate by placing the outlet tubing into a small graduated cylinder for 15 s. Multiply volume of water collected by 4 to determine drip rate (mL/min). Record drip rate in field book. Put the measured injectate back into the carboy.

4. Place weighted inlet tubing into carboy and remove air bubbles in tubing.

5. Start pumping injectate into the stream and record the exact time you begin the release (i.e., start stopwatch).

6. At the downstream station, monitor the rise in conductivity and wait for plateau. Conductivity will increase as the injectate reaches the downstream station. Plateau has been reached when conductivity no longer increases but remains stable.

7. Once conductivity stabilizes, take PLT water samples and PLT conductivity readings at each station moving from downstream to upstream.

8. Measure the injectate drip rate again and record in field book. This time, make sure to empty injectate from graduated cylinder into the stream. Turn off the pump and record the exact stop time and total release minutes, as well as the final volume of injectate in the carboy.

9. Collect a sample of the injectate and put into labeled bottle. Keep this bottle away from other samples to avoid contamination (e.g., place bottle in a separate ziploc bag or whirl-pak).

10. Place water samples in a cooler for transport to the laboratory; on return place all samples into a freezer until ready for N analyses.

11. Collect data on stream widths after release is completed. Stream widths should be measured every 5−10 m from the injection point for ∼20 measurements. Width measurements will be used to calculate N uptake parameters.

12. *Tips*:

    a. Wear gloves and use care when handling injectate to avoid contamination.

    b. It is very important that the injectate completely mixes by the first sampling station. Dripping injectate into a riffle area (either natural or constructed) will help with mixing.

    c. Rinse the pump well after releases because salt solutions are highly corrosive. Pull weighted inlet tube out of carboy and place just upstream of the drip point and rinse tubing thoroughly (∼5 min) with clean stream water.

31.3.2.3 Laboratory Procedures

Analysis of Water Samples

1. Analyze water samples for nutrient concentrations using standard methods (APHA, 2012). If using Br^- as a conservative tracer, analyze for that ion as well.
2. *Tips*:
    a. When analyzing water samples, be sure to analyze your samples from low to high concentrations to avoid contamination (i.e., run BKD samples first, then PLT samples moving from downstream samples toward injection point).
    b. Different methods currently exist for analysis of different solutes. For example, two methods are commonly used to quantify NH_4^+-N: the indophenol blue method for samples above 10 µg/L (APHA, 2012) and the fluorometric method for low concentrations (Holmes et al., 1999). When choosing an appropriate method, make sure to check the most recent literature for any refinements of methods used for quantifying your specific solute of interest. Please see Chapter 36 in this volume for the detailed fluorometric protocol to quantify NH_4^+-N.

Calculating Nutrient Uptake Length (S_w)

1. Enter all BKD and PLT concentrations measured for each station into a spreadsheet and calculate the mean PLT concentration at each station.
2. Plot $\ln\left(\text{mean PLT} - \text{BKD } NH_4^+\text{-N}\right)/(\text{PLT} - \text{BKD conductivity})$ versus distance downstream (see Fig. 31.1).
3. Using the regression equation for the plotted line, S_w (m) is the inverse of the absolute value of the slope.
4. Calculate uptake velocity, V_f, and whole-stream uptake, U, using S_w (see General Design section).

Tip: In some systems, nitrification can be estimated from NH_4^+ releases by monitoring the change in NO_3^--N concentrations at sampling stations, with increasing NO_3^--N concentration as one moves downstream indicative of reach-scale nitrification (Bernhardt et al., 2002).

31.3.2.4 Modifications and Enhancements

Although the short-term nutrient release approach, using minimum enrichment of background nutrient concentrations, works well for quantifying nutrient uptake in small streams, a pulse-based approach can also be used to quantify nutrient uptake in a stream (Payn et al., 2008; Chapter 30). Two pulse-based methods have recently been used to estimate nutrient uptake in small streams and larger streams and rivers. The tracer additions for spiraling curve characterization (TASCC) approach (Covino et al., 2010) involves pulsing a concentrated solution of reactive and conservative tracer into a stream and then collecting water chemistry samples at a single downstream station over the course of the breakthrough curve (BTC) when the pulse is moving past the sampling station. Using the TASCC approach, the goal is to saturate uptake and estimate ambient uptake rates by quantifying uptake on each sample taken over the course of the BTC (Covino et al., 2010). The pros and cons of the TASCC and short-term nutrient addition approaches have been compared recently, and the appropriate method choice largely depends on site-specific conditions and the context of research questions (Trentman et al., 2015).

In nonwadeable streams and rivers, one can also use the minimal pulse addition technique (mPAT), which represents a combination of the short-term nutrient release approach and TASCC; this approach has been used successfully in much larger rivers ($Q > 80,000$ L/s) than possibly using the short-term nutrient additions (Tank et al., 2008). The mPAT involves pulsing a concentrated solution of reactive and conservative tracer into a stream or river, and then sampling across the BTC at multiple sampling stations downstream ($n = 5$). After analyzing reactive and conservative tracer in BTC samples ($n = 40$ for each BTC), the total mass of reactive and conservative tracer to pass by each station is quantified by integrating the area under each BTC. The relative mass of reactive versus conservative tracer is then regressed against distance downstream to calculate k and S_w, similar to the calculations presented for short-term nutrient additions. While each approach has its strengths and weaknesses, we chose to focus on the short-term nutrient release approach due to its ease of use and flexibility for application across a wide range of field sites.

31.3.3 Advanced Method 2: Short-Term ^{15}N Stable Isotope Tracer Addition

31.3.3.1 General Preparations

Isotope Purchase and Site Selection

1. Purchase the stable isotope well in advance of your release (~3−6 months). We recommend a target enrichment for the stream of $\delta^{15}N = 5000$‰. If cost is an issue, target enrichment can be decreased, but reduced target enrichment results in less label incorporation in the selected biomass compartment.

2. Locate an analytical laboratory that can conduct the mass spectrometry on organic solid samples, either at your own institution or one that does contract work. Check the web and see information below.
3. Select your stream reach using the same criteria as for Advanced Method 1, bearing in mind that isotope costs are directly proportional to stream discharge and N concentrations (i.e., NO_3^- flux). Small or low-N streams are the most cost-effective.

Presampling and Preliminary Information for Calculations

1. Collect preaddition water samples and discharge measurements from a minimum of five points along the length of the stream reach. Because these data will be used to calculate how much isotope to add, precise measurements of N concentration and discharge are imperative.
2. See Table 31.4 for calculation of ^{15}N needed. Amount of ^{15}N to add to the stream channel is based on known variables including stream discharge, stream NO_3^- concentration, desired enrichment, enrichment of isotope salt, and experiment duration. Thus, these variables should be determined prior to calculation of ^{15}N needed. It is assumed that the amount of ^{15}N added to the carboy is the amount needed for the entire release. Drip rate of the pump should be adjusted so the solute is almost completely depleted over the course of the release and should be calculated prior to the release.
3. See Advanced Method 1, Eq. (31.6), and Table 31.3 for calculation of conservative tracer salt needed. To calculate isotope needed, first calculate the background ^{15}N flux:

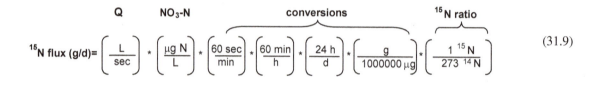

$$^{15}N \text{ flux (g/d)} = \left(\frac{L}{sec}\right) * \left(\frac{\mu g\ N}{L}\right) * \left(\frac{60\ sec}{min}\right) * \left(\frac{60\ min}{h}\right) * \left(\frac{24\ h}{d}\right) * \left(\frac{g}{1000000\ \mu g}\right) * \left(\frac{1\ ^{15}N}{273\ ^{14}N}\right) \qquad (31.9)$$

4. To calculate g $K^{15}NO_3$ to add to the 20-L carboy, incorporate the calculated ^{15}N flux into the equation below:

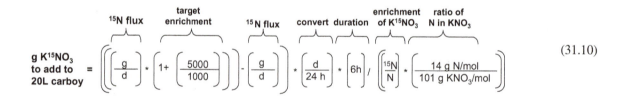

$$\text{g } K^{15}NO_3 \text{ to add to 20L carboy} = \left(\left(\left(\frac{g}{d}\right) * \left(1 + \frac{5000}{1000}\right)\right) - \left(\frac{g}{d}\right)\right) * \left(\frac{d}{24\ h}\right) * \left(6h\right) / \left(\left(\frac{^{15}N}{N}\right) * \left(\frac{14\ g\ N/mol}{101\ g\ KNO_3/mol}\right)\right) \qquad (31.10)$$

5. Note that the target enrichment of 5000 and the duration of the release of 6 h are suggested values and can be modified to suit specific conditions.
6. *Tips*:
   a. After you calculate the quantity of isotope and conservative tracer (Cl^- or Br^-) needed, weigh them into containers that are well-sealed to avoid contamination.
   b. Because of the potential for contamination, all salts weighed out should be clearly labeled and double-bagged.
   c. Be extremely careful when weighing out isotope salts. "This should be done away from any sample processing." Even a miniscule amount of salt can contaminate an entire laboratory and ruin all current and future experiments where isotopes would be used!
   d. Prepare an area of the laboratory designated for isotope sample processing. Clear the area, wipe down with ethanol, and lay down laboratory paper. Signs should be placed around the area indicating that enriched samples are being processed.

TABLE 31.4 Advanced Method 2: Calculation of ^{15}N-labeled salt to add to stream channel over the course of the experiment.

Date Stream	(A) Discharge (L/s)	(B) NO$_3^-$-N (µg/L)	(C) Target δ^{15}N (‰)	(D) NO$_3^-$-N Flux (µg/day)	(E) NO$_3^-$-N Flux (g/day)	(F) ^{15}N Flux (g/day)	(G) ^{15}N Needed for Target (g/day)	(H) Total ^{15}N to Add (g/h)	(I) Drip Duration (h)	(J) Total ^{15}N Needed for Release (g)	(K)[a] Enrichment of K^{15}NO$_3$	(L) Total K^{15}NO$_3$ Needed for Release (g)	(M)[b] Cost (US$)
A	10	500	5000	432,000,000	432	1.58	9.50	0.33	6.00	1.98	0.10	143	
B	10	1000	5000	864,000,000	864	3.16	18.99	0.66	6.00	3.96	0.10	286	
C	10	1000	5000	864,000,000	864	3.16	18.99	0.66	6.00	3.96	0.99	29	852
D	50	1000	5000	4,320,000,000	4320	15.82	94.95	3.30	6.00	19.78	0.99	144	4261
Calculation				A × B × 60 × 60 × 24	D/1,000,000	E/273	F × [1 + (C/1000)]	(G − F)/24		I × H		H/[K^{15}(14/101.2)]	
Explanation				Convert to days	Convert to grams	Multiply by background ratio of ^{15}N:^{14}N	Calculating ^{15}N needed for target enrichment	Subtract ^{15}N natural abundance and convert to hrs		Multiply g ^{15}N/h by drip duration	Proportion of ^{14}N to ^{15}N in KNO$_3$	Divide total ^{15}N (g/h) by ratio of molecular weight N:KNO$_3$ multiplied enrichment of K^{15}NO$_3$	Multiply g needed by cost

[a] Isotope (e.g., K^{15}NO$_3$) can be purchased in various degrees of ^{15}N enrichment (10% and 99% ^{15}N-enrichment are most common but be sure to check isotope bottle label).
[b] Costs are based on an estimate of $84.50 per gram of 99% enriched K^{15}NO$_3$.

31.3.3.2 Field Procedures

Before Turning on the Dripper

1. Measure a 200–300 m stream reach and mark 10 stations along the reach with flagging tape (if $Q > 50$ L/s, extend the stream reach and mark additional stations).
2. Choose stations based on expected uptake (e.g., if the stream is replete with algae, uptake will be high and stations should be concentrated closer to the injection point with only a couple further downstream).
3. Collect background samples of your rapid-turnover compartment (i.e., filamentous green algae or epilithic biofilm scraped from rock surface) at each sampling station for determination of the natural abundance $\delta^{15}N$ signal. See below for laboratory processing details.
4. Make a copy of Fig. 31.5 for the field datasheet and gather items on checklist.

Carboy Preparation

1. Fill the carboy with ~ 18 L of stream water and set up pump at injection point.
2. While wearing gloves, add a small amount of water to the vial containing the ^{15}N salt to dissolve just prior to adding it to the carboy.
3. While wearing gloves, add isotope and conservative tracer salt to carboy. Rinse vials a few times into carboy after addition to remove all the salts.
4. While wearing gloves, mix injectate by capping and shaking carboy or stirring with a stick after bringing volume to 20 L.
5. *Tips*:
   a. Record the mass of $K^{15}NO_3$ salt added to the injection carboy and the initial volume of injectate.
   b. Be sure to wear disposable gloves when handling the ^{15}N salt or solution and carefully discard the gloves when finished. Contamination by ^{15}N is a major concern.

Adding Injectate to Stream and Sampling at Plateau

1. Suspend tubing in the center of the stream channel slightly above the surface, keeping the rest of the tubing out of the stream. Drip injectate into stream at a constant rate (rate should be set so that the 20-L carboy is almost completely emptied over the duration of the experiment) for ~ 6 h to label the rapid-turnover compartment with ^{15}N.
2. Record the start time in which the injection began and pump flow rate (determined using timed drip into graduated cylinder; see Advanced Method 1 for details). Make sure to empty injectate from graduated cylinder into the stream.

```
Stream name_____ Date_____
Crew_____
Injectate release rate checks_____
Start time_____ g Cl- added_____
Stop time_____ g K15NO3 added_____
Total drip time (min)_____ Carboy volume_____
```

Station	Distance from injection (m)	BKD Conductivity	PLT Conductivity
1			
2			
3			
4			
5			

Field checklist
- ☐ Collect BKD compartment samples and measure conductivity at all stations
- ☐ Turn pump on
- ☐ Set injectate release rate
- ☐ Start ^{15}N/Cl- release
- ☐ Check injectate release rate
- ☐ Collect PLT compartment samples and measure conductivity at all stations
- ☐ Re-check injectate release rate
- ☐ Turn pump off
- ☐ Rinse samples thoroughly
- ☐ Measure stream width at 5-10 m intervals

FIGURE 31.5 Datasheet and checklist for short-term ^{15}N tracer release (Advanced Method 2; *BKD*, background; *PLT*, plateau).

3. After 6 h of continuous injection, collect samples of your fast-turnover compartment (e.g., algae) at each station (moving from downstream to upstream stations) for later ^{15}N analysis. Make sure you have a sample from before the drip begins for background stable isotope analysis of biomass compartment.
4. Place each sample in a small ziploc and label clearly with date, stream, site, and "^{15}N sample."
5. At the same time, to determine the effects of dilution on isotope labeling, take water samples for conservative tracer concentrations (as described in Advanced Method 1 above) or measure conductivity at each station.
6. After all plateau isotope and conservative tracer samples have been collected, turn off the injection pump. Rinse tubing thoroughly by pumping clean water from upstream of the injection point for ∼10 min.
7. Record the final (unused) volume of injectate in your field book.
8. If using filamentous algae as your biomass compartment, to avoid contamination of samples by the ^{15}N in the water column where you sampled, go upstream of the dripper and rinse samples thoroughly before placing samples on ice for transport back to the laboratory. Rinse samples starting with lowest enrichment, those taken from the most downstream site, and process moving upstream to avoid contamination. If you are collecting biofilm, allow biofilm slurry to settle out in sample container, pour off supernatant, replace with clean water, and repeat a few times to isolate only particulates.
9. Collect data on stream widths after the release is completed. Stream widths should be measured every 5−10 m from the injection point for ∼20 measurements. Width measurements will be used to calculate uptake parameters.
10. *Tips*:
    a. Check the injection rate every hour. After a rate check, be sure to put graduated cylinder contents into the stream. Adjust injection rate if necessary.
    b. It is very important that there be complete mixing of the injection by the first sampling station. Dripping injectate into a riffle area (either natural or constructed) will help with mixing.
    c. A composite sample of each biomass compartment should be collected at each station. For example, take a little bit of algae from several places at each station and combine for ^{15}N sample.

31.3.3.3 Laboratory Procedures

Sample Processing for Selected Rapid-Turnover Compartment (e.g., Filamentous Green Algae)

1. To avoid contamination, sort samples from lowest to highest enrichment for processing (i.e., downstream to upstream).
2. Place (or pour in case of biofilm slurry) each sample into its own labeled aluminum pan for drying.
3. Dry each sample at 55°C for 72 h.
4. Grind samples into a fine powder and place into labeled scintillation vials for storage until analysis via mass spectrometry. Make sure to clean your grinding instrument between samples to minimize isotope carryover.
5. Prepare each sample for analysis of ^{15}N by mass spectrometry by weighing a small amount of ground sample (using a microbalance) into a 4.6-mm tin capsule (Costech Laboratories) and then press into a sealed "bullet shaped" capsule.
6. Analyze samples on an isotope-ratio mass spectrometer, in a reputable university or commercial analytical laboratory capable of analyzing ^{15}N samples.
7. *Tips*:
    a. Thoroughly rinse all field equipment with tap water to ensure that no ^{15}N contaminates the laboratory.
    b. Keep all ^{15}N contaminated materials and samples separate and clear of other equipment in the laboratory. Clean all processing materials thoroughly between samples.
    c. Prepared sample capsules can be placed into a 96-well plate and covered with a well cap for transport to the mass spectrometry laboratory.
    d. If cost is limiting, analyze samples from every other station (i.e., stations 1, 3, 5, etc.) first to determine where peak ^{15}N enrichment occurred, and then analyze more samples if necessary.
    e. Ground ^{15}N samples in scintillation vials will contain enough material for multiple analytical replicates if needed.

31.3.3.4 Potential Modifications to Protocols

1. Either $^{15}NH_4^+$ or $^{15}NO_3^-$ can be used as an isotopic tracer.
2. In areas with high-background stream water conductivity, Br^- should be used as the conservative tracer instead of Cl^-. Bromide can be measured using ion chromatography in the laboratory (see details in Advanced Method 1).
3. If the selected stream reach does not have an abundance of filamentous green algae, alternative substrata can be used for the rapid-turnover compartment including biofilm and flocculent fine benthic organic matter. But these compartments may have spatially heterogeneous N uptake rates, depending on colonization/deposition dynamics, resulting in slower

incorporation of the ^{15}N tracer. The more active the compartment (i.e., rapidly cycling), the more isotopic tracer will be incorporated, resulting in a higher tracer δ^{15}N signal.

4. If your stream has a lack of stable substrate (e.g., a sand-dominated, low-gradient stream) unglazed ceramic tiles can be colonized ahead of time with stream biofilm by placing them in the stream 2−4 weeks prior to the short-term isotope addition (see Chapter 12). Tiles are then harvested after the 6-h isotope release and analyzed in the same manner as previously described.

31.3.3.5 Calculation of ^{15}N Uptake Length

1. Uptake length is calculated similarly as in Advanced Method 1, Eq. (31.2). However, ^{15}N uptake length (S_w) is calculated using the inverse of the slope of the regression of ln(tracer δ^{15}N compartment/conservative tracer concentration) versus distance from the injection point in meters (Fig. 31.6). "Tracer" is defined in Fig. 31.1. Uptake velocity (V_f) and whole-stream uptake (U) are then calculated as described above (Eqs. 31.5 and 31.6).

31.3.3.6 Modifications and Enhancements

The outlined method focuses on a short-term ^{15}N isotope release and quantifying a single specific uptake compartment, algae, which has fast turnover (i.e., quick isotopic labeling). Longer releases of ^{15}N lasting days to weeks allow for quantification of the multitude of N-transformations occurring within a stream (Peterson et al., 2001; Mulholland et al., 2002), including various assimilatory pathways (Sobota et al., 2012) and dissimilatory transformations (Mulholland et al., 2008). After adding a sufficient amount of ^{15}N to a stream, samples of the entire food web can be collected to trace N throughout the ecosystem and establish the ultimate fate of N removed from the water column. For example, Mulholland et al. (2008) found that the vast majority of NO_3^--N that is removed from the water column enters assimilatory pathways, with only ∼10−20% being denitrified and permanently removed from the ecosystem. The added ^{15}N can be tracked through all possible assimilatory compartments within a stream, and the important pathways appear to be driven by light availability (Hall et al., 2009).

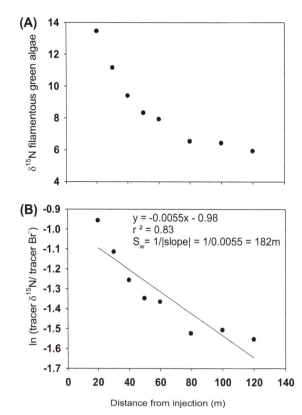

FIGURE 31.6 (A) ^{15}N Enrichment (background-corrected) of filamentous green algae in a headwater stream in southwest Michigan, USA, after a short-term ^{15}N tracer release and (B) same data plotted for calculation of uptake length, S_w, which is the inverse slope of ln(^{15}N:Br$^-$ concentration) over distance from the injection point.

31.4 QUESTIONS

31.4.1 Basic Method: Nutrient Diffusing Substrata

1. Was each stream limited by N or P? Did colimitation occur? What factors may explain your results?
2. When would you expect N limitation to occur? When would you expect P limitation to occur? When would you expect colimitation to occur?
3. Why is it important to understand what nutrients limit algal growth in streams?

31.4.2 Advanced Method 1: Short-Term Nitrogen Release

4. What was the N uptake length (S_w), uptake velocity (V_f), and uptake rate (U) calculated for each stream?
5. How do spiraling metrics compare to other stream ecosystems? How does areal uptake compare to terrestrial ecosystems? Note: Measured uptake metrics can be contrasted with other streams using supplemental material available from Hall et al. (2013).
6. Would you expect NO_3^- or NH_4^+ uptake rates to be higher? How would P uptake compare?

31.4.3 Advanced Method 2: Short-Term ^{15}N Tracer Release

7. What was the N uptake length (S_w), uptake velocity (V_f), and uptake rate (U) calculated for each stream?
8. How would N uptake compare if different biotic compartments were sampled for ^{15}N enrichment? Would the estimate be similar? Why?
9. How did N uptake rate determined using the short-term nutrient addition in Advanced Method 1 compare to the ^{15}N uptake rate determined with Advanced Method 2 and why might they differ? Are relative patterns similar for S_w and V_f?
10. Sometimes N uptake cannot be detected within a stream reach using either a short-term nutrient release or a short-term ^{15}N tracer release. What factors would yield these results for each method?

31.5 MATERIALS AND SUPPLIES

31.5.1 Basic Method

Field Materials
  L-shaped stakes
  Parafilm
  Field books
  Cooler with ice
  Polycon cups (1oz; Madan Plastics #1's) with nutrient diffusing substrate
  Gray plastic L-bars (US Plastics- #45031)
  Fritted glass discs (i.e., glass crucible covers, Leco #528-042)
  Small 4″ cable ties
  Rubber silicon glue
  Needle-nose pliers
  Plastic wrap
  Clear waterproof tape
Laboratory Materials
  YSI oxygen meter
  Extra-small stir bar
  Small Rubbermaid
  Shaker table
  60-mL disposable syringes
  1-L beaker
  Paper towels
  2-L Erlenmeyer flask
  Stir/heat plates
  Granulated agar

Ammonium Chloride (NH$_4$Cl)
Potassium Phosphate (KH$_2$PO$_4$)
Sodium or Potassium Nitrate (NaNO$_3$ or KNO$_3$)
Ziplok bags, snack size
35-mL centrifuge tubes
1-L Nalgene bottles

31.5.2 Advanced Method 1

Field Materials
   Carboy
   Bottle for filling carboy
   Injection Pump
   Pump tubing with weight
   Charged battery
   Graduated cylinder
   60-mL water bottles
   Disposable gloves
   Ion-specific probe
   Field books
   Weighed salts
   Meter stick
   Meter tape
   Flow meter
   60-mL disposable syringes
   Syringe filters
   Cooler with ice
   Stopwatch
Laboratory Materials
   Ammonium Chloride (NH$_4$Cl)
   Potassium Phosphate (KH$_2$PO$_4$)
   Sodium or Potassium Nitrate (NaNO$_3$ or KNO$_3$)
   Disposable Gloves

31.5.3 Advanced Method 2

Field Materials
   Carboy
   Bottle for filling carboy
   Injection Pump
   Pump tubing with weight
   Charged battery
   Graduated cylinder
   60-mL water bottles
   Whirl-pak bags
   Disposable Gloves
   Sieve
   Ion-specific probe
   Field books
   Weighed salts
   Meter stick
   Meter tape
   Flow meter
   Cooler with ice

Stopwatch
60-mL disposable syringes
Syringe filters
Laboratory Materials
Coffee grinder
Scintillation vials
96-well plates
Drying oven
Isotope ($K^{15}NO_3$)
Ammonium Chloride (NH_4Cl)
Potassium Phosphate (KH_2PO_4)
Sodium or Potassium Nitrate ($NaNO_3$ or KNO_3)
Disposable Gloves
4×6 mm tin capsules (Costech)
Microbalance

REFERENCES

Alexander, R.B., Smith, R.A., Schwarz, G.E., 2000. Effect of stream channel size on the delivery of nitrogen to the Gulf of Mexico. Nature 403, 758−761.

Alexander, R.B., Smith, R.A., Schwarz, G.E., Boyer, E.W., Nolan, J.V., Brakebill, J.W., 2008. Differences in phosphorus and nitrogen delivery to the Gulf of Mexico from the Mississippi River Basin. Environmental Science and Technology 42, 822−830.

American Public Health Association (APHA), 2012. Standard Methods for the Examination of Water and Wastewater, twenty-second ed. American Public Health Association, Washington, DC.

Bechtold, H.A., Marcarelli, A.M., Baxter, C.V., Inouye, R.S., 2012. Effects of N, P, and organic carbon on stream biofilm nutrient limitation and uptake in a semi-arid watershed. Limnology and Oceanography 57, 1544−1554.

Bernhardt, E.S., Likens, G.E., 2004. Controls on periphyton biomass in heterotrophic streams. Freshwater Biology 49, 14−27.

Bernhardt, E.S., Hall Jr., R.O., Likens, G.E., 2002. Whole-system estimates of nitrification and nitrate uptake in streams of the Hubbard Brook Experimental Forest. Ecosystems 5, 419−430.

Bernot, M.J., Dodds, W.K., 2005. Nitrogen retention, removal, and saturation in lotic ecosystems. Ecosystems 8, 442−453.

Burgin, A.J., Hamilton, S.K., 2008. NO_3^--driven SO_4^{2-} production in freshwater ecosystems: implications for N and S cycling. Ecosystems 11, 908−922.

Capps, K.A., Booth, M.T., Collins, S.M., Davison, M.A., Moslemi, J.M., El-Sabaawi, R.W., Simonis, J.L., Flecker, A.S., 2011. Nutrient diffusing substrata: a field comparison of commonly used methods to assess nutrient limitation. Journal of the North American Benthological Society 30, 522−532.

Carpenter, S.R., Caraco, N.F., Correll, D.L., Howarth, R.W., Sharpley, A.N., Smith, V.H., 1998. Nonpoint pollution of surface waters with phosphorus and nitrogen. Ecological Applications 8, 559−568.

Cordell, D., White, S., 2011. Peak phosphorus: clarifying the key issues of a vigorous debate about long-term phosphorus security. Sustainability 3, 2027−2049.

Corkum, L.D., 1996. Responses of chlorophyll-a, organic matter, and macroinvertebrates to nutrient additions in rivers flowing through agricultural and forested land. Archiv für Hydrobiologie 136, 391−411.

Costello, D.M., Rosi-Marshall, E.J., Shaw, L.E., Grace, M.R., Kelly, J.J., 2016. A novel method to assess effects of chemical stressors on natural biofilm structure and function. Freshwater Biology 61, 2129−2140.

Covino, T.P., McGlynn, B.L., McNamara, R.A., 2010. Tracer additions for spiraling curve characterization (TASCC): quantifying stream nutrient uptake kinetics from ambient to saturation. Limnology and Oceanography: Methods 8, 484−498.

David, M.B., Gentry, L.E., Kovacic, D.A., Smith, K.M., 1997. Nitrogen balance in and export from an agricultural watershed. Journal of Environmental Quality 26, 1038−1048.

David, M.B., Gentry, L.E., 2000. Anthropogenic inputs of nitrogen and phosphorus and riverine export for Illinois, USA. Journal of Environmental Quality 29, 494−508.

Davis, J.C., Minshall, G.W., 1999. Nitrogen and phosphorus uptake in two Idaho (USA) headwater wilderness streams. Oecologia 119, 247−255.

Diaz, R.J., Rosenberg, R., 2008. Spreading dead zones and consequences for marine ecosystems. Science 321, 926−929.

Dodds, W.K., 1997. Distribution of runoff and rivers related to vegetative characteristics, latitude, and slope: a global perspective. Journal of the North American Benthological Society 16, 162−168.

Dodds, W.K., Bouska, W.W., Eitzmann, J.L., Pilger, T.J., Pitts, K.L., Riley, A.J., Schloesser, J.T., Thornbrugh, D.J., 2009. Eutrophication of US freshwaters: analysis of potential economic damages. Environmental Science and Technology 43, 12−19.

Dube, M.G., Culp, J.M., Scrimgeour, G.J., 1997. Nutrient limitation and herbivory: processes influenced by bleached kraft pulp mill effluent. Canadian Journal of Fisheries and Aquatic Sciences 54, 2584−2595.

Duff, J.H., Triska, F.J., 2000. Nitrogen biogeochemistry and surface-subsurface exchange in streams. In: Jones, J.B., Mulholland, P.J. (Eds.), Streams and Groundwaters. Academic Press, San Diego, CA, pp. 197–220.

Elser, J.J., Bracken, M.E.S., Cleland, E.E., Gruner, D.S., Harpole, W.S., Hillebrand, H., Ngai, J.T., Seabloom, E.W., Shurin, J.B., Smith, J.E., 2007. Global analysis of nitrogen and phosphorus limitation of primary producers in freshwater, marine, and terrestrial ecosystems. Ecology Letters 10, 1135–1142.

Fairchild, G.W., Lowe, R.L., Richardson, W.B., 1985. Algal periphyton growth on nutrient-diffusing substrates: an in situ bioassay. Ecology 66, 465–472.

Fenn, M.E., Poth, M.A., Aber, J.D., Baron, J.S., Bormann, B.T., Johnson, D.W., Lemly, A.D., McNulty, S.G., Ryan, D.F., Stottlemyer, R., 1998. Nitrogen excess in North American ecosystems: predisposing factors, ecosystem responses, and management strategies. Ecological Applications 8, 706–733.

Fewtrell, L., 2004. Drinking-water nitrate, methemoglobinemia, and global burden of disease: a discussion. Environmental Health Perspectives 112, 1371–1374.

Francoeur, S.N., Biggs, B.J.F., Smith, R.A., Lowe, R.L., 1999. Nutrient limitation of algal biomass accrual in streams: seasonal patterns and a comparison of methods. Journal of the North American Benthological Society 18, 242–260.

Francoeur, S.N., 2001. Meta-analysis of lotic nutrient amendment experiments: detecting and quantifying subtle responses. Journal of the North American Benthological Society 20, 358–368.

Galloway, J.N., Dentener, F.J., Capone, D.G., Boyer, E.W., Howarth, R.W., Seitzinger, S.P., Asner, G.P., Cleveland, C.C., Green, P.A., Holland, E.A., Karl, D.M., Michaels, A.F., Porter, J.H., Townsend, A.R., Vörösmarty, C.J., 2004. Nitrogen cycles: past, present, and future. Biogeochemistry 70, 153–226.

Gregory, S.V., Swanson, F.J., McKee, W.A., Cummins, K.W., 1991. An ecosystem perspective of riparian zones. BioScience 41, 540–551.

Gruber, N., Galloway, J.N., 2008. An Earth-system perspective of the global nitrogen cycle. Nature 451, 293–296.

Hall, R.O., Tank, J.L., Sobota, D.J., Mulholland, P.J., O'Brien, J.M., Dodds, W.K., Webster, J.R., Valett, H.M., Poole, G.C., Peterson, B.J., Meyer, J.L., McDowell, W.H., Johnson, S.L., Hamilton, S.K., Grimm, N.B., Gregory, S.V., Dahm, C.N., Cooper, L.W., Ashkenas, L.R., Thomas, S.M., Sheibley, R.W., Potter, J.D., Niederlehner, B.R., Johnson, L.T., Helton, A.M., Crenshaw, C.M., Burgin, A.J., Bernot, M.J., Beaulieu, J.J., Arango, C.P., 2009. Nitrate removal in stream ecosystems measured by ^{15}N addition experiments: total uptake. Limnology and Oceanography 54, 653–665.

Hall, R.O., Baker, M.A., Rosi-Marshall, E.J., Tank, J.L., Newbold, J.D., 2013. Solute-specific scaling of inorganic nitrogen and phosphorus uptake in streams. Biogeosciences 10, 7323–7331.

Harpole, S.H., Ngai, J.T., Cleland, E.E., Seabloom, E.W., Borer, E.T., Bracken, M.E.S., Elser, J.J., Gruner, D.S., Hillebrand, H.H., Shurin, J.B., Smith, J.E., 2011. Nutrient co-limitation of primary producer communities. Ecology Letters 14, 852–862.

Helton, A.M., Poole, G.C., Meyer, J.L., Wollheim, W.M., Peterson, B.J., Mulholland, P.J., Bernhardt, E.S., Stanford, J.A., Arango, C., Ashkenas, L.R., Cooper, L.W., Dodds, W.K., Gregory, S.V., Hall, R.O., Hamilton, S.K., Johnson, S.L., McDowell, W.H., Potter, J.D., Tank, J.L., Thomas, S.M., Valett, H.M., Webster, J.R., Zeglin, L., 2011. Thinking outside the channel: modeling nitrogen cycling in networked river ecosystems. Frontiers in Ecology and the Environment 9, 229–238.

Hill, B.H., Elonen, C.M., Seifert, L.R., May, A.A., Tarquinio, E., 2012. Microbial enzyme stoichiometry and nutrient limitation in US streams and rivers. Ecological Indicators 18, 540–551.

Hoellein, T.J., Tank, J.L., Kelly, J.J., Rosi-Marshall, E.J., 2010. Seasonal variation in nutrient limitation of microbial biofilms colonizing organic and inorganic substrata in streams. Hydrobiologia 649, 331–345.

Hoellein, T.J., Tank, J.L., Rosi-Marshall, E.J., Entrekin, S.A., Lamberti, G.A., 2007. Controls on spatial and temporal variation of nutrient uptake in three Michigan headwater streams. Limnology and Oceanography 52, 1964–1977.

Holmes, R.M., Aminot, A., Kerouel, R., Hooker, B.A., Peterson, B.J., 1999. A simple and precise method for measuring ammonium in marine and freshwater ecosystems. Canadian Journal of Fisheries and Aquatic Sciences 56, 1801–1808.

Hood, J.M., McNeely, C., Finlay, J.C., Sterner, R.W., 2014. Selective feeding determines patterns of nutrient release by stream invertebrates. Freshwater Science 33, 1093–1107.

Howarth, R.W., Billen, G., Swaney, D., Townsend, A., Jaworski, N., Lajtha, K., Downing, J.A., Elmgren, R., Caraco, N., Jordan, T., Berendse, F., Freney, J., Kudeyarov, V., Murdoch, P., Zhao-Liang, Z., 1996. Regional nitrogen budgets and riverine N & P fluxes for the drainages to the North Atlantic Ocean: natural and human influences. Biogeochemistry 35, 75–139.

Johnson, L.T., Tank, J.L., Dodds, W.K., 2009. The influence of land use on stream biofilm nutrient limitation across eight North American ecoregions. Canadian Journal of Fish and Aquatic Sciences 66, 1081–1094.

Kemp, M.J., Dodds, W.K., 2001. Spatial and temporal patterns of nitrogen in pristine and agriculturally influenced streams. Biogeochemistry 53, 125–141.

Kemp, M.J., Dodds, W.K., 2002. The influence of variable ammonium, nitrate, and oxygen concentrations on uptake, nitrification, and denitrification rates. Limnology and Oceanography 47, 1380–1393.

Lansdown, K., McKew, B.A., Whitby, C., Heppell, C.M., Dumbrell, A.J., Binley, A., Olde, L., Trimmer, M., 2016. Importance and controls of anaerobic ammonium oxidation influenced by riverbed geology. Nature Geoscience 9, 357–360.

Lehman, J.T., 1976. Ecological and nutritional studies on *Dinobryon* Ehrenb.: seasonal periodicity and the phosphate toxicity problem. Limnology and Oceanography 21, 646–658.

Lutgens, F.K., Tarbuck, E.J., 1992. The Atmosphere: An Introduction to Meteorology. Prentice Hall, Upper Saddle River, NJ.

Mulholland, P.J., Tank, J.L., Webster, J.R., Bowden, W.B., Dodds, W.K., Gregory, S.V., Grimm, N.B., Hamilton, S.K., Johnson, S.L., Marti, E., McDowell, W.H., Merriam, J.L., Meyer, J.L., Peterson, B.J., Valett, H.M., Wollheim, W.M., 2002. Can uptake length in streams be determined by nutrient addition experiments? Results from an interbiome comparison study. Journal of the North American Benthological Society 21, 544–560.

Mulholland, P.J., Helton, A.M., Poole, G.C., Hall, R.O., Hamilton, S.K., Peterson, B.J., Tank, J.L., Ashkenas, L.R., Cooper, L.W., Dahm, C.N., Dodds, W.K., Findlay, S.E.G., Gregory, S.V., Grimm, N.B., Johnson, S.L., McDowell, W.H., Meyer, J.L., Valett, H.M., Webster, J.R., Arango, C.P., Beaulieu, J.J., Bernot, M.J., Burgin, A.J., Crenshaw, C.L., Johnson, L.T., Niederlehner, B.R., O'Brien, J.M., Potter, J.D., Sheibley, R.W., Sobota, D.J., Thomas, S.M., 2008. Stream denitrification across biomes and its response to anthropogenic nitrate loading. Nature 452, 202−205.

Mulholland, P.J., Hall, R.O., Sobota, D.J., Dodds, W.K., Findlay, S.E.G., Grimm, N.B., Hamilton, S.K., McDowell, W.H., O'Brien, J.M., Tank, J.L., Ashkenas, L.R., Cooper, L.W., Dahmn, C.N., Gregory, S.V., Johnson, S.L., Meyer, J.L., Peterson, B.J., Poole, G.C., Valett, H.M., Webster, J.R., Arango, C.P., Beaulieu, J.J., Bernot, M.J., Burgin, A.J., Crenshaw, C.L., Helton, A.M., Johnson, L.T., Niederlehner, B.R., Potter, J.D., Sheibley, R.W., Thomas, S.M., 2009. Nitrate removal in stream ecosystems measured by ^{15}N addition experiments: denitrification. Limnology and Oceanography 54, 666−680.

Newbold, J.D., Elwood, J.W., O'Neill, R.V., Van winkle, W., 1981. Measuring nutrient spiraling in streams. Canadian Journal of Fisheries and Aquatic Sciences 38, 860−863.

O'Brien, J.M., Dodds, W.K., Wilson, K.C., Murdock, J.N., Eichmiller, J., 2007. The saturation of N cycling in Central Plains streams: ^{15}N experiments across a broad gradient of nitrate concentrations. Biogeochemistry 84, 31−49.

Payn, R.A., Gooseff, M.N., Benson, D.A., Cirpka, O.A., Zarnetske, J.P., Bowden, W.B., McNamara, J.P., Bradford, J.H., 2008. Comparison of instantaneous and constant-rate stream tracer experiments through non-parameteric analysis of residence time distributions. Water Resources Research 44, W06404.

Peterson, B.J., Wollheim, W.M., Mulholland, P.J., Webster, J.R., L Meyer, J., Tank, J.L., Marti, E., Bowden, W.B., Valett, H.M., Hershey, A.E., McDowell, W.H., Dodds, W.K., Hamilton, S.K., Gregory, S., Morrall, D.D., 2001. Control of nitrogen export from watersheds by headwater streams. Science 292, 86−90.

Pringle, C.M., Paaby-Hansen, P., Vaux, P.D., Goldman, C.R., 1986. In situ nutrient assays of periphyton growth in a lowland Costa Rican stream. Hydrobiologia 134, 207−213.

Rabalais, N.N., Turner, R.E., Wiseman Jr., W.J., 2002. Gulf of Mexico hypoxia, a.k.a. "the dead zone". Annual Review of Ecology and Systematics 33, 235−263.

Reisinger, A.J., Tank, J.L., Dee, M.M., 2016. Regional and seasonal variation in nutrient limitation of river biofilms. Freshwater Science 35, 474−489.

Rosi-Marshall, E.J., Kincaid, D.W., Bechtold, H.A., Royer, T.V., Rojas, M., Kelly, J.J., 2013. Pharmaceuticals suppress algal growth and microbial respiration and alter bacterial communities in stream biofilms. Ecological Applications 23, 583−593.

Sabater, F., Butturini, A., Marti, E., Munoz, I., Romani, A., Wray, J., Sabater, S., 2000. Effects of riparian vegetation removal on nutrient retention in a Mediterranean stream. Journal of the North American Benthological Society 19, 609−620.

Scavia, D., Allan, J.D., Arend, K.K., Bartell, S., Beletsky, D., Bosch, N.S., Brandt, S.B., Briland, R.D., Daloğlu, I., DePinto, J.V., Dolan, D.M., Evans, M.A., Farmer, T.M., Goto, D., Han, H., Höök, T.O., Knight, R., Ludsin, S.A., Mason, D., Michalak, A.M., Richards, R.P., Roberts, J.J., Rucinski, D.K., Rutherford, E., Schwab, D.J., Sesterhenn, T., Zhang, H., Zhou, Y., 2014. Assessing and addressing the re-eutrophication of Lake Erie: central basin hypoxia. Journal of Great Lakes Research 40, 226−246.

Stream Solute Workshop, 1990. Concepts and methods for assessing solute dynamics in stream ecosystems. Journal of the North American Benthological Society 9, 95−119.

Schlesinger, W.H., Bernhardt, E.S., 2013. Biogeochemistry: An Analysis of Global Change, third ed. Academic Press, San Diego, CA.

Smil, V., 2000. Phosphorus in the environment: natural flows and human interferences. Annual Reviews of Energy and the Environment 25, 53−88.

Sobota, D.J., Johnson, S.L., Gregory, S.V., Ashkenas, L.R., 2012. A stable isotope tracer study of the influences of adjacent land use and riparian condition on fates of nitrate in streams. Ecosystems 15, 1−17.

Subalusky, A.L., Dutton, C.L., Rosi-Marshall, E.J., Post, D.M., 2015. The hippopotamus conveyor belt: vectors of carbon and nutrients from terrestrial grasslands to aquatic systems in sub-Saharan Africa. Freshwater Biology 60, 512−525.

Tanaka, T., Kawasaki, K., Daimon, S., Kitagawa, W., Yamamoto, K., Tamaki, H., Tanaka, M., Nakatsu, C.H., Kamagata, Y., 2014. A hidden pitfall in the preparation of agar media undermines microorganism cultivability. Applied and Environmental Microbiology 80, 7659−7666.

Tank, J.L., Webster, J.R., 1998. Interactions of substrate and nutrient availability on wood biofilm processes in streams. Ecology 79, 151−162.

Tank, J.L., Dodds, W.K., 2003. Responses of heterotrophic and autotrophic biofilms to nutrients in ten streams. Freshwater Biology 48, 1031−1049.

Tank, J.L., Rosi-Marshall, E.J., Baker, M.A., Hall, R.O., 2008. Are rivers just big streams? A pulse method to quantify nitrogen demand in a large river. Ecology 89, 2935−2945.

Trentman, M.T., Dodds, W.K., Fencl, J.S., Gerber, K., Guarneri, J., Hitchman, S.M., Peterson, Z., Rüegg, J., 2015. Quantifying ambient nitrogen uptake and functional relationships of uptake versus concentration in streams: a comparison of stable isotope, pulse, and plateau approaches. Biogeochemistry 125, 65−79.

Vitousek, P.M., Aber, J.D., Howarth, R.W., Likens, G.E., Matson, P.A., Schindler, D.W., Schlesinger, W.H., Tilman, D.G., 1997. Human alteration of the global nitrogen cycle: sources and consequences. Ecological Applications 7, 737−750.

Vitousek, P.M., Howarth, R.W., 1991. Nitrogen limitation on land and in the sea. How can it occur? Biogeochemistry 13, 87−115.

Ward, M.H., Mark, S.D., Cantor, K.P., Weisenburger, D.D., Correa-Villaseñor, A., Zahm, S.H., 1996. Drinking water nitrate and the risk of non-Hodgkin's lymphoma. Epidemiology 7, 465−471.

Winterbourn, M.J., 1990. Interactions among nutrients, algae and invertebrates in a New Zealand mountain stream. Freshwater Biology 23, 463−474.

Wold, A.P., Hershey, A.E., 1999. Spatial and temporal variability of nutrient limitation in 6 North Shore tributaries to Lake Superior. Journal of the North American Benthological Society 18, 2−14.

Webster, J.R., Patten, B.C., 1979. Effects of watershed perturbation on stream potassium and calcium dynamics. Ecological Monographs 49, 51−72.

Chapter 32

Nitrogen Transformations

Walter K. Dodds[1], Amy J. Burgin[2], Amy M. Marcarelli[3] and Eric A. Strauss[4]

[1]*Division of Biology, Kansas State University;* [2]*Ecology and Evolutionary Biology and Environmental Studies, Kansas Biological Survey and The University of Kansas;* [3]*Department of Biological Sciences, Michigan Technological University;* [4]*Department of Biology, University of Wisconsin—La Crosse*

32.1 INTRODUCTION

Nitrogen is an important limiting element in streams (Dodds and Smith, 2016), as well as a pollutant that stimulates unwanted algal growth. Further, ammonia and nitrate can both be toxic to vertebrates, including humans, at high concentrations (Thurston et al., 1981; Carmago et al., 2005). Transport of nitrogen through streams can lead to water quality problems in coastal areas (Alexander et al., 2000), yet rivers and streams retain and process substantial amounts of nitrogen. Streams are also a globally important source of nitrous oxide (Beaulieu et al., 2011), which is one of the most potent greenhouse gasses leading to global warming.

Biotic and abiotic factors influence transformations of nitrogen into different forms in the environment. The major forms of nitrogen (N) are organic N (including proteins, nucleic acids, urea, and various other compounds) and inorganic N (Dodds and Whiles, 2010). Major gaseous forms of inorganic N include N_2 gas as a major component of the atmosphere, nitrous oxide (N_2O), and ammonia (NH_3). Dissolved inorganic ionic forms of nitrogen include ammonium (NH_4^+), nitrite (NO_2^-), and nitrate (NO_3^-).

32.1.1 Nitrogen Fluxes

All organisms require nitrogen and must take it into their cells to maintain viability and grow. Nitrogen that is taken from the environment and converted into biological molecules in the cell is referred to as assimilated. Nitrogen can be assimilated in some organic forms (e.g., many heterotrophic organisms require amino acids from their diet to synthesize proteins). Plants and microbes can also assimilate inorganic nitrogen. The pathway to assimilate inorganic nitrogen requires it to be converted to NH_4^+ before it can be converted to amino acids and other organic compounds. However, other forms can be converted to NH_4^+ before assimilation (Fig. 32.1).

In addition to *assimilatory* pathways, *dissimilatory* pathways exist where some types of nitrogen are converted to others in energy-yielding pathways. Each of the forms of nitrogen has different reduction—oxidation (or redox, the relative

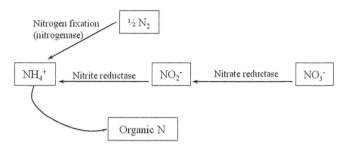

FIGURE 32.1 Assimilatory pathways of nitrogen uptake. All forms of inorganic N must be converted to ammonium. Note that use of nitrate requires more energy than ammonium and use of N_2 even more energy. *After Dodds and Whiles (2010).*

Methods in Stream Ecology. http://dx.doi.org/10.1016/B978-0-12-813047-6.00010-3

availability of free electrons) states, and relative to the environment they are in, can yield or require energy when they are converted among forms. Specifically, the more a compound differs in oxidation—reduction state from the overall chemistry of the water it is in, the greater the potential energy it has. So, converting this compound to the same potential energy as the rest of the chemicals dissolved in the water will yield energy. Conversely, converting a compound to a form that is vastly different from the redox state of the rest of the chemicals in solution requires potential energy.

The inorganic nitrogen compounds can be ordered by redox potential, aligned from most oxidized to most reduced $NO_3^- > NO_2^- > N_2O > N_2 > NH_4^+$. This order is not too difficult to remember as the more oxygen atoms a compound has and the fewer hydrogen atoms, the more oxidized it is. Dissolved oxygen (O_2) is a key determinant of oxidation—reduction state of the water in which compounds are dissolved—if O_2 is absent, then redox is low and electrons are plentiful (because the O_2 does not react with them). Thus, the presence of O_2 in large part dictates which nitrogen transformations dominate. Given the order of oxidation—reduction state, the conditions under which various nitrogen transformations will occur can be predicted (Fig. 32.2).

The final major class of elemental flux is excretory, which is mineralization or, in the absence of O_2, sometimes referred to as ammonification. Mineralization occurs when heterotrophs break down organic N compounds that are N-rich and they need to excrete the excess nitrogen. Generally this nitrogen is excreted in the form of NH_4^+ by aquatic organisms. It is important to understand this flux because it is what keeps all the nitrogen in a system from building up as organic nitrogen and in part controls transformation dynamics. Some of the fluxes in Figs. 32.1 and 32.2 are not very important in freshwaters (e.g., ANNAMOX, Schubert et al., 2006), and so they will not be considered in detail here. For more detail on methods to measure these fluxes, please refer to Huygens et al. (2013).

When measuring rates of these N cycle processes, we distinguish between potential and actual rates as well as between gross and net rates. A potential rate is the rate in the absence of limiting factors. In the natural environment, many biogeochemical processes can be limited by any of several things, including the amount of substrate available for a re-action, the amount of enzyme present to drive a reaction, temperature, light, and other factors. So, for an individual sample, if we optimize all conditions and measure the maximum rates possible given the enzymes that are present, these are referred to as potential rates. In contrast, we can mimic the natural environmental conditions as closely as possible and these are referred to as actual rates. Any dissolved N pool will have transformations that feed into that pool and those that use up the pool. The change in the entire size of the pool (net transformation rate) will be dictated by the total rate of the reaction transforming the N out of the pool (gross rate) minus the rate of the process supplying the pool.

This chapter will cover the basic methods of (1) *actual and potential denitrification*, (2) gross and net *nitrification*, and (3) *nitrogen fixation*. These constitute standard enzyme assays or net flux estimates. We then present advanced methods that take advantage of stable isotopes as N tracers and direct measurements of changes in gas concentrations to quantify fluxes that are more difficult to trace with standard enzyme assays, in the form of direct measures of (1) nitrogen fixation and (2) *dissimilatory nitrate reduction to ammonia* (DNRA).

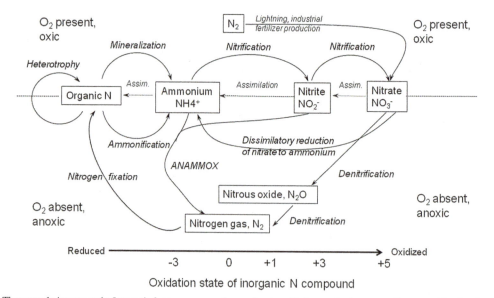

FIGURE 32.2 The general nitrogen cycle. Inorganic forms are arranged according to oxidation—reduction state with most oxidized at the right-hand side of the diagram. Some fluxes predominate in oxic conditions (top half of diagram) and others in anoxic (bottom half of diagram). *After Dodds and Whiles (2010).*

32.1.2 Small-Scale Assays for Fluxes

In this chapter, we cover common small-scale assays of various N transformation rates. These are generally "bottle" assays and may or may not scale to whole-system rates (Schindler, 1998). For a more detailed discussion of some whole-system rate measures, see Chapter 31 in this book. Readers are cautioned that bottle assays can lead to rate estimates that are different from those in the natural environment because samples must be disturbed to make such measurements, but for determining relative rates in comparative experiments, these estimates can be useful (Fig. 32.3A and B). Small incubations can diverge from in situ rates because the effect of the walls of the container reduces flow, light, and water exchange and can alter temperature. Generally incubations should be carried out in as short a time period as possible to get a reliable signal, and researchers often use a series of measures over time to verify that rates are consistent during longer incubations. Larger incubation vessels are more likely to capture heterogeneity but become more challenging to control and replicate as size increases and the more closely the operator is attempting to replicate in situ conditions (e.g., flow, temperature, light, replenishment of water).

In general, samples used to measure fluxes will be taken for the major types of substrata present in streams. Note that in this chapter we refer to the reactants that enzymes act on as substrates, but that solid materials on the bottom of the stream are substrata or substratum. Incubations in small vials are ideal for conducting many measurements in a short period of time and for comparing among different substrata (e.g., sediment, epiphytes, water column), but can be problematic for biofilms because they may require separating periphyton (see Chapter 12) from their growth substrata. Chambers can be used to measure biofilms attached to larger substrata such as wood, rocks, and organic matter, but also restrict the size the types of substrata that can be incubated and can only be run with a limited number of replicates at a time. Construction of recirculating chambers has been addressed in several papers (e.g., Dodds and Brock, 1998; Rüegg et al. 2015). Nonrecirculating chambers such as the polycarbonate containers depicted in Fig. 32.3D are ideal for incubating periphyton attached to their growth substrata and can be modified to create flow in the chambers using stir plates and bars or by attaching small pumps via Tygon tubing that circulates water across the growth substrata (Fig. 32.3C). Recirculating chambers such as that depicted in Fig. 32.3E provide the most realistic flow conditions, but they are expensive and may not be available to all researchers. Researchers are urged to consider their options for incubation containers at the onset of an experiment and select the one that is most appropriate for their substratum of interest, budget, and need for precise measurements versus replication.

32.2 GENERAL DESIGN

32.2.1 Basic Method 1: Denitrification—Determining Unamended Denitrification and Denitrification Enzyme Activity Rates

Denitrification is a series of dissimilatory microbial reductions of inorganic nitrogen, beginning with NO_3^- and ending with the production of N_2 gas. These reactions are essentially anaerobic respiration pathways that a wide range of organisms (from the archaea, bacteria, and fungi) are capable of performing when O_2 is not available as an electron acceptor. Determining rates of denitrification in streams and/or comparing denitrification among systems or treatments is often of great interest because NO_3^- is usually the most abundant form of inorganic N in streams and because the process can ultimately result in a loss of N from a stream as N_2 gas. Thus, denitrification is a critical process in regulating N availability and removal from ecosystems (Seitzinger et al., 2006).

Direct measurement of N_2 gas or use of isotopic techniques is often recommended to achieve actual unbiased rates of denitrification, but these methods are complex and require sophisticated equipment (see Section 32.4 on Advanced Methods). For a detailed review of different techniques used to measure denitrification, see Groffman et al. (2006). In this section, we will describe two similar methods to estimate denitrification that use acetylene gas as an inhibitor of the final step in the denitrification pathway—the reduction of N_2O to N_2. If this final step is inhibited, it is only necessary to measure the accumulation rate of N_2O in a sample to estimate denitrification. Incubation conditions can artificially stimulate synthesis of enzymes for denitrification (e.g., increase anoxic conditions or increase substrate availability), so chloramphenicol (an enzyme synthesis inhibitor) is also added to samples during incubation to limit the denitrification rate to that based on enzymes already in the collected sample.

The unamended denitrification method measures the accumulation of N_2O in samples that are incubated under anoxic conditions without substrate amendments. One criticism of this method has been that this method underestimates denitrification because acetylene also inhibits nitrification (NO_3^- production). If denitrifiers are limited by the availability of NO_3^-, a reduction in NO_3^- production can have a negative effect on the measured denitrification rate.

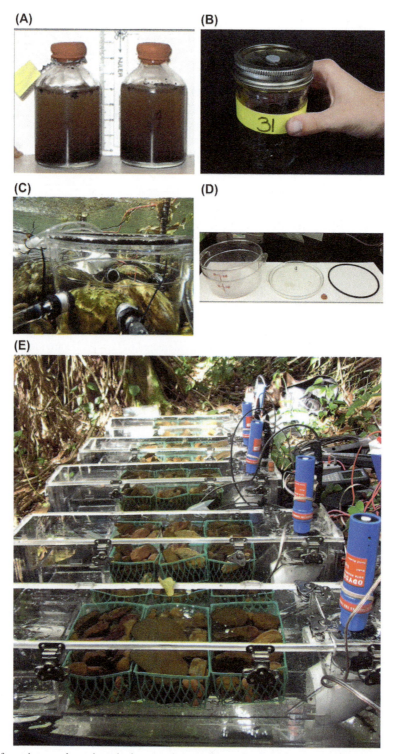

FIGURE 32.3 Examples of containers used to estimate in situ rates where gas flux needs to be determined. Serum vials (A) and canning jars (B) are ideal to incubate many replicates and/or small volumes of water or sediment, but may require separating periphyton from their growth substrata. Chambers can be circulating (C, E) or noncirculating (D). Nonrecirculating chambers such as the polycarbonate containers in (D) are ideal for incubating periphyton attached to their growth substrata and can be modified to create flow in the chambers using stir plates and bars or by attaching small pumps via Tygon tubing that circulate water across the growth substrata (C). Recirculating chambers in (E) can provide the most realistic flow conditions, but are expensive and can only be run with a limited number of replicates at a time. These chambers are being run streamside, and each has a light probe and oxygen probe attached. They can be fitted with septa for gas sampling and incubated submerged for temperature control (see Rüegg et al., 2015). *Photo credits: (A) A. Marcarelli; (B) E. Strauss; (C) M. Schenk; (D) E. Eberhard; (E) L. Koenig.*

In contrast, denitrification enzyme activity (DEA) measures denitrification potential in the presence of added substrates (i.e., glucose and NO_3^-). Thus, DEA measures full denitrification expression, that is, the potential of a sample given the extant level of denitrification enzymes (Groffman et al., 1999). DEA measurements alone are useful to show relative denitrification rates among sites or treatments. More specifically, DEA measurements can be used to identify denitrification hotspots within a stream or to compare denitrification potentials among streams. Even though DEA rates are considered "potentials," they do not always exceed ambient rates measured with more sophisticated techniques (Findlay et al., 2011). Rates measured with the DEA method should not be considered actual rates of denitrification nor be used to estimate N budgets because of the added NO_3^- and organic carbon. However, efforts have been made to combine measurements of NO_3^- availability, nitrification, unamended denitrification, and DEA to estimate actual NO_3^- flux in riverine ecosystems (Richardson et al., 2004).

32.2.2 Basic Method 2: Nitrification—Determining Gross and Net Nitrification Rates

Nitrification is a key microbial two-step transformation in the nitrogen cycle because it is the only natural pathway whereby nitrate is produced within a system. The energy gain from this aerobic chemoautotrophic process is relatively low, and rates are generally low compared to other nitrogen cycle processes. However, because the process can operate at low rates even with relatively low ammonium concentrations (Dodds and Jones, 1987), it occurs in many environments. Naturally low nitrification rates consequently necessitate longer incubation times compared to measuring rates of other processes, especially if isotopes are not used. Even though the rates are low, studies have shown nitrification rates to be influenced by important environmental factors including light, temperature, O_2, ammonium availability, pH, organic carbon availability, and C:N ratio (Strauss and Lamberti, 2000; Strauss et al., 2002). Nitrifying bacteria are generally attached to substrata within the stream, including sediment, fine particulate organic matter, and algae, and thus measurement of nitrification rates will involve incubating one or more of these materials. The methods described below can easily be modified to test the effect of different environmental conditions.

Gross nitrification is the absolute amount of ammonium converted to nitrate under oxic conditions. Gross rates can be estimated as the difference in ammonium concentrations between incubations in which nitrification was inhibited with the chemical nitrapyrin (2-chloro-6-[trichloromethyl]-pyridine) and those in which nitrification was allowed to occur. It is assumed that mineralization and ammonium assimilation are uninhibited in both incubations and that the ammonium increases in the incubations containing nitrapyrin is a result of inhibited ammonium oxidation. The method here describes a laboratory incubation of sediment and stream water placed in flasks. Modifications can be made to substitute other substrata depending on the research objective. When using sediment, studies often only use the uppermost 5 cm of sediment from the stream bottom.

Net nitrification is the change in nitrate availability through time and is the difference between gross nitrification and assimilatory/dissimilatory nitrate reduction. The net nitrification method below is similar, yet simpler, than the method for gross nitrification. Net nitrification can be measured simply as the change in nitrate concentration within a single incubation of substratum in stream water. As with gross nitrification, the method presented here will use sediment as the substratum but other substrata can be substituted depending on the objective.

32.2.3 Basic Method 3: Nitrogen Fixation

Nitrogen fixation (N_2 fixation), the transformation of nitrogen gas (N_2) to ammonium, can only be performed by certain heterotrophic bacteria, archaea, and cyanobacteria (Raymond et al., 2004). All of these organisms catalyze this reduction reaction using the enzyme nitrogenase. Even with this enzyme, the N_2 fixation reaction is energetically expensive, and nitrogenase is strongly inhibited by the presence of O_2, creating a unique challenge for N_2-fixing organisms that must expend energy and resources carrying out the fixation reaction as well as protecting the enzyme from O_2. In streams, N_2 fixation is primarily carried out by autotrophic cyanobacteria, particularly *Nostoc*, *Anabaena*, and *Calothrix* (Whitton, 2012). Species within these genera all form heterocysts, which are specialized, thick-walled cells where the N_2 fixation reaction occurs and the enzyme is protected from O_2. Although free-living, unicellular, nonheterocystous cyanobacteria have not been observed fixing N_2 in any stream to date, diatoms of the order Rhopalodiales, which host N_2-fixing cyanobacterial endosymbionts related to the unicellular *Cyanothece* sp. (Prechtl et al., 2004) are commonly found growing as epiphytes and on hard substrata in many streams.

The *acetylene reduction* assay is an indirect method for estimating N_2 fixation by measuring the activity of the nitrogenase enzyme (Stewart et al., 1967; Flett et al., 1976). The assay works because the nitrogenase enzyme recognizes the triple bond between C atoms in a molecule of acetylene (C_2H_2; $H-C \equiv C-H$) as equivalent to the triple covalent bond between N atoms in an N_2 molecule. Nitrogenase will break one of the bonds between C atoms in acetylene, in the process of converting the molecule to ethylene (C_2H_4). Nitrogenase enzyme activity is estimated in the acetylene reduction assay by introducing acetylene to an airtight container or chamber along with the N_2 fixer of interest and measuring the amount of ethylene produced over a known time period. The rate at which ethylene gas is produced is related to the potential N_2 fixation rate.

Nitrogen fixation rates are dependent on a variety of environmental conditions that limit either the rate of the process itself (e.g., supply of enzyme cofactors, presence of O_2) or the energy available to the organisms that can be dedicated to carry out the reaction. Any of the factors that limit or constrain primary production in streams (e.g., light, nutrient supply, flow conditions) can also limit or constrain rates of N_2 fixation for autotrophic N_2 fixers. Because N_2 fixation rates in streams are particularly light sensitive, most researchers have opted to conduct acetylene reduction assays in the field, where light and temperature conditions can be easily maintained by conducting incubations in containers submerged in the study streams.

The acetylene reduction assay must be carried out in an airtight arena, and a variety of options have been used to conduct these measurements in streams, from simple to complex (Fig. 32.3). The simplest approach is to seal the substratum or biofilm of interest suspended in water into a small, glass container such as a serum vial (Fig. 32.3A). If the study is focused on autotrophic N_2 fixers such as cyanobacteria, then the chambers must be transparent and attention should be paid to the wavelengths of light transmitted through the selected material, as well as its ability to bind organic gasses such as acetylene and ethylene. Glass is ideal for light transmission and lack of reaction with organic gasses, but is not practical for large chambers. Plexiglas and polycarbonate are not reactive with organic gasses and transmit most visible light, but both absorb various wavelengths of infrared and ultraviolet light (although some ultraviolet-transmissive Plexiglas is available to build chambers; Dodds and Brock, 1998). Polyethylene and other plastics should be avoided as they can be very reactive with organic gasses such as acetylene and ethylene. The chamber or vial selected must be outfitted with at least one sampling port where septa can be placed to allow collection of gas samples using a syringe and needle.

To conduct the assay, substrata are enclosed in the chamber either suspended in or with a similar volume of overlying water. A headspace equal to about 10% of the container volume composed primarily of acetylene is introduced and the container is agitated to dissolve the acetylene in the water in equilibrium with the headspace, and then an initial headspace sample is collected to document conditions at the start of the incubation. The method used to introduce headspace can vary depending on the total volume of the chamber—for small volumes headspaces can easily be introduced using a syringe, while others have used a small balloon sealed in the chamber and then popped through a septa to introduce larger volumes of acetylene (e.g., Grimm and Petrone, 1997). Following several hours of incubation, the chamber is again agitated to equilibrate gas between the water and headspace, after which a headspace sample is collected to measure the ethylene produced during the incubation. The rate of ethylene production is typically linear for 6–8 h and then declines as the supply of acetylene is depleted. Therefore, most researchers conduct their incubations for 2–4 h, although it is strongly suggested that researchers conduct their own time-course incubations to determine ideal incubation duration to detect production of ethylene with confidence while avoiding the decline in production that occurs over long time periods. Gas samples are stored in gastight containers and subsequently analyzed for initial and final concentrations of ethylene using a gas chromatograph (GC).

32.2.4 Advanced Methods: Isotopes for Flux Rate Measurement

Additional power to detect biogeochemical processes and more refined estimates of transformation rates can be gained by incorporating the stable isotope ^{15}N into bottle rate measurements. Indeed, some N cycling processes—most notably DNRA—can only be measured using ^{15}N tracers. The advantage of increased resolution to detect biogeochemical changes using ^{15}N tracers is balanced by the challenges associated with the high costs of isotopically labeled chemicals, accessibility to instruments to measure isotopes (e.g., isotope ratio mass spectrometer, IRMS), and the need for additional training necessary to understand optimal quality control measures and contamination prevention.

Membrane inlet mass spectrometry (MIMS) is another specialized technique using a mass spectrometer (mass spec) often employed by nitrogen biogeochemists to gain additional power in measuring transformation rates, most notably for denitrification (N_2 production). MIMS can be used to also measure N_2 fixation, and most recently a method has been validated for using MIMS to measure DNRA (ox-MIMS; Yin et al., 2014). The principle of the MIMS measurement is based on how gases dissolve into water; therefore, the instrument also often referred to as a dissolved gas analyzer (available from Bay Instruments, Kana et al., 1994). In essence, gases dissolve into water based on known physics of solubility related to temperature, salinity, and relative humidity. Sample water with these dissolved gases is passed through gas-permeable tubing (silicon membrane) housed in a vacuum-tight chamber, so gases pass through the membrane into a vacuum and are then transferred to the mass spec. The mass spec collects signals corresponding to different masses. The signals of specific atomic mass units (amu) can be related to different gaseous components, including $4 =$ helium; $18 =$ water vapor; $28-30 =$ dinitrogen (using isotopes $^{14,14}N$, $^{14,15}N$, $^{15,15}N$); $32 =$ oxygen; $40 =$ argon. The amplitude of the signal is proportional to the amount of the gas present, and by comparing it to a standard of known concentration (and knowing the linearity of the instrument), the mass of a certain gas present in a given sample can be calculated. Some gases are biologically inert (e.g., Ar) and can therefore be used as conservative tracers to correct for physical processes and compared to changes in biologically active gases (e.g., N_2). Once you have access to a MIMS, samples are relatively easy and cheap to collect, but can be time-consuming to process (~ 5 min per sample and we know of no autosampler for this equipment).

32.3 SPECIFIC METHODS

32.3.1 Basic Method 1: Denitrification

32.3.1.1 General Preparation

1. Gas chromatograph: Both denitrification methods described here require measuring N_2O gas concentrations with a GC. Instrument configurations vary widely depending on manufacturer, detectors, columns, and sampler, but the basic N_2O analysis configuration would be a GC equipped with a ^{63}Ni electron capture detector at $350°C$ and a GC column similar to Porapak Q 80/100. Consultation with your GC manufacturer is advisable for ideal configuration.

2. Laboratory gases: Measurement of unamended denitrification and DEA will require the use of several gases in the laboratory. Gas purity is important as noted here.

   a. GC carrier gas: Usually 95% argon—5% methane (P5) or ultrahigh purity N_2 are used as the GC carrier gas for N_2O analysis.

   b. Gas standards: Calibration of GC with gas standards of known N_2O concentration will be necessary. A good series of standards to have on hand would be 1, 10, 100, and 1000 ppm$_v$ N_2O (balance gas is N_2). Small cylinders of these high-quality N_2O gas standards can be purchased from many specialty gas companies. Intermediate standard concentrations can be made via syringe dilutions with atmospheric air. For example, to produce a 50 ppm$_v$ N_2O gas standard, draw 2.5 mL of atmospheric air into a gas syringe already containing 2.5 mL of 100 ppm$_v$ N_2O. The first time you run this analysis you should use a wide range of standards from 0 to 1000 ppm$_v$. Eventually, you might be able to adjust the actual standard concentrations based on the level of denitrification activity in your system.

   c. Acetylene: High-quality contaminant-free acetylene gas is used to inhibit the reduction of N_2O during the incubation period. The preferred type of acetylene gas is atomic absorption grade which can be purchased from specialty gas companies. This gas can be used immediately without further purification. Industrial grade, such as welding grade acetylene can also be used but acetone and other contaminants must first be removed by bubbling gas through a concentrated H_2SO_4 trap and then a distilled/deionized water trap prior to use.

   d. Anoxic environment gas: To ensure an anoxic environment during the incubation, the samples are flushed with an O_2-free gas. Oxygen-free Ar, N_2, or He is routinely used for this purpose.

3. DEA solution: The goal of DEA analysis is to measure the full expression of denitrification enzymes present in the sample at the time of collection. This DEA solution will be added to samples before incubation to supply denitrifiers with an available source of high-quality organic carbon and NO_3^- to alleviate substrate limitation. The solution also contains chloramphenicol to inhibit new enzyme synthesis. Dissolve 1.01 g of KNO_3, 0.30 g of glucose, and 1.00 g of chloramphenicol in a 1000-mL volumetric flask with about 900 mL of deionized water. Wrap the flask with aluminum foil to protect it from light, place the flask with a Teflon stir bar on a stir plate, and stir until the chloramphenicol completely dissolves (usually requires stirring overnight). Bring the solution up to volume with deionized water and transfer the solution to a 1-L amber bottle. Store in the refrigerator. This solution may also be made without glucose and/or NO_3^- to test for C and/or N limitation. A chloramphenicol-only solution will be used for measuring unamended denitrification.

4. Incubation vessels: Just about any airtight container with septa access should be suitable. Many studies have used Erlenmeyer flasks or media bottles. An inexpensive option is canning jars sealed with standard canning rings and lids (Fig. 32.3). The canning lids will need to be hole-punched and fitted with a septum. Heating the canning lids in hot water prior to use may help soften the plastisol sealant and improve the airtight seal. It is also possible to incubate an intact sediment core in a stoppered plastic tube as long as there is septa access to the airtight headspace. If incubating intact cores, diffusion of acetylene and DEA solution can be limited. If the vessel is not glass, N_2O adsorption to the vessel should be tested before using. Regardless of vessel used, the actual headspace volume will need to be known and should be between 100 and 300 mL.

32.3.1.2 Denitrification Rate Procedures (Laboratory)

Incubation Vessel Setup

1. To each incubation vessel, add 25 mL sediment and 20 mL stream water. If using sediment cores, the uppermost 5 cm of sediment from a 2.54-cm ID core would be 25.3 mL. The sediment and water volumes listed are for a vessel volume of approximately 200–400 mL; if a larger or smaller container is used, scale the materials added proportionally. If sediment and/or water volumes are changed, the volume of amendment solution added in step 2/3 will also need to be changed proportionally. Record the number of each vial and contents on a data sheet (Table 32.1).
2. For unamended denitrification samples, add 5 mL of 1 g chloramphenicol/L solution to each vessel.
3. For DEA measurement, instead add 5 mL of DEA solution.
4. Seal lid on incubation vessel. It is important that the seal is airtight and capable of withstanding both positive and negative pressure.
5. Insert a needle (c. 22 gauge) attached to vacuum line/pump into the vessel and evacuate for 90 s (Fig. 32.3).
6. Add anoxic environment gas to the vessel to a small positive pressure (c.30 kPa).
7. Repeat evacuation and gas flushing for a total of three cycles.
8. Bring vessel to atmospheric pressure by venting excess gas through a water-filled syringe (Fig. 32.3).
9. Add acetylene to sample headspace. The volume added should equal 10% of headspace. Note the time acetylene is added to each vessel; this is the start of the incubation period.

Incubation, Sampling, and Analyses

1. Place samples on an orbital shaker (175 rpm) and incubate at desired temperature for 6 h (3 h for DEA).
2. Collect and store a time series of gas samples from each vessel.
3. Using a 5-mL disposable syringe and needle (c. 22 gauge), collect a 3-mL gas sample at 1, 2, 4, and 6 h for unamended denitrification samples. For DEA samples, collect samples at 0.5, 1.0, 1.5, and 3 h. Insert needle into vessel and pump syringe $3\times$ before collecting gas sample. *Note*: if total headspace volume of vessel is less than 100, 3 mL of acetylene should be returned to the vessel after sampling to maintain positive pressure inside the vessel.
4. Place gas sample into an evacuated 2-mL serum bottle crimp-sealed with a butyl rubber septa.
5. If N_2O concentrations will not be determined within 24 h, N_2O standards for the standard curve should also be stored in evacuated serum bottles and stored with samples. This will help account for any container effects on gas samples. For example, N_2O can bind to certain types of septa material.
6. Using a GC calibrated with N_2O gas standards, determine N_2O concentration (ppm_v) in the gas samples.

Calculation of Denitrification Rate

1. For each gas sample collected and analyzed on the GC, convert N_2O concentration from volume units to mass units. That is, convert ppm_v N_2O to μg N_2O-N/L:

$$C_m = \frac{C_v \cdot M \cdot P}{R \cdot T} \tag{32.1}$$

where C_m = mass/volume concentration (μg N_2O-N/L) in headspace; C_v = ppm_v or volume/volume concentration in μL/L; M = mole weight of nitrogen in N_2O (28 g/mole of N_2O); P = pressure in atmospheres (ATM = 760 mmHg); R = universal gas constant (0.0820575 L $\cdot$ ATM/K $\cdot$ mole); T = room temperature in Kelvin (K = °C + 273.15).
2. Calculate the total mass of N_2O-N per sample, accounting for N_2O dissolved in water and in gas phase:

$$C_T = C_m \cdot (V_g + V_l \cdot \beta) \tag{32.2}$$

where C_T = total mass of N_2O-N in vessel; V_g = headspace volume (L) in vessel; V_l = liquid volume (L) in vessel; β = Bunsen coefficient = $1.2407 - 0.0398 \cdot temp + 0.0005 \cdot temp^2$; temp = incubation temperature in °C.

TABLE 32.1 Data sheet for recording data necessary for calculating denitrification rate. See also Online Worksheet 32.1.

Date:

Project:

Investigators:

Sample	Incubation Vessel ID	Treatment	Sediment Volume (mL)	Liquid Volume (mL)	Head-space Volume (mL)	Incubation Temperature (°C)	Incubation Start Time	T_1 Time	T_1 N_2O (ppm$_v$)	T_2 Time	T_2 N_2O (ppm$_v$)	T_3 Time	T_3 N_2O (ppm$_v$)	T_4 Time	T_4 N_2O (ppm)

TABLE 32.2 Data sheet for recording data necessary for calculating gross nitrification rate. See also Online Worksheet 32.1.

Date:

Project:

Investigators:

Sample	Flask ID	Nitrapyrin or DMSO	Sediment Volume (mL)	Liquid Volume (mL)	Incubation Temperature (°C)	Incubation Start Time	T_0 Time	T_0 NH_4^+ (mg N/L)	T_F Time	T_F NH_4^+ (mg N/L)
1		N								
1		D								
2		N								
2		D								
3		N								
3		D								
4		N								
4		D								

3. Calculate denitrification rates in μg N₂O-N/h:

$$\text{Denitrification Rate} \left(\mu g \ N_2O\text{-}Nh^{-1} \ S^{-1} \right) = \frac{C_{Tf} - C_{Ti}}{\Delta t \cdot S} \tag{32.3}$$

where C_{Tf} = total final mass of N₂O-N (~4 h sample for unamended denitrification or ~1.5 h sample for DEA); C_{Ti} = total initial mass of N₂O-N (2 h sample for unamended denitrification or 0.5 h sample for DEA); Δt = change in incubation time (h); S = sediment characteristic in which to express rate (e.g., surface area calculated from core cross section, dry mass of sediment, ash-free dry mass (AFDM), or sediment volume).

Tips and Notes

1. Measuring a minimum of five replicate incubations per experimental unit (e.g., five incubations per site within a stream) is recommended because of heterogeneity.
2. Gas samples collected at different times may be used when determining denitrification rate; however it should be established that the increase in N₂O is linear over the time interval used. An alternative technique to calculate denitrification rate is to plot N₂O concentration against sample time and calculate denitrification rate as the slope of the line.

32.3.2 Basic Method 2: Nitrification

32.3.2.1 Gross Nitrification Procedure

Day 0 Flask Setup

1. For each sample, you will need two 125-mL Erlenmeyer flasks. One flask will contain sediment, stream water, and the nitrapyrin nitrification inhibitor solution (this flask will hereafter be referred to as the "N flask"). The other flask, the "D flask," will contain sediment, stream water, and dimethyl sulfoxide (DMSO, the solvent in the inhibitor solution). Record each flask number and contents on a data sheet (Table 32.2).
2. For each flask, place 25 mL of sediment into a labeled 125-mL Erlenmeyer flask using a plastic powder funnel. If using sediment cores, the uppermost 5 cm of sediment from a 2.54 cm ID core would be 25.3 mL as calculated for the volume of a cylinder.
3. Rinse funnel with exactly 81 mL of site water, catching all rinse water in the flask (i.e., flask should contain only 25 mL sediment plus 81 mL site water).

4. To all N flasks add 20 μL of nitrapyrin solution (1 g/20 mL dissolved in DMSO) and to all D flasks add 20 μL of DMSO. *Note*: be careful when handling DSMO;[1] while the DMSO is not toxic, it is a "super solvent" that will allow toxic chemicals dissolved in it to easily pass into the skin.
5. Tightly cover flask openings with Parafilm and shake to thoroughly mix the contents.
6. Immediately remove the Parafilm and pipette 6 mL of the slurry into a labeled 15-mL plastic centrifuge tube already containing 6 mL of 2 N KCl. This step is required to mobilize positively charged NH_4^+ ions that are ionically bound to negatively charged sediment material.
7. Cover centrifuge tubes with Parafilm. Allow NH_4^+ to extract from the samples for approximately 60 min, inverting to mix samples at least every 10 min.
8. During the NH_4^+ extraction step, initiate the nitrification incubation. Place the N and D flasks on an orbital shaker at 175 rpm. To discourage microbial contamination and facilitate aeration, loosely cover each flask with a piece of aluminum foil but do not recover flasks with Parafilm. Incubate samples for 3 days in the dark at ambient temperature.
9. Returning to the NH_4^+ extracted samples in the centrifuge tubes, filter each sample through a glass fiber filter (GFF) (e.g., Pall Type A/E or Whatman GF/C or GF/F). Brief centrifugation of samples before filtration will settle much of the sediment and ease filtration. Store filtered sample at 4°C in a clean 20-mL scintillation vial or other small bottle until analyzed for NH_4^+. If analysis will not be completed within 24 h, add 200 μL of 10% H_2SO_4 to each sample.

Day 3 Laboratory Procedures

1. After the 3-day incubation, remove N and D flasks from the shaker.
2. Collect Day 3 samples, extract NH_4^+, filter, and prepare sample as described for Day 0 samples.
3. Analyze samples for NH_4^+-N concentration using the method of your choice. For a detailed fluorometric protocol to quantify NH_4^+-N, please see Chapter 36 in this volume.
4. If samples were acidified and stored, neutralize the acid with 200 μL of 3.6 M NaOH before NH_4^+ analysis.
5. Regardless of the method, be sure your standards account for the DMSO and KCl in the samples (i.e., all standards should be made in a matrix solution of 10 μL DMSO/100 mL 1.0 N KCl).

Calculation of Nitrification Rate

1. Express the rate in appropriate units (e.g., μg N · mL sediment^{-1} d^{-1}, or g N m^{-2} d^{-1}). Various units are obtained by using mass versus moles of nitrogen nitrified, different time units, and different sediment/substratum characteristics.
2. In general, nitrification rate can be calculated as:

$$\text{Gross Nitrification Rate} = \frac{(\Delta N - \Delta D)}{t \cdot S} \tag{32.4}$$

where ΔN = change (final−initial) in NH_4^+-N mass in the N flask during the incubation period; mass can be calculated by multiplying the concentration of NH_4^+ (mg N/L) by the volume of the sample (0.1 L); ΔD = change in NH_4^+-N mass in the D flask during the incubation period; t = actual time of incubation (e.g., days or hours); and S = sediment characteristic in which to express the nitrification rate. Examples may include surface area calculated from core cross section, dry mass of sediment, AFDM, or volume of sediment.

Tips and Notes

1. Measuring a minimum of five replicate incubations per experimental unit is recommended because of heterogeneity.
2. An incubation duration of 3 days has been shown to produce linear changes in NH_4^+-N concentration in samples from numerous stream ecosystems (Strauss et al., 2002). However, collecting more samples throughout the incubation period and regressing NH_4^+-N concentration against time for linearity over the incubation is prudent.
3. Be cautious of contamination of control flasks (D flasks) with nitrapyrin (e.g., pipette tips, bottle caps, etc.). Nitrapyrin is highly effective at inhibiting nitrification and even trace amounts can introduce significant error.

32.3.2.2 Net Nitrification Procedure

Day 0 Flask Setup

1. For each flask, place 25 mL of sediment into a labeled 125-mL Erlenmeyer flask using a plastic powder funnel.
2. Rinse funnel with exactly 85 mL of site water, catching all rinse water in the flask

---

1. Wear personal protective equipment including lab coat, eyewear, and gloves.

3. Tightly cover flask openings with Parafilm and shake to thoroughly mix the contents.
4. Immediately remove the Parafilm and pipette 10 mL of the slurry into a labeled 15-mL plastic centrifuge tube.
5. Place the flasks on an orbital shaker at 175 rpm. To discourage microbial contamination and facilitate aeration, loosely cover each flask with a piece of aluminum foil but do not recover flasks with Parafilm. Incubate samples for 3 days in the dark at ambient temperature.
6. Returning to the samples in the centrifuge tubes, filter each sample through a GFF (e.g., Pall Type A/E or Whatman GF/C or GF/F). Brief centrifugation of samples before filtration will settle much of the sediment and ease filtration. Store filtered sample at 4°C in a clean 20-mL scintillation vial until analyzed for NO_3^-. If analysis will not be completed within 24 h, add 200 μL of 10% H_2SO_4 to each sample.

Day 3 Laboratory Analyses

1. After the 3-day incubation, remove the flasks from shaker.
2. Collect Day 3 samples, filter, and prepare sample as described for Day 0 samples.
3. Analyze samples for NO_3^--N concentration using the method of your choice. If samples were acidified and stored, neutralize the acid with 200 μL of 3.6 M NaOH before NO_3^- analysis.

Calculation of Nitrification Rate

Net nitrification rate can be calculated as:

$$\text{Net Nitrification Rate} = \frac{\Delta NO_3^--N}{t \cdot S} \tag{32.5}$$

where $\Delta N =$ change (final−initial) in NO_3^--N mass in the flask during the incubation period; mass can be calculated by multiplying the concentration of NO_3^- (mg N/L) by the volume of the sample (0.1 L); $t =$ actual time of incubation; and $S =$ substratum characteristic in which to express the nitrification rate (i.e., per unit mass, area, or volume).

Tips and Notes

1. Measuring a minimum of five replicate incubations per experimental unit is recommended because of inherent heterogeneity.
2. Since this method intentionally examines the combined effect of several processes and does not single out a particular process (e.g., nitrification alone), one should use caution in interpreting the measured rates.
3. You can also measure the change in NH_4^+-N concentration using this method as an estimate of net N mineralization. If you do this though, you will need to extract the NH_4^+ bound to sediment material before measuring NH_4^+ concentration as described in the gross nitrification method.

32.3.3 Basic Method 3: Nitrogen Fixation

32.3.3.1 General Preparation

Gas Chromatograph Configuration

1. A flame ionization detector (FID)−equipped GC is needed to analyze ethylene. Many configurations are available and it is recommended that the manufacturer of the specific brand of machine used be consulted for optimal setup.
2. Several columns can be used to analyze ethylene—the two most common are Poropak T and Hayesep T. Example GC configurations with a Hayesep T column are ultrahigh purity helium carrier at 30 psi, ultrahigh purity hydrogen to FID at 35 psi, column temperature 40°C, FID temperature 180°C. These settings result in clear separation of ethylene and acetylene peaks with retention times of about 1.5 and 3.8 min, respectively.
3. Although you do not need to know the size of the acetylene peak to calculate fixation rates, you do need to ensure that all of the acetylene gas has moved off the column before running the next sample. Ramping up the column temperature after the ethylene peak is recorded can move the acetylene through the column more quickly and shorten the analysis time per sample.
4. Similar to denitrification, you will also want to purchase high-purity, premixed standards for calculating concentrations of ethylene; 100 or 1000 ppm ethylene in He are useful concentrations that can be easily diluted to create lower concentration standards depending on rates observed in your study systems.

Site Selection, Preparation, and Incubation Decisions

1. Select the desired stream reach for the experiment, and determine the dominant substrata type(s) for the incubation. Make sure the reach has an area with water depths adequate to submerge incubating samples with light conditions/shading typical of the overall study reach.

2. Select the enclosure to be used for the incubation and determine the total volume of that container. Decide on target volumes to be used for the measurements, such that (1) substratum volume equals 30–45% of total volume, (2) water volume equals 45–60% of total volume, and (3) headspace equals 10% of total volume.
3. Procure enough acetylene for all assays. For discussion on acetylene purity see information under preparations for denitrification.

32.3.3.2 Field Procedures

Setting up a Run

1. Fixation should be run as close to the middle of the day as possible, since rates may be light-dependent and closely related to patterns in primary production. Alternately, the incubation could be repeated over a 24-h period to characterize how rates change with light availability.
2. Fill the enclosure with substratum to be measured (e.g., rocks or wood), water, and a balloon filled with acetylene, if using for a larger enclosure (e.g., chamber). Record number and contents of each container on a data sheet (Table 32.3).
3. Place enclosure in the water and seal.
4. Repeat for the desired number of replicates per site. Also set up several "blanks" with water only to account for any enclosure effects or impurities in acetylene.

Initializing the Assay

1. If using chambers, measure the temperature in the chamber by inserting a thermometer into the sampling port. Record the temperature separately for each chamber; this number is necessary to calculate gas concentrations. With smaller incubation enclosures, measure stream temperature.
2. Replace the septa underwater, making sure that there are no air bubbles.
3. Pop the balloon in the chamber by inserting a needle through the sampling port, or add the headspace by withdrawing the desired volume of water and replacing with the acetylene (may require two injection ports to pull water from one while pushing gas into another).
4. Agitate the enclosure for 20 s to equilibrate the gas dissolved in the water with that in the headspace (not necessary for recirculation chambers).

TABLE 32.3 Data sheet for in situ acetylene reduction assays to measure N_2 fixation. See also Online Worksheet 32.1.

Date:

Project:

Investigators:

Chamber #	Initial Temp (°C)	T_0 Time	Final Temp (°C)	T_F Time	Headspace Volume (mL)	Water Volume (mL)	Substrate Volume (mL)

5. Collect an initial sample by bringing the headspace to the sampling port and withdrawing a sample (volume should be proportional to overall headspace volume—e.g., with a 2-L chamber and 200-mL headspace, we collect about a 10-mL sample; for a 50-mL headspace, we collect a 2- or 3-mL sample).
6. Record the initial time of sample (Table 32.3).
7. Store the initial gas sample in a gastight syringe, evacuated serum vial, or Exetainer (Labco, UK) until analysis.
8. Incubate chamber in situ for desired incubation time, typically 2−4 h depending on expected rates.

Terminating the Assay

1. At the end of the desired incubation time, collect a final gas sample by repeating steps 4−7 above.
2. After the final sample is collected, measure the temperature in each enclosure as in step 1 above.
3. Open the enclosure and measure volume of water and sediment in the chamber. If using a hard substratum, you will also want to estimate surface area to allow scaling of rates per unit area.
4. If you wish to express rates per unit biomass, process substrata for standing crop biomass (see chlorophyll and AFDM protocols in Chapter 12).

32.3.3.3 Laboratory Procedures

Analysis of Headspace Samples via Gas Chromatography

1. Ethylene concentration can be analyzed on any GC equipped with an FID.
2. On each day that the GC samples are analyzed, also run standard concentration samples of ethylene. Run standard samples before any unknown samples each day to ensure that the GC is functioning properly, then continue to analyze a standard every 10 samples to correct/monitor for instrument drift.

Calculating N$_2$ Fixation Rates (Based on Capone, 1993)

1. Determine the solubility correction (SC) for ethylene in aqueous phase (SC) as:

$$SC = 1 + \left(\alpha \cdot \frac{A}{B} \right) \tag{32.6}$$

where α = Bunsen coefficient for ethylene at the incubation temperature, A = water volume in the chamber, and B = headspace volume; α can be determined from a chemistry handbook such as Dean (1992).

2. Calculate the quantity of ethylene in the sample as:

$$ethylene (nmol) = \left(\frac{Peak\ Height_{sample}}{Peak\ Height_{standard}} \right) \cdot C_{standard} \cdot B \cdot SC \tag{32.7}$$

where Peak Height$_{sample}$ = ethylene peak height in sample, Peak Height$_{sample}$ = ethylene peak height in a standard, $C_{standard}$ = concentration of that standard in nmol mL^{-1}, B = headspace volume in mL (volume of headspace in assay vessel in mL), and SC = solubility correction as described above.

3. Calculate the rate of fixation in the chambers as:

$$Ethylene\ fixation\ rate \left(\frac{nmol}{t} \right) = \frac{(ethylene_{final} - ethylene_{initial})}{t} \tag{32.8}$$

where ethylene$_{final}$ and ethylene$_{initial}$ = ethylene concentrations in the chamber at the beginning and end of the incubation and t = duration of the incubation period (commonly reported in hours).

4. Convert the ethylene fixation rate to the N$_2$ fixation rate assuming a ratio of 3 moles of ethylene produced for every 1 mole of N$_2$ gas potentially fixed (but see N-fixation literature for discussions of the appropriateness of this conversion factor and alternates).
5. Scale the fixation rate per substratum area or unit biomass as desired.

Tips and Notes

A time series of incubations is suggested for early runs in a particular system. This will allow for determination of the minimum incubation time to detect activity as well as an incubation length where rates remain linear.

32.3.4 Advanced Method 1: Using ^{15}N to Measure DNRA

32.3.4.1 General Preparation

Analytical Considerations (Prepare >1 Month Prior to Experiment)

1. Plan to make $^{15}NO_3^-$ solution at an appropriate concentration for your field site. Solution can be made from $K^{15}NO_3$ or $Na^{15}NO_3$.[2] It is easiest to dose out the appropriate amount of ^{15}N as a solution (by volume), as opposed to measuring out very small masses.

   a. Nitrate concentrations are often difficult to impossible to measure in sediments due to the low concentrations and fast removal rates. Therefore, addition of any $^{15}NO_3^-$ may stimulate removal rates, as well as DNRA or denitrification rates. The additional $^{15}NO_3^-$ should be added at concentrations small enough to minimize this stimulation, but large enough to get measurable signal. In general, the labeled nitrate should not boost the concentration of the ambient nitrate by more than 10%.

   b. Isotopes are always expressed relative to a standard, which is a notation that takes some getting used to. If you are not familiar with using isotopes, we highly recommend a general reference to start (e.g., Kendall and Caldwell, 1998; Michener and Lajtha, 2008; see also Chapter 23). Isotopic enrichment is usually expressed one of two ways, as atom percent or as delta notation. Enrichment of an experiment is calculated as (from Steingruber et al., 2001):

$$\varepsilon = [NO_3^-]_a - \frac{[NO_3^-]_b}{[NO_3^-]_a} \tag{32.9}$$

where $[NO_3^-]$ = nitrate concentration of the solution after (a) and before (b) the $^{15}NO_3^-$ tracer addition. Enrichment of ^{15}N can be expressed two ways:

   i. Atom percent = (mass of heavy isotope)/(mass of all isotopes). In the case of N:

$$\text{Atom \% } ^{15}\text{N} = [^{15}\text{N}/(^{14}\text{N} + ^{15}\text{N})] \cdot 100 \tag{32.10}$$

   ii. Delta notation (δ, expressed in ‰):

$$\delta‰ = (R_{sample}/R_{standard} - 1) \cdot 1000 \tag{32.11}$$

where R = ratio of the heavy to light isotope in the sample or standard (see also Chapter 23).

2. Gather syringes of various sizes (1, 5, 10, 20, 60 mL) depending on the volume of your assay bottle and the amount of tracer you need to add. You will also need assay bottles (e.g., 60-mL Nalgene or 120-mL Wheaton bottles). It is best to use bottles that do not have any potential for background ^{15}N contamination and create an airtight seal. You will also need access to a shaker table large enough to hold all of your assay bottles and diffusion bottles.

3. After your assay is complete, quickly process the resulting samples to measure for $^{15}NH_4^+$ and $^{15}NO_3^-$. Quick turnaround depends on having the following materials assembled and ready to use:

   a. Diffusion bottles—these are a separate set of bottles in which your diffusions will be done (see Section 32.3.4.3 step 7). Diffusions require a vessel with a tight-fitting lid, which can withstand some pressure buildup without gas escape.

   b. Acid traps—see details in Section 32.3.4.3 step 7b.

4. Finally, determine ahead of time where you plan to send your samples for analysis by IRMS. It is best to contact the contract lab well ahead of time to ask about their backlog and turnaround times. It is also good to ask other researchers about the reliability of the laboratory chosen for analyses. Pick one that has analyzed similar samples before, has well established quality control and quality assurance procedures, and can meet your time requirements for sample turnaround. This often is not the least expensive analytical laboratory (see also Chapter 23 for recommendations).

32.3.4.2 Field Procedures

1. If you collect substrata directly into your assay bottles, you need to take steps to make the bottles anoxic in the field (e.g., adding water to the sediments so they do not become oxidized, or exchanging the bottle headspace with an anoxic gas such as He or N_2).

2. Alternatively, you can collect intact cores from your study site and carefully transport them back to the laboratory. Ideally cores are broken down under anaerobic conditions (e.g., Coy anaerobic chamber or "glove box"). If a glove box is not available, portable glove bags can be used and are less expensive (e.g., Aldrich AtmosBag, Sigma—Aldrich).

---

2. The 98 atom% $Na^{15}NO_3^-$ can be purchased from Sigma—Aldrich (item # 364606) at ~US $100/g, which can be enough for a large number of bottle assays. Stable isotopes can be backordered for months, so be sure to place your order well ahead of the experiment.

32.3.4.3 Laboratory Procedures

Adding ^{15}N to Incubations

1. Determine how much of your substratum to add to an assay bottle (e.g., 10 grams of wet sediment). Record the wet mass and also be sure to estimate the dry weight by drying a subsample and calculating the percentage of water in the substratum. Keep the substratum as anoxic as possible, either by putting it in a glove box or by adding deoxygenated water (e.g., site water sparged with an inert gas such as He) to the sample immediately after weighing it out.

2. Add enough tracer to each assay bottle to reach a desired predetermined final NO_3^- concentration. The decision of how much ^{15}N to add depends on the goals of the study and the estimated rates. For example, if your goal is to measure as close to ambient rates as possible and you are working with a substratum you expect to be very anoxic, you would want to add as little ^{15}N as necessary to meet the minimum detection limits of the instrument that will analyze your samples, which varies by lab and by instrument (e.g., a MIMS can measure ^{15}N, but is much less sensitive than an IRMS and thus requires greater enrichment). Nitrate is often undetectable in anoxic pore waters; thus, adding any nitrate will artificially elevate the rates being measured.

3. Incubate bottles for a predetermined time period. Again, how long to incubate the bottles depend on the predicted rates. If you expect you are working with a substratum that will yield high rates, it is best to incubate for as short of a period as possible. If your rates are completely unknown, we strongly advise running a "pilot" batch of samples wherein you incubate assays for different time periods to determine the shortest possible incubation time that will yield measurable results for the IRMS. While this may sound time intensive, it can be time well spent, as too short of an incubation could result in an entirely unusable (and expensive) data set. Incubate the bottles under as similar conditions to the field collection as possible (e.g., light and temperature); this often requires environmental chambers.

4. Once the incubation is complete, separate the water and the sediment fractions. This is often done by centrifugation. The $^{15}NH_4^+$ will be in both the water fraction and sorbed to the sediments; $^{15}NH_4^+$ can be released from the sediments by exchanging it with 2 M KCl (as a final concentration; be sure to account for the volume of water in your sediment that may dilute a straight 2 M solution of KCl; Robertson et al., 1999).

5. Measure the NH_4^+ concentration of the KCl solution and of the water fraction. If analyzing a KCl sample, be sure your standards are also prepared using KCl as a matrix.

6. Once you know the concentration of NH_4^+ in your samples, you can estimate the volume you need to reach the minimum mass requirements of the IRMS (or other mass spectrophotometer) used to analyze the samples. It is advisable to not diffuse your entire sample volume. Samples can be frozen indefinitely. If working with contract labs, samples can go missing (in the mail or through other means) or be dropped, and thus, if you have diffused your entire sample, you may need to redo the experiment. Once you know the volume of the samples you will need to diffuse onto acid traps, you can proceed to the diffusion method. Again, the volume will likely vary by sample, so be sure to calculate the volume for the full range of NH_4^+ concentrations in your samples.

   a. Example calculation: Your bottle contains 50 mL of sample at 0.5 mg/L NH_4^+. The total mass of NH_4^+ is 0.2 mg (0.156 mg as N or 156 µg N). The IRMS requires a minimum of 20 µg N per sample. Diffusing 10 mL (1/5th) of sample would result in 31.2 µg N from the resulting IRMS sample, well above the minimum detection limit, but also leaving four-fifths of the sample remaining should it need to be reprocessed.

7. The final step before sending samples off for IRMS analysis involves transferring the dissolved NH_4^+ onto an acidified filter "trap" so that it can be analyzed as a solid. The alkaline headspace diffusion method increases the pH of the sample water with the addition of magnesium oxide (MgO), thereby converting the dissolved NH_4^+ into gaseous ammonia, which can be trapped onto an acidified filter either floating on or suspended above the water. The dried filter containing trapped ammonia is packed into combustion capsules for analysis on an IRMS interfaced with an elemental analyzer (EA). Diffusion vessels require a vessel with a tight-fitting lid, which can withstand some pressure buildup without gas escape.

   a. Record volumes and concentrations of samples to determine the efficiencies of the conversion and trapping processes. Also use a "standard" of NO_3^- solution at known concentration, similar to the concentration of the samples (to calculate conversion efficiency), blanks (deionized water), and controls (deionized water with reagents to check for reagent contamination).

   b. Add a filter packet to each diffusion bottle. Filter packets are made from two pieces of 2.54-cm Teflon tape with an acidified (25 µL of 2.5 M $KHSO_4$) Whatman GF/D GFF (1 cm diameter) sandwiched between the tape. The tape covers are then pressed together where they overlap to make a seal using an item with a diameter slightly larger than 1 cm (e.g., a scintillation vial). A weak seal can allow the alkaline solution to contact the filter during the diffusion process, thereby neutralizing the acid and compromising the sample (Fillery and Recous, 2001). Alternatively, filter packets can be suspended above the water in the headspace by a small hook glued to the inside lid of the bottle. This requires more preparation (e.g., building the diffusion hooks), and care should be taken to not use materials that off-gas NH_4^+, as may be the case with some plastics or glue.

c. After the addition of a filter packet, add 50 g of NaCl and 3 g of MgO/L of sample directly into the diffusion vessel, sealing the vessel immediately. Both salts should be precombusted (450−500°C). Shake for 2 weeks at 40°C or 3 weeks at room temperature to ensure the complete trapping of ammonia on the filter.

d. Recovery is checked using standards of an appropriate concentration to match the samples. Also, to correct for chemical contaminants, include reagent blanks containing only the NaCl and MgO and, if appropriate, Devarda's alloy.

e. The measurement of $^{15}NO_3^-$ is a slight variation on the above diffusion method. The only difference is that the NH_4^+ is first driven off and the remaining $^{15}NO_3^-$ is reduced to NH_4^+ with Devarda's alloy.

   i. Calculate sample volume based on desired mass, as in step 6a above.

   ii. Add 5 g of NaCl and 3 g of MgO to the sample, boil briefly with a stir bar until volume is ∼ 100 mL. Heating and stirring will speed the conversion of NH_4^+ to NH_3 (gas) and drive it out of the water phase.

   iii. Once the NH_4^+ conversion is complete, only NO_3^- remains in the sample. Add 0.5 g of Devarda's alloy to the sample, along with another 0.5 g of NaCl and 0.5 g of MgO. Immediately place a filter packet in the bottle and seal it. Cap the bottle tightly.

   iv. Place bottles at 60°C for 48 h. Remove from oven and shake for 7 days to facilitate full transfer of NH_4^+ (from the reduced NO_3^-) onto the filter packet. Remove filter and process for IRMS analysis.

f. The methods summarized here were used by the Lotic Intersite Nitrogen eXperiment (LINX) group. The LINX protocol is more detailed and is available for download at http://andrewsforest.oregonstate.edu/data/abstract.cfm?dbcode=AN006. Several variants of this method are published and none has clearly emerged as superior (e.g., Holmes et al., 1998; Herman et al., 1995; Diaconu et al., 2005). New methods to measure $^{15}NH_4^+$ continue to be proposed (e.g., Gardner et al., 1995; Yin et al., 2014) and may eventually replace the diffusion method.

8. Once filters are dried (in a desiccator for ∼3−4 days), pack filters into tins suitable for elemental analysis. Place into a 96-well plate or similar tray and make a label key.

9. Send samples to an IRMS lab. Include sample, standards, controls, and blanks.

Calculating DNRA Rates

1. Convert isotope ratios into mole fractions (MF)

   a. The data file you receive from the contract lab should contain the delta value ($\delta^{15}N$, expressed as ‰) and the mass of sample (μg N).

$$^{15}N\ MF\ =\ \frac{[((\delta^{15}N/1000) + 1)\cdot 0.0036765]}{[1\ +\ ((\delta^{15}N/1000) + 1)\cdot 0.0036765]} \tag{32.12}$$

   This conversion will result in a very small number. For 0 per mil (0‰) enrichment, the calculation should result in a $^{15}N\ MF = 0.003663$.

2. Multiple MF by N pool sizes

   a. As an example, we have a sample at 50‰ enrichment, which converts to an MF = 0.003845. We measured the N concentration in this sample as 200 μg NH_4^+/L or 156 μg N/L, which converts to 11.14 μmol N/L. Therefore, 11.14 μmol N/L * 0.003845 = 0.04 μmol ^{15}N/L. This can be multiplied by the appropriate volume to convert from concentration to mass of ^{15}N in the assay bottle.

3. Convert to a rate (flux)

   a. Divide by the incubation time to convert the mass to a rate. The final units will be in μmol ^{15}N/h.

4. More complete calculations are available in the LINX2 protocols, referenced above. These describe how to correct for the reagent blanks and for any addition of nitrate to the sample, as may be needed to reach the detection threshold of the IRMS (known as "spikes").

Tips and Notes

1. A time series of incubation is suggested for early runs in a particular system. This will allow for determination of the minimum incubation time to detect activity as well as an incubation length where rates remain linear.

2. Practice making diffusion filter packets prior to use. These should be inspected well with a light from behind to ensure there are no holes but also that there is a good seal (Teflon tape pressured together is clear).

32.3.5 Advanced Method 2: Using MIMS to Measure Net N$_2$ Flux

Net N$_2$ flux is a balance between producing processes (denitrification) and consuming processes (N fixation). Net N$_2$ flux can be quantified by measuring N$_2$/Ar ratios in natural environments (Laursen and Seitzinger, 2002; McCutchan et al., 2003) or in experimental enclosures (e.g., assay bottles) using MIMS. MIMS allows for simple, rapid, and selective analysis of dissolved air gas concentrations dissolved into water samples. MIMS can also be paired with ^{15}N enrichment to more specifically measure N cycling processes, including denitrification, particularly when paired with the isotope pairing technique (e.g., Risgaard-Petersen et al., 2003). For simplicity, we will focus on the easier-to-measure net N$_2$ flux in a bottle assay. Since the goal is to measure potentially small changes in N$_2$ concentration against a *very* large background pool of atmospheric N$_2$ (78%), contamination is a major obstacle and concern with any MIMS method. Excessive caution in guarding against any air entry, once an experiment is started, is the surest way to produce useable results. Remember—air exposure or bubbles are your worst enemy! Complete the incubations and transfer steps underwater, as is possible, to help ensure that samples are not compromised by contact with the atmosphere.

32.3.5.1 Assay Procedure

1. Set up bottle assays as described for other protocols in this chapter, noting the mass of substratum and volume of water added to each bottle. Fill each bottle to full capacity—eliminate any air bubbles in the bottles. You also need a set of "blank" or "control" bottles that only contain water and do not have any substrata.
   a. As with all bottle assays, the timing of this assay is highly dependent on the expected rate of the process; if you expect very "fast" net N$_2$ flux, you will incubate these for a shorter period of time.
   b. If you want to speed up the reaction or are more interested in "potential" rates as opposed to near-ambient rates, the addition of NO$_3^-$ to the assay bottles will likely enhance N$_2$ fluxes by increased denitrification.
   c. If you are not going to artificially stimulate N$_2$ fluxes, we highly recommend running some pilot tests prior to your full-scale experiment to determine target incubation times and rates.
2. Sampling times: Samples are collected before and after the assay.
   a. Collect $t = 0$ sample (beginning sample) from your control and experimental assay bottles. Ideally, since these are starting from the same source water, they should be identical in gas composition. Incubate bottles on a shaker table for the predetermined time.
   b. Collect your $t =$ final samples from the control and experimental assay bottles. Record the temperature of the water bath at the end of the experiment.
3. Sample collection: MIMS samples are typically collected into 12-mL Exetainer vials. When filling the vials, use a small diameter piece of Tygon (or Viton) tubing to extend down to the bottom of the Exetainer, allowing you to fill the Exetainer from the bottom up.
   a. This action minimizes entrainment of atmospheric gasses (potential source of contamination), which can disrupt the dissolved gas signatures.
   b. Overflow the tube $\sim 3\times$, and slowly lift the tubing out of the vial taking care to replace the volume of the tubing with water so there is no headspace in the vial.
   c. Once the tubing is out, there should be a semicircle of water over the rim of the vial. Additional detail and pictures are available in Burgin et al. (2013).
4. Sample preservation: Preserve sample with 200 µL of a 50% zinc chloride (ZnCl$_2$) solution (50% W/V; dissolve 50 g in 100 mL deionized water). Add the aliquot of ZnCl$_2$ to the Exetainer placing the tip of the pipette under the water surface. Quickly cap the Exetainer and flip the vial upside down to mix the heavy ZnCl$_2$ solution into the water. You should also check for bubbles that may have formed around the cap seal. Any samples with bubbles should be discarded and resampled.
   a. Caution: ZnCl$_2$ is corrosive and acutely toxic, so it should be handled with extreme caution and wearing appropriate personal protective equipment! The reaction dissolving zinc chloride in H$_2$O is exothermic, so beware of heating in containers. ZnCl$_2$ is also harmful to aquatic environments so be extra cautious when using it in the field; waste bottles should not be discarded in the field. All needles used during the procedures should be disposed off following standard sharps disposal procedures.
5. Sample storage: Once samples are collected, keep them as cool as possible, preferably in a refrigerator at 4°C. Keeping gases cooler than the temperature at which they are collected helps to keep the gases in solution. If the samples are allowed to warm up, significant bubble formation can occur, resulting from degassing. As an added protection, if samples are to be shipped or will be stored for long periods, it is best to keep the samples underwater. Keeping samples underwater

limits extreme temperature fluctuations and also limits any potential evaporation that can occur near the seal of the cap. In general, ~40−50 samples can be placed into a 1-L Nalgene bottle, or similar container. If samples are placed in water, sample IDs should be written with a thick Sharpie and covered over with Parafilm so that the sample ID is not rubbed off in transit. Alternatively, white electrical tape works well in place of standard label tape and will stay on underwater.

   a. Place all bottles in a large water-filled container (e.g., plastic box), ensuring that the bottles are completely submerged, particularly at areas where gas might leak (e.g., seals or lids). Place a lid on the box to keep water from splashing out. Measure the temperature of the water bath; ideally, keep a recording thermometer in the water bath for the duration of the experiment. If that is not available, record the temperature whenever samples are collected. Temperature is a key driver of gas solubility and is necessary for determining the dissolved N_2 and Ar.

6. Experimental design: As described, this design employs a simple two-point rate calculation ($t = 0$ and $t =$ final). More complicated and likely more accurate rates can be measured by (1) collecting multiple samples from the same assay bottle over time or (2) destructively harvesting replicate assay bottles at different time points in an experiment.

   a. If the first approach is used, an assay bottle with a septum needs to be employed so that water can be injected into the assay bottle to remove a sample for MIMS analysis. The gas signature of the water injected into the ongoing assay should also be accounted for in subsequent calculations.

   b. If the second approach is used, you should take care that the assay bottles are all as homogeneous as possible and incubations are started at the exact same time, as minor differences between bottles can cause complications for the eventual rate calculations.

32.3.5.2 Running MIMS Samples

1. Each MIMS is uniquely built from different components. If running your samples in-house, follow the instructions specific to your instrument.

2. If sending samples to another lab for analysis, use the following precautions:

   a. Keep Exetainers underwater to decrease risk of contamination.

   b. Keep Exetainers and water bath at a cooler temperature than at which they were collected to decrease risk of degassing due to warming. It is best to add multiple ice packs to the cooler and ship the samples overnight.

32.3.5.3 Calculations (Adapted From Kana et al., 1998)

1. Calculate N_2 and Ar concentrations:

$$N_2 \text{ concentration} = \frac{\text{MIMS signal}_{sample}}{\text{MIMS signal}_{standard}} \cdot \text{solubility of } N_2 \qquad (32.13)$$

where "MIMS signal" = mass signal output from the mass spectrometer for the standard water and the solubility of N_2 is for the temperature, salinity, and barometric pressure of the standard, determined from the solubility tables, such as those in Colt (2012). The same formula is used for Ar.

2. Calculate the N_2 flux from the N_2:Ar:

$$N_2 \text{ flux} = (N_2 : \text{Ar } T_0 - N_2 : \text{Ar } T_f) \cdot \text{Ar}_{pred} \qquad (32.14)$$

where $T_0 =$ starting time point, $T_f =$ final time point, and $\text{Ar}_{pred} =$ predicted Ar concentration at saturation, which can be found in a solubility table, as in Colt (2012).

3. Calculate Net N_2 flux:

$$\text{Flux} \left(\mu\text{mol/area}^2 \text{ time} \right) = (N_2 \text{ flux} \cdot \text{water volume}) / (\text{surface area} \cdot \text{incubation time}) \qquad (32.15)$$

Units should be specific to how you measure surface area and time for the incubation. Net fluxes can be either positive (denitrification driven) or negative (N fixation driven), but cannot tell you which process is dominant or larger than the other.

32.4 QUESTIONS

32.4.1 Denitrification

1. Compare the unamended and the potential rates of denitrification. Did they differ? If so, why?

2. How do the denitrification rates you measured compare to those in other streams and rivers using similar methods? For comparison, see rates in Martin et al. (2001), Richardson et al. (2004), and Findlay et al. (2011).
3. What other factors besides availability of nitrate and organic carbon might affect denitrification rates?
4. Denitrification is often considered a beneficial process in relation to coastal eutrophication. Why would that be true?
5. In some systems, high rates of denitrification might decrease overall N:P ratios. Why might this be a concern in some ecosystems?

32.4.2 Nitrification

6. How do the nitrification rates you measured compare to those in other streams and rivers measured using similar methods? For comparison, see rates in Kemp and Dodds (2002), Strauss et al. (2002, 2004), and Starry et al. (2005).
7. The incubation period for nitrification is much longer than that for denitrification. Why is this so? What can be inferred about the biology of nitrifying bacteria based on the long incubation time?
8. We provided methods for measuring net and gross nitrification. Why would you want to measure one or the other process?
9. How can nitrification (NO_3^- production) limit denitrification when one process is aerobic and the other is anaerobic? In other words, how can both processes be occurring in the same place?

32.4.3 Nitrogen Fixation

10. What was the N_2 fixation rate that you measured for your study reach?
11. Which taxa do you think were responsible for the nitrogen fixation in your reach?
12. How do the N_2 fixation rates you measured compare to those in other streams? How about those measured in other ecosystems? For comparison, see tables in Grimm and Petrone (1997) and Marcarelli et al. (2008).
13. How might changing the nutrient or light availability alter rates of N_2 fixation? What other environmental factors could be important for controlling rates of this process?
14. Why are isotopic methods necessary when there are small pools of the measured compound that turn over quickly?
15. Why do ^{15}N methods make it easier to detect rates of denitrification when there is so much N_2 in the atmosphere and dissolved in most streams?

32.4.4 DNRA and N_2 Flux

16. How do the DNRA rates you measured compare to rates measured in other studies? See Nogaro and Burgin (2014) or Washbourne et al. (2011) for comparison.
17. What are the processes that contribute to net N_2 fluxes, as measured using the MIMS method described herein? How can you tell which process is dominant in your system?
18. How do the net N_2 fluxes you measured compare to other studies that do not use isotopic enrichment (e.g., Grantz et al., 2012; Deemer et al., 2011)? How do the net N_2 fluxes compare to studies that measure denitrification using ^{15}N tracers (e.g., Mulholland et al., 2008)?

32.5 MATERIALS AND SUPPLIES

32.5.1 Denitrification

Field Materials
Sediment corer and utensils
Plastic bags for sediment
Plastic bottles for stream water
Thermometer
Cooler with ice
Labeling tape and permanent marker
Data sheet, clipboard, pencil(s)
Lab Materials
Data sheet

Incubation vessels with septa (see details in General Preparation)
Labeling tape and permanent marker
Plastic powder funnel
Adjustable pipettes and tips
Graduated cylinders
Disposable syringe (5 mL)
Syringe needles (c. 22 gauge)
Vacuum pump
Tubing
Orbital shaker
Glass serum bottles (2 mL) with butyl rubber septa and aluminum seals
Crimper
Gases (see details in General Preparations)
Volumetric flasks
Potassium nitrate (KNO_3)
Glucose ($C_6H_{12}O_6$)
Chloramphenicol (Sigma−Aldrich CAS # 56-75-7)

32.5.2 Nitrification

Field Materials
Sediment corer and utensils
Plastic bags for sediment
Plastic bottles for stream water
Thermometer
Cooler with ice
Labeling tape and permanent marker
Data sheet, clipboard, pencil(s)
Lab Materials
Data sheet
Labeling tape and permanent marker
Glass Erlenmeyer flasks (125 mL)
Adjustable pipettes and tips
Graduated cylinders
Orbital shaker
Plastic centrifuge tube (15 mL)
Centrifuge
Parafilm
Aluminum foil
Syringe filter apparatus
Glass fiber filters (e.g., Pall Type A/E or Whatman GF/C or GF/F)
Glass scintillation vial (20 mL)
Volumetric flasks
Potassium chloride (KCl)
Dimethyl sulfoxide (DMSO)
Nitrapyrin (2-chloro-6-[trichloromethyl]-pyridine; Sigma−Aldrich CAS # 1929-82-4)
Sulfuric acid (H_2SO_4)
Sodium hydroxide (NaOH)

32.5.3 Nitrogen Fixation

Data sheet and clipboard, pencil(s)
Chambers and septa for injection port(s)
Balloons or syringes as needed to introduce acetylene headspace

Acetylene gas—either in a compressed air tank or generated from calcium carbide
Syringes and needles for collecting gas samples
Thermometer
Gas storage vials or Exetainers (two per chamber)
Large (2 L minimum) graduated cylinder for measuring volume of water and substratum
Rite-in-the-rain paper for tracing surface areas of substrata
Other assorted field materials—sharpies, label tape, etc.

32.5.4 Dissimilatory Nitrate Reduction to Ammonia

Field Materials
Sediment corer and utensils
Plastic bags for sediment
Plastic bottles for stream water
Thermometer
Cooler with ice
Labeling tape and permanent marker
Data sheet, clipboard, pencil(s)

Lab Materials
Data sheet or lab notebook
Labeling tape and permanent marker
Wheaton or Nalgene bottles for assay
$^{15}NO_3^-$ solution
Adjustable pipettes and tips
Syringes (various sizes)
Orbital shaker
Plastic centrifuge tube (15 or 50 mL)
Centrifuge
Diffusion bottles
Acid trap filter packs
Glass scintillation vials (20 mL)
Potassium chloride (KCl)
Sodium chloride (NaCl)
Magnesium oxide (MgO)
Sulfuric acid (H_2SO_4)
EA tins and 96-well plate for organizing/storing samples

32.5.5 N_2 Flux

Field Materials
Sediment corer and utensils
Plastic bags for sediment
Plastic bottles for stream water
Thermometer
Cooler with ice
Labeling tape and permanent marker
Data sheet, clipboard, pencil(s)

Lab Materials
Data sheet or lab notebook
Labeling tape and permanent marker
Wheaton or Nalgene bottles for assay
Exetainers
Tygon or Viton tubing for sample transfer between assay bottle and Exetainer
Luer-lok syringes (various sizes)

Luer-lok connectors for creating tubing transfer component
Zinc chloride (as sample preservative)
Pipette and tips
Gloves and appropriate waste disposal
Electrical tape or Parafilm for preserving sample labels
Storage containers with good watertight seals (Lock & Lock is a good brand)

REFERENCES

Alexander, R.B., Smith, R.A., Schwarz, G.E., 2000. Effect of stream channel size on the delivery of nitrogen to the Gulf of Mexico. Nature 403, 758−761.

Beaulieu, J.J., et al., 2011. Nitrous oxide emission from denitrification in stream and river networks. Proceedings of the National Academy of Sciences of the United States of America 108, 214−219.

Burgin, A.J., Hamilton, S.K., McCarthy, M.J., Gardner, W., 2013. Methods for measuring NO_3^- reduction processes in wetlands. In: Megonigal, P., DeLaune, R. (Eds.), Methods in Biogeochemistry of Wetlands. Soil Science Society of America Book Series, Madison, WI, USA, pp. 519−533.

Camargo, J.A., Alonso, A., Salamanca, A., 2005. Nitrate toxicity to aquatic animals: a review with new data for freshwater invertebrates. Chemosphere 58, 1255−1267.

Capone, D.G., 1993. Determination of nitrogenase activity in aquatic samples using the acetylene reduction procedure. In: Kemp, P.F., Sherr, B.F., Sherr, E.B., Cole, J.J. (Eds.), Handbook of Methods in Aquatic Microbial Ecology. Lewis Publishers, Boca Raton, Florida, USA, pp. 621−631.

Colt, J., 2012. Dissolved Gas Concentration in Water, second ed. Elsevier, London, UK.

Dean, J.A., 1992. Lange's Handbook of Chemistry, fifteenth ed. McGraw-Hill, Inc., New York, New York, USA.

Deemer, B.R., Harrison, J.A., Whitling, E.W., 2011. Microbial dinitrogen and nitrous oxide production in a small eutrophic reservoir: an in situ approach to quantifying hypolimnetic process rates. Limnology and Oceanography 56, 1189−1199.

Diaconu, C., Brion, N., Elskens, M., Baeyens, W., 2005. Validation of a dynamic ammonium extraction technique for the determination of ^{15}N at enriched abundances. Analytica Chimica Acta 554, 113−122.

Dodds, W.K., Brock, J., 1998. A portable flow chamber for in situ determination of benthic metabolism. Freshwater Biology 39, 49−59.

Dodds, W.K., Jones, R.D., 1987. Potential rates of nitrification and denitrification in an oligotrophic fresh-water sediment system. Microbial Ecology 14, 91−100.

Dodds, W.K., Smith, V.H., 2016. Nitrogen, phosphorus, and eutrophication in streams. Inland Waters 6, 155−164.

Dodds, W.K., Whiles, M.R., 2010. Freshwater Ecology: Concepts and Environmental Applications of Limnology, second ed. Academic Press, Burlington, Massachusetts, USA.

Fillery, I.R., Recous, S., 2001. Use of enriched ^{15}N sources to study soil N transformations. In: Unkovich, M.J., Pate, J.S., McNeill, A., Gibbs, J. (Eds.), Stable Isotope Techniques in the Study of Biological Processes and Functioning of Ecosystems. Kluwer Academic Publishers, Dordrecht, Netherlands, pp. 167−194.

Findlay, S.E.G., Mulholland, P.J., Hamilton, S.K., Tank, J.L., Bernot, M.J., Burgin, A.J., Crenshaw, C.L., Dodds, W.K., Grimm, N.B., McDowell, W.H., Potter, J.D., Sobota, D.J., 2011. Cross-stream comparison of substrate-specific denitrification potential. Biogeochemistry 104, 381−392.

Flett, R., Hamilton, R., Campbell, N., 1976. Aquatic acetylene-reduction techniques: solutions to several problems. Canadian Journal of Microbiology 22, 43−51.

Gardner, W.S., Bootsma, H.A., Evans, C., John, P.A.S., 1995. Improved chromatographic analysis of ^{15}N: ^{14}N ratios in ammonium or nitrate for isotope addition experiments. Marine Chemistry 48, 271−282.

Grantz, E.M., Kogo, A., Scott, J.T., 2012. Partitioning whole-lake denitrification using in situ dinitrogen gas accumulation and intact sediment core experiments. Limnology and Oceanography 57, 925−935.

Grimm, N.B., Petrone, K.C., 1997. Nitrogen fixation in a desert stream ecosystem. Biogeochemistry 37, 33−61.

Groffman, P.M., Altabet, M.A., Böhlke, J., Butterbach-Bahl, K., David, M.B., Firestone, M.K., Giblin, A.E., Kana, T.M., Nielsen, L.P., Voytek, M.A., 2006. Methods for measuring denitrification: diverse approaches to a difficult problem. Ecological Applications 16, 2091−2122.

Groffman, P.M., Holland, E.A., Myrold, D.D., Robertson, G.P., Zou, X.M., 1999. Denitrification. In: Robertson, G.P., Coleman, D.C., Bledsoe, C.S., Sollins, P. (Eds.), Standard Soil Methods for Long Term Ecological Research. Oxford University Press, New York, New York, USA, pp. 272−288.

Herman, D., Brooks, P., Ashraf, M., Azam, F., Mulvaney, R., 1995. Evaluation of methods for nitrogen-15 analysis of inorganic nitrogen in soil extracts. II. Diffusion methods. Communications in Soil Science and Plant Analysis 26, 1675−1685.

Holmes, R.M., McClelland, J.W., Sigman, D.M., Fry, B., Peterson, B.J., 1998. Measuring $^{15}N - NH_4^+$ in marine, estuarine and fresh waters: an adaptation of the ammonia diffusion method for samples with low ammonium concentrations. Marine Chemistry 60, 235−243.

Huygens, D., Trimmer, M., Rütting, T., Müller, C., Heppell, C.M., Lansdown, K., Boeckx, P., 2013. Biogeochemical nitrogen cycling in wetland ecosystems: nitrogen-15 isotope techniques. In: DeLaune, R., Reddy, K., Richardson, C., Megonigal, J. (Eds.), Methods in Biogeochemistry of Wetlands. Soil Science Society of America Book Series, Madison, WI, USA, pp. 553−591.

Kana, T.M., Darkangelo, C., Hunt, M.D., Oldham, J.B., Bennett, G.E., Cornwell, J.C., 1994. Membrane inlet mass spectrometer for rapid high-precision determination of N_2, O_2, and Ar in environmental water samples. Analytical Chemistry 66, 4166−4170.

Kana, T.M., Sullivan, M.B., Cornwell, J.C., Groszkowski, K.M., 1998. Denitrification in estuarine sediments determined by membrane inlet mass spectrometry. Limnology and Oceanography 43, 334−339.

Kemp, M.J., Dodds, W.K., 2002. Comparisons of nitrification and denitrification in prairie and agriculturally influenced streams. Ecological Applications 12, 998–1009.

Kendall, C., Caldwell, E., 1998. Fundamentals of isotope geochemistry. In: Kendall, C., McDonnell, J.J. (Eds.), Isotope Tracers in Catchment Hydrology. Elsevier, Amsterdam, Netherlands, pp, 51–58.

Laursen, A.E., Seitzinger, S.P., 2002. Measurement of denitrification in rivers: an integrated, whole reach approach. Hydrobiologia 485, 67–81.

Marcarelli, A.M., Baker, M.A., Wurtsbaugh, W.A., 2008. Is in-stream N-2 fixation an important N source for benthic communities and stream ecosystems? Journal of the North American Benthological Society 27, 186–211.

Martin, L.A., Mulholland, P.J., Webster, J.R., Valett, H.M., 2001. Denitrification potential in sediments of headwater streams in the southern Appalachian Mountains, USA. Journal of the North American Benthological Society 40, 505–519.

McCutchan, J.H., Saunders, J.F., Pribyl, A.L., Lewis, W.M., 2003. Open-channel estimation of denitrification. Limnology and Oceanography: Methods 1, 74–81.

Michener, R., Lajtha, K., 2008. Stable Isotopes in Ecology and Environmental Science. John Wiley & Sons, Oxford, UK.

Mulholland, P.J., Helton, A.M., Poole, G.C., Hall, R.O., Hamilton, S.K., Peterson, B.J., Tank, J.L., Ashkenas, L.R., Cooper, L.W., Dahm, C.N., Dodds, W.K., Findlay, S.E.G., Gregory, S.V., Grimm, N.B., Johnson, S.L., McDowell, W.H., Meyer, J.L., Valett, H.M., Webster, J.R., Arango, C.P., Beaulieu, J.J., Bernot, M.J., Burgin, A.J., Crenshaw, C.L., Johnson, L.T., Niederlehner, B.R., O'Brien, J.M., Potter, J.D., Sheibley, R.W., Sobota, D.J., Thomas, S.M., 2008. Stream denitrification across biomes and its response to anthropogenic nitrate loading. Nature 452, 202–205.

Nogaro, G., Burgin, A.J., 2014. Influence of bioturbation on denitrification and dissimilatory nitrate reduction to ammonium (DNRA) in freshwater sediments. Biogeochemistry 120, 279–294.

Prechtl, J., Kneip, C., Lockhart, P., Wenderoth, K., Maier, U.-G., 2004. Intracellular spheroid bodies of *Rhopalodia gibba* have nitrogen-fixing apparatus of cyanobacterial origin. Molecular Biology and Evolution 21, 1477–1481.

Raymond, J., Siefert, J.L., Staples, C.R., Blankenship, R.E., 2004. The natural history of nitrogen fixation. Molecular Biology and Evolution 21, 541–554.

Richardson, W.B., Strauss, E.A., Bartsch, L.A., Monroe, E.M., Cavanaugh, J.C., Vingum, L., Soballe, D.M., 2004. Denitrification in the Upper Mississippi River: rates, controls, and contribution to nitrate flux. Canadian Journal of Fisheries and Aquatic Sciences 61, 1102–1112.

Risgaard-Petersen, N., Nielsen, L.P., Rysgaard, S., Dalsgaard, T., Meyer, R.L., 2003. Application of the isotope pairing technique in sediments where anammox and denitrification coexist. Limnology and Oceanography: Methods 1, 63–73.

Robertson, G.P., Coleman, D.C., Bledsoe, C.S., Sollins, P., 1999. Standard Soil Methods for Long-Term Ecological Research. Oxford University Press, Oxford, England.

Rüegg, J., Brant, J.D., Larson, D.M., Trentman, M.T., Dodds, W.K., 2015. A portable, modular, self-contained recirculating chamber to measure benthic processes under controlled water velocity. Freshwater Science 34, 831–844.

Schindler, D.W., 1998. Replication versus realism: the need for ecosystem-scale experiments. Ecosystems 1, 323–334.

Schubert, C.J., Durisch-Kaiser, E., Wehrli, B., Thamdrup, B., Lam, P., Kuypers, M.M., 2006. Anaerobic ammonium oxidation in a tropical freshwater system (Lake Tanganyika). Environmental Microbiology 8, 1857–1863.

Seitzinger, S., Harrison, J.A., Böhlke, J.K., Bouwman, A.F., Lowrance, R., Peterson, B., Tobias, C., Drecht, G.V., 2006. Denitrification across landscapes and waterscapes: a synthesis. Ecological Applications 16, 2064–2090.

Starry, O.S., Valett, H.M., Schreiber, M.E., 2005. Nitrification rates in a headwater stream: influences of seasonal variation in C and N supply. Journal of the North American Benthological Society 24, 753–768.

Steingruber, S.M., Friedrich, J., Gächter, R., Wehrli, B., 2001. Measurement of denitrification in sediments with the ^{15}N isotope pairing technique. Applied and Environmental Microbiology 67, 3771–3778.

Stewart, W., Fitzgerald, G., Burris, N., 1967. In situ studies on N$_2$ fixation using the acetylene reduction technique. Proceedings of the National Academy of Sciences of the United States of America 58, 2071–2078.

Strauss, E.A., Lamberti, G.A., 2000. Regulation of nitrification in aquatic sediments by organic carbon. Limnology and Oceanography 45, 1854–1859.

Strauss, E.A., Mitchell, N.L., Lamberti, G.A., 2002. Factors regulating nitrification in aquatic sediments: effects of organic carbon, nitrogen availability, and pH. Canadian Journal of Fisheries and Aquatic Sciences 59, 554–563.

Strauss, E.A., Richardson, W.B., Bartsch, L.A., Cavanaugh, J.C., Bruesewitz, D.A., Imker, H., Heinz, J.A., Soballe, D.M., 2004. Nitrification in the Upper Mississippi River: patterns, controls, and contribution to the NO_3^- budget. Journal of the North American Benthological Society 23, 1–14.

Thurston, R.V., Russo, R.C., Vinogradov, G., 1981. Ammonia toxicity to fishes. Effect of pH on the toxicity of the unionized ammonia species. Environmental Science and Technology 15, 837–840.

Washbourne, I.J., Crenshaw, C.L., Baker, M.A., 2011. Dissimilatory nitrate reduction pathways in an oligotrophic aquatic ecosystem: spatial and temporal trends. Aquatic Microbial Ecology 65, 55–64.

Whitton, B.A., 2012. Ecology of Cyanobacteria II: Their Diversity in Space and Time. Springer Science & Business Media, Dordrecht, Netherlands.

Yin, G., Hou, L., Liu, M., Liu, Z., Gardner, W.S., 2014. A novel membrane inlet mass spectrometer method to measure $^{15}NH_4^+$ for isotope-enrichment experiments in aquatic ecosystems. Environmental Science and Technology 48, 9555–9562.

Chapter 33

Phosphorus Limitation, Uptake, and Turnover in Benthic Stream Algae

Alan D. Steinman[1] and Solange Duhamel[2]

[1]Annis Water Resources Institute, Grand Valley State University; [2]Lamont—Doherty Earth Observatory, Columbia University

33.1 INTRODUCTION

Increased loading of nutrients into streams and lakes has become one of the major environmental problems facing society today. Non—point source pollution, associated with changing land use patterns and practices, has resulted in increased impairments to water bodies (e.g., Jordan et al., 1997; Carpenter et al., 1998; Allan, 2004; Dodds, 2006; Kaushal et al., 2011; Michalak et al., 2013). These impairments include cultural eutrophication, harmful algal blooms, thermal pollution, increased sedimentation, and increased loadings of contaminants, such as metals, pesticides, oil, and grease. The ability of an aquatic ecosystem to assimilate these stressors without exhibiting impairment depends largely on its biology (i.e., species composition and abundance, elemental stoichiometry), chemistry (i.e., nutrient quantity and quality), geology (i.e., underlying lithology), and geomorphology (i.e., constrained or unconstrained valley floor).

In stream ecosystems, benthic algae and bacteria represent a potentially important biotic sink for pollutants, such as excess nutrients (Mulholland and Rosemond, 1992; Bernhardt et al., 2003; Cardinale, 2011), although adsorption to stream sediments also can be a major sink, especially for phosphorus (Mulholland, 1996; Stutter et al., 2010). Determining the rates at which nutrients are taken up and released can provide important information in assessing how large a nutrient load a stream, lake, or estuary can process before its integrity is negatively impacted (cf. Dodds, 2003; Withers and Jarvie, 2008; Steinman and Ogdahl, 2015).

The nutrient that we focus on in this chapter is phosphorus (P). Inorganic P is commonly considered the element most likely to limit primary production in freshwater ecosystems (Schindler, 1977; Hecky and Kilham, 1988; Hudson et al., 2000; but see below). Although P concentrations in healthy plants are relatively low, usually ranging from 0.1% to 0.8% of dry mass (Raven et al., 1981), P is an essential element. Some of the more important functions played by P in plants include being a structural component of "high-energy" phosphate compounds [e.g., ADP and adenosine triphosphate (ATP)], nucleic acids, several essential coenzymes, and cell membrane constituents (phospholipids), as well as being involved in the phosphorylation of sugars.

Although a significant number of stream studies have indicated that P limits the growth of benthic algae (e.g., Stockner and Shortreed, 1978; Elwood et al., 1981; Peterson et al., 1983; Bothwell, 1989; Dodds et al., 1997), the phenomenon is not universal (McCall et al., 2014), and P is by no means the only limiting nutrient in lotic ecosystems. Nitrogen (N) has been found to be the limiting nutrient in some streams (Grimm and Fisher, 1986; Hill and Knight, 1988; Lohman et al., 1991; Tank and Dodds, 2003), whereas other lotic systems can be colimited by N and P (Rosemond et al., 1993; Perrin and Richardson, 1997; Francoeur, 2001; Tank and Dodds, 2003), micronutrients (Pringle et al., 1986), or light (Hill et al., 1995; Carey et al., 2007; Johnson et al., 2009).

In this chapter, three different aspects of P utilization by benthic algae will be covered: (1) assessment of P limitation, (2) measurement of P uptake rates, and (3) determination of the release rate of P (expressed as the turnover rate). We note two caveats regarding this chapter. First, we focus exclusively on inorganic P; it is likely that dissolved organic phosphorus plays an important, albeit relatively undefined role, in the nutrient dynamics of freshwater algae (cf. Hwang et al., 1998; Pant et al., 2002), especially in P-limited environments (Karl and Björkman, 2015). However, treatment of

Methods in Stream Ecology. http://dx.doi.org/10.1016/B978-0-12-813047-6.00011-5

this topic is beyond the scope of this chapter (but see Chapter 24). Second, although we use the term benthic algae throughout the chapter, it should be noted that the benthic algae attached to submerged substrata in streams usually exist as part of a complex assemblage variously referred to as periphyton, *aufwuchs*, or biofilm. This assemblage usually consists of algae, bacteria, fungi, and meiofauna (see Chapters 9–11, 14) that exist within a mucilaginous, poly-saccharide matrix (Lock et al., 1984), and each biotic group has different affinities for P. Indeed, even within a group, P uptake and cycling may be influenced by the abundance of different species and growth forms (cf. Steinman et al., 1992; Davies and Bothwell, 2012).

33.1.1 Assessment of P Limitation

Nutrient limitation in algae can be assessed in several different ways, including elemental composition of biomass, nutrient enrichment bioassays, enzymatic activities, and physiological responses. Elemental composition can suggest nutrient limitation because the proportions of carbon (C), N, and P, while confined to a relatively narrow range in algae (Hall et al., 2005), nonetheless vary in response to both ambient nutrient concentrations in the water and ambient light conditions (Finlay et al., 2011; Drake et al., 2012). The ratios of C:N:P have profound ecological implications, as nutrient stoichiometry at the base of trophic food webs can influence or be influenced by trophic level interactions, population dynamics, taxonomic structure at the community level, and ecosystem level processes such as nutrient limitation and cycling (Hillebrand and Kahlert, 2001; Frost et al., 2002; Stelzer and Lamberti, 2002; Hillebrand et al., 2008). Kahlert (1998), in a review of the literature, found that the optimal (i.e., conditions without nutrient limitation or surplus) C:N:P ratio of freshwater benthic algae was 158:18:1 (molar), which deviates from both the Redfield ratio of 106:16:1 (Redfield, 1958; but see Geider and La Roche, 2002) derived from mixed phytoplankton populations and the ratio of 119:17:1 obtained for marine benthic microalgae (Hillebrand and Sommer, 1999). It is likely that carbon-rich detritus (cf. Cross et al., 2003), the inclusion of macroalgae (cf. Hillebrand and Sommer, 1999), and the carbon content of the mucilaginous biofilm matrix account, at least in part, for the higher C:P ratio of Kahlert's benthic algae compared to Redfield's planktonic algae. When C, N, or P become limiting in the environment, this condition can be reflected in a lower level of nutrient present in the algal cell. For example, if P concentration becomes growth limiting in a stream, tissue C:P and N:P ratios would be expected to increase because the algae make more efficient use of the P incorporated into cells. Moreover, certain microorganisms have evolved mechanisms to reduce their P requirements. For example, some phytoplankton species can partially substitute sulfur for P in membrane lipids (Van Mooy et al., 2009); more controversial has been the suggestion that an aquatic bacterium may be able to partially substitute arsenic for P in nucleic acids (Wolfe-Simon et al., 2011), although this work has been challenged (e.g., Reaves et al., 2012). For freshwater planktonic algae, C:P values >129 and N:P values >22 (as opposed to Redfield ratios of 106 and 16, respectively) suggest at least moderate P deficiency in algae (Hecky et al., 1993). However, these ratios increase considerably for freshwater benthic algae, where P deficiency is suggested if C:P values exceed 369 and N:P values exceed 32 (Kahlert, 1998). See Chapter 36 for more information and exercises on elemental composition.

Nutrient enrichment bioassays involve the addition of nutrients to a stream, either in the form of diffusing substrata (see Chapter 31), fertilizer pellets (Steinman et al., 2011), or solute injections (see Chapters 30 and 31). The enrichment lasts for some designated period of time, and its effect is evaluated by change in biomass (see Chapter 12) or primary productivity (see Chapter 34) in enriched compared to unenriched systems.

An enzymatic assay that has proven to be a reliable indicator of P limitation is whole-community *phosphatase* activity (PA). The phosphatase enzyme hydrolyzes phosphate ester bonds, thereby releasing orthophosphate (PO_4) from organic P compounds. PA is quantified by measuring the amount of hydrolysis produced after the phosphatase enzyme comes in contact with an added organic P substrate, thereby releasing PO_4. The most common type of phosphatase assayed in freshwater systems is alkaline phosphatase, which hydrolyzes phosphomonoesters. In contrast, phosphodiesterase hydrolyzes phosphodiesters, while adenosine triphosphatase hydrolyzes ATP. PA is regulated by P concentration present in the water (Dyhrman and Ruttenberg, 2006; Duhamel et al., 2010); PA is generally repressed as inorganic P concentrations increase. Conversely, PA generally increases as inorganic P concentrations decline in aquatic ecosystems (Healey, 1973; Wetzel, 1981; Currie et al., 1986; Espeland et al., 2002; Scott et al., 2009; Ellwood et al., 2012). Thus, PA has been used to infer P limitation for aquatic microflora (Healey and Hendzel, 1979; Burkholder and Wetzel, 1990; Hernandez et al., 2002; Newman et al., 2003). Based upon their results from phytoplankton culture studies, Healey and Hendzel (1979) suggested that phosphatase levels above 0.003 mmol P mg chlorophyll a^{-1} h^{-1} indicate moderate P deficiency and levels above 0.005 mmol P mg chlorophyll a^{-1} h^{-1} indicate severe P deficiency. It should be noted that these values are specific to the environmental conditions and substrate that they used: pH 8.5, 35°C, and 10 µM *o*-methylfluorescein phosphate (which they found to support saturated PA). In systems with complex organic P substrates (e.g., dystrophic systems), PA can lead to biased conclusions if the assays do not include the appropriate phosphatases (cf. Pant et al., 2002); in addition, normalizing to chlorophyll may inflate algal-derived phosphatase rates as it does not account for bacterial biomass, which can be substantial in some periphyton mats.

The application of enzyme-labeled fluorescence (ELF) has opened up research avenues in the use of PA to detect P limitation. Although the details of this method are beyond the scope of this chapter, a brief review helps illustrate possible future directions of this field. Rengefors et al. (2001) used ELF to differentiate PA at the species-specific scale for phytoplankton. ELF results in a fluorescent product when it reacts with alkaline phosphatase, which precipitates at the site of enzymatic activity (i.e., cell surface). This method makes it possible to determine not only which species contains PA but also the relative quantity of PA (Nedoma et al., 2003; Duhamel et al., 2009). ELF also has been used to (1) differentiate whole-community PA from that of individual bacterial cell PA in wetland periphyton biofilms (Espeland et al., 2002), (2) determine the contribution of microbes to overall phosphatase production in periphyton mats inhabiting *Utricularia* traps (Sirová et al., 2009), and (3) localize AP within freshwater macroalgal epiphyte assemblages (Young et al., 2010). ELF labeling of both intact and homogenized periphyton mats may result in interesting insights regarding where P limitation is occurring within the biofilm matrix (Sharma et al., 2005).

33.1.2 P Uptake Rates

The relationship between the nutrient concentration in the water and the rate at which nutrients are taken up by algae can be described by a hyperbolic function (Fig. 33.1). The Michaelis–Menten equation for enzyme kinetics is often used to describe this function:

$$V = V_m(S/(K_s + S)) \tag{33.1}$$

where V = nutrient uptake rate, V_m = maximum nutrient uptake rate, S = concentration of the nutrient, and K_s = half-saturation constant (or nutrient concentration at which nutrient uptake is one-half the maximal uptake rate). From a biological perspective, there are two critical considerations in Fig. 33.1. First, nutrient uptake rates become saturated as nutrient concentration increases. Empirical studies have shown that saturation of P uptake can occur at very low concentrations in both individual benthic diatom cells (<1 µg/L; Bothwell, 1989) and whole streams (<10 µg/L; Mulholland et al., 1990). Thus, investigations examining P uptake in benthic algae must consider the possibility that saturation will influence uptake kinetics even at relatively low concentrations. Second, the constant K_s provides a useful index of a cell's affinity for a nutrient—a lower K_s suggests a greater affinity for the nutrient, which can confer a competitive advantage when the nutrient is present at low concentrations. Generally, taxa that have low K_s values have a competitive advantage at low nutrient concentrations. However, K_s values appear to be fixed and do not appear to vary much under different environmental conditions. Rather, the physiological reason why nutrient-limited algae often increase their short-term nutrient uptake rates when exposed to elevated nutrient concentrations is because of an increase in V_m and not a change in K_s

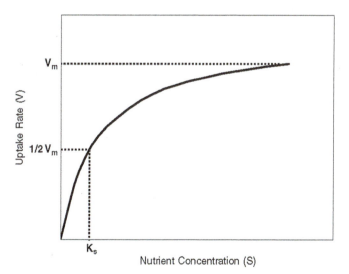

FIGURE 33.1 Relationship between nutrient concentration (*S*) and nutrient uptake rate (*V*). V_{max}, the maximum nutrient uptake rate; K_s, the half-saturation constant or the nutrient concentration at which the uptake rate is one-half of the maximum uptake rate.

(Darley, 1982; Lohman and Priscu, 1992). However, multiphasic P uptake systems have been documented in unicellular algae (Rivkin and Swift, 1982; Jansson, 1993), suggesting that K_s values are not static. From an energetic perspective, increasing V_m makes more sense, as a change in K_s requires the alteration of existing enzyme structures or the induction of an alternative enzyme, whereas increasing V_m requires only the activation or additional synthesis of an existing enzyme (Rivkin and Swift, 1982). Of course, over the long term, elevated nutrient levels may lead to an altered community structure, resulting in dominance by new algal or bacterial species, which may have greater V_m or different K_s values, thereby changing nutrient kinetics in the benthic community.

It is also important to distinguish between nutrient-limited *uptake* rates (above) and nutrient-limited *growth* rates. The relationship between nutrient concentration and algal growth can be modeled using either the Monod model or the Droop model. The Monod model relates algal growth to the external concentration of nutrients in the water, whereas the Droop model relates algal growth to internal (cellular) concentration of nutrients. For additional details on these models, see Droop (1974), Rhee (1978), Kilham and Hecky (1988), and Borchardt (1996).

33.1.3 P Turnover Rates

This portion of the chapter is designed to examine P turnover rates in benthic stream algae. P turnover may provide an index of internal cycling in the algal community. Once an algal cell takes up P from the external medium, the P can be incorporated into structural elements, maintained in a labile pool, or excreted from the cell. Cells that are phosphorus limited may be less likely to release the P they have taken up (back to the external medium) than cells that are phosphorus replete (but see Cembella et al., 1984; Borchardt et al., 1994). Thus, the P turnover rate in algae (i.e., loss of P from algal cell relative to total algal P) may be lower in P-limited cells than P-saturated cells, assuming both the P-limited and P-replete cells have similar metabolic activities and are exposed to similar grazing pressures (Steinman et al., 1995). One way to measure P turnover in algae is to label the cells with a P *radioisotope* (e.g., ^{32}P or ^{33}P) in the laboratory, place the algae back into the natural or a controlled environment, and then measure the amount of radioactive P present in the cells over time. This gives an apparent P turnover rate, as turnover is being estimated from the entire periphyton matrix and not from individual cells.

33.1.4 Overview of Chapter

This chapter examines P limitation and uptake in benthic algae collected from a relatively low- and a relatively high-P stream. In theory, the benthic algae growing in the low- and high-P streams should have adapted to the different ambient conditions and exhibit different ecological attributes (Fig. 33.2). Specifically, if the algae in the low-P stream are P-limited, they should have lower biomass and greater C:P ratios than algae collected from the high-P stream, all else being equal (although intense internal nutrient cycling within the periphyton matrix can compensate for low external nutrient concentrations; Dodds, 2003; Mulholland et al., 1994). In addition, algae in the low-P stream should have greater phosphatase activities, lower K_s values, lower total P uptake rates (i.e., computed as mass per unit time), and lower apparent P turnover rates (greater retention) than algae from the high-P stream (Fig. 33.2).

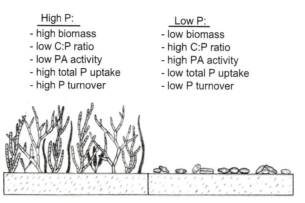

High P:
- high biomass
- low C:P ratio
- low PA activity
- high total P uptake
- high P turnover

Low P:
- low biomass
- high C:P ratio
- high PA activity
- low total P uptake
- low P turnover

FIGURE 33.2 Phosphorus-related attributes for hypothetical attached algal communities exposed to high-P (left side) and low-P (right side) conditions.

33.2 GENERAL DESIGN

This chapter describes the methodology to measure P limitation, uptake, and turnover in benthic algae. Although valuable information will be gleaned from any of these methods in isolation, we recommend combining them when possible to gain a broader understanding of P-related processes in streams.

Site selection should include two streams, one with relatively high P (e.g., >20 µg PO_4-P/L, if available) and one with relatively low P (e.g., <5–10 µg PO_4-P/L, if available). If differences in algal response are to be detected, it is critical that the algae be exposed to ecologically meaningful differences in nutrient concentration. We recommend using an undisturbed stream (if available) for the "low-P" system, where there are few obvious impacts (e.g., point source inputs, unnatural absence of riparian vegetation, livestock in streams). For the "high-P" system, use streams receiving either agricultural runoff or point sources containing high P (e.g., sewage effluent) or clarifying tanks at sewage treatment facilities (Davis et al., 1990). If all streams in the region have low levels of P, then it may be possible to enrich a stream with P for a sustained period of time (e.g., >4 weeks) to create high-P conditions (e.g., Steinman, 1994; Cross et al., 2003) or, alternatively, enrich a stream with N to force P limitation. These changes can be done through the use of nutrient-diffusing substrata (see Chapter 31) or solute additions (see Chapter 30) if permitted by local regulatory agencies. Regardless of which streams are selected, collect algae from sites in the two streams that are comparable in terms of other environmental conditions (e.g., irradiance level, current velocity, discharge, temperature, grazer density) to the greatest extent practicable.

33.2.1 Basic Method 1: Phosphatase Activity

This method consists of two parts: an assay of PA, followed by normalizing activity to an estimate of biomass (see Chapter 12). Normalizing PA to Chlorophyll fails to account for bacteria in the production of PA, likely resulting in an overestimation of PA. Alternatively, PA can be normalized to ash-free dry mass (AFDM) (see Chapter 12), which does account for bacteria (and other organic matter). However, this approach may underestimate PA, as AFDM includes materials such as fungi and detritus, which are not directly involved in PA. PA also can be normalized to surface area, but in that case, rates may be more reflective of how much (or how little) active biomass is present than the P concentrations in the water column.

Phosphatase acts on a variety of organic P compounds. Synthetic substrate analogs are used to measure PA in natural samples. After hydrolysis of the PO_4 moiety, these substrates release a product that can be assayed colorimetrically (e.g., 5-bromo-4-chloro-3-indolyl phosphate) or fluorometrically [e.g., methylumbelliferyl phosphate (MUF-P), 3-*o*-methylfluorescein (MFP)]. While fluorogenic substrates are considered to be the most sensitive, colorimetric assays are preferred in environments with high biomass as they are easier to perform (Hernandez et al., 2002; Hoppe, 2003; Jansson et al., 1988; Manafi et al., 1991). In this method, we add a commercially available compound, *para*-nitrophenyl phosphate (*p*-NPP), to determine the PA present. When PO_4 is hydrolyzed from *p*-NPP, *p*-nitrophenol (*p*-NP) is formed, which can be measured spectrophotometrically.

33.2.2 Basic Method 2: Chemical Composition (C:P Ratio in Algal Tissue)

This method consists of three components: measurement of algal AFDM (and conversion to C), acid digestion of combusted matter to obtain dissolved P leached from ashed algal tissue (Solórzano and Sharp, 1980), and then measurement of inorganic P in oxidized algal material according to standard methods. Ideally, the C concentration in algae would be measured with an elemental analyzer (see Chapter 36). However, this instrument is not always available. Consequently, in this method we present an alternative approach to estimate C, based on measurement of AFDM (see Chapter 12), which is easy to perform but less accurate than elemental analysis. We assume C is 53% of AFDM, a reasonably accurate assumption for most algal communities (Wetzel, 1983).

33.2.3 Basic Method 3: Net Nutrient Uptake—Stable Phosphorus

This procedure involves the measurement of net loss of dissolved inorganic phosphorus from the water in which the algae are growing. Because certain classes of organic P compounds are partially hydrolyzed by the standard molybdenum blue assay used to measure dissolved inorganic P concentration (see below), results are typically reported as soluble reactive phosphorus (SRP). The method consists of three components: sample water during the incubation, measure SRP in water samples, and measure algal biomass. Water is sampled at the start of the incubation, and thereafter at 30 and 60 min, and analyzed for SRP according to standard methods (APHA et al., 1995). Large changes in the biomass:water volume ratio resulting from water sampling during the incubation period should be avoided by minimizing sample volumes or number of samples. If the water volume in the incubation chambers is low, the 30 min samples can be omitted. We recommend that volumetric change be limited to <10% of initial volume during sampling. Ideally, chambers attached to pumps that could recirculate water during the incubation would be used (Fig. 33.3), as water velocity will influence the uptake rate of P in benthic stream algae (Whitford and Schumacher, 1964). However, if chambers and pumps are not available, the method can still be completed by using large (2 L) glass chambers and stir plates. An open petri dish is glued to the bottom of the chamber, into which is placed a stir bar. Then coarse-meshed screening (e.g., chicken wire) is placed over the petri dish, thereby creating a shelf onto which are placed the substrata with attached algae. The rotation speed of the stir bar is varied until it matches approximately the current velocity in the sampled streams.

The use of stable (i.e., nonradioactive) elements to measure *net nutrient uptake* rate is dependent on the initial nutrient concentration in the ambient water being high enough to measure the remaining nutrient at the end of the incubation period. For example, if P concentrations are low at the beginning of the incubation, they may be below the detection limit at the end. Another potential problem with this measurement is that if nutrient regeneration rates from algae are similar to nutrient uptake rates (i.e., the community is at steady state with respect to nutrient dynamics), then no net uptake will be measured, even if total uptake rates are appreciable. An alternative approach is to add nutrients to stream water. This elevates nutrient concentrations above ambient levels, ensuring that concentrations at the end of the incubation will still be high enough to be measured, and temporarily increases nutrient uptake rates above rates of nutrient regeneration. However, this approach measures only nutrient uptake potential at the higher concentration and may be an overestimate of ambient uptake rate depending on the degree of enrichment (see Mulholland et al., 1990, 2002). Alternatively, a nutrient addition

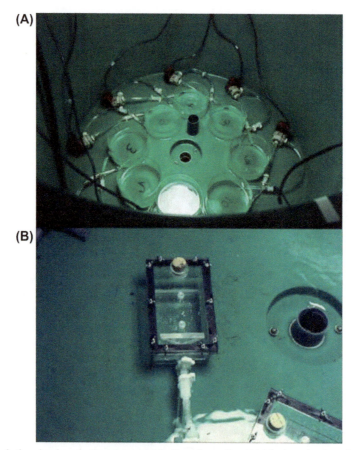

FIGURE 33.3 Examples of incubation chambers that have been used for P uptake studies. (A) 2-L glass chambers fitted with adapters to accept tubing attached to submersible pumps. Pumps circulate water within the chambers. Placing adjustable clamps on tubing line can reduce flow rate, if so desired. (B) 1-L plexiglass chamber with detachable lid. Lid attaches to main body of chamber with wing nuts; gaskets provide a leak-proof seal. Chambers are attached to submersible pumps. Note the large port (far end with cork) in lid, which allows an oxygen meter to be placed directly in the chamber to measure metabolism. The two small ports allow for injection of radioisotope into the chamber.

approach that involves multiple levels of nutrient enrichment can be used to approximate total uptake rate at ambient nutrient concentration (Payn et al., 2005; see also Chapter 30). In this chapter, we provide instructions for measuring net uptake rates at ambient nutrient concentrations (i.e., without enrichment).

Although we recommend the measurement of SRP in this analysis, we note that SRP may overestimate the true concentration of phosphate in the water sample. That is because the soluble portion of SRP may include colloidal P in the filtrate, while the reactive portion of SRP may include organic forms of P that react with the reagents (Rigler, 1966; Hudson et al., 2000). The difference between SRP and phosphate concentration is most distinct in waters with very low-P concentrations but becomes less noticeable as ambient P concentration increases.

33.2.4 Advanced Method 1: Gross Nutrient Uptake (Phosphorus Radiotracer)

This procedure involves measuring the loss of $^{33}PO_4$ added to the water in which the algae are growing.[1] Stable and radioisotopic tracers are important tools to monitor and quantify microbial growth and metabolism, as well as the cycling of elements. Because P has only one stable isotope (^{31}P), the use of radioactive isotopic tracers of phosphorus (i.e., ^{32}P or ^{33}P) can be critical in quantifying P fluxes in aquatic environments. P radioisotopes are particularly useful as they can be added to the sample at trace levels, and thus not perturb the steady-state phosphate concentration. This is particularly relevant in P-limited environments with very low PO_4 concentrations, where P radioisotopes allow rigorous estimation of the uptake kinetic parameters and the turnover time of phosphate and organic P substrates (e.g., ^{33}P-glucose-6-phosphate, ^{33}P-ATP).

Recent studies also have used variations in the natural abundance of the stable ^{18}O isotope bound to P ($\delta^{18}O_P$) to study P transformations and determine P sources in aquatic systems (Colman et al., 2005; McLaughlin et al., 2006; Elsbury et al., 2009; Young et al., 2009). P is strongly bound to oxygen (e.g., orthophosphate, polyphosphate), and in many aquatic ecosystems the oxygen isotopic composition of $\delta^{18}O_P$ is not in isotopic equilibrium with ambient water, and may thus reflect the phosphate sources to water bodies. These analyses require sophisticated instrumentation and are not included in this chapter.

Uptake rates will be calculated in this method from algae growing in high-P and low-P streams. This method should use algae growing on small artificial substrata [e.g., unglazed ceramic cylinders (Steinman et al., 1991b) or unglazed tiles placed in the streams for a period long enough to acquire an algal community similar to natural substrata]. The use of small artificial substrata allows AFDM to be measured directly on the substratum without it being physically removed, thereby minimizing contact with radioactive P in the algae. In addition, if the turnover option is to be completed (see below), the P in the algae on these substrata can be extracted with relative ease.

33.2.5 Advanced Method 2: Phosphorus Turnover

This method involves measuring the rate at which radiolabeled P, incorporated into algal biomass, is lost from the algal assemblage over time.[2] Ideally, this procedure will be piggybacked on the prior method of measuring P uptake rates using ^{33}P. The method consists of four parts: radiolabeling of algae, oxidation of labeled algae, extraction of P from ash, and measuring radioactivity in subsamples of extracted material.

After the ^{33}P uptake method is completed, substrata are either returned to the high- and low-P streams if allowed, or placed into lab-based recirculating chambers or aquaria containing either high or low concentrations of P. Four substrata are sampled from each stream on four occasions over a 10-day period. The algae on the substrata are oxidized, and ^{33}P is extracted from the ash. A subsample of this extract is diluted, placed into scintillation cocktail, and assayed for radioactivity using liquid scintillation spectroscopy.

P turnover for each stream is calculated as the first-order rate constant of the decline in ^{33}P activity over time (slope of relationship between ln [^{33}P] in algae vs. time). A mean activity is calculated on each date from the four substrata collected and used in the regression with time. For the purposes of this method, we recommend normalizing ^{33}P content to unit area

---

1. Extreme caution must be exercised when using radioisotopes. Users must consult with the radiation safety officer at their institution. We recommend the use of ^{33}P, instead of ^{32}P, because of its lower energy, although it is more expensive. Even with the relatively low maximum energy of ^{33}P (0.248 MeV), the small amount of radioactivity used (0.5 mCi/L), and the short half-life of the isotope (25.4 days), all handling of the isotope must be done with extreme care.

2. Extreme caution needs to be exercised when using radioisotopes. See the prior cautionary note. In addition, because this exercise involves potentially placing radiolabeled algae back into the natural environment, users must consult with the radiation safety officer at their institution regarding restrictions or other potential concerns about this protocol. If this option is not viable, labeled algae can be placed into chambers or aquaria filled with water of high and low-phosphorus concentration, to evaluate the influence of P concentration on turnover.

of substratum, as opposed to biomass, although this approach assumes relatively similar biomass levels among substrata or that sufficient samples are collected on each date to take into account the natural variability of biomass in the system. If ^{33}P content is expressed per unit biomass, it becomes necessary to introduce a growth correction factor to account for any net growth during the period of the experiment (because the amount of radioactivity per unit biomass in the sample will decline due to dilution by the accrual of new, nonlabeled biomass). Also, if the extracted ^{33}P samples are counted on different days over the period of the turnover experiment, they must be corrected for radioactive decay from the start of the experiment; because of the short half-life of ^{33}P (25 days), some of the decline in ^{33}P content in algae will be the result of radioactive decay. As a solution to this problem, all of the ^{33}P extract samples for the entire turnover study can be assayed on the scintillation counter at the same time at the end of the study, thereby minimizing the need to correct for decay.

33.3 SPECIFIC METHODS

33.3.1 Basic Method 1: Phosphatase Activity

33.3.1.1 Preparation Protocol

1. At least 1 month, and preferably 3–6 months, prior to the experiment, place approximately 50 of the 3 cm × 3 cm unglazed ceramic tiles or ceramic cylinders (Steinman et al., 1991b; Du-Co Ceramics Co., Saxonburg, PA: http://www.ceramics.com/duco/) in riffle habitats of the two streams to be sampled (50 tiles allows several teams of students to replicate this analysis); riffles should have similar physical characteristics (i.e., current velocity, flow, irradiance, substrate). We recommend substrates of small size because this allows them to be placed directly into an extraction jar at the time of sampling, without having to remove algae from the surface. If unglazed tiles are used, and are purchased attached to each other in sheets (as opposed to individual tiles), place the entire sheet in the stream, which minimizes the likelihood of tiles being lost if high discharge occurs.
2. At least five tiles per stream should be analyzed (four for PA and one control per stream). Alternatively, small rocks can be used but they must be small enough to fit in the incubation jars and be submersed in a small volume of water.
3. Label two acid-washed plastic containers (c. 30 cm × 30 cm = 900 cm^2) by stream name or type (high-P; low-P).
4. Label 10 wide-mouth glass incubation jars (30 mL or large enough to contain the substratum) by stream type (high-P or low-P stream): one tile per jar for each of the 4 PA tiles and one control per stream type (10 total).
5. Prepare 150 mM *p*-NPP solution (add 2.78 g of *para*-nitrophenyl phosphate to 50 mL of ultrapurified deionized water).

33.3.1.2 Field Collection Protocol

1. Collect tiles and filter water [filter about 500 mL of stream water into an acid-washed 1-L plastic bottle using a hand pump and a Whatman GFF or Gelman Type A/E glass fiber filter (GFF)] from each stream.
2. Fill the two acid-washed plastic containers (one labeled high P and the other labeled low P) with the appropriate stream water, and place five tiles (one extra per stream for PA in case of loss; control treatment has no tile) from each stream inside the container. Attach the lids to completely water-filled containers (which minimizes tile movement), and place the containers in a cooler to be transported back to the laboratory.

33.3.1.3 Laboratory Protocol

1. Using an automatic pipette, transfer 20 mL of filtered stream water (use more water if needed to completely submerse substratum) to each of the 10 incubation jars labeled for PA (5 jars for each stream type). In the laboratory, separate the sheet of tiles into individual tiles if necessary (ignore any glue that may remain attached to individual tiles following separation), and place one tile into each incubation jar. Leave one jar per stream without a tile (control).
2. Using an adjustable volume 1-mL pipette (set to 0.4 mL), transfer 0.4 mL of the *p*-NPP solution into the water (units of mmol/L) in each of the incubation jars (or proportionately more if water volume is >20 mL), cap the jar, and gently mix. Incubate the jars at room temperature for 30 min, gently hand swirling the jars every 3–5 min.
3. After 30 min, filter the water in each jar by removing the water using a 25-mL plastic syringe and filtering it through a 0.45 µm pore size syringe-mounted filter (e.g., Syrfil-MF, Costar Corp., Cambridge, MA) and collecting 10 mL of filtrate in a labeled glass scintillation vial.
4. Add 0.05 mL of 1N NaOH to each vial containing the 10 mL of filtrate from each incubation jar to bring the pH up to ~10 (for maximum color development of nitrophenol). Measure the absorbance of each filtrate at 410 nm against deionized water using a dual-beam spectrophotometer and a 1-cm pathlength cuvette.

5a. Biomass as chlorophyll *a*: Remove the tile from each jar, rinse it by immersing it into filtered, unamended stream water, and place it in a small plastic jar or centrifuge tube containing a known volume of 90% acetone that is sufficient to cover the substratum if measuring chlorophyll; follow the procedures in Chapter 12 for chlorophyll analysis.

5b. Biomass as AFDM: Remove the tile from each jar, rinse it by immersing it into filtered, unamended stream water, and scrape the material off the tile following the procedures for measuring AFDM described in Chapter 12.

33.3.1.4 Data Analysis

1. PA is calculated from the absorbance of the *p*-NPP solution as follows:

$$PA = (Abs_{sample} - Abs_{blank}) \times 58 \times Volume_{(inc)} \tag{33.2}$$

where Abs_{sample} = absorbance reading of sample at 410 nm, Abs_{blank} = absorbance reading of control at 410 nm (filtered stream water only, to correct for natural phosphatase in water), $Volume_{(inc)}$ = volume of stream water in which each algal sample is incubated (in L); [if 20 mL is used (as described in this method), this value will be 0.02]. Use Table 33.1 for data entry and calculations. The value 58 in Eq. (33.2) is the specific absorbance (at pH > 10) of nitro-phenol, which is the hydrolysis product of *p*-NPP.

2. PA of the periphyton community on each cylinder is computed as the amount of NPP hydrolyzed [mmol of nitrophenyl (NP) produced] per unit biomass for each sample, to obtain chlorophyll-normalized or AFDM-normalized PA (units of mmol mg chlorophyll a^{-1} 0.5 h^{-1} or mmol mg $AFDM^{-1}$ 0.5h^{-1}, respectively). If phosphatase levels are very low, the incubation period can be extended to 1 h, and the values are reported per hour. Alternatively, PA could be normalized by tile surface area to obtain area-specific PA (units of mmol cm^{-2} 0.5 h^{-1}).

33.3.2 Basic Method 2: Chemical Composition

33.3.2.1 Preparation Protocol

1. If tiles are to be used for this experiment (in lieu of rocks), place approximately 50 of the 3 cm × 3 cm unglazed ceramic tiles or ceramic cylinders (Steinman et al., 1991b) in selected high-P and low-P streams to be sampled (50 tiles allows several teams of students to replicate this analysis). Tiles should be placed in streams at least 1 month, and preferably 3–6 months, prior to the experiment.

2. Label two acid-washed plastic containers (30 cm × 30 cm) by stream name or type (high-P; low-P).

3. Combust six acid-washed 10-mL glass beakers at 500°C for 1 h. Cut out 6 pieces of aluminum foil that are large enough to cover the top of each beaker. Lightly etch the sample number onto the foil with a pointed object (do not write it on the beaker because it will burn off during combustion, potentially contributing to dry mass and leaving one unable to track individual samples).

TABLE 33.1 Sample data sheet for determination of phosphatase activity (PA). See also Online Worksheet 33.1.

Date:				Stream:					
Sample	Absorbance (410 nm)	Net Abs. (Absorbance Minus Sample Blank)	Volume (L)	PA (mmol ½h^{-1})	Chlorophyll *a* (mg)	Chl-Specific PA (mmol mg^{-1} ½h^{-1})	Ash-Free Dry Mass (AFDM) (mg)	AFDM-Specific PA (mmol mg^{-1} ½h^{-1})	
Blank									
1									
2									
.									
.									
n									

33.3.2.2 Field Collection Protocol

1. Collect rocks or ceramic tiles from each stream.
2. Fill the two plastic containers (one labeled high P and the other labeled low P) with the appropriate stream water, and place three small rocks/tiles from each stream inside the container. Attach the lids to completely water-filled containers (which minimizes rock/tile movement) and place the containers in a cooler to be transported back to the laboratory.

33.3.2.3 Laboratory Protocol

1. Follow the general procedures outlined in Chapter 12 for determination of AFDM, including the following modifications. After the algae are removed from each rock/tile, add the slurry (make sure the volume is less than 10 mL) to the bottom of the precombusted, acid-washed 10-mL tared, glass beaker. Cover the top of the beaker with numbered aluminum foil. Dry the beaker to constant weight at 105°C (c. 24–48 h). Remove the beakers from the drying oven and transfer them to desiccators until weighing.
2. After the beakers have been weighed, place them in a muffle furnace at 500°C for at least 4 h (make certain the oven is at 500°C before timing), remove, and allow beakers to cool to room temperature in a desiccator and reweigh.
3. Using a 5-mL pipette, add 5 mL of 2N HCl to the beaker, label the beaker with the sample number, and replace the aluminum foil with parafilm over the beaker to prevent evaporation. Acid extraction of ashed material should last at least 24 h. Place beakers in the laboratory hood during the extraction period.
4. After extraction, transfer contents of each beaker to a 500-mL volumetric flask. Rinse the beaker with deionized water and pour rinse water into the volumetric flask as well. Bring the total volume in the volumetric flask to 500 mL by adding ultrapurified deionized water (this will result in a leachate of 0.02N HCl).
5. Pour each sample into separate acid-washed plastic bottles, label accordingly, and analyze using standard methods for analysis of P in water (APHA et al., 1995; see below).[3]

33.3.2.4 Data Analysis

1. Calculate the amount of carbon in the sample by multiplying the AFDM by 0.53. Carbon content is estimated by assuming that 53% of AFDM is composed of carbon (Wetzel, 1983). Although this value may vary slightly among algal groups and environmental conditions, the variance is low ($\pm 5\%$) compared to other cellular constituents. Alternatively, you may follow the guidance in Chapter 36, which calls for estimating carbon content by assuming that 30% of DM is composed of carbon. Use Table 33.2 for data entry and calculations.
2. Calculate the concentration of P in each sample by comparing its absorbance against a standard curve developed from the standards analyzed. The total amount of P (in mg) is then calculated by multiplying the P concentration by 0.5 (because the total volume of diluted leachate is 0.5 L).
3. The C:P ratio is calculated by dividing the total C by the total P in each sample (converted to the same mass units) and then multiplying by 2.58 (to convert to a molar basis). Compare the ratio to the Redfield ratio (106:1) and analyze the differences between the high-P and low-P streams.

33.3.3 Basic Method 3: Net Nutrient Uptake (Stable Phosphorus)

33.3.3.1 Preparation Protocol

1. At least 1 month, and preferably 3–6 months, prior to the experiment, place approximately 50 of the 3 cm × 3 cm unglazed ceramic tiles or ceramic cylinders (Steinman et al., 1991b) in selected high-P and low-P streams to be sampled. If tiles were purchased in sheets, and uptake is to be normalized by AFDM, then tiles should be preashed to remove attached glue, which otherwise would be included in the AFDM measurement. If tiles are purchased individually or uptake is to be normalized by chlorophyll *a*, then no preashing is necessary.
2. Label two acid-washed plastic containers (30 cm × 30 cm) by stream name or type (high P; low P).
3. Label six acid-washed 50-mL collection bottles according to treatment (high P vs. low P) and time ("initial," "30 min," and "60 min").

---

3. Standards for P analysis must be made in 0.02 N HCl to be comparable to that of samples. Make sure to use personal protective equipment when handling reagents.

TABLE 33.2 Sample data sheet for determination of chemical composition (italicized letters in formulae refer to column letter). See also Online Worksheet 33.1.

Date: Stream:

Sample	A Beaker + Dried Material	B Beaker + Ashed Material	C Ash-Free Dry Mass (AFDM) = A − B	D Carbon (mg) (AFDM × 0.53)	E Phosphorus (mg) From Digestion and SRP Analysis	F Molar C:P [(D/E) × 2.58]
1						
2						
3						
.						
.						
.						
n						

4a. Label 20 aluminum weigh boats by etching the bottom of the boat with a sharp edge to designate sample number, if measuring biomass by AFDM or

4b. Label 20 small plastic jar or centrifuge tube to designate sample number, if measuring biomass by chlorophyll *a*.

33.3.3.2 Field Collection Protocol

1. Collect tiles and filter water (filter 1 L of stream water into an acid-washed 1-L plastic bottle using a hand pump and a Whatman GFF or Gelman Type A/E GFF) from each stream.
2. Place 10 tiles into a labeled plastic container (high P or low P) per team, which is filled with appropriate stream water. Attach the lids to completely water-filled containers (which minimizes tile movement), and place the containers in a cooler to be transported back to the laboratory.

33.3.3.3 Laboratory Protocol

1. Transfer the 1 L of filtered stream water and tiles into one incubation chamber per stream (the number of tiles placed in each chamber will be dependent on the amount of biomass attached to the substratum; a general rule of thumb would be at least 10 tiles if biomass is low and 5 to 10 tiles if it is high). Ideally there would be multiple chambers with 5 to 10 tiles per stream, not just one chamber per stream type, to have true replication but that can become prohibitively expensive. As a consequence, we default here to just one chamber per treatment stream.
2. Using a 30-mL automatic pipette, remove 30 mL of stream water from each chamber and transfer to the sample bottle labeled "initial." Filter the 30-mL water samples through a 0.45 μm pore size syringe filter (e.g., Syrfil-MF, Costar Corp., Cambridge, MA). Start either the pumps or the stir bar in the chamber.
3. Remove 30 mL of stream water at 30 and 60 min after the start of the incubation, and transfer the water to the appropriately labeled bottle. Filter the samples as in step 2. If the water samples are not going to be analyzed for SRP within a few hours, place the bottles in the refrigerator (for storage up to 1 week) or freezer (for storage >1 week) until they can be analyzed for SRP levels (APHA et al., 1995; see below).
4. After 60 min, remove the tiles from the chamber. If normalizing uptake by AFDM, see step 5; if normalizing uptake by chlorophyll *a*, see step 6.
5a. Place the tiles in an appropriately labeled aluminum weigh boat and dry the tiles to constant weight at 105°C (c. 24−48 h). Remove the weigh boats from the drying oven and transfer them to desiccators until weighing.
5b. After the weigh boats have been weighed, place them in muffle furnace at 500°C for at least 4 h (make certain the ovens are at 500°C before timing), remove, and allow them to cool to room temperature in a desiccator and reweigh.

Calculate AFDM as the difference between the dry mass and the combusted mass, and sum the AFDM of all substrates in each chamber.

6. Place each tile in a small plastic jar or centrifuge tube containing a known volume of 90% acetone that is sufficient to cover the substratum if measuring chlorophyll; follow the procedures in Chapter 12 for chlorophyll analysis.
7. Analyze water samples for SRP (see below).

33.3.3.4 Soluble Reactive Phosphorus Analysis (Adapted From APHA et al., 1995)

1. Make up appropriate reagents:
   a. H_2SO_4 solution: Add 140 mL concentrated sulfuric acid to 900 mL of ultrapurified deionized water.
   b. Ammonium molybdate solution: Dissolve 15 g of ammonium molybdate in 500 mL of ultrapurified deionized water (store in polyethylene bottle in the dark).
   c. Ascorbic acid solution: Dissolve 2.7 g of ascorbic acid in 50 mL of ultrapurified deionized water. Make immediately before using or keep frozen.
   d. Antimony potassium tartrate solution: Dissolve 0.34 g of antimony potassium tartrate in 250 mL ultrapurified deionized water.
   e. Mixed reagent: Combine 25 mL of sulfuric acid solution, 10 mL of ammonium molybdate solution, 5 mL of antimony potassium tartrate solution, and 5 mL of ascorbic acid solution. Use within 6 h of preparation.
   f. Phosphorus standards: (1) Stock solution—Dissolve 0.2197 g of anhydrous K_2HPO_4 in 1 L of ultrapurified deionized water (1.00 mL = 50 μg P/L); (2) Prepare four standard curve solutions by diluting from stock solution—a typical range is 5—1000 μg/L but this may be extended in either direction depending on the concentrations in the selected streams and sensitivity of instrumentation; (3) Develop a standard curve of absorbance versus SRP concentration.
2. Add 3.0 mL of mixed reagent to 30 mL of standard and all samples and mix thoroughly.
3. Wait for at least 20 min, but not longer than 1 h, and measure absorbance of solution at 885 nm against distilled water on a spectrophotometer using 10-cm pathlength cuvettes (shorter pathlengths can be used, but this is not recommended as analytical sensitivity is reduced).
4. Calculate SRP concentration (μg/L) by comparing absorbance of sample against the standard curve.

33.3.3.5 Data Analysis

1. Plot the SRP concentration versus time to determine whether or not the relationship appears to be linear. Calculate the net P uptake rate using the following formula:

$$V = ([C_o - C_f] \times L)/t \tag{33.3}$$

where V = net uptake rate (μg P/h), C_o = initial SRP concentration, C_f = final SRP concentration, L = incubation volume (in L), and t = time period of incubation (h). The net P uptake rate should then be normalized to either total biomass in the incubation (e.g., AFDM or chlorophyll a) or total substratum surface area. Use Table 33.3 for data entry and calculations.

TABLE 33.3 Sample data sheet for determination of net P uptake. See also Online Worksheet 33.1.

Date:	Stream:
Time (min)	Soluble Reactive Phosphorus Concentration (μg/L)
0	
30	
60	

Calculated uptake rate:
Total ash-free dry mass (AFDM) or chlorophyll a in sample:
Uptake per unit AFDM (μg P mg AFDM^{-1} min^{-1}):
Uptake per unit chlorophyll a (μg P mg chlorophyll a^{-1} min^{-1}):

33.3.4 Advanced Method 1: Gross Nutrient Uptake (Phosphorus Radiotracer)

(Review cautionary notes on use of radioisotopes described previously.)

33.3.4.1 Preparation Protocol

1. At least 1 month and preferably 3−6 months, prior to the experiment, place approximately 100 small unglazed ceramic tiles or ceramic cylinders in selected high-P and low-P streams to be sampled. If tiles are used, they should be preashed to remove attached glue, which otherwise would be included in the AFDM measurement.
2. Label acid-washed plastic containers (30 cm × 30 cm) by stream name or type (high-P; low-P).
3. Each team assigned to a stream should have six 25-mL scintillation vials, each containing 15 mL of scintillation cocktail, labeled according to treatment (high P or low P) and time (background, 10, 20, 30, 45, and 60 min).

33.3.4.2 Field Collection Protocol

1. Collect 5 (uptake only) or 16 (uptake and turnover) tiles and water (filter 1 L of stream water into an acid-washed 1-L plastic bottle using a hand pump and a Whatman GFF or Gelman Type A/E GFF) from each stream.
2. Place the 5 (uptake only) or 16 (uptake and turnover) tiles per stream into labeled acid-washed plastic containers (high P or low P), which are filled with stream water. Attach the lids to completely water-filled containers (which minimizes tile movement) and place containers in a cooler to be transported back to the laboratory.

33.3.4.3 Laboratory Protocol

1. Transfer the 1 L of filtered stream water and 5 (uptake only) or 16 (uptake and turnover) tiles into each incubation chamber. An extra tile will be collected from each stream but remain unlabeled by isotope; this tile will serve as a background control (see Section 33.3.5.2).
2. Transfer approximately 50 mL of the filtered stream water to a 60-mL acid-washed plastic bottle, which will be analyzed for SRP concentration (see Section 33.3.3.4).
3. Remove 1 mL of water from each chamber just prior to the ^{33}P injection.[4] Transfer this water to the appropriately labeled scintillation vial (background) and mix thoroughly.
4. Inject 0.5 mCi of carrier-free ^{33}PO$_4$ (as either orthophosphoric acid or phosphate salt dissolved in water) with a micropipette into each chamber. The micropipette tip will be extremely radioactive, so it should be removed immediately from the pipette after use and discarded in a radioactive waste bin.
5. Remove 1 mL of water from each chamber at 10, 20, 30, 45, and 60 min after the start of the incubation. Transfer the water to the appropriately labeled scintillation vial and mix thoroughly.[5]
6. After the 60 min sample is collected, carefully remove the tiles from the chamber using forceps or tongs (**remember, the material attached to the tiles will be radioactive and should be handled with great care**). These tiles will then be processed for AFDM measurement, if turnover is not to be measured. For AFDM measurement, follow the procedures outlined below (step 7), making sure to avoid touching the radioactive material (keeping the tiles inside the beaker at all times minimizes this risk). If P turnover is to be measured, it is recommended that the tiles remain in the radiolabeled chamber water for an additional 5 h (6 h total) to allow for greater incorporation of ^{33}P by the algae. Radiolabeled tiles are then transported back out to the streams (if permitted) or returned to recirculating chambers/aquaria if turnover is to be measured. **Cautionary notes**: Wear gloves, safety glasses, and lab coats at all times when handling radioactive samples (consult the local radiation safety officer at your institution for guidance and specific regulations associated with your site). Store the radioactive water from the incubation in sealed and labeled carboys until the radioactivity decays to background levels. It is recommended to store the water for at least 10 half-lives before disposal (the half-life of ^{33}P is 25.4 days).
7. Finish measuring AFDM according to the methods outlined above, with the important modification of *not* removing the algae from the substratum. Instead, weigh the substratum with attached algae, and calculate AFDM as the difference in mass before and after combustion. Double-bag, seal, label, and store the radioactive waste until the radioactivity decays

---

4. If there are obvious signs of seston in the chamber water, it will be necessary to filter the subsamples before they are added to the scintillation vials (to remove radioactively labeled particulate material). This can be done by removing approximately 3 mL of water from each chamber with a 15-mL syringe, filtering the water through a 0.45-μm pore size syringe filter (e.g., Syrfil-MF, Costar Corp., Cambridge, MA) into a small beaker and then pipetting 1 mL of this filtrate into the scintillation vial.

5. The first sample is not taken until 10 min to allow complete mixing of radioisotope within the chamber.

to background levels (10 half-lives or ~250 days). Place the radioactive chambers in an appropriately labeled containment bin until reuse.

8. Count each scintillation vial for 10 min on a liquid scintillation counter (the counting efficiency for ^{33}P is generally >90%, and because the sample matrix is the same for all samples, no correction for counting efficiency is needed). No decay correction is needed if all samples are counted within a few hours of each other.

9. SRP concentration of the initial stream water will be measured according to the methods outlined above (Section 33.3.3.4).

33.3.4.4 Data Analysis

1. Gross P uptake rate is measured using the first-order rate coefficient of radiotracer depletion in the water (k), the concentration of SRP in the stream water, and the water volume during the incubation (Steinman et al., 1991a). This procedure consists of three steps:
   a. Calculate k by regressing the ln-normalized scintillation count data (minus the background value determined from the sample collected just prior to ^{33}P injection) against time. Use Table 33.4 for data entry and calculations (also see Fig. 33.4).
   b. Gross P uptake rate is then estimated by multiplying k by the SRP concentration and by the water volume in the incubation chamber. Based on the data from the sample data sheet and depicted in Fig. 33.4, k (-0.0038/min $= -0.0228$/h) is multiplied by 6.2 (SRP concentration) and 1.0 (L of water in chamber). This rate is in units of μg P/h.
   c. Gross P uptake rate should then be normalized to the biomass in the chamber (μg P mg AFDM^{-1} h^{-1}) or by surface area of substrata in the chamber (μg P cm^{-2} h^{-1}). The gross uptake rate is divided either by the total AFDM in the chamber (e.g., 70.4 mg based on the sample data sheet) or total substrata surface area in chamber (e.g., 160 cm^2 based on the sample data sheet) to obtain a normalized uptake rate.

33.3.5 Advanced Method 2: Phosphorus Turnover

33.3.5.1 Preparation Protocol

1. Combust 40 acid-washed 10-mL glass beakers at 500°C for 1 h. Cut out 40 pieces of aluminum foil that are large enough to cover the top of each beaker. Gently etch the following information onto the foil with a pointed object (do not write it on the beaker because it will burn off during combustion, potentially contributing to dry mass and leaving one unable to track individual samples): stream type (high-P or low-P), sampling date (Day 0, 2, 5, or 10), and replicate (a–d) (i.e., 16 beakers/stream type).

TABLE 33.4 Sample data sheet for determination of gross phosphorus uptake (radiotracer). See also Online Worksheet 33.1.

Date: Stream:

Time (min)	Counts per Minute (CPM ^{33}P)	CPM Minus Background	ln (CPM Minus Background)
0 (Background)	49.4		
10	2009.8	1960.4	7.581
20	1939.1	1889.7	7.544
30	1841.7	1792.3	7.491
45	1761.8	1712.4	7.446
60	1669.4	1620	7.390

Calculated uptake rate constant (k): -0.0038/min
Stream water soluble reactive phosphorus concentration (μg/L): 6.2
Calculated uptake rate (μg P min^{-1}): 0.024
Ash-free dry mass (AFDM) (mg) or surface area (cm^2) in sample:
AFDM = 70.4 mg; surface area = 160 cm^2
Uptake per unit AFDM (μg P mg AFDM^{-1} min^{-1}): 0.000341
Uptake per unit surface area (μg P cm^{-2} min^{-1}): 0.00015

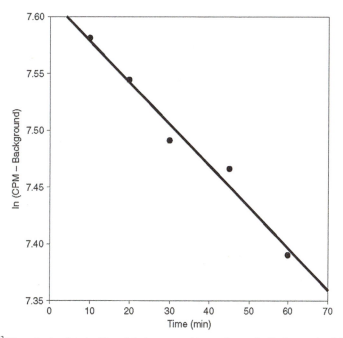

FIGURE 33.4 Radioactivity of ^{33}P in water incubated with periphyton exposed to grazing snails. Background activity of ^{33}P has been subtracted from measured activity and the data ln-normalized. Data used to generate the figure are based on real experiments and are included in the total P uptake sample data sheet (Table 33.3). *CPM*, counts per minute.

33.3.5.2 Field Placement and Collection (Option 1)

This option is to be followed if permission is obtained by the appropriate authorities to place radiolabeled material in the selected streams. If permission cannot be obtained, follow the steps in Section 33.3.5.3 (Option 2, below).

1. After completion of the radioisotopic uptake study (Section 33.3.4.3), carefully remove the tiles with forceps or tongs from each chamber, place them into plastic containers, and transport them to the high-P and low-P streams.
2. Place tiles in locations that have similar current velocity, irradiance level, temperature, and grazer density in both streams, if possible. Remove four radioactive tiles from each stream at 1 h [the 1 h incubation allows "day 0" samples to be rinsed by "cold" (i.e., nonradioactive) stream water prior to sampling to wash off adsorbed residual ^{33}P] and again at 2, 5, and 10 days, and place them in appropriately labeled and tared 50-mL glass beakers. One additional unlabeled tile should be processed from each stream; any radioactivity associated with the algae on this tile will be subtracted off all other counts, as it represents the naturally occurring background radioactivity (or the background activity associated with the scintillation counting of samples). The laboratory procedures associated with this option are included in Section 33.3.5.4 (below).

33.3.5.3 Chamber/Aquarium Placement and Collection (Option 2)

For this option, radiolabeled tiles are placed in either a high-P or low-P recirculating chamber or aquarium with unlabeled and filtered (0.45 µm) water from each stream. Ideally, the water in each chamber should be changed daily to minimize released ^{33}P from being taken up again. However, this option is time-consuming and can generate a considerable volume of contaminated waste. If water is not changed, it should be recognized that the calculated turnover rates will represent underestimates of true turnover.

1. After completion of the radioisotopic uptake study (Section 33.3.4.3), carefully remove the 17 tiles (16 labeled; 1 unlabeled) with forceps or tongs from each chamber, place them into plastic containers, and transport them to the unlabeled high-P or low-P chambers. If possible, chambers should be exposed to light and temperature regimes that are similar to ambient conditions and be fitted with a means to circulate or move the water (e.g., aeration or stirring).
2. Remove four tiles from each chamber at 1 h (the 1-h incubation allows "day 0" samples to be rinsed by stream water prior to sampling to wash off adsorbed residual ^{33}P) and again at 2, 5, and 10 days, and place them in appropriately labeled, tared 50-mL glass beakers. An additional unlabeled tile also should be processed from each stream; any radioactivity associated with the algae on this tile will be subtracted off all other counts, as it represents the naturally occurring background radioactivity.

TABLE 33.5 Sample data sheet for determination of P turnover. See also Online Worksheet 33.1.

Date:			Stream:		
		ln (Counts per Minute)			
Replicate	Background	Day 0 Minus Background	Day 2 Minus Background	Day 5 Minus Background	Day 10 Minus Background
A					
B					
C					
D					

33.3.5.4 Laboratory Protocol (Applicable for Both Options 1 and 2)

1. On the day of collection (days 0, 2, 5, and 10), place the beakers in a drying oven and dry the tiles to constant weight at 105°C (usually 24–48 h).
2. Once the tiles reach a constant dry mass, weigh the dried tiles, and combust them for a minimum of 4 h at a full 500°C. Remove tiles, allow them to cool to room temperature in a desiccator, and reweigh.
3. Using a 10-mL pipette, add 10 mL of 2N HCl to the beaker (make sure this is enough to cover all the periphyton), label the beaker with the appropriate sample designation, and place Parafilm over the beaker. Leaching of ashed material should last at least 24 h. Place beakers in the laboratory hood.
4. After a minimum exposure of 24 h to the acid, add 10 mL of deionized water to each beaker (to reduce acidity to 1N), swirl the beaker gently to mix thoroughly, and pipette 1 mL of the diluted leachate to a scintillation vial containing 15 mL of scintillation cocktail.
5. Count each sample on a liquid scintillation counter for 10 min. It is recommended that all samples from the turnover experiment be counted during the same run (within several hours of each other) at experiment's end to preclude the need to apply a radioactive decay correction factor. The counts are used to determine turnover rate.

33.3.5.5 Data Analysis

1. P turnover rate is computed by linear regression of ln [^{33}P] counts in algae versus time in stream (in days). P turnover rate is therefore expressed as a first-order turnover rate constant (days^{-1}). The background radioactivity associated with unlabeled tiles should be subtracted from each sample count prior to performing the ln-transformation and the regression, and the regression should be based on a mean value derived from the four substrata sampled on each day. Use Table 33.5 for data entry and calculations.
2. It should be emphasized that this determination of P turnover rate may not be an accurate physiological index of total algal P turnover rate. This is because not all P pools within the algal cells will have reached isotopic equilibrium during the 6 h of ^{33}P exposure during the uptake part of the experiment (cf. Scinto and Reddy, 2003). However, the approach described should provide a reasonable basis for comparing turnover rates in the more rapidly cycling P pools between different streams.

33.4 QUESTIONS

33.4.1 Limitation: Phosphatase Activity

1. Was phosphatase activity greater in the low-P stream, as hypothesized? If not, what might explain this result?
2. What other factors besides P concentration and biomass might influence the PA in the two streams?
3. Why is it important to normalize the PA data by an index of biomass?
4. Phosphatase is an inducible enzyme. That is, it is synthesized upon metabolic demand, as opposed to a constitutive enzyme, which is always present. What advantage is there to an organism in maintaining phosphatase as an inducible enzyme?

33.4.2 Limitation: Chemical Composition

5. C:P ratios substantially greater than 106:1 (on a molar basis) suggest P deficiency in planktonic algae (the Redfield ratio). Why is the C:P ratio suggesting P deficiency different in plankton compared to benthic algae (>360:1)?

6. Some algal species have greater carbon demands than others because of more carbon-based compounds in their cell walls. How would this type of demand influence the interpretation of the C:P ratio?

7. Many algal species exhibit "luxury uptake" of phosphorus (Rier et al., 2016), whereby they take up excessive amounts of P when it is available (e.g., during high-P conditions) and then store the P intracellularly (in polyphosphate bodies). How would luxury uptake of P influence the chemical composition ratio of benthic algae?

33.4.3 Net Uptake: Stable Phosphorus

8. Were the net uptake rates similar or different between the two streams? If they were different, what might account for this difference?

9. Sometimes, no net uptake is measured during an incubation (i.e., the amount of P measured at the start of the experiment is the same as that at the end of the experiment). Assuming that the algae are biologically active and actively taking up phosphorus, what might explain this result?

10. Would you expect the same P uptake rates by algae irrespective of which method of measuring uptake was used (i.e., stable vs. radioisotopic P)? If so, why? If not, how might the rates differ and why?

33.4.4 Total Uptake: Radiolabeled Phosphorus

11. Were the gross uptake rates similar or different between the two streams? If they were different, what might account for this difference? In which stream do you expect to measure the greater uptake of ^{33}P? Why?

12. By keeping the incubation time short in this exercise, you minimize the possibility that any radioactive P that was taken up could be rereleased within the incubation period (i.e., minimize the possibility of recycling). Thus, the radioactive P removed from the water is assumed to represent the total or *gross nutrient uptake* rate. How does this differ from net uptake rate (i.e., which measure should be greater)? Why?

33.4.5 Turnover

13. Were the P turnover rates similar or different between the two streams? If they were different, what might account for this difference?

14. If the ^{33}P was allowed to come to complete isotopic equilibrium within the algae during the uptake part of the experiment, would you expect measured P turnover rates to be greater or lower than those measured? Why?

15. How might the thickness of the periphyton matrix influence turnover rates? What about grazing activity?

33.5 MATERIALS AND SUPPLIES

Letters in parentheses indicate in which method (A = Basic Method 1, B = Basic Method 2, C = Basic Method 3, D = Advanced Method 1, or E = Advanced Method 2) the item is used.[6]

Field Materials
    1-L polyethylene bottles (A, C, D)
    Cooler (A, B, C, D)
    Hand pumps with GFF (or equivalent) filters (A, C, D)
    Holder to transport beakers to and from field (E)
    Tupperware containers to accommodate tiles or rocks (A, B, C, D, E)
    Unglazed ceramic tiles (e.g., tiles measuring 3 cm × 3 cm or ceramic cylinders) (A, C, D, E)
Laboratory Materials (excludes SRP analysis; see Section 33.3.3)
    1N NaOH (A)

---

6. Any use of isotope requires specific laboratory protocols. These protocols are available from the Safety and/or Risk Management Department at the institution or laboratory licensed for isotope use. These protocols must be followed carefully. It is essential that gloves (we recommend double gloving, using vinyl gloves directly over the hands and disposable gloves over the vinyl ones), lab coats, and safety glasses be worn at all times.

2N HCl (B, E)

10-mL glass beakers (B, D)

25-mL plastic syringes with syringe holders; 0.45 μm pore size (A, C, D)

25-mL scintillation vials (A, D, E)

50-mL glass beakers (E)

50-mL polyethylene bottles (C)

60-mL polyethylene bottles (D)

100 mL polyethylene bottles (B)

100-mL volumetric cylinders (B)

150 mM *para*-nitrophenyl phosphate solution (add 2.78 g of *para*-nitrophenyl phosphate to 50 mL of ultrapurified deionized water) (A)

90% acetone (90 parts acetone with 10 parts saturate magnesium carbonate solution)

Aluminum foil (B, E)

Aluminum weigh boats (C, D)

Carrier-free ^{33}P isotope (0.5 mCi/chamber; order from PerkinElmer: http://www.perkinelmer.com/Catalog/Family/ID/Phosphorus33%20Radionuclide%20Orthophosphoric%20Acid%20in%201mL%20HClfree%20Water) (D)

Course-bristled toothbrushes (B)

EcoLume Scintillation Cocktail (ICN Scientific, Costa Mesa, CA) (D, E)

Large pans or trays (B)

Parafilm (B, E)

Plastic jars or centrifuge tubes (A)

Reagents for SRP analysis (B, D)

Wide-mouth glass incubation jars (30 mL or larger for larger substrata) (A)

Laboratory Equipment

Analytical balance (A, B, C, D, E)

Automatic pipettes (1 mL; 10 mL; 30 mL) (A, B, C, D, E)

Desiccators (B, C)

Drying oven (B, C, D, E)

Liquid Scintillation Counter (D, E)

Muffle furnace (B, C, D, E)

Recirculating chambers (either with pumps or stirrers to circulate water) (C, D)

Spectrophotometer (narrow band width: 0.5−2.0 nm) and cuvettes (1 and 10 cm pathlength) (A, B, C, D)

ACKNOWLEDGMENTS

This chapter is devoted to the memory of Pat Mulholland, who was the coauthor on previous editions and whose influence spirals throughout this chapter. May his memory serve as a blessing. The authors thank Mary Ogdahl and Gary Lamberti for their constructive comments on the chapter and Eric Nemeth for his artistic assistance. These experiments, and several of the methods, were developed during research supported by the National Science Foundation's Ecosystem Program, the US Department of Energy's Office of Science, Office of Biological and Environmental Research, and the Michigan Department of Environmental Quality.

REFERENCES

Allan, J.D., 2004. Landscapes and riverscapes: the influence of land use on stream ecosystems. Annual Review of Ecology, Evolution, and Systematics 35, 257−284.

APHA, AWWA, WEF, 1995. Standard Methods for the Examination of Water and Wastewater, nineteenth ed. American Public Health Association, Washington, DC, USA.

Bernhardt, E.S., Likens, G.E., Buso, D.C., Driscoll, C.T., 2003. In-stream uptake dampens effects of major forest disturbance on watershed nitrogen export. Proceedings of the National Academy of Sciences of the United States of America 100, 10304−10308.

Borchardt, M.A., 1996. Nutrients. In: Stevenson, R.J., Bothwell, M.L., Lowe, R.L. (Eds.), Algal Ecology. Academic Press, San Diego, CA, USA, pp. 183−227.

Borchardt, M.A., Hoffmann, J.P., Cook, P.W., 1994. Phosphorus uptake kinetics of *Spirogyra fluviatilis* (Charophyceae) in flowing water. Journal of Phycology 30, 403−412.

Bothwell, M.L., 1989. Phosphorus-limited growth dynamics of lotic periphytic diatom communities: areal biomass and cellular growth rate responses. Canadian Journal of Fisheries and Aquatic Sciences 46, 1293−1301.

Burkholder, J.M., Wetzel, R.G., 1990. Epiphytic alkaline phosphatase on natural and artificial plants in an oligotrophic lake: re-evaluation of the role of macrophytes as a phosphorus source for epiphytes. Limnology and Oceanography 35, 736−747.

Cardinale, B.J., 2011. Biodiversity improves water quality through niche partitioning. Nature 472, 86−89.

Carey, R.O., Vellidis, G., Lowrance, R., Pringle, C.M., 2007. Do nutrients limit algal periphyton in small blackwater coastal plain streams? Journal of the American Water Resources Association 43, 1183−1193.

Carpenter, S.R., Caraco, N.F., Correll, D.L., Howarth, R.W., Sharpley, A.N., Smith, V.H., 1998. Nonpoint pollution of surface waters with phosphorus and nitrogen. Ecological Applications 8, 559−568.

Cembella, A.D., Antia, N.J., Harrison, P.J., 1984. The utilization of inorganic and organic phosphorus compounds as nutrients by eukaryotic microalgae: a multidisciplinary perspective: Part 1. Critical Reviews in Microbiology 10, 317−391.

Colman, A.S., Blake, R.E., Karl, D.M., Fogel, M.L., Turekian, K.K., 2005. Marine phosphate oxygen isotopes and organic matter remineralization in the oceans. Proceedings of the National Academy of Sciences of the United States of America 102, 13023−13028.

Cross, W.F., Benstead, J.P., Rosemond, A.D., Wallace, J.B., 2003. Consumer-resource stoichiometry in detritus-based streams. Ecology Letters 6, 721−732.

Currie, D.J., Bentzen, E., Kalff, J., 1986. Does algal-bacterial phosphorus partitioning vary among lakes? A comparative study of orthophosphate uptake and alkaline phosphatase activity in fresh water. Canadian Journal of Fisheries and Aquatic Sciences 43, 311−318.

Darley, W.M., 1982. Algal Biology: A Physiological Approach. Blackwell Scientific, Oxford, UK.

Davies, J.-M., Bothwell, M.L., 2012. Responses of lotic periphyton to pulses of phosphorus: P-flux controlled growth rate. Freshwater Biology 57, 2602−2612.

Davis, L.S., Hoffmann, J.P., Cook, P.W., 1990. Seasonal succession of algal periphyton from a wastewater treatment facility. Journal of Phycology 26, 611−617.

Dodds, W.K., 2003. The role of periphyton in phosphorus retention in shallow freshwater aquatic systems. Journal of Phycology 39, 840−849.

Dodds, W.K., 2006. Eutrophication and trophic state in rivers and streams. Limnology and Oceanography 51, 671−680.

Dodds, W.K., Smith, V.H., Zander, B., 1997. Developing nutrient targets to control benthic chlorophyll levels in streams: a case study of the Clarke Fork River. Water Research 31, 1738−1750.

Drake, W.M., Scott, J.T., Evans-White, M., Haggard, B., Sharpley, A., Rogers, C.W., Grantz, M.E., 2012. The effect of periphyton stoichiometry and light on biological phosphorus immobilization and release in streams. Limnology 13, 97−106.

Droop, M.R., 1974. The nutrient status of algal cells in continuous culture. Journal of the Marine Biological Association of the United Kingdom 54, 825−855.

Duhamel, S., Gregori, G., Van Wambeke, F., Nedoma, J., 2009. Detection of extracellular phosphatase activity at the single-cell level by enzyme-labeled fluorescence and flow cytometry: the importance of time kinetics in ELFA labeling. Cytometry. Part A: The Journal of the International Society for Analytical Cytology 75A, 163−168.

Duhamel, S., Dyhrman, S.T., Karl, D.M., 2010. Alkaline phosphatase activity and regulation in the North Pacific Subtropical Gyre. Limnology and Oceanography 55, 1414−1425.

Dyhrman, S., Ruttenberg, K.C., 2006. Presence and regulation of alkaline phosphatase activity in eukaryotic phytoplankton from the coastal ocean: implications for dissolved organic phosphorus remineralization. Limnology and Oceanography 51, 1381−1390.

Ellwood, N.T.W., Di Pippo, F., Albertano, P., 2012. Phosphatase activities of cultured phototrophic biofilms. Water Research 46, 378−386.

Elsbury, K.E., Paytan, A., Ostrom, N.E., Kendall, C., Young, M.B., McLaughlin, K., Rollog, M.E., Watson, S., 2009. Using oxygen isotopes of phosphate to trace phosphorus sources and cycling in Lake Erie. Environmental Science and Technology 43, 3108−3114.

Elwood, J.W., Newbold, J.D., Trimble, A.F., Stark, R.W., 1981. The limiting role of phosphorus in a woodland stream ecosystem: effects of P enrichment on leaf decomposition and primary producers. Ecology 62, 146−158.

Espeland, E.M., Francoeur, S.N., Wetzel, R.G., 2002. Microbial phosphatase in biofilms: a comparison of whole community enzyme activity and individual bacterial cell-surface phosphatase expression. Archiv für Hydrobiologie 153, 581−593.

Finlay, J.C., Hood, J.M., Limm, M.P., Power, M.E., Schade, J.D., Welter, J.R., 2011. Light-mediated thresholds in stream-water nutrient composition in a river network. Ecology 92, 140−150.

Francoeur, S.N., 2001. Meta-analysis of lotic nutrient amendment experiments: detecting and quantifying subtle responses. Journal of the North American Benthological Society 20, 358−368.

Frost, P.C., Stelzer, R.S., Lamberti, G.A., Elser, J.J., 2002. Ecological stoichiometry of trophic interactions in the benthos: understanding the role of C:N:P ratios in lentic and lotic habitats. Journal of the North American Benthological Society 21, 515−528.

Geider, R., La Roche, J., 2002. Redfield revisited: variability of C:N:P in marine microalgae and its biochemical basis. European Journal of Phycology 37, 1−17.

Grimm, N.B., Fisher, S.G., 1986. Nitrogen limitation in a Sonoran Desert stream. Journal of the North American Benthological Society 5, 2−15.

Hall, S.R., Smith, V.H., Lytle, D.A., Leibold, M.A., 2005. Constraints on primary producer N:P stoichiometry along N:P supply ratio gradients. Ecology 86, 1894−1904.

Healey, F.P., 1973. Inorganic nutrient uptake and deficiency in algae. Critical Reviews in Microbiology 3, 69−113.

Healey, F.P., Hendzel, L.L., 1979. Fluorometric measurement of alkaline phosphatase activity in algae. Freshwater Biology 9, 429−439.

Hecky, R.E., Kilham, P., 1988. Nutrient limitation of phytoplankton in freshwater and marine environments: a review of recent evidence on the effects of enrichment. Limnology and Oceanography 33, 796−822.

Hecky, R.E., Campbell, P., Hendzel, L.L., 1993. The stoichiometry of carbon, nitrogen, and phosphorus in particulate matter of lakes and oceans. Limnology and Oceanography 38, 709−724.

Hernandez, I., Niell, F.X., Whitton, B.A., 2002. Phosphatase activity of benthic marine algae. An overview. Journal of Applied Phycology 14, 475–487.

Hill, W.R., Knight, A.W., 1988. Nutrient and light limitation of algae in two northern California streams. Journal of Phycology 24, 125–132.

Hill, W.R., Ryon, M.G., Schilling, E.M., 1995. Light limitation in a stream ecosystem: responses by primary producers and consumers. Ecology 76, 1297–1309.

Hillebrand, H., Kahlert, M., 2001. Effect of grazing and nutrient supply on periphyton biomass and nutrient stoichiometry in habitats of different productivity. Limnology and Oceanography 46, 1881–1898.

Hillebrand, H., Sommer, U., 1999. The nutrient stoichiometry of benthic microalgal growth: redfield proportions are optimal. Limnology and Oceanography 44, 440–446.

Hillebrand, H., Frost, P., Liess, A., 2008. Ecological stoichiometry of indirect grazer effects on periphyton nutrient content. Oecologia 155, 619–630.

Hoppe, H.G., 2003. Phosphatase activity in the sea. Hydrobiologia 493, 187–200.

Hudson, J.J., Taylor, W.D., Schindler, D.W., 2000. Phosphate concentrations in lakes. Nature 406, 54–56.

Hwang, S.-J., Havens, K.E., Steinman, A.D., 1998. Phosphorus kinetics of planktonic and benthic assemblages in a shallow subtropical lake. Freshwater Biology 40, 729–745.

Jansson, M., 1993. Uptake, exchange, and excretion of orthophosphate in phosphate-starved *Scenedesmus quadricauda* and *Pseudomona* K7. Limnology and Oceanography 38, 1162–1178.

Jansson, M., Olsson, H., Pettersson, K., 1988. Phosphatases; origin, characteristics and function in lakes. Hydrobiologia 170, 157–175.

Johnson, L.T., Tank, J.L., Dodds, W.K., 2009. The influence of land use on stream biofilm nutrient limitation across eight North American ecoregions. Canadian Journal of Fisheries and Aquatic Sciences 66, 1081–1094.

Jordan, T.E., Correll, D.L., Weller, D.E., 1997. Effects of agriculture on discharges of nutrients from coastal plain watersheds of Chesapeake Bay. Journal of Environmental Quality 26, 836–848.

Kahlert, M., 1998. C:N:P ratios of freshwater benthic algae. Archiv für Hydrobiologie, Special Issues: Advances in Limnology 51, 105–114.

Karl, D.M., Björkman, K., 2015. Dynamics of dissolved organic phosphorus. In: Hansell, D.A., Carlson, C.A. (Eds.), Biogeochemistry of Marine Dissolved Organic Matter. Academic Press, Burlington, Ontario, Canada, pp. 233–334.

Kaushal, S.S., Groffman, P.M., Band, L.E., Elliott, E.M., Shields, C.A., Kendall, C., 2011. Tracking nonpoint source nitrogen pollution in human-impacted watersheds. Environmental Science and Technology 45, 8225–8232.

Kilham, P., Hecky, R.E., 1988. Comparative ecology of marine and freshwater phytoplankton. Limnology and Oceanography 33, 776–795.

Lock, M.A., Wallace, R.R., Costerton, J.W., Ventullo, R.M., Charlton, S.E., 1984. River epilithon: toward a structural-functional model. Oikos 42, 10–22.

Lohman, K., Priscu, J.C., 1992. Physiological indicators or nutrient deficiency in *Cladophora* (Chlorophyta) in the Clark Fork of the Columbia River, Montana. Journal of Phycology 28, 443–448.

Lohman, K., Jones, J.R., Baysinger-Daniel, C., 1991. Experimental evidence for nitrogen limitation in a northern Ozark stream. Journal of the North American Benthological Society 10, 14–23.

Manafi, M., Kneifel, W., Bascomb, S., 1991. Fluorogenic and chromogenic substrates used in bacterial diagnostics. Microbiological Reviews 55, 335–348.

McCall, S.J., Bowes, M.J., Warnaars, T.A., Hale, M.S., Smith, J.T., Warwick, A., Barrett, C., 2014. Phosphorus enrichment of the oligotrophic River Rede (Northumberland, UK) has no effect on periphyton growth rate. Inland Waters 4, 121–132.

McLaughlin, K., Kendall, C., Silva, S.R., Young, M., Paytan, A., 2006. Phosphate oxygen isotope ratios as a tracer for sources and cycling of phosphate in North San Francisco Bay, California. Journal of Geophysical Research: Biogeosciences 111, G03003.

Michalak, A.M., Anderson, E.J., Beletsky, D., Boland, S., Bosch, N.S., Bridgeman, T.B., Chaffin, J.D., Chog, K., Confesor, R., Daloğlu, I., DePinto, J.V., Evans, M.A., Fahnenstiel, G.L., He, L., Ho, J.C., Jenkins, L., Johengen, T.H., Kuo, K.C., LaPorte, E., Liu, X., McWilliams, M.R., Moore, M.R., Posselt, D.J., Richards, R.P., Scavia, D., Steiner, A.L., Verhamme, E., Wright, D.M., Zagorski, M.A., 2013. Record-setting algal bloom in Lake Erie caused by agricultural and meteorological trends consistent with expected future conditions. Proceedings of the National Academy of Sciences of the United States of America 110, 6448–6452.

Van Mooy, B.A., Fredricks, H.F., Pedler, B.E., Dyhrman, S.T., Karl, D.M., Koblížek, M., Lomas, M.W., Mincer, T.J., Moore, L.R., Moutin, T., Rappé, M.S., Webb, E.A., 2009. Phytoplankton in the ocean use non-phosphorus lipids in response to phosphorus scarcity. Nature 458, 69–72.

Mulholland, P.J., 1996. Role of nutrient cycling in streams. In: Stevenson, R.J., Bothwell, M.L., Lowe, R.L. (Eds.), Algal Ecology. Academic Press, San Diego, CA, USA, pp. 609–633.

Mulholland, P.J., Rosemond, A.D., 1992. Periphyton response to longitudinal nutrient depletion in a woodland stream: evidence of upstream-downstream linkage. Journal of the North American Benthological Society 11, 405–419.

Mulholland, P.J., Steinman, A.D., Elwood, J.W., 1990. Measurement of phosphorus uptake length in streams: comparison of radiotracer and stable PO_4 releases. Canadian Journal of Fisheries and Aquatic Sciences 47, 2351–2357.

Mulholland, P.J., Steinman, A.D., Marzolf, E.R., Hart, D.R., DeAngelis, D.L., 1994. Effect of periphyton biomass on hydraulic characteristics and nutrient cycling in streams. Oecologia 98, 40–47.

Mulholland, P.J., Tank, J.L., Webster, J.R., Bowden, W.B., Dodds, W.K., Gregory, S.V., Grimm, N.B., Hamilton, S.K., Johnson, S.L., Martí, E., McDowell, W.H., Merriam, J.L., Meyer, J.L., Peterson, B.J., Valett, H.M., Wollheim, W.M., 2002. Can uptake length in streams be determined by nutrient addition experiments? Results from an interbiome comparison study. Journal of the North American Benthological Society 21, 544–560.

Nedoma, J., Štrojsová, A., Vrba, J., Komárková, J., Šimek, K., 2003. Extracellular phosphatase activity of natural plankton studied with ELF97 phosphate: fluorescence quantification and labelling kinetics. Environmental Microbiology 5, 462–472.

Newman, S., McCormick, P.V., Backus, J.G., 2003. Phosphatase activity as an early warning indicator of wetland eutrophication: problems and prospects. Journal of Applied Phycology 15, 45–59.

Pant, H.K., Reddy, K.R., Dierberg, F.E., 2002. Bioavailability of organic phosphorus in a submerged aquatic vegetation-dominated treatment wetland. Journal of Environmental Quality 31, 1748–1756.

Payn, R.A., Webster, J.R., Mulholland, P.J., Valett, H.M., Dodds, W.K., 2005. Estimation of stream nutrient uptake from nutrient addition experiments. Limnology and Oceanography: Methods 3, 174–182.

Perrin, C.J., Richardson, J.S., 1997. N and P limitation of benthos abundance in the Nechako River, British Columbia. Canadian Journal of Fisheries and Aquatic Sciences 54, 2574–2583.

Peterson, B.J., Hobbie, J.E., Corliss, T.L., Kriet, D., 1983. A continuous flow periphyton bioassay: tests of nutrient limitation in a tundra stream. Limnology and Oceanography 28, 582–595.

Pringle, C.M., Paaby-Hansen, P., Vaux, P.D., Goldman, C.R., 1986. In situ assays of periphyton growth in a lowland Costa Rica stream. Hydrobiologia 134, 207–213.

Raven, P.H., Evert, R.F., Curtis, H., 1981. Biology of Plants, third ed. Worth Publishers, New York, USA.

Reaves, M.L., Sinha, S., Rabinowitz, J.D., Kruglyak, L., Redfield, R.J., 2012. Absence of detectable arsenate in DNA from arsenate-grown GFAJ-1 cells. Science 337, 470–473.

Redfield, A.C., 1958. The biological control of chemical factors in the environment. American Scientist 46, 205–221.

Rengefors, K., Pettersson, K., Blenckner, T., Anderson, D.M., 2001. Species-specific alkaline phosphatase activity in freshwater spring phytoplankton: application of a novel method. Journal of Plankton Research 23, 435–443.

Rhee, G.Y., 1978. Effects of N:P atomic ratios and nitrate limitation on algal growth, cell composition, and nitrate uptake. Limnology and Oceanography 23, 10–24.

Rier,, S.T., Kinek,, K.C., Hay, S.E., Franceoeur, S.N., 2016. Polyphosphate plays a vital role in the phosphorus dynamics of stream periphyton. Freshwater Science 35, 490–502.

Rigler, F.H., 1966. Radiobiological analysis of inorganic phosphorus in lakewater. Verhandlungen der Internationalen Vereinigung für Theoretische und Angewandte Limnologie 16, 465–470.

Rivkin, R.B., Swift, E., 1982. Phosphate uptake by the oceanic dinoflagellate *Pyrocystis noctiluca*. Journal of Phycology 18, 113–120.

Rosemond, A.D., Mulholland, P.J., Elwood, J.W., 1993. Top-down and bottom-up control of stream periphyton: effects of nutrients and herbivores. Ecology 74, 1264–1280.

Schindler, D.W., 1977. The evolution of phosphorus limitation in lakes. Science 195, 260–262.

Scinto, L.J., Reddy, K.R., 2003. Biotic and abiotic uptake of phosphorus by periphyton in a subtropical freshwater wetland. Aquatic Botany 77, 203–222.

Scott, J.T., Lang, D.A., King, R.S., Doyle, R.D., 2009. Nitrogen fixation and phosphatase activity in periphyton growing on nutrient diffusing substrata: evidence for differential nutrient limitation in stream periphyton. Journal of the North American Benthological Society 28, 57–68.

Sharma, K., Inglett, P.W., Reddy, K.R., Ogram, A.V., 2005. Microscopic examination of photoautotrophic and phosphatase-producing organisms in phosphorus-limited Everglades periphyton mats. Limnology and Oceanography 50, 2057–2062.

Sirová, D., Borovec, J., Černá, B., Rejmánková, E., Adamec, L., Vrba, J., 2009. Microbial community development in the traps of aquatic *Utricularia* species. Aquatic Botany 90, 129–136.

Solórzano, L., Sharp, J.H., 1980. Determination of total and dissolved phosphorus and particulate phosphorus in natural waters. Limnology and Oceanography 25, 574–578.

Steinman, A.D., 1994. The influence of phosphorus enrichment on lotic bryophytes. Freshwater Biology 31, 53–63.

Steinman, A.D., Ogdahl, M.E., 2015. TMDL reevaluation: reconciling phosphorus load reductions in a eutrophic lake. Lake and Reservoir Management 31, 115–126.

Steinman, A.D., Mulholland, P.J., Kirschtel, D.B., 1991a. Interactive effects of nutrient reduction and herbivory on biomass, taxonomic structure, and P uptake in lotic periphyton communities. Canadian Journal of Fisheries and Aquatic Sciences 48, 1951–1959.

Steinman, A.D., Mulholland, P.J., Palumbo, A.V., Flum, T.F., DeAngelis, D.L., 1991b. Resilience of lotic ecosystems to a light-elimination disturbance. Ecology 72, 1299–1313.

Steinman, A.D., Mulholland, P.J., Hill, W.R., 1992. Functional responses associated with growth form in stream algae. Journal of the North American Benthological Society 11, 29–243.

Steinman, A.D., Mulholland, P.J., Beauchamp, J.J., 1995. Effects of biomass, light, and grazing on phosphorus cycling in stream periphyton communities. Journal of the North American Benthological Society 14, 371–381.

Steinman, A.D., Ogdahl, M.E., Wessell, K., Biddanda, B., Kendall, S., Nold, S., 2011. Periphyton response to simulation nonpoint source pollution in the lower Muskegon River watershed. Aquatic Ecology 45, 439–454.

Stelzer, R.S., Lamberti, G.A., 2002. Ecological stoichiometry in running waters: periphyton chemical composition and snail growth. Ecology 83, 1039–1051.

Stockner, J.G., Shortreed, K.R., 1978. Enhancement of autotrophic production by nutrient addition in a coastal rainforest stream on Vancouver Island. Journal of the Fisheries Research Board of Canada 35, 28–34.

Stutter, M.I., Demars, B.O.L., Langan, S.J., 2010. River phosphorus cycling: separating biotic and abiotic uptake during short-term changes in sewage effluent loading. Water Research 44, 4425–4436.

Tank, J.L., Dodds, W.K., 2003. Nutrient limitation of epilithic and epixylic biofilms in ten North American streams. Freshwater Biology 48, 1031–1049.

Wetzel, R.G., 1981. Long-term dissolved and particulate alkaline phosphatase activity in a hardwater lake in relation to lake stability and phosphorus enrichments. Verhandlungen der Internationalen Vereinigung für Theoretische und Angewandte Limnologie 21, 369–381.

Wetzel, R.G., 1983. Limnology, second ed. Saunders College Publishing, Philadelphia, Pennsylvania, USA.

Whitford, L.A., Schumacher, G.J., 1964. Effect of a current on respiration and mineral uptake in *Spirogyra* and *Oedogonium*. Ecology 45, 168–170.

Withers, P.J.A., Jarvie, H.P., 2008. Delivery and cycling of phosphorus in rivers: a review. Science of the Total Environment 400, 379–395.

Wolfe-Simon, F., Blum, J.S., Kulp, T.R., Gordon, G.W., Hoeft, S.E., Pett-Ridge, J., Stolz, J.F., Webb, S.M., Weber, P.K., Davies, P.C.W., Anbar, A.D., Oremland, R.S., 2011. A bacterium that can grow by using arsenic instead of phosphorus. Science 332, 1163–1166.

Young, M.B., McLaughlin, K., Kendall, C., Stringfellow, W., Rollog, M., Elsbury, K., Donald, E., Paytan, A., 2009. Characterizing the oxygen isotopic composition of phosphate sources to aquatic ecosystems. Environmental Science and Technology 43, 5190–5196.

Young, E.B., Tucker, R.C., Pansch, L.A., 2010. Alkaline phosphatase in freshwater *Cladophora*-epiphyte assemblages: regulation in response to phosphorus supply and localization. Journal of Phycology 46, 93–101.

Chapter 34

Stream Metabolism

Robert O. Hall, Jr. [1] and **Erin R. Hotchkiss** [2,3]

[1]*Department of Zoology and Physiology, University of Wyoming;* [2]*Département des sciences biologiques, Université du Québec à Montréal;* [3]*Department of Biological Sciences, Virginia Polytechnic Institute and State University*

34.1 INTRODUCTION

Ecosystem metabolism governs the fixation and mineralization of organic carbon (C) in streams. Metabolism comprises (1) total fixation of inorganic to organic C from all photoautotrophs, called *gross primary production* (GPP), and (2) the mineralization of organic to inorganic C by all organisms in a stream reach, called *ecosystem respiration* (ER). The balance between GPP and ER is *net ecosystem production* (NEP) where:

$$\text{NEP} = \text{GPP} + \text{ER} \tag{34.1}$$

Here we refer to ER as a negative flux measured as the consumption of dissolved O_2, which means we limit our definition of ER to aerobic respiration. Thus, NEP is the sum of a positive GPP value and a negative ER value. NEP can be positive, which means that over the time scale of measurement a stream is fixing more organic C than it is respiring and thus is storing or exporting some amount of organic C (i.e., the stream is *autotrophic*) (Lovett et al., 2006). Alternatively, NEP can be negative meaning that the stream is mineralizing more C than it is fixing and must therefore be receiving external organic C or respiring stored organic C (i.e., the stream is *heterotrophic*). The relationship between GPP and ER can also be represented as a ratio, where a stream with GPP/ER <1 is heterotrophic and GPP/ER >1 is autotrophic. Respiration in heterotrophic streams is often subsidized by external inputs of terrestrially derived organic C (fixed by primary producers on land), thus making most streams heterotrophic and net sources of CO_2 to the atmosphere (Duarte and Prairie, 2005; Marcarelli et al., 2011; Hotchkiss et al., 2015).

Metabolism is a fundamental property of ecosystems. Measurements of whole-ecosystem GPP, ER, and NEP using diel O_2 data integrate all aerobic organisms (autotrophs, heterotrophs) and habitats (benthic, planktonic, and hyporheic zones) that contribute to the fixation, transformation, and availability of organic matter in streams and rivers. GPP and ER, and changes in the balance between the two, reflect some combination of watershed, upstream, and internal processes. Consequently, changes in light, nutrients, and organic matter can alter rates of metabolism (Roberts et al., 2007; Bernot et al., 2010). In turn, GPP and ER can strongly influence levels of dissolved oxygen, the availability of organic matter for consumers, and other water quality parameters. Given the links between metabolism, water quality, animal production, and anthropogenic changes to landscapes and waterways, measurements of GPP and ER are increasingly being used as a metric to assess ecosystem health (Young et al., 2008). Estimates of GPP and ER can also increase our understanding of controls on food web energy fluxes, the capacity of streams to transform and remove excess nutrients, and how streams may respond to environmental change.

ER can be further split into two components: respiration by autotrophs (R_a; *autotrophic respiration*) and respiration by heterotrophs (R_h; *heterotrophic respiration*):

$$\text{ER} = R_a + R_h \tag{34.2}$$

The dominant autotrophs in streams are algae, bryophytes (nonvascular plants; mosses and liverworts), cyanobacteria, and macrophytes (vascular plants). Because autotrophs respire some proportion of the organic C fixed during photosynthesis, the fixed C that remains and is available to consumers is *net primary production* (NPP):

$$\text{NPP} = \text{GPP} + R_a \tag{34.3}$$

Methods in Stream Ecology. http://dx.doi.org/10.1016/B978-0-12-813047-6.00012-7

In contrast to metabolism estimates in terrestrial ecosystems that may include calculations of both NEP and NPP, direct measurements of net production in freshwaters are largely limited to NEP because in aquatic ecosystems with high rates of autotrophic biomass turnover and low standing stocks of autotrophic biomass, scientists lack good methods to estimate the mass of autotrophic C remaining after R_a. Some fraction of GPP is also leached from autotrophs as dissolved organic C (Baines and Pace, 1991; Hotchkiss and Hall, 2015), removing fixed C from autotrophic biomass that is not respired by the autotrophs themselves. Additionally, autotrophs in streams coexist with heterotrophs, often interacting on small spatial scales in complex biofilm structures, hindering our ability to isolate R_a from the respiration of coexisting heterotrophs. One exception is the ability to estimate the NPP of macrophytes or bryophytes, but these methods are often destructive, such as physically removing known areas of macrophytes for biomass estimates (Fisher and Carpenter, 1976) and will still not fully capture the total NPP of most streams with metabolically active biofilms. It is possible to statistically analyze daily metabolism data to estimate the proportion of GPP immediately respired; a review of >20 streams estimated that autotrophs and associated heterotrophs respired 44% of GPP soon after fixation (Hall and Beaulieu, 2013). While not an estimate of R_a alone, the fraction of GPP not immediately respired will largely represent the remaining autotrophic organic C, which is an essential food source for heterotrophs in stream and river food webs (Minshall, 1978; Thorp and Delong, 2002).

Estimates of stream metabolism have a long history. Borrowing the ideas from research on coral reefs (Sargent and Austin, 1949; Odum and Odum, 1955), H.T. Odum outlined the procedure for estimating GPP and ER in an open channel (Odum, 1956). His procedure formed the basis of our methods today and was a breakthrough in the sense that his method allowed ecologists to consider C transformations at the scale of a stream reach and then measure these rates at the same scale. In addition to open channel methods, much work has used chamber and bottle incubation methods in streams to estimate metabolism, both through (1) measuring changes in dissolved gases pre- and postincubation (Bott et al., 1997; Dodds and Brock, 1998) and by (2) using radioactive isotopes to trace the conversion of inorganic to organic C (Steeman Nielsen, 1952; Bott and Ritter, 1981), which, depending on incubation length, provides an estimate between gross and net primary production (Peterson, 1980). These methods are necessary when one is interested in partitioning metabolism from different components of the ecosystem, such as planktonic versus benthic or pool versus riffle, but incubations render scaling metabolism to the reach scale difficult because of spatial heterogeneity. Further, incubations miss certain habitats such as hyporheic metabolism (Fellows et al., 2001) and can create artificial environments through isolation from the stream reach that produce less accurate estimates of metabolism. The use of open-channel methods greatly enhances our ability to consider streams as a whole ecosystem, where changes in O_2 within a stream reach integrate the metabolism of water column, benthic, and hyporheic components.

Since Odum's time, metabolism methods have greatly improved, most notably advanced by the development and increasing affordability of dissolved oxygen (O_2) sensors with high-temporal frequency logging capabilities. Consequently, open-channel measurements of O_2 have become a standard method for estimating aquatic ecosystem metabolism, where the change in O_2 over time is a function of GPP, ER, and air–water gas exchange fluxes (G):

$$\frac{(dO_2)}{dt} = GPP + ER + G \tag{34.4}$$

Open-channel metabolism methods assume measurements of O_2 can be used as a proxy for C cycling given the relationship between O_2 consumption and production, aerobic organic C mineralization and fixation, and CO_2 production and consumption during ER and GPP, respectively. We use measurements of O_2 primarily because CO_2 dissolved in water is not simply CO_2—dissolved inorganic C largely occurs as bicarbonate (HCO_3^-), with smaller fractions of free CO_2 and carbonate (CO_3^{2-}); the fraction of each dissolved inorganic C species will shift with changes in pH (Drever, 1988).

Physical processes (e.g., gas exchange, groundwater inflow) alter O_2 dynamics in streams and must be considered when estimating metabolism. The influence of air–water gas exchange (G in Eq. 34.4) on dissolved O_2 can be substantial, especially in highly turbulent streams with low rates of GPP and ER. The gas exchange flux, G, is the product of a gas exchange rate (K, d^{-1}) times the saturation deficit of dissolved O_2. This rate, K, is a difficult part of measuring metabolism and we will revisit ways to measure and model K in the general design and detailed methods sections of this chapter. We note that inputs of groundwater to streams, which are often depleted in O_2, can dilute stream water O_2 and can be mistaken for rates of O_2 consumption and consequently bias metabolism estimates (Hall and Tank, 2005). The methods presented in this chapter will not focus on ways to account for groundwater inputs of low-O_2 water (but see Chapter 8), but keep in mind groundwater may be an important factor governing O_2 and other water chemistry parameters in your study streams.

Here we cover the general design and more specific methods needed to estimate whole-stream, open-channel metabolism. We focus on methods using diel measurements of O_2 and temperature at a single station (i.e., one-station methods), which are suitable for stream reaches without dams or substantial groundwater and tributary inputs. We do not cover the use of ^{14}C tracers or light/dark bottle and chamber methods, largely because the procedures for these measurements have not changed much through the years and have been covered well by others (e.g., Bott, 2006). Moreover, we believe that open-channel metabolism integrates the appropriate spatial scale needed to measure and monitor stream ecosystem function. The following sections cover methods needed to collect good quality O_2 data as well as the additional parameters needed for metabolism calculations. We also cover the methods used to make the metabolism calculations themselves—primarily, (1) a basic direct calculation method and (2) a more advanced inverse modeling technique using maximum likelihood to estimate GPP and ER (and sometimes K). Given the importance of knowing K for accurate estimates of ER, we also devote time to discussing methods used to measure and model K, and when these different methods might be most appropriate. Our objective is to provide the central knowledge needed to estimate stream metabolism as well as an understanding of the assumptions behind many of the common methods. Instead of trying to cover many different types of metabolism modeling methods, most of which are still rapidly evolving as we write this chapter, we have chosen to focus on the fundamentals and refer you to additional sources in the literature and in open-source code for more complex measurement and modeling options.

34.2 GENERAL DESIGN

34.2.1 Basic Method: Stream Metabolism

34.2.1.1 Site Selection and Data Collection

The study reach can be chosen from a range of stream types, but a few characteristics will make estimating metabolism easier. The study stream should have a relatively low slope and therefore low gas exchange rates. Low gas exchange permits estimating gas exchange from the O_2 data themselves and reduces errors in ER and GPP because diel O_2 swings are far from saturation. High light and thus some GPP will also assist in estimating gas exchange. Steep, forested streams are difficult to estimate metabolism and we suggest not using those for class exercises. One-station metabolism estimates assume that the stream is relatively homogenous over a length roughly equal to three times the turnover length of O_2 ($3v/K$, where v is stream velocity) (Chapra and Di Toro, 1991; Reichert et al., 2009). Thus one-station methods do not work immediately below big discontinuities such as dams, sewage outfalls, and waterfalls. Two-station methods are needed in these situations, but will not be discussed here. Procedures for two-station metabolism can be found elsewhere (Halbedel and Büttner, 2014; Hall et al., 2016).

Metabolism estimates require collection of dissolved O_2 and water temperature data in the field at high-frequency time intervals (5—15 min) for ≥ 24 h. Light data are needed and can be measured directly or estimated from solar models. At minimum, stream physical attributes needed are mean depth, which can be measured directly or from discharge, width, and velocity by assuming hydraulic continuity (see Chapter 3; Eq. 34.8 below). Depending on the method used to estimate air-water gas exchange (see further), one or more gas tracer and conservative solute additions in the field may also be needed.

34.2.2 The Fundamental Metabolism Equation

The fundamental equation used to describe the primary biological and physical processes that change dissolved O_2 over time (as in Odum, 1956) is given as:

$$\frac{dO}{dt} = \frac{GPP}{z} + \frac{ER}{z} + K_O(O_{sat} - O)$$

(34.5)

which can be discretized and solved via an Euler numeric solution to:

$$O_i = O_{i-\Delta t} + \left(\frac{GPP}{z} + \frac{ER}{z} + K_O\left(O_{sat(i-\Delta t)} - O_{i-\Delta t}\right)\right)\Delta t$$

(34.6)

where O_2 at time i (O_i) (g O_2/m^3) is equal to O_2 at the previous time ($O_{i-\Delta t}$) plus time step-specific rates of GPP, ER, and air-water gas exchange (based on the O_2 gas exchange rate K_O (1/d) and the difference between dissolved O_2 and O_2 at saturation (O_{sat}) for a given temperature and barometric pressure). In this model, ER is a negative O_2 flux because O_2 is being consumed. z is mean stream depth (m). Δt (d) is the measurement interval of logged O_2 data for a one-station metabolism model in streams, lakes, and rivers. This equation does not include an adjustment for groundwater entering a reach (but see Hall and Tank, 2005). Symbols and units for all metabolism parameters are listed in Table 34.1.

TABLE 34.1 Parameter symbols and units for ecosystem metabolism models.

Parameter Symbol	Parameter Description	Parameter Units
O	Measured O_2	g O_2/m^3 (= mg/L)
mO	Modeled O_2	g O_2/m^3
GPP	Gross primary production	g O_2 m^{-2} d^{-1}
ER	Ecosystem respiration	g O_2 m^{-2} d^{-1}
met$_i$	Instantaneous net metabolism	g O_2 m^{-2} d^{-1}
Δt	Measurement time step	d
z	Mean stream depth	m
v	Mean stream velocity	m/s
K_O	O_2 gas exchange rate	1/d
K_{600}	General gas exchange rate	1/d
O_{sat}	O_2 at saturation	g O_2/m^3
PAR	Photosynthetically active radiation	μmol m^{-2} s^{-1}
bp$_{abs}$	Absolute barometric pressure	mm Hg
bp$_{std}$	Barometric pressure standardized to sea level	mm Hg or inches Hg

34.2.3 Air-Water Gas Exchange

A central unknown in measuring open channel metabolism is gas exchange between the water and the atmosphere. This flux is the product of the saturation deficit of O_2 (O_{sat} − O) times the gas exchange rate K (1/d) for O_2. The saturation deficit is easily measured, but the gas exchange rate is much more difficult. Ecologists have used several ways to estimate gas exchange in streams and rivers. These cover three approaches: (1) use gas exchange measured elsewhere to infer gas exchange in a focal stream. Such approaches may use theory of gas exchange based on physical models to predict gas exchange in a focal stream (Tsivoglou and Neal, 1976). Another way is to empirically predict gas exchange based on many past tracer experiments (Raymond et al., 2012). These techniques work well to describe the mean or median flux from a population of streams, but they have high prediction error for any one stream. For example, predictions using equations in Raymond et al. (2012) will have about a fivefold prediction error, which therefore gives a fivefold error on ER when estimating metabolism. For this reason, most practitioners use approaches (2 and 3): generating empirical estimates of gas exchange from their study stream. Two ways to estimate K are to (2) use an experimental tracer gas addition and (3) estimate gas exchange from the diel O_2 data themselves. Tracer gas additions greatly improved ecologists' ability to measure metabolism in small streams, and in particular, those with high rates of gas exchange (Wanninkhof et al., 1990; Marzolf et al., 1994). If O_2 exchange is low, it may not be possible to detect a measureable decline in tracer added. Fortunately in that situation, it may be much easier to measure gas exchange using the O_2 data themselves. This chapter covers methods for calculating gas exchange using either tracer additions or O_2 data because we believe these are the most appropriate for generating accurate estimates of GPP and ER.

Given the different approaches to measuring gas exchange, you may be asking, "Which one should I use?" This question is open at the time of writing this chapter (Demars et al., 2015; Holtgrieve et al., 2016). One key assumption of the tracer addition method is that converting K from tracer gas to O_2 follows theory (Asher et al., 1997), although this has received little empirical testing in streams (Holtgrieve et al., 2016). We suggest that the type of stream will in part determine which method to use. In streams with high K and low GPP, it will be harder to estimate gas exchange using O_2 data than using a tracer addition. Conversely, tracers may work poorly when gas exchange is low, and this is the situation where estimating K from the O_2 data works best. For a 1-day course exercise in metabolism, adding tracer gas may be too much work for a small return of potential increase in accuracy of GPP and ER. On the other hand, if one is installing a sensor to measure long-term metabolism in a stream that may have low GPP or high gas exchange, we strongly suggest using a tracer gas addition. At a minimum, a tracer estimate of K can be used to provide prior information for K in a Bayesian modeling approach (Holtgrieve et al., 2010, 2016). We currently use both empirical approaches: we often use a combination of tracer additions and modeling using O_2 data (Hotchkiss and Hall, 2014, 2015) but tracers are often required to estimate K in low-productivity streams. In some rivers estimating K from the O_2 data themselves can work well (Hall et al., 2016; Genzoli and Hall, 2016).

34.2.4 Advanced Analyses: Inverse Modeling of Ecosystem Metabolism

The goal of inverse modeling is to (1) model known values using a process equation that includes at least one parameter with an unknown value and (2) minimize the difference between modeled values and data by solving for unknown parameter values until a given set of parameters provides the best fit between modeled and measured data (Hilborn and Mangel, 1997). Inverse modeling of ecosystem metabolism is based on modeling water column O_2 using best model estimates of GPP and ER and minimizing the difference between modeled and measured O_2 by solving for the most likely values of GPP, ER, and sometimes K (Holtgrieve et al., 2010; Hotchkiss and Hall, 2014).

While past methods for estimating ecosystem metabolism often involved exact solutions for GPP and ER using O_2 data (Mulholland et al., 2001; Bernot et al., 2010; and many others), inverse modeling does not use any O_2 data in Eq. (34.7). The O_2 data are only used to test how well model estimates of GPP and ER can reproduce diel O_2 concentrations. A maximum likelihood estimation (MLE) or Bayesian parameter estimation from Markov chain Monte Carlo sampling (MCMC) iteratively finds parameter estimates that provide the best fit of modeled and measured O_2. For example, we can use MLE or MCMC to estimate GPP and ER as a function of modeled O_2 (mO) and parameters GPP, ER, and air-water O_2 exchange:

$$mO_i = mO_{i-\Delta t} + \frac{GPP \times PAR_t}{z \times \sum PAR} + \frac{ER}{z}\Delta t + K_O\left(O_{sat(i-\Delta t)} - mO_{i-\Delta t}\right)\Delta t \qquad (34.7)$$

This model uses the relative amount of light $\left(\frac{PAR_t}{\sum PAR}\right)$ at each time step to turn on GPP during the day. A best-fit solution will always exist for GPP, ER, and K given a set of O_2 data, but inverse modeling allows for a visual comparison of modeled versus measured O_2 as well as a quantification of the goodness of fit with modeled parameter estimates.

34.3 SPECIFIC METHODS

34.3.1 Basic Method: Stream Metabolism

34.3.1.1 In the Field: Site Parameters and Diel O_2 Data

1. Estimate stream physical parameters. Metabolism estimates require measurement of hydraulic and geomorphic parameters of the study reach. The key parameter needed for metabolism is mean water depth, z (m). It is possible to measure mean depth from many transects, but given the great heterogeneity of stream depths within and among transects, the most accurate method is to calculate mean depth based on measured estimates of discharge (Q, m^3/s), mean velocity (v, m/s), and wetted width (w, m), that is:

$$z = \frac{Q}{w \times v} \qquad (34.8)$$

Q can be measured from a gage, the velocity-area method, or via dilution gaging of a salt slug (see Chapter 3). We use the dilution gaging method whenever possible because it allows us to estimate nominal travel time and v (distance downstream ÷ nominal travel time from the salt slug addition as well (see Chapter 30). Wetted width is simply measured using a tape or laser range finder at ~ 20 transects. If the river is too big to do a salt slug release, it is possible to estimate velocity by knowing Q, w, and slope, from which depth can be calculated using the Manning's equation (see Chapter 3). For longer-term metabolism measurements, you will need to account for changes in mean depth over time by developing site-specific relationships between mean depth and water level.

2. Measure or model diel light. If you use inverse approaches to calculate metabolism, you will need estimates of light input to the stream. These light data can either be estimated based on time and geographical location (see Supplemental File 34.3) or measured directly using recording light sensors (see Chapter 7). If using light sensors, place them in a location near the stream representative of stream channel light conditions and record at the same intervals as O_2. Because the inverse modeling approach we present below assumes GPP is a linear function of light, it is unnecessary to record light in units of photosynthetically active radiation (PAR, μmol photons m^{-2} s^{-1}); radiometric units (e.g., W m^{-2} s^{-1} will work fine.

3. Program and calibrate O_2 sensor. A variety of these are available ranging from multiparameter sondes (e.g., YSI, Hydrolab, In Situ, Eureka, and others) to smaller, less expensive sensors that only measure and record O_2 and temperature (e.g., PME and Onset). All brands use new optical sensor technology and work fine for metabolism. Older sondes use electrode sensors, which work well for short-term deployments, but are less precise and will drift over time. You also have the option to record O_2 through a diel cycle using a standard O_2 meter and sensor, but we find that the recording sensors are much easier, despite that you will miss the camaraderie of an all-nighter (J.L. Meyer, pers. comm.). Program the sensor to record O_2 at 15-min intervals or less; we use 5- or 10-min intervals. If you plan to use the nighttime regression method to estimate K from O_2 data, 5-min intervals are the best option. When programming the sensor, it is important to ensure that time is accurate and time zone is known. If using calculated light, Standard Time is preferable to Daylight Time to avoid 1-h offsets in light and O_2 data.

   Accurate calibration of sensors is essential. The degree of undersaturation at night (Eq. 34.13) defines ER; thus, if true nighttime saturation is 90% and you are 5% high on calibration, ER will be underestimated by half. There are several ways to calibrate. One is to put the sensor in wet atmosphere and assume saturation. Sensors measure the partial pressure of O_2, and dissolved O_2 at saturation is the same partial pressure as O_2 in a wet atmosphere. If you use this method, it is necessary that the O_2 sensor and temperature sensor read the exact same temperature; we often submerse sondes, sealed in a wet air cap, underwater to achieve this calibration. Another means, and one that requires more gear, is to vigorously bubble a pot of water with air and immerse the sensors. A bubbling pot is very near saturation (although the bubbling equipment and pot may read slightly high and need correction; Hall et al., 2016). Small sensors, such as PME MiniDOTs, come precalibrated from the factory. We find these calibrations to be within 2% the true value, but you should check yours before first deployment, and certainly check for changes in calibration over time. Depending on the logging sensor you have chosen, you may have the opportunity to rescale O_2 measurements to calibrated values so that the sensor itself is calibrated before deployment. For sensors that you cannot recalibrate yourself, you will need to record the difference between logged values and calibration chamber values to correct the logged O_2 data after deployment.

4. Measure barometric pressure. It is necessary to know the barometric pressure of both your calibration location and study stream. If you have a handheld barometer that reads absolute barometric pressure, use that. Otherwise you can calculate barometric pressure by knowing your elevation above sea level and standardized barometric pressure (bp) from a nearby weather station using the following equation (i.e., the barometric formula; Colt, 2012).

$$bp_{abs} = bp_{st} \times 25.4 \times e^{\dfrac{-9.80665 \times 0.0289644 \times alt}{8.31447 \times 288.15}} \tag{34.9}$$

   where bp_{abs} = absolute barometric pressure in mm Hg, bp_{st} = standardized bp in inches of Hg, and alt = altitude above sea level in meter. We use inches for bp_{st}, because that is how you see it from any US weather station. Delete the 25.4 if you live where bp_{st} is reported in mm of Hg. Do not use the bp_{st} from a weather station to directly calibrate a sensor unless you are within 30 m of sea level; you will nearly always need to correct for elevation.

5. Deploy O_2 sensor and download data. To measure and log O_2 for metabolism estimates, find a site near the bank in an area where there is enough downstream flow to move water over the sensor (so not a backwater or pool) but slow enough to avoid damaging the sensor and to allow safe installation. Secure the sensor with rope or cable to something immovable to ensure that the sensor will stay submerged at the deployment site. For long-term installations you will need to consider a deployment and anchoring scheme that will protect and keep the sensor in place during high flow. Bolting sensor housing to boulders or bridges may be needed. For most short deployments, we tie the sensor to a riparian tree. If you deploy the sensor in an area with high human activity, we suggest hiding the sensor as much as possible to avoid theft. Deploy the sensor for at least the night before, the day during, and the night after the focal metabolism day.

   Following deployment, download the data following the manufacturer's instructions. Save the data as a text file (we use.csv) for easy import into Excel, R, or the analysis program of your choice. Depending on the length of deployment, you may also want to recheck sensor calibration after the measurement period.

34.3.1.2 In the Field: Gas Exchange From Tracer Additions

Tracer additions provide a direct measure of gas exchange in a stream reach. To measure gas exchange in the field requires adding a tracer gas such as propane, sulfur hexafluoride (SF_6), or argon. Each has its benefits and costs. Propane is cheap and readily available from gas stations or hardware stores. Large amounts are required to measure longitudinal declines due to gas exchange, so this technique works best in smaller streams. SF_6 is easily measured because gas chromatographs with electron capture are highly sensitive and therefore require very small amounts of SF_6. But SF_6 is expensive and a particularly intense greenhouse gas, with 23,900 times the forcing of CO_2. A 4.5-kg tank of SF_6 contains the greenhouse forcing of driving a mid-size car about 120,000 km. Argon is also cheap (it is used for welding), but it so far has only been used in small streams and requires a membrane-inlet mass spectrometer to accurately measure when near atmospheric partial pressures.

The gas tracer addition procedure is almost exactly as if for nutrient uptake experiments (see Chapters 30 and 31): gas and a conservative tracer are pumped into a stream, and the decay rate is calculated as if the gas is a reactive tracer. We refer the reader to the above nutrient uptake chapters for details on setting up a drip experiment with a reactive solute and conservative tracer (e.g., NaCl or bromide), conservative tracer sampling, and correcting reactive solute concentrations for any dilution of conservative tracer downstream of the release site.

1. Establish tracer measurement stations along study reach. Establish two or more (ideally at least four) measurement stations at known distances downstream of the tracer addition site. We note that the mean travel distance of O_2 in streams is $0.7v/K$ (v = water velocity in m/min; K = gas exchange rate in min^{-1}), and a gas exchange reach much smaller than that may not represent the processes controlling O_2 in streams. While background concentrations of propane and SF_6 should be near-zero, it is best to collect several replicate "pre" samples before hooking up the tracer gas to the regulator and starting the tracer addition (see sampling protocol further for volume and methods).

2. Add tracer. Pump gas at a continuous rate throughout the continuous conservative tracer release at the same release site; ideally in a constrained part of the stream with rapid mixing. If using propane, add the gas through a large diffuser, e.g., a Pentair AS23S airstone. SF_6 flow rates can be very low (100 mL/min is fine for a stream with 50−200 L/s discharge) and requires only a small diffuser; a fine bubble aquarium airstone will work well. If adding NaCl as your conservative tracer, a handheld conductivity meter may be used to confirm when the tracers have approached plateau at your downstream sampling stations.

3. Sample tracer. After the conservative tracer reaches plateau at the downstream sites, sample the conservative solute and tracer gas at each site by collecting 3 or more replicate samples of 45 mL of water in a 60-mL plastic syringe with a stopcock. Syringes should be prerinsed 3× with stream water immediately before sampling. Take care to keep the water bubble-free during and after collecting the sample to avoid contaminating dissolved gases with air. Close the stopcock after ensuring the 45 mL sample is bubble free. Turn off the gas and stop pumping conservative solute after sampling (or, for reaches with longer travel time, while still at plateau concentration at the downstream-most reaches). Keep all samples at a similar temperature to prevent differences in measured gas concentrations due to changes in solubility.

4. Transfer tracer samples to gas-tight containers. Go far away from the stream before transferring your sample from syringe to gas vial to avoid contamination with any tracer gas degassing from the stream. Open the stopcock, add 15 mL air ("headspace") to each syringe, and reclose the stopcock. Shake syringes vigorously for 5−10 min (same length of time for all samples) to transfer the tracer gas dissolved in the water sample into the headspace. Attach a needle to the stopcock, open the stopcock between syringe and vial, and inject the headspace from the sample syringe into a 10−12 mL gas-tight serum vial, taking care to keep the stopcock closed to the atmosphere at all times during transfer. The vial can either be preevacuated or, as you push in the gas sample, gently pull out the same amount of air from the vial with an empty syringe and needle. Transfer the remaining water into sample bottles for analysis of conservative solute.

The best sample size and ratio of water/gas for headspace extraction may vary depending on the elevation of your sample site, sensitivity of your gas chromatograph (GC) to tracer gas, size of your study stream, and release rate of tracer gas into the stream. We have used the above method for both propane and SF_6 with good results. There are variations of these basic tracer gas addition techniques used by different research groups, such as bubbling tracer gas into a container of stream water with conservative solute for a single drip addition of water highly concentrated with tracer gas and conservative solute (Tobias et al., 2009), storing water samples in airtight containers until injecting headspace with a GC autosampler, and many others.

34.3.1.3 In the Laboratory: Gas Exchange From Tracer Additions

1. Measure the tracer gas peak areas using an appropriate GC. It is beyond the scope of this chapter to provide detail on these measurements because the recommendations would be specific to each GC. Propane requires a GC with flame ionization detector. SF_6 can be estimated using a GC with electron capture detector. Because propane and SF_6 have near-zero background concentrations in streams, and we measure the relative decline in the gas downstream from the addition site, it is not necessary to measure the concentrations of either by fitting a standard curve.

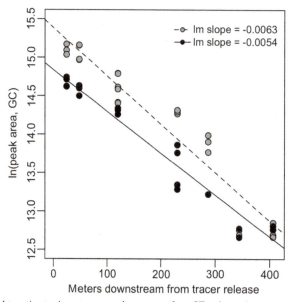

FIGURE 34.1 Gas tracer additions used to estimate air-water gas exchange rates from SF_6 releases in an upper reach of French Creek, Wyoming, USA, on September 30 and October 01, 2012. *Lines* are linear model fits to data using the *lm()* function in R. Slopes from the linear model (*lm* slope) are K_{SF_6} in units of 1/m. Multiplying $K_{SF_6,d}$ × stream velocity (6998 m/d) as in Eq. (34.10) gives $K_{SF_6,t} = 38$ and 44 (1/d). Using temperature-specific Schmidt number scaling to convert $K_{SF_6,t}$ to K_{600} with the *K600fromSF6* function in the supplemental online R code gives $K_{600} = 68$ and 79 (1/d).

2. Calculate K from GC data. Once you have concentrations or peak areas, calculate the per meter rate of gas evasion (K_{gas}) using Eq. (31.2) from Chapter 31 (see also Fig. 34.1). Convert this per length ($K_{gas,d}$; m^{-1}) rate to a per time ($K_{gas,t}$; s^{-1}) rate as:

$$K_{gas,t} = K_{gas,\,d} \times v \qquad (34.10)$$

where v = mean velocity of the stream (m/s). Gases diffuse at different rates depending on the type of gas and the temperature. We use the ratio of their Schmidt numbers to convert gas exchange among gases and temperatures. It is typical to report gas exchange at a common Schmidt number of 600 (K_{600}) and we will do that here:

$$\frac{K_{600}}{K_{gas,t}} = \left(\frac{600}{Sc_{gas}}\right)^{-n} \qquad (34.11)$$

where Sc_{gas} = Schmidt number for the tracer gas (i.e., propane, SF_6, argon) and n = Schmidt number exponent (usually assumed to be 0.5; Jähne et al., 1987). For example, K_{600} can be calculated from $K_{propane,t}$ as:

$$K_{600} = \left(\frac{600}{2864 - (154.14T) + (3.791T^2) - (0.0379\,T^3)}\right)^{-0.5} \times K_{propane,t} \qquad (34.12)$$

Eq. (34.12) includes the temperature-specific calculation of $Sc_{propane}$ given stream temperature (T; °C) during the tracer addition (Raymond et al., 2012). Temperature-specific Schmidt numbers calculations for O_2, CO_2, SF_6, Ar, and many other gases can be found in Raymond et al. (2012).

34.3.1.4 On the Computer: Gas Exchange From O_2 Data

1. Calculate gas exchange rates from nighttime regression of O_2 data. The nighttime regression method is an older method developed by Hornberger and Kelly (1975). This method requires that O_2 be out of equilibrium at the start of nighttime due to the day's photosynthetic activity. The rate of return to equilibrium is a function of the gas exchange rate. Thus at night with no GPP, metabolism is:

$$\frac{\Delta O}{\Delta t} = ER + K_O(O_{sat} - O) \qquad (34.13)$$

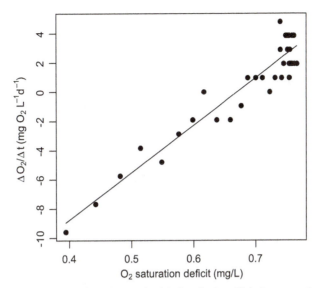

FIGURE 34.2 Estimating air-water gas exchange rates from changes in nighttime O_2 data. Nighttime regression output is for September 22, 2012 in French Creek, Wyoming, USA. *Line* is least-squares regression fit to data and was generated using function *nightreg()* provided in the supplemental online R code. *Slope* of the line is 32/d (K_O), which converts to a K_{600} of 43/d.

This equation is that of a line where $\Delta O/\Delta t$ is on the Y axis, the DO deficit ($O_{sat} - O$) is on the X axis, ER is the Y intercept, and K_O is the slope (Fig. 34.2). This method works best in streams with high GPP and low rates of gas exchange. If gas exchange is too high, then $\Delta O/\Delta t$ approaches 0 soon after nightfall and there are not enough data to conduct a regression. Given that the return of $\Delta O/\Delta t$ to equilibrium is a function of K, then length of time throughout the night to conduct nighttime regression will vary. We suggest starting right at nightfall and iteratively extending the time through the night until points cluster around $\Delta O/\Delta t \sim 0$. K can also be estimated as a free parameter in a model for metabolism. We describe this approach in detail in the next section.

34.3.1.5 On the Computer: Estimating Metabolism

In the next sections, we will step through the process to calculate metabolism from oxygen data. One needs to know how to use a spreadsheet (e.g., Excel) or a programming language (e.g., R, Matlab) to calculate metabolism. Hand calculations are tedious and error-prone. We much prefer a programming language to a spreadsheet and we have provided R code in the online supplements to this chapter. Excellent R tutorials and websites are available to help students and researchers become familiar with the R coding environment, basic commands, and troubleshooting (e.g., Zuur et al., 2009; R Development Core Team, 2016). The annotated R code we have provided will work for all calculations in this chapter, including the nonlinear minimization method described further. If you are not familiar with R, the Excel worksheet that we have provided in the online supplements will work just as well, but for the direct calculation only. We suggest that you do not use Excel for any nonlinear function minimization.

Direct Calculation

The simplest way to calculate one-station metabolism is via direct calculation of metabolism at each time step. This method requires an independent measure of gas exchange, either from a gas tracer addition or the nighttime regression method. We have provided as online supplements both an Excel spreadsheet (see Online Worksheet 34.1) and R code (see Supplemental File 34.2) to do this calculation. To calculate net metabolism at any time point (met$_i$), one simply rearranges Eq. (34.6):

$$\text{met}_i = \left(\frac{O_i - O_{i-\Delta t}}{\Delta t} - K_{O(i)}(O_{sat,i} - O_i) \right) z \tag{34.14}$$

1. Calculate the change in O_2 over time. The first term in Eq. (34.14), $O_i - O_{i-1}/\Delta t$, is easily calculated from sensor data.
2. Calculate the gas exchange flux given K_O and O_{sat}. The second term, gas exchange flux, must vary with temperature for both K_O and O_2 saturation. We use Schmidt number scaling to convert a normalized K with Schmidt number of

600 (K_{600}) to K_O using the temperature-specific Schmidt number for O_2 (Sc_O; Wanninkhof, 1992) following the same approach as Eq. (34.12) above:

$$K_O = \frac{K_{600}}{(600/(1800.6 - 120.10T + 3.7818T^2 - 0.047608T^3))^{-0.5}} \quad (34.15)$$

Next it is necessary to calculate oxygen saturation (O_{sat}) as a function of T and barometric pressure, bp (mm Hg) at each time step. Based on Garcia and Gordon (1992) and Colt (2012), we estimate

$$O_{sat} = e^{a0 + a1 \times ts + a2 \times ts^2 + a3 \times ts^3 + a4 \times ts^4 + a5 \times ts^5} \times \frac{bp - u}{760 - u} \times 1.42905 \quad (34.16)$$

given

$$ts = \ln\left(\frac{(298.15 - T)}{(273.15 + T)}\right) \quad (34.17)$$

and

$$u = 10^{\left(8.10765 - \frac{1750.3}{(235+T)}\right)} \quad (34.18)$$

where $a0 = 2.00907$, $a1 = 3.22014$, $a2 = 4.0501$, $a3 = 4.94457$, $a4 = -0.256847$, $a5 = 3.88767$.

3. Calculate net metabolism (met_i) for each time step.
4. Estimate GPP and ER from met_i and light. Determine sunrise and sunset times for your stream; these data are easily found for your site from a local weather station or online (e.g., http://www.esrl.noaa.gov/gmd/grad/solcalc/). At night met_i is instantaneous ER, and in day it is NEP. Daily ER is the mean of met_i before sunrise and after sunset (Fig. 34.3). GPP is the area under the curve between met_i and ER during the day. This direct calculation can be implemented in Excel, R or any other environment. Because GPP and ER take place at the same time during the day, it is difficult to isolate these two processes and calculate both accurately with only O_2 data to estimates two unknowns. Consequently, scientists often assume that ER at night (measured easily because photosynthesis is not taking place) is equal to ER during the day, despite evidence that ER during the day may indeed be higher than nighttime ER (Tobias et al., 2007; Hotchkiss and Hall, 2014).

34.3.1.6 Advanced Method: Metabolism Modeling

The advance method of inverse metabolism modeling with R requires several of the Basic Method steps outlined earlier: measuring site-specific physical parameters, collecting and downloading well-calibrated diel O_2 data, measuring or

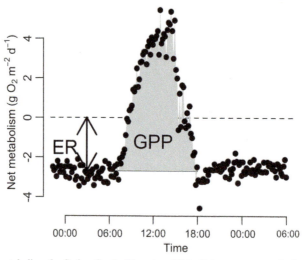

FIGURE 34.3 Direct calculation of metabolism for Spring Creek, Wyoming, USA. *Points* are net metabolism calculated via Eq. (34.14). K_{600} was 25.7/d and was chosen to be the same as that calculated from the inverse model approach. *Shaded area* is GPP. *Arrow* is instantaneous ecosystem respiration. Daily ER is instantaneous ER scaled to 1 d. GPP = 1.95 g O_2 m^{-2} d^{-1}, ER = -2.68 g O_2 m^{-2} d^{-1}.

modeling light, and potentially estimating K from tracer additions and/or nighttime regression methods. The R code described further (see Supplemental File 34.3) includes functions to model light, convert tracer-specific K rates to K_{600}, and estimate K from nighttime regression of O_2 data.

Inverse Modeling

The inverse modeling approach requires two steps. First is identifying a suitable model of O_2 dynamics. We will use the model from Eq. (34.7), recognizing that there are many possible models. Next is computing the best-fit values of the parameters given the data. There are multitude of ways to do this step (Bolker et al., 2013). We will focus on nonlinear minimization of the $-$log likelihood of the data, which assumes normally distributed errors and provides the same results as sums of squares minimization. Likelihood is advantageous because it provides a first step in understanding uncertainty in the GPP and ER estimates.

1. Estimate GPP, ER (and potentially K) using inverse modeling in R. Using the annotated R code we have provided (see Supplemental File 34.3), estimate GPP, ER, and K as free (unknown) parameters (function *rivermetabK* in Part 7 of Supplemental File 34.3). If you have an independent estimate of K (e.g., from a gas tracer addition) then use the code where K is entered as a fixed, and not free, parameter (function *rivermetab* in Part 8 of Supplemental File 34.3). Note that you will need to convert any tracer-specific K rates to K_{600} before running the *rivermetab* model (see K functions in Part 4 of Supplemental File 34.3). We have provided code to estimate K_{600} from nighttime O_2 regressions (function *nightreg* in Part 5 of Supplemental File 34.3). You may use prior information you have about K at your site (e.g., from tracer additions or nighttime regressions) as a starting value for K in the *rivermetabK* model. If you lack light data, we have provided code to estimate solar insolation as a function of geographic location, time of day, and day of year (function *lightest* in Part 6 of Supplemental File 34.3). We have provided some data (see Supplemental Files 34.4 and 34.5) to practice using with the code in Supplemental File 34.3 and that can be checked against example model output in Supplemental File 34.6.

2. Critically evaluate model results. A key consideration when using any inverse modeling approach is to check the model fit with the data! The plotting routine in R (part of *rivermetabK* and *rivermetab* functions in Supplemental File 34.3) will show the modeled O_2 and the O_2 data given model estimates of GPP and ER (and K for *rivermetabK*). Visually observe the fit; Fig. 34.4 shows such a fit for French Creek, Wyoming, USA, from the data we provided in Supplemental File 34.3. As with any calculation exercise, is important to check that the modeled estimated parameters (GPP, ER, and maybe K) make sense: Is GPP positive and ER negative? Are GPP, ER, and K within the range of rates reported by others in the literature?

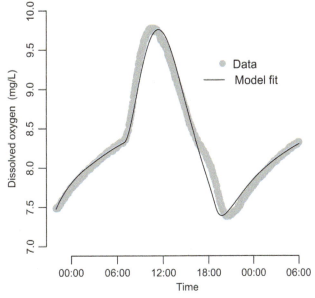

FIGURE 34.4 Inverse model fit of metabolism for French Creek, Wyoming, USA. *Points* are oxygen data. *Line* is a model fit using Eq. (34.9) and is derived from data and code supplied in Supplemental File 34.3. GPP $= 3.4$ g O_2 m^{-2} d^{-1}, ER $= -2.6$ g O_2 m^{-2} d^{-1} and estimate of K_{600} is 33.9/d.

It is important to note that even having a good fit of modeled O_2 to O_2 data or a low −log likelihood does not necessarily mean that you know the parameter estimates with low uncertainty. It is possible that a range of values of the parameter estimates (in particular ER and K) can give an equally good fit of the model to the data; this phenomenon is called equifinality. Techniques for accurately estimating parameter uncertainty within any one day require Bayesian methods and are beyond the scope of this chapter. Such methods are still being developed and tested and require statistical procedures that account for both process and observation error in the O_2 model (Clark, 2007; Hall et al., unpublished data). One way to test the amount of parameter uncertainty would be to simply collect a week's worth of data at constant stream flow, calculate metabolism for each day, and look for day-to-day variation in K. If that variation is large (say, varying by 50%) it is likely that uncertainty in metabolism parameters is high.

Many other ways, both from modeling and statistical perspectives, exist to estimate metabolism parameters. The approaches we outline here are simple and suitable for learning the methods and for classroom instruction. Varying model structures include accounting for the role of variable flow and turbidity in driving rates of ecosystem metabolism. Hanson et al. (2008) found that Eq. (34.4) provided similar estimates of GPP and ER compared with other more complex models that included temperature-driven ER or more complex light functions. Statistically, it is advantageous to use Bayesian methods to account for parameter error and to develop multilevel models of multiday time series of metabolism. At the time of writing this chapter such statistical methods are in development. However, these computing tools (e.g., streamMetabolizer, https://github.com/USGS-R/streamMetabolizer) are available as open-source code and this code will evolve with time. We advise researchers specializing in metabolism estimation to use these or analogous approaches to calculate metabolism for long time series.

34.4 QUESTIONS

1. Some countries and a few U.S. states have started to include metabolism estimates in their environmental monitoring and health assessment programs. Which environmental changes could influence rates of GPP, ER, and NEP? Sketch a graph of your predictions for how one or more changing environmental variables (biotic or abiotic) might influence metabolism in streams.
2. Consider your study stream: (1) Where do you see the highest densities of primary producers and organic material? Would you predict spatially homogenous or heterogeneous contributions to reach-scale GPP and ER? Why? (2) Most streams and rivers are heterotrophic. What types of terrestrial organic matter might be supporting ER in your stream? How might this change over seasons?
3. If a stream is autotrophic, what do you think is the eventual fate of this fixed C?
4. Using the example data provided in Supplemental Files 34.4 and 34.5 and/or metabolism estimates for your own study site(s), compare your estimates of GPP and ER with the range of rates published in the literature (Fig. 34.5). (1) Was

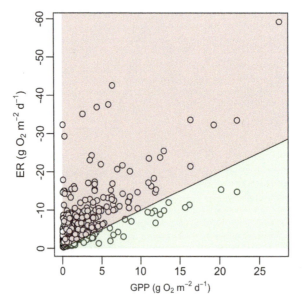

FIGURE 34.5 Rates of GPP and ER from streams and rivers across the globe. The line is a 1:1 line. Median GPP is 1.6 g O_2 m^{-2} d^{-1} (25%−75% quartiles = 0.49, 3.86). Median ER is −4.7 g O_2 m^{-2} d^{-1} (25%−75% quartiles = −2.49, −9.12). *Updated from Hall et al. (2016).*

this stream autotrophic or heterotrophic on the measurement day(s)? (2) How do rates of GPP and ER compare to those measured at other sites? (3) What might be common characteristics of sites occupying different parts of Fig. 34.5?

5. Assuming 10 streams have the same ER of -10 g O_2 m^{-2} d^{-1}, but widely different rates of gas exchange, (say $K_O = 5$, 10, 20, 50/d), use Eq. (34.13) to plot the relationship between K_O and nighttime O_2 concentrations. How does gas exchange control equilibrium O_2 concentration at night?

6. If you miscalibrate an oxygen sensor to 20% above the true calibration, what will happen to estimates of ER?

7. Write a metabolism model using Eq. (34.7) that plots diel O_2 data based on GPP, ER, and z. (1) Given GPP of 4 g O_2 m^{-2} d^{-1}, ER of -4 g O_2 m^{-2} d^{-1}, and z of 0.3 m, synthetic light data for your location, and constant temperature of 15 C, plot the variation in the O_2 diel patterns across a range of gas exchange rates (e.g., K_O from 5 to 50/d). You can use R or Excel for this exercise. Adapting the R code in the Supplemental File 34.3 would be the easiest. How does gas exchange affect the shape of the diel curve? Note: It is this change in shape that allows the inverse modeling procedure to solve for GPP, ER, and K simultaneously for some streams. (2) Further simulations: How might the magnitude of K's influence on O_2 values change in sites with lower or higher rates of GPP and/or ER?

34.5 MATERIALS AND SUPPLIES

Multiparameter sonde (e.g., YSI, Hydrolab, In Situ, Eureka) or Recording oxygen sensor (e.g., PME-Minidot, Onset HOBO)

Rope or cable to secure sensor to an immovable object

Light logger, if measuring (e.g., Onset Pendant)

Metric measuring tapes (e.g., 100 m) or laser range finder

Calibration equipment (large bucket to hold sondes, air pump, air stone, barometer)

Equipment for propane or SF$_6$ additions (see Chapters 30 and 31)

Computer with Excel and R

ACKNOWLEDGMENTS

Alison Appling commented on an earlier draft of this chapter. NSF grant EPS-1208909 partially supported this work.

REFERENCES

Asher, W., Karle, L., Higgins, B., 1997. On the differences between bubble-mediated air-water transfer in freshwater and seawater. Journal of Marine Research 55, 813–845.

Baines, S.B., Pace, M.L., 1991. The production of dissolved organic matter by phytoplankton and its importance to bacteria: patterns across marine and freshwater systems. Limnology and Oceanography: Methods 36, 1078–1090.

Bernot, M.J., Sobota, D.J., Hall, R.O., Mulholland, P.J., Dodds, W.K., Webster, J.R., Tank, J.L., Ashkenas, L.R., Cooper, L.W., Dahm, C.N., Gregory, S.V., Grimm, N.B., Hamilton, S.K., Johnson, S.L., McDowell, W.H., Meyer, J.L., Peterson, B.J., Poole, G.C., Valett, H.M., Arango, C.P., Beaulieu, J.J., Burgin, A.J., Crenshaw, C., Helton, A.M., Johnson, L.T., Merriam, J., Niederlehner, B.R., O'Brien, J.M., Potter, J.D., Sheibley, R.W., Thomas, S.M., Wilson, K., 2010. Inter-regional comparison of land-use effects on stream metabolism. Freshwater Biology 55, 1874–1890.

Bolker, B.M., Gardner, B., Maunder, M., Berg, C.W., Brooks, M., Comita, L., Crone, E., Cubaynes, S., Davies, T., Valpine, P., 2013. Strategies for fitting nonlinear ecological models in R, AD Model Builder, and BUGS. Methods in Ecology and Evolution 4, 501–512.

Bott, T.L., 2006. Primary productivity and community respiration. In: Hauer, F.R., Lamberti, G.A. (Eds.), Methods in Stream Ecology, second ed. Academic Press, San Diego, CA, pp. 533–556.

Bott, T.L., Ritter, F.P., 1981. Benthic algal productivity in a Piedmont stream measured by ^{14}C and dissolved oxygen change procedures. Journal of Freshwater Ecology 1, 267–278.

Bott, T.L., Brock, J.T., Baattrup-Pedersen, A., Chambers, P.A., Dodds, W.K., Himbeault, K.T., Lawrence, J.R., Planas, D., Snyder, E., Woolfardt, G.M., 1997. An evaluation of techniques for measuring periphyton metabolism in chambers. Canadian Journal of Fisheries and Aquatic Sciences 54, 715–725.

Chapra, S.C., Di Toro, D.M., 1991. Delta method for estimating primary production, respiration, and reaeration in streams. Journal of Environmental Engineering 117, 640–655.

Clark, J.S., 2007. Models for Ecological Data: An Introduction. Princeton University Press, Princton, NJ.

Colt, J., 2012. Dissolved Gas Concentrations in Water, second ed. Elsevier, London, UK.

Demars, B., Thompson, J., Manson, J.R., 2015. Stream metabolism and the open diel oxygen method: principles, practice, and perspectives. Limnology and Oceanography: Methods 13, 356–374.

Dodds, W.K., Brock, J.T., 1998. A portable flow chamber for in situ determination of benthic metabolism. Freshwater Biology 39, 49–59.

Drever, J.I., 1988. The Geochemistry of Natural Waters. Prentice-Hall, Englewood Cliffs, NJ.

Duarte, C.M., Prairie, Y.T., 2005. Prevalence of heterotrophy and atmospheric CO_2 emissions from aquatic ecosystems. Ecosystems 8, 862–870.

Fellows, C.S., Valett, M.H., Dahm, C.N., 2001. Whole-stream metabolism in two montane streams: contribution of the hyporheic zone. Limnology and Oceanography: Methods 46, 523–531.

Fisher, S.G., Carpenter, S.R., 1976. Ecosystem and macrophyte primary production of the Fort River, Massachusetts. Hydrobiologia 49, 175–187.

Garcia, H., Gordon, L., 1992. Oxygen solubility in seawater: better fitting equations. Limnology and Oceanography 37, 1307–1312.

Genzoli, L., Hall, R.O., 2016. Shifts in Klamath River metabolism following a reservoir cyanobacterial bloom. Freshwater Science 35, 795–809.

Halbedel, S., Büttner, O., 2014. MeCa, a toolbox for the calculation of metabolism in heterogeneous streams. Methods in Ecology and Evolution 5, 971–975.

Hall, R.O., Beaulieu, J.J., 2013. Estimating autotrophic respiration in streams using daily metabolism data. Freshwater Science 32, 507–516.

Hall, R.O., Tank, J.L., Baker, M.A., Rosi-Marshall, E.J., Hotchkiss, E.R., 2016. Metabolism, gas exchange, and carbon spiraling in rivers. Ecosystems 19, 73–86.

Hall, R.O., Tank, J.L., 2005. Correcting whole-stream estimates of metabolism for groundwater input. Limnology and Oceanography: Methods 3, 222–229.

Hanson, P.C., Carpenter, S.R., Kimura, N., Wu, C., Cornelius, S.P., Kratz, T.K., 2008. Evaluation of metabolism models for free-water dissolved oxygen methods in lakes. Limnology and Oceanography: Methods 6, 454–465.

Hilborn, R., Mangel, M., 1997. The Ecological Detective: Confronting Models with Data. Princeton University Press, Princeton, NJ.

Holtgrieve, G.W., Schindler, D.E., Branch, T.A., A'mar, Z.T., 2010. Simultaneous quantification of aquatic ecosystem metabolism and reaeration using a Bayesian statistical model of oxygen dynamics. Limnology and Oceanography 55, 1047–1063.

Holtgrieve, G.W., Schindler, D.E., Jankowski, K., 2016. Comment on Demars et al. 2015, "Stream metabolism and the open diel oxygen method: principles, practice, and perspectives". Limnology and Oceanography: Methods 14, 110–113.

Hornberger, G.M., Kelly, M.G., 1975. Atmospheric reaeration in a river using productivity analysis. Journal of the Environmental Engineering Division 101, 729–739.

Hotchkiss, E.R., Hall, R.O., 2014. High rates of daytime respiration in three streams: use of $\delta^{18}O_{O2}$ and O_2 to model diel ecosystem metabolism. Limnology and Oceanography: Methods 59, 798–810.

Hotchkiss, E.R., Hall, R.O., 2015. Whole-stream ^{13}C tracer addition reveals distinct fates of newly fixed carbon. Ecology 96, 403–416.

Hotchkiss, E.R., Hall, R.O., Sponseller, R.A., Butman, D., Klaminder, J., Laudon, H., Rosvall, M., Karlsson, J., 2015. Sources of and processes controlling CO_2 emissions change with the size of streams and rivers. Nature Geoscience 8, 696–699.

Jähne, B., Münnich, K.O., Dutzi, R.A., Huber, W., Libner, P., 1987. On the parameters influencing air-water gas exchange. Journal of Geophysical Research 92, 1937–1949.

Lovett, G.M., Cole, J.J., Pace, M.L., 2006. Is net ecosystem production equal to ecosystem carbon accumulation? Ecosystems 9, 152–155.

Marcarelli, A.M., Baxter, C.V., Mineau, M.M., Hall, R.O., 2011. Quantity and quality: unifying food web and ecosystem perspectives on the role of resource subsidies in freshwaters. Ecology 92, 1215–1225.

Marzolf, E.R., Mulholland, P.J., Steinman, A.D., 1994. Improvements to the diurnal upstream-downstream dissolved oxygen change technique for determining whole-stream metabolism in small streams. Canadian Journal of Fisheries and Aquatic Sciences 51, 1591–1599.

Minshall, G., 1978. Autotrophy in stream ecosystems. BioScience 28, 767–770.

Mulholland, P.J., Fellows, C.S., Tank, J.L., Grimm, N.B., Webster, J.R., Hamilton, S.K., Martí, E., Ashkenas, L., Bowden, W.B., Dodds, W.K., McDowell, W.H., Paul, M.J., Peterson, B.J., 2001. Inter-biome comparison of factors controlling stream metabolism. Freshwater Biology 46, 1503–1517.

Odum, H.T., 1956. Primary production in flowing waters. Limnology and Oceanography 1, 102–117.

Odum, H.T., Odum, E.P., 1955. Trophic structure and productivity of a windward coral reef community on Eniwetok Atoll. Ecological Monographs 25, 291–320.

Peterson, B.J., 1980. Aquatic primary productivity and the $^{14}C-CO_2$ method: a history of the productivity problem. Annual Review of Ecology and Systematics 11, 359–385.

R Development Core Team, 2016. R: A Language and Environment for Statistical Computing. R Foundation for Statistical Computing, Vienna, Austria.

Raymond, P.A., Zappa, C.J., Butman, D., Bott, T.L., Potter, J., Mulholland, P., Laursen, A.E., McDowell, W.H., Newbold, J.D., 2012. Scaling the gas transfer velocity and hydraulic geometry in streams and small rivers. Limnology and Oceanography: Fluids & Environments 2, 41–53.

Reichert, P., Uehlinger, U., Acuña, V., 2009. Estimating stream metabolism from oxygen concentrations: effect of spatial heterogeneity. Journal of Geophysical Research - Biogeosciences 114, G03016.

Roberts, B.J., Mulholland, P.J., Hill, W.R., 2007. Multiple scales of temporal variability in ecosystem metabolism rates: results from 2 years of continuous monitoring in a forested headwater stream. Ecosystems 10, 588–606.

Sargent, M.C., Austin, T.S., 1949. Organic productivity of an Atoll. Transactions of the American Geophysical Union 30, 245–249.

Steeman Nielsen, E., 1952. The use of radio-active carbon (C^{14}) for measuring organic production in the sea. Journal du Conseil Permanent International pour l'Exploration de la Mer 18, 117–140.

Thorp, J.H., Delong, M.D., 2002. Dominance of autochthonous autotrophic carbon in food webs of heterotrophic rivers. Oikos 96, 543–550.

Tobias, C.R., Böhlke, J.K., Harvey, J.W., 2007. The oxygen-18 isotope approach for measuring aquatic metabolism in high-productivity waters. Limnology and Oceanography 52, 1439–1453.

Tobias, C.R., Böhlke, J.K., Harvey, J.W., 2009. A simple technique for continuous measurement of time-variable gas transfer in surface waters. Limnology and Oceanography: Methods 7, 185–195.

Tsivoglou, E.C., Neal, L.A., 1976. Tracer measurement of reaeration: III. Predicting the reaeration capacity of inland streams. Journal of the Water Pollution Control Federation 48, 2669—2689.

Wanninkhof, R., 1992. Relationship between wind speed and gas exchange over the ocean. Journal of Geophysical Research 97, 7373—7382.

Wanninkhof, R., Mulholland, P.J., Elwood, J.W., 1990. Gas exchange rates for a first-order stream determined with deliberate and natural tracers. Water Resources Research 26, 1621—1630.

Young, R.G., Matthaei, C.D., Townsend, C.R., 2008. Organic matter breakdown and ecosystem metabolism: functional indicators for assessing river ecosystem health. Journal of the North American Benthological Society 27, 605—625.

Zuur, A.F., Ieno, E.N., Meesters, E., 2009. A Beginner's Guide to R. Springer-Verlag, New York.

Chapter 35

Secondary Production and Quantitative Food Webs

Arthur C. Benke and Alexander D. Huryn

Department of Biological Sciences, University of Alabama

35.1 INTRODUCTION

Secondary production is the formation of heterotrophic biomass *through time* (e.g., Benke, 1993). Annual secondary production, for example, is the sum of all biomass produced ("production") by a population during one year. This includes biomass remaining at the end of the year, and all production lost from the aquatic environment during this period. This chapter focuses on aquatic invertebrates, but the same principles and methods can apply to vertebrates as well. When working with invertebrates, losses may include mortality (e.g., disease, parasitism, cannibalism, predation), loss of tissue reserves (e.g., molting, silk, starvation), emigration, and emergence. The relationship between production and related bioenergetic parameters can be represented by the familiar equations:

$$I = A + F \tag{35.1}$$

and

$$A = P + R + U \tag{35.2}$$

where I = ingestion, A = assimilation, F = food that is defecated (egestion), P = production, R = respiration, and U = excretion (e.g., Calow, 1992). Food that is assimilated thus contributes to production (P), respiration (R), and excretion (U). Each of these is fluxes (or flows) of materials or energy, with units of mass or energy area^{-1} time^{-1}. At the level of the individual, P represents growth, whereas at the level of the population it represents the collective growth of all individuals in the population. Obviously, how much an organism grows depends on how much it eats, but growth also depends on how efficiently that food is converted to new tissue. Two characteristics of an organism's bioenergetics determine this efficiency: assimilation efficiency (AE = A/I) and net production efficiency (NPE = P/A). The product of AE and NPE is gross production efficiency (GPE), which converts ingestion to production in one step. Among stream macroinvertebrates, assimilation efficiency is likely to be the most variable term, ranging from less than 5% for detritivores to almost 90% for carnivores (Benke and Wallace, 1980). NPE for macroinvertebrates shows less variation and is often close to 50%. Thus, a detritivore might convert only 2% −3% ($\approx 0.05 \times 0.5$) of its food to production, whereas, a predator might convert as much as 45% ($\approx 0.9 \times 0.5$).

Historically, different kinds of units have been used to represent secondary production (e.g., kcal m^{-2} year^{-1}, grams wet mass m^{-2} year^{-1}, grams dry mass m^{-2} year^{-1}). However, most studies in recent decades have used mass units. For studies of macroinvertebrates in particular, dry mass [or ash-free dry mass (AFDM)] is the norm. Carbon units, as in primary production studies (Chapter 34) are rarely used. Nonetheless, standard conversions are available. For example, Waters (1977) suggested using 1 g dry mass $\approx$ 6 g wet mass $\approx$ 0.9 g AFDM $\approx$ 0.5 g C $\approx$ 5 kcal $\approx$ 21 kJ. More recently, Benke et al. (1999) presented data showing that 1 g dry mass ranges from 0.91 to 0.96 g AFDM among major insect orders, but values for mollusks and decapods were higher. They also suggested that 1 g AFDM (rather than 1 g dry mass) $\cong$ 0.5 g C. Mass units can also be converted into elemental units besides C such as P, N, and heavy metals (e.g., Cross et al., 2007; Singer and Battin, 2007; Chapter 36).

There are now many estimates of annual production for entire communities of stream macroinvertebrates (see reviews of Benke, 1993; Benke and Huryn, 2010). These range from $\sim$2 to >100 g dry mass m^{-2} year^{-1}, with the majority of values being $\leq$20 g dry mass m^{-2} year^{-1} (Fig. 35.1).

Methods in Stream Ecology. http://dx.doi.org/10.1016/B978-0-12-813047-6.00013-9

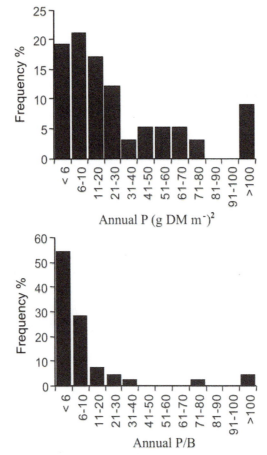

FIGURE 35.1 Frequency distribution of annual production and production/biomass (*P/B*s) for 58 estimates of benthic macroinvertebrate production for streams and rivers worldwide. Sources and actual values for these estimates are provided in Benke (1993).

35.1.1 Biomass Turnover and the *P/B* Concept

To appreciate the concept of secondary production, it is important to understand the relationship between production and biomass. Biomass (*B*) is a measurement of how much living tissue mass for a population is present at one instant in time (or averaged over several periods of time), and its units are mass (or energy) per unit area (e.g., g/m^2) (Benke, 1993). Production, on the other hand, is a flow (e.g., g m^{-2} year^{-1}). Production divided by biomass (*P/B*) is therefore a rate, with units of inverse time (e.g., year^{-1}). Since any unit of time can be selected for estimating a rate, we can calculate annual *P/B*, weekly *P/B*, daily *P/B*, etc. *P/B* is essentially a weighted mean value of biomass growth rates of all individuals in the population. Alternatively, a *cohort P/B* is defined as production of a population over its *life span* divided by the mean biomass over this same time period. A convenient property of the cohort *P/B* is that it has a relatively constant value of about 5 (range usually 3–8). Because it is calculated over a variable period of time (i.e., life span), it is a ratio (unitless) rather than a rate.

Annual P/B values of benthic invertebrates were once thought to vary from only about 1–10 (Waters, 1977), and this is probably true for several groups. For example, a population with a life span of 1 year (i.e., univoltine) will have an annual *P/B* of about 5, almost identical to the cohort *P/B*. A bivoltine population will have an annual *P/B* of about 10. However, much higher values (approaching or exceeding 100) have now been shown for at least some of the dipterans and mayflies which have very short development times (e.g., Jackson and Fisher, 1986; Benke, 1998; Ramírez and Pringle, 2006). High *P/B* values are also possible for meiofauna (Tod and Schmidt-Araya, 2009; also see Chapter 14). In contrast, organisms with life spans >5 years, such as some fishes, crayfishes, and mollusks can have *P/B* values <1. Annual *P/B* values are thus a direct function of the development time of a population and values for individual populations have been shown to vary from <0.1 to >200 (see Table 1 in Huryn and Wallace, 2000). Thus, knowing a population's *P/B* value is almost as important as knowing production itself. Annual *P/B* values estimated for entire communities of stream macroinvertebrates have almost as wide a range from <1 to >100, with most being <6 (Fig. 35.1).

35.1.2 Utility of Secondary Production in Ecosystem Studies

Understanding factors determining levels and limits of ecosystem production is a central goal in ecosystem ecology. It should therefore not be surprising that studies of secondary production of stream invertebrates have figured prominently in the development of stream ecosystem theory. Since the pioneering monograph by Allen (1951) and the seminal review by Waters (1977), studies of secondary production in streams have focused on a diversity of ecological questions (e.g., Benke and Huryn, 2010; Dolbeth et al., 2012). These studies fall into three general categories: those simply documenting levels of production of populations and communities (reviews by Waters, 1977; Benke, 1984, 1993; Huryn and Wallace, 2000), those attempting to determine physical and biological factors controlling levels and limits to production (Benke, 1984; Huryn and Wallace, 1987; Huryn, 1998), and those that have used estimates of production as a metric for assessing some aspect of the bioenergetic performance of a population or its interactions with other members of its community (e.g., Benke and Wallace, 1980, 1997; Ross and Wallace, 1981, 1983; Georgian and Wallace, 1983; Wallace and O'Hop, 1985; Short et al., 1987; Plante and Downing, 1989; Benke and Jacobi, 1994).

Much of the research on secondary production has involved empirically based inductive approaches. Within the last three decades, however, production has been used with increasing frequency as a response variable in stream ecosystem experiments (e.g., Wallace and Gurtz, 1986; Lugthart and Wallace, 1992; Peterson et al., 1993; Hall et al., 2000, 2011; Cross et al., 2006, 2007; Hannesdóttir et al., 2013; Ledger et al., 2011, 2013; Wallace et al., 2015) and in studies of the effects of land use on stream communities (Sallenave and Day, 1991; Shieh et al., 2002, 2003; Carlisle and Clements, 2003; Woodcock and Huryn, 2007, 2008; Gücker et al., 2011; Johnson et al., 2013b). Estimates of secondary production have been particularly effective in these applications because it integrates several other components of ecological performance—density, biomass, individual growth rate, reproduction, survivorship, and development time (Benke, 1993).

35.2 GENERAL DESIGN

Studies of secondary production in streams usually encompass the habitat or reach scale. In the first case, a specific habitat within a study reach is usually sampled (e.g., snag or riffle), and the units of production are reported per area of habitat. In the second, all major habitats within a reach are sampled, and units are reported per area of reach. The appropriate reach length will vary depending upon the purpose of a study, but generally depends on two considerations. The first is habitat structure—all major habitats should be represented in repeated and discrete patches so that variability among habitat patches will be incorporated into the sampling design. Second, the reach length should be long enough to ensure that migration and emigration of individuals during the study will be minimal. For most studies of invertebrate production in wadeable streams, reaches in the range of 50–500 m in length are probably sufficient. However, reach length must be considered more carefully in systems where species show migratory behavior, such as freshwater shrimp in the neotropics.

35.2.1 Population Density

Estimates of secondary production—regardless of approach—require accurate measurement of population density and size structure. Sampling methods used to estimate density are usually based on some form of quadrat sampling. Depletion removal methods have also been used for crayfish (Rabeni et al., 1997; Whitmore and Huryn, 1999). The most appropriate type of sampler will depend on substratum type. A Surber or Hess sampler might be used in a cobble area or on a flat bedrock habitat. A petite ponar grab or corer might work best in shallow gravel or sand (e.g., Ogeechee River corer, Gillespie et al., 1985, manufactured by Wildco, Inc.). As with any quantitative sampling, replication is necessary to obtain accurate density estimates. The distribution of stream biota is extraordinarily patchy, and this will usually be the greatest contributor to both imprecision and inaccuracy of production estimates. A sufficient number of samples (i.e., from 4 to 6 samples typically have been used) should thus be taken to ensure accuracy and to maintain the statistical power of the study to an appropriate level.

Study designs range from completely randomized sampling (habitat and reach scale), to sampling stratified by habitat (reach scale). In the latter case, reach-scale estimates of production can be obtained by calculating production separately for each habitat, quantifying the relative cover of the different habitats, weighting habitat-specific production by the relative area of the habitat, and summing these estimates (e.g., Huryn and Wallace, 1987; Smock et al., 1992). This latter approach requires fewer samples to attain a given level of precision than completely randomized designs, but requires accurate identification and delineation of habitats.

Most methods used to estimate production require repeated sampling of density over the entire developmental cycle of the target population. Samples are taken monthly in most studies; a schedule that is logical for invertebrates with annual life cycles. This schedule may be a useful compromise when estimating community production for temperate streams as well as for many taxa in tropical streams. Monthly sampling will result in poor resolution of the population dynamics of organisms with short life cycles, however. For studies focusing on such taxa (e.g., *Siphlonisca aerodromia*, Huryn, 2002), samples taken at weekly intervals may be required. In cases where growth and development are not synchronous, the sampling schedule is of less concern if steady state biomass can be assumed (see noncohort methods for further considerations for the analysis of such taxa). At the other end of the spectrum, seasonal or even annual sampling may be adequate for long-lived invertebrate taxa (>1 year; e.g., snails, Huryn et al., 1995; crayfish, Whitmore and Huryn, 1999), as has sometimes been employed for fish production studies (Waters et al., 1990).

In temperate streams, short intervals between sampling in spring and summer and long intervals between sampling during winter may be advisable because higher growth rates, and often the bulk of production, typically occur during the warmer months (e.g., Benke and Jacobi, 1994; Johnson et al., 2000). Caution is required for community studies; however, because some important taxa are bioenergetically active only during winter and early spring (Benke and Jacobi, 1994). Probably the best approach for accurate estimates of community production is to combine several sampling approaches that are optimal for populations suspected to be major contributors to total system production. Ideally, a thorough knowledge of the life histories of different taxa within a community will allow the planning of a sampling regime that will provide the best accuracy for a given effort, but such information is often not available in advance. If you do not know this in advance, then you can obtain general knowledge of expected life histories by searching the literature.

35.2.2 Population Size Structure

Population size structure refers to the density of individuals within different size classes of a population. For the purpose of estimating production, breaking down a population into size classes is essential for applying methods used in estimating growth and the loss of individuals over time due to mortality, as well as providing a convenient way for estimating biomass. Size classes can be defined arbitrarily on the basis of body length or head capsule width, or they can be based on criteria such as instar or developmental indicators (e.g., appearance of histoblasts, etc.). The use of length classes is both effective and convenient, however. Length can be measured very precisely using an ocular micrometer or less precisely using a sheet of 1 mm graph paper placed directly on the microscope stage. The latter approach allows the rapid sorting of individuals into length classes that are suitable for most methods used to estimate production.

35.2.3 Individual and Population Biomass

To calculate production by any method, it is essential that biomass is determined. The product of length-specific mass (mg/individual) and density (number of individuals/m^2) within a length size class yields an estimate of size-specific biomass (mg/m^2). The sum of biomass for all size groups is population biomass.

The relationship between individual length and mass for a given taxon is often obtained from a length–mass relationship. Nonpreserved (fresh) animals collected for this purpose provide the best results since preservation (especially in ethanol) results in shrinkage of soft body parts and losses of dry mass by leaching. Animals preserved in a formalin solution will provide estimates comparable to nonpreserved specimens. The procedure involves measuring the lengths of individual animals from a wide range of size categories under a dissecting microscope. The eyepiece must be fitted with a micrometer so that lengths can be measured to at least 0.1 mm. Subsequently, the measured individuals are dried, usually in a drying oven for a minimum of 24 h at 60°C, cooled in a desiccator, and weighed on an analytical balance with acceptable precision. It is best to have at least 20 measurements. A linear regression is then developed of the form

$$\ln W = \ln a + b \ln L \tag{35.3}$$

where W = individual mass, L = length, a = a constant, and b = slope of the regression. This equation is the linear equivalent of a power curve, $W = aL^b$. Since we expect a cubic relationship between L and W, b should be reasonably close to 3 (Benke et al., 1999).

In the absence of time or equipment to determine a length–mass relationship, one can use literature values to obtain length-specific mass. Benke et al. (1999) updated and added to the useful equations of Smock (1980) in summarizing relationships for benthic insects, crustaceans, and mollusks from North America and usually to the genus or species level. Order-level equations from Benke et al. are presented in Table 35.1. Additional equations for aquatic insects are presented by Johnston and Cunjak (1999) for northeastern North America; Meyer (1989), Beerstiller and Zwick (1995), and Burgherr and Meyer (1997) for Europe; Towers et al. (1994) for New Zealand; and Miyasaka et al. (2008) for Japan.

TABLE 35.1 Mean values of coefficients *a* and *b* from length—mass regressions for major insect and crustacean (Decapoda and Amphipoda) orders using total length, except for Decapoda (carapace length). W = dry mass (mg); L = body length (mm); n = number of equations from which mean *a* and *b* were obtained. Values of *a* and *b* did not significantly differ among insect orders and Amphipoda due to interspecific variability within orders; Decapoda values of *a* and *b* differed significantly from all others. Note that all values of *b* are relatively close to 3.

Order	*n*	*a*	*b*
Decapoda	9	0.0147	3.626
Amphipoda	7	0.0058	3.015
Coleoptera	9	0.0077	2.910
Diptera	43	0.0025	2.692
Ephemeroptera	54	0.0071	2.832
Hemiptera	4	0.0108	2.734
Megaloptera	7	0.0037	2.838
Odonata	18	0.0078	2.792
Plecoptera	36	0.0094	2.754
Trichoptera	34	0.0056	2.839

Modified from Table 2 in Benke et al. (1999).

35.3 SPECIFIC METHODS

The methods for estimating production can be divided into two basic categories: cohort and noncohort (Waters, 1977; Benke, 1984, 1993). Cohort techniques may be used when it is possible to follow a cohort (i.e., individuals that hatch from eggs within a reasonably short time span and grow at about the same rate) through time. When a population's life history is more complex, a noncohort technique often must be used. Other approaches described below deal with using short cuts, applying statistical methods, and developing quantitative food webs. All of these techniques require quantitative collection (i.e., number per square meter) of the macroinvertebrate species for which estimates are made. See Chapter 15 and Merritt et al. (2008) for quantitative collection techniques.

35.3.1 Cohort Techniques

As a cohort develops through time, a general decrease in density (N), due to mortality, and an increase in individual mass (W), due to growth, occurs (Fig. 35.2). Interval production (i.e., time between two sampling dates) is easily calculated directly from field data by the *increment-summation method* as the product of the mean density between two sampling dates $(\overline{N})$ and the increase in individual mass ΔW (i.e., $\overline{N} \times \Delta W$). Assuming there is only one generation per year, annual production is calculated as the sum of all interval estimates, plus the initial biomass:

$$P = B_{\text{initial}} + \sum \overline{N} \Delta W \qquad (35.4)$$

The initial biomass (B_{initial}) represents an approximation of production that has accumulated before the first sampling date.

However, if one wants to examine production patterns throughout the year, mean daily production for an interval can be calculated by dividing each $\overline{N} \Delta W$ by the days in the interval. This converts interval production into a true flow (i.e., g m^{-2} day^{-1}). A study of the stream caddisfly *Brachycentrus spinae* provides an especially clear example for illustrating the calculation of production using a cohort method, such as the increment-summation method (Table 35.2, modified from Ross and Wallace, 1981). Ross and Wallace (1981) used the *instantaneous growth method* (see below), which provides production estimates very similar to those in Table 35.2. Note that the cohort P/B ratio is close to 5 (i.e., 5.96).

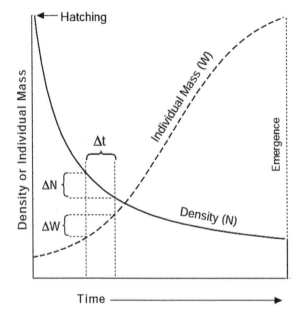

FIGURE 35.2 Hypothetical cohort of a stream insect showing curves of individual growth in mass (*W*) and population mortality (*N*). *Adapted from Benke (1984)*.

In addition to the increment-summation method, there are three closely related ways of calculating production using cohort data that should give very similar results (e.g., Waters, 1977; Gillespie and Benke, 1979; Benke, 1984). The *removal-summation method* is most similar to the increment-summation method, but calculates production *lost* during the sampling interval as the product of the decrease in density (ΔN, Fig. 35.2) and the mean individual mass $(\overline{W})$ over the interval (i.e., $\overline{W} \times \Delta N$ rather than $\overline{N} \times \Delta W$). Adding the increase in biomass (ΔB) between sampling dates to the production loss equals interval production as calculated above (i.e., $\overline{W}\Delta N + \Delta B = \overline{N}\Delta W$ for any interval). The *Allen curve method* is a graphical approach (Fig. 35.3) in which the area under a curve of density versus mean individual mass approximates total production of a cohort (Allen, 1951; Waters, 1977; Gillespie and Benke, 1979; Benke, 1984). The Allen curve also illustrates the relationship between biomass (*B*) and the changes in numbers (ΔN) and individual mass (ΔW) over sampling intervals that are used in the tabular methods (e.g., Table 35.2). For example, $Y + Z$ (Fig. 35.3) $\cong \overline{N}\Delta W$ in the increment-summation method (Table 35.2). The *instantaneous growth method* can also be used to calculate production during a sampling interval (see below).

35.3.2 Noncohort Techniques: Size—Frequency Method

When a population cannot be followed as a cohort from field data, it is necessary to use a noncohort method to estimate production. These methods require independent approximations of either development time or biomass growth rates. The *size—frequency method* (Hynes and Coleman, 1968; Hamilton, 1969; Benke, 1979) assumes that a mean size—frequency distribution determined from samples collected throughout the year approximates a mortality curve for an *average cohort*. A study of the stream mayfly *Tasmanocoenis tonnoiri* provides a good illustration of this method (Table 35.3, modified from the data of Marchant, 1986). The decrease in density (ΔN) from one size (i.e., length) category to the next is multiplied by the mean mass between size categories $(\overline{W})$, using the same rationale as for the removal-summation method. Before summing the products (i.e., $\overline{W}\Delta N$) for each size class, each value should be multiplied by the total number of size classes (Table 35.3, final column). This is done because it is assumed that there is a total development time of one year, and that there is the same number of cohorts during the year as size classes (see Hamilton, 1969 or Benke, 1984 for a more complete rationale). Cohort *P/B* is equal to the sum of the final column (i.e., production assuming a 1-year life span) divided by the sum of the biomass column. In this particular case, the cohort *P/B* (9.5) is considerably higher than usually expected (5), due to the fact that a very small fraction of the population survived to the larger size classes.

If development time is much different than a year, it is necessary to apply a correction factor to the basic size—frequency calculation; sum of final column of Table 35.3. This involves multiplication by 365/CPI where CPI (i.e., cohort production interval) is the mean development time in days from hatching to final size (Benke, 1979). In the example of

TABLE 35.2 Calculation of annual and daily production of *Brachycentrus spinae* using the increment-summation method.

Date	Density (No./m²) N	Individual Mass (mg) W	Biomass (mg/m²) $N \times W$	Individual Growth (mg) $\Delta W = W_2 - W_1$	Mean $\overline{N}$ (No./m²) $(N_1 + N_2)/2$	Interval P (mg/m²) $\overline{N}\Delta W$	Daily P (mg m⁻² day⁻¹) $\overline{N}\Delta W/\Delta t$
May 18	282.9	0.021	5.9				
				0.036	254.9	9.17	0.66
June 01	226.8	0.057	12.9				
				0.031	204.4	6.33	0.53
June 13	181.9	0.088	16.0				
				0.085	160.2	13.62	0.85
June 29	138.5	0.173	24.0				
				0.179	123.9	22.17	1.58
July 13	109.2	0.352	38.4				
				0.588	98.4	57.83	4.45
July 26	87.5	0.940	82.3				
				0.266	73.9	19.64	0.89
August 17	60.2	1.206	72.6				
				0.590	54.3	32.01	2.46
August 30	48.3	1.796	86.7				
				0.025	42.6	1.06	0.07
September 15	36.8	1.821	67.0				
				1.378	32.0	44.03	2.45
October 03	27.1	3.199	86.7				
				0.358	20.1	7.18	0.17
November 14	13.0	3.557	46.2				
				1.074	10.9	11.71	0.51
December 07	8.8	4.631	40.8				
				2.222	6.2	13.83	0.27
January 27	3.8	6.853	26.0				
				1.624	3.1	5.08	0.24
February 17	2.6	8.477	22.0				
				3.071	2.2	6.60	0.28
March 13	1.7	11.548	19.6				
				3.252	0.9	2.76	0.06
April 27	0.0	14.800	0.0				
				Annual P	$= 5.9 +$	253.03	$= 258.93$
		Cohort B	$= 43.4$		Cohort P/B	$= 5.96$	
		Annual B	$= 39.8$		Annual P/B	$= 6.51$	

B = mean biomass; $\overline{N}$ = mean density between two consecutive dates; P = production. Annual production is calculated by adding the sum of the interval production column and the biomass estimated on the first sampling date. Mean biomass was estimated from monthly means since the sampling regime involved both monthly and bimonthly samples. Thus, mean cohort biomass was for 11 months and mean annual biomass for 12 months. Data from Ross and Wallace (1981).

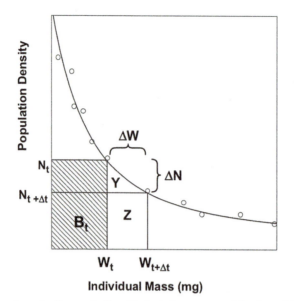

FIGURE 35.3 Hypothetical Allen curve for estimating production. *Circles* indicate means of density and individual mass from samples. Curve is smoothed to provide an approximate fit to the points. Production is equal to the area under the curve. Note that W_t, N_t, ΔW, ΔN, and B_t correspond to the same terms in Table 35.2.

Table 35.3, Marchant estimated a mean CPI of 5 months based upon his interpretation of life histories from size—frequency histograms. Annual production is thus calculated, using months rather than days, as $352.5 \times 12/5 = 846.1$ mg m^{-2} year^{-1}, with an annual P/B of 22.9. These estimates are somewhat different than that found by Marchant since he used a geometric rather than a linear calculation of mean individual mass between size categories (mass at loss, sixth column of Table 35.3). Some investigators argue that geometric means provide more accurate estimates of individual biomass over a given time interval because growth is usually exponential rather than linear. The use of geometric versus linear means is usually a matter of preference, however, because the former generally provides only slightly lower values than the latter. Shorter CPIs (e.g., 30 days) require even greater corrections (i.e., $365/30 \cong 12$). If it is not possible to approximate CPI from field data, as done by Marchant, it is necessary to obtain this information from populations reared in the laboratory or in the field. A final point is that CPI is inversely related to biomass turnover rates (i.e., daily or annual P/B). For example, if CPI = 30 days (a relatively short time), then annual P/B = cohort $P/B \times 365/\text{CPI} \approx 5 \times 365/30 \approx 60$ (a relatively high value). Benke (1993) noted that the *size—frequency* method has been used more than any other production method for stream invertebrates and this is probably still true.

35.3.3 Noncohort Techniques: Instantaneous Growth Rate Method

The second noncohort technique is the *instantaneous growth rate method*. It involves the calculation of a daily instantaneous growth rate:

$$g = \frac{\ln(W_{t+\Delta t}/W_t)}{\Delta t} \tag{35.5}$$

where W_t = mean mass of an individual at time t, $W_{t+\Delta t}$ = mean mass of an individual at time $t + \Delta t$, and Δt = length of the time interval. Daily production (P_d) is calculated as:

$$P_d = g \times \overline{B} \tag{35.6}$$

where $\overline{B}$ = mean population biomass for two consecutive dates in units of g/m^2 (e.g., Benke and Parsons, 1990; Benke and Jacobi, 1994). Unlike the size—frequency method, the instantaneous growth rate method is valuable for tracking changes in production over time (e.g., Georgian and Wallace, 1983; Benke, 1998).

When applying the instantaneous growth rate method as a *cohort* approach, g may be estimated directly from changes in average cohort biomass between sampling dates using Eq. (35.5). The effect of sample error on estimates of growth rate between sampling intervals may result in negative values for g if growth rates are low. The problem can be eliminated by regressing mean individual mass against days since hatching and using a continuous exponential model to estimate W_t and

TABLE 35.3 Calculation of annual production of *Tasmanocoenis tonnoiri* using the size–frequency method.

Length (mm)	Density (No./m²)	Individual Mass (mg)	No. Lost (No./m²)	Biomass (mg/m²)	Mass at Loss (mg)	Biomass Lost (mg/m²)	Times No. Size Classes
	N	W	ΔN	$N \times W$	$\overline{W} = W_1 + W_2/2$	$\overline{W}\Delta N$	$\overline{W}\Delta N \times 6$
0.5	706.0	0.001	−142.0	0.71	0.011	−1.491	(−8.95)[a]
1.5	848.0	0.02	730.0	16.96	0.050	36.5	219.00
2.5	118.0	0.08	72.0	9.44	0.130	9.36	56.16
3.5	46.0	0.18	42.0	8.28	0.265	11.13	66.78
4.5	4.0	0.35	3.7	1.40	0.435	1.61	9.66
5.5	0.3	0.52[b]	0.3	0.16	0.520[b]	0.16	0.94
			Biomass	= 36.94		Production (Uncorrected)	= 352.5
			Cohort *P/B*	= 9.5			
			Annual *P/B*	= 22.9		Annual *P*[c] (Prod. × 12/5)	= 846.1

The density column (the average cohort) is the mean value from samples taken throughout the year. B = mean biomass; P = production; $\overline{W}$ = mean individual mass between two size classes.

[a] Negative value at top of table (right column) disregarded since it is probably an artifact caused by inefficient sampling of smallest size class or rapid growth through size interval. If negative values are found below a positive value (not shown in example), they should be included in the summation.

[b] Final "mass at loss" should be equal to individual mass of the largest size class.

[c] Annual production is calculated by multiplying "uncorrected production" by a cohort production interval (CPI) correction factor (12 mo/CPI), where CPI = 5 mo (see text).

Data from Marchant (1986).

$W_{t+\Delta t}$. Another potential source of error when using this approach occurs late in cohort development of aquatic insects when apparent mean size decreases due to the early emergence of large individuals. Unless this latter source of error is accounted for, the sample method results in underestimates of g which can lead to large errors in production because population biomass is often greatest shortly before emergence.

When using the instantaneous growth method as a *noncohort* approach, g is estimated from animals grown in the laboratory or in the field (e.g., Huryn and Wallace, 1986; Hauer and Benke, 1987, 1991; Cross et al., 2005). For very large invertebrates, such as snails and crayfish, growth rates can be measured using mark and recapture of free-ranging individuals (branding, tagging, tattooing; Huryn et al., 1995; Whitmore and Huryn, 1999; Venarsky et al., 2014). An alternative procedure that can be used for many different taxa is through the use of in situ growth chambers. Larvae of various sizes are confined in chambers, which can be made from short lengths of plastic tubing (i.d. = 7.7 cm) capped with 63–500 μm mesh. The mesh can be attached by gluing or with cable ties. The fine mesh has been used for chironomid larvae (Huryn and Wallace, 1986), the larger for mayflies (Leptophlebiidae, Siphlonuridae; Huryn, 1996a, 2002). Chambers may be anchored directly to the stream bottom. In habitats where oxygen may reach low levels (e.g., floodplain swamps), they should be supported on foam floats in such a way that a portion of the mesh on either end of the chamber will be submerged regardless of fluctuating water levels (Huryn, 2002). Chambers may be loosely packed with conditioned detritus or pebbles coated with biofilm to provide food (Huryn, 1996a, 2002). Alternatively, the chambers may be deployed for 2–3 weeks prior to stocking to allow biofilm to grow on their walls. All of these approaches have been used successfully in both streams and wetlands (Huryn and Wallace, 1986; Huryn, 1996a, 2002). Once chambers are prepared, individual larvae or groups of even-sized larvae, representing the range available for a given taxon, are measured and placed into the growth chambers. At appropriate intervals (e.g., weekly), larvae are removed, their lengths recorded, and new individuals placed in the chambers. g is calculated using Eq. (35.5) and further equations estimating g as a function of water temperature and individual mass may be derived using regression models (e.g., Huryn and Wallace, 1986; Hauer and Benke, 1987, 1991). In cases where both the cohort and noncohort approaches can be applied simultaneously, the accuracy of growth rates obtained from confined individuals may be assessed by comparing plots of predicted growth trajectories based on the regression equations, to size–frequency data from the field (Huryn, 2002).

The most accurate estimates of production will probably be obtained when it is possible to determine size-specific growth rates due to differences in growth rates between size classes. Production of the ith size class is $P_i = g_i \times \overline{B}_i$, where g_i and $\overline{B}_i$ are the growth rate and mean biomass of the ith size class, respectively, and total daily production (P_d) is the sum of the production of all size classes,

$$P_d = g_1 B_1 + g_2 B_2 + \ldots + g_i B_i \tag{35.7}$$

35.3.4 "Shortcut" Approaches

Given the amount of labor required to directly measure invertebrate production, it is not surprising that shortcut approaches have been developed. Of the several offered (Benke, 1984), we believe that three are particularly useful. The first is based on the annual P/B, which is a rough estimator of the annual biomass growth rate of a population. Waters (1969) showed that the cohort P/B values for invertebrate populations fell within a relatively narrow range (2–8), and he suggested that *cohort production* might be estimated as the product of biomass and a suitable P/B (5 is usually suggested). Thus, if a population is known to be univoltine, an annual P/B of 5 can be used as an approximation. When entire benthic communities are considered, however, the range of P/Bs for *annual production* varies from <1 to >100 because cohorts of different taxa may require periods of several weeks to several years (Benke, 1993). The range of expected annual P/Bs can be narrowed by considering only small temperate streams (e.g., mean discharge ∼0.1–1.0 m³/s and mean annual temperature 5–10°C), where annual P/B's range from 2.2 to 8.7 (10 streams) or from 4.2 to 7.9 (7 streams) (Benke, 1993). On the basis of these prior studies, a rough but reasonable estimate of the expected range of secondary production for a stream with similar characteristics can be obtained as the product of community biomass and a range of annual P/Bs from 2.2 to 8.7 or even from 4.2 to 7.9.

A second useful shortcut approach is based upon meta-analysis using multiple regression models to estimate invertebrate production as a function of more readily measured variables, usually temperature, population biomass, and maximum body size (Morin and Bourassa, 1992; Benke, 1993; Morin and Dumont, 1994; Benke et al., 1998). Empirical models now seem to be used quite frequently for marine benthos (Brey, 1990; Tumbiolo and Downing, 1994; Beukema and Dekker, 2013). However, this approach has been criticized because it is prone to imprecision and inaccuracy, particularly when used to estimate production for single species at a single location (Benke, 1993). On the other hand, the estimation of production for entire communities or groups of species (e.g., functional groups) as summed composites of estimates for single populations may increase the accuracy of this approach (Benke et al., 1998). Indeed, Benke (1993) and Webster et al. (1995) showed that production of invertebrate functional feeding groups (sensu Cummins, 1973) may be estimated as various linear and polynomial functions of stream size (as indicated by mean annual discharge). Although the latter models are not appropriate for estimating production per se, the testing and refinement of meta-analytical approaches merits further research because they may provide biologically reasonable ranges of production at large spatial scales required for multiscale studies of stream ecosystem processes (e.g., among general categories of streams).

A third shortcut method is the use of the biomass of emergent aquatic insects as an indicator of total larval production. A meta-analysis by Statzner and Resh (1993) revealed a statistically significant relationship between emerging biomass and benthic secondary production for 18 streams in Europe ($r^2 = 0.81$, $p < .0001$). Their analysis indicated that adult emergence represented $\sim 24.3\%$ of benthic insect production for these streams. A useful "rule of thumb" for estimating stream insect production for the region represented in the analysis would thus be the product of biomass of emerging insects and a coefficient of "4.1" (i.e., 1/0.243). A subsequent study of a lake outlet stream in Germany by Poepperl (2000) indicated that the mean ratio of emerging insect biomass to larval production was 18.3%, or a coefficient of 5.5, providing some support for the value of 4.1 suggested by Statzner and Resh (1993). Recently, Johnson et al. (2013a) used Poepperl's (2000) 18.3% to estimate community production from emergence in assessing effects of mountaintop coal mining in West Virginia.

35.3.5 Statistical Approaches

The quantification of the uncertainty of production estimates has been a long-standing problem due to the fact that a study of a population or community of a single ecosystem will ultimately provide a single value. The uncertainty of this value, and thus the ability to objectively compare it to values estimated for other ecosystems, however, will be unknown. This is because methods used to estimate such uncertainty require replication. In most studies of lotic secondary production, the appropriate replicate is the stream itself. Unfortunately replication at this spatial scale will be impossible in many, if not most, cases. Several attempts to produce algorithms estimating variance for production estimates have been suggested (Krueger and Martin, 1980; Newman and Martin, 1983; Morin et al., 1987). The most flexible method, requiring the fewest assumptions, however, is bootstrapping—a nonparametric resampling technique (Effron and Tibshirani, 1993).

Bootstrapping is used to estimate the uncertainty of variables with unknown or complex frequency distributions and for situations in which logistical constraints do not allow replication. At minimum, it provides an estimate of the uncertainty inherent in a particular data set. If the data are unbiased and of sufficient coverage, however, bootstrapping will provide an estimate of the true probability distribution underlying any given parameter (Effron and Tibshirani, 1993). It is important to be aware that, as for any statistical approach, bootstrapping requires a solid foundation of data for meaningful results—it also requires the same philosophical and methodological rigor as other statistical approaches.

Bootstrapping has been used to estimate confidence intervals (CIs) for production estimates for both populations and communities calculated by both the size–frequency and instantaneous growth methods (Morin et al., 1987; Huryn, 1996b, 1998). Estimates of CIs are derived by randomly resampling each of the original data sets used to estimate production, with replacement, until a predetermined number of bootstrap data sets are produced (usually 500 or 1000). The mechanical process of randomizing and iteratively resampling the original data sets to produce bootstrap data sets is readily accomplished using Microsoft Excel spreadsheet functions plus Visual Basic or R. The bootstrap data sets are then combined to estimate production which ultimately yields a vector of bootstrap production values (usually 500 or 1000). A mean and approximate 95% CI—as an example—can then be produced from this vector by discarding the upper and lower 2.5% of bootstrap values (Fig. 35.4) or by using an alternative approach such as the bias-corrected percentile method (Meyer et al., 1986). Differences between two vectors of bootstrap production values can be assessed by comparing the degree of overlap of CIs or by using an approach such as the two-sample randomization test (Fig. 35.4; Manly, 1991). In cases where more than two vectors are involved, preplanned orthogonal comparisons can be assessed using a matrix of probabilities estimated using the two-sample randomization test. Family-wise error rate can be controlled using the Bonferroni correction (Keppel, 1982).

35.3.6 Quantification of Food Webs

One application of production analysis is in quantification of food webs, a subject of considerable interest in ecology (Benke and Wallace, 1997; Woodward et al., 2005; Benke, 2011; Cross et al., 2013). There are two types of food webs for which production is particularly useful: a flow web and an *I/P* web (Woodward et al., 2005; Benke, 2011). A *flow web* quantifies the ingestion flow of each prey to each predator (e.g., mg m^{-2} year^{-1}) (Benke and Wallace, 1997). An *I/P web* (ingestion/production) estimates the fraction of each prey's production ingested by each predator and can sum all predatory mortality for each prey species (Woodward et al., 2005; Benke, 2011). To construct each food web, the investigator first

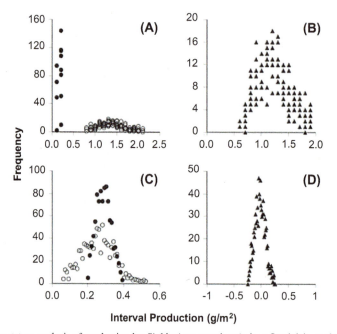

FIGURE 35.4 (A) Results of a bootstrap analysis of production by *Siphlonisca aerodromia* in a floodplain wetland in Maine, USA, in 1997 (*filled circles*) and 1998 (*open circles*, Huryn, 2002 and unpublished). The frequency distribution of the bootstrap data set for each year is shown. Only points falling within the 95% confidence intervals (CIs) are included (i.e., the highest 25 values and the lowest 25 values of 1000 total are not shown). The lack of overlap between 95% CIs indicates that production was significantly different between years. (B) Same as A except that the data are for *Eurylophella* and production between years is not significantly different. (C) Application of the two-sample randomization test to data shown in A. The bootstrap vector for production in 1997 was subtracted from 1998 and the 95% CI of the resulting frequency distribution is shown. The 95% CI does not contain zero, indicating that production between years was significantly different. (D) Same as C except that the data are for *Eurylophella* and the 95% CI contains zero, indicating that production is not significantly different between years.

needs to estimate production of each consumer. The *I/P* web also requires production of basal food types if fraction of basal resources consumed is of interest. Also required is the fraction of each food type eaten by a consumer using *quantitative* gut analyses (i.e., the biomass fraction of each food type consumed by each species). Finally, estimates of ecological efficiencies are needed (e.g., assimilation efficiency) and are usually obtained from the literature. These two types of food webs are illustrated in Fig. 35.5 for a hypothetical stream community with diatoms and detritus as basal food resources, grazers and shredders as primary consumers, and an insect predator.

The sequence of calculations required in building these food webs is illustrated in Table 35.4. The first step is to determine the relative amount (RA_{ij}) of each food type i assimilated by each taxon j given the proportion of food type in the diet and assimilation efficiency as

$$RA_{ij} = G_{ij} \times AE_i \tag{35.8}$$

where G_{ij} = proportion of a food type i in consumer j's diet from gut analyses and AE_i = assumed assimilation efficiency (assimilation/ingestion) of food type i. The relative amount of food type i assimilated by consumer j has little value by itself and must be converted into the fractional amount assimilated as

$$FA_{ij} = \frac{RA_{ij}}{\sum_{i=1}^{n} RA_{ij}} \tag{35.9}$$

The actual amounts of each food type assimilated (AA_{ij}) and contributing to consumer j's production (trophic basis of production, Benke and Wallace, 1980) is determined by multiplying the fractional amount assimilated (FA_{ij}) by consumer j's production (P_j) or

$$AA_{ij} = FA_{ij} \times P_j \tag{35.10}$$

Ingestion flows can be easily determined by dividing the production of consumer j attributed to each food type (AA_{ij}) by the GPE for that food type (Table 35.4). GPE is simply the product of NPE and AE. NPE is that fraction of assimilation

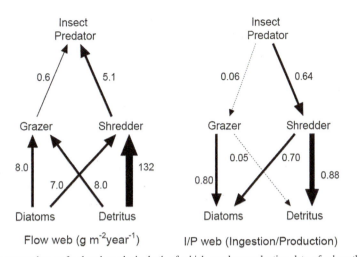

FIGURE 35.5 Two contrasting approaches to food web analysis, both of which employ production data of a hypothetical stream insect assemblage (Table 35.4): flow web (left) and ingestion/production (*I/P*) web (right). In the flow web, *arrows* of different thickness entering each insect correspond to flow (via ingestion) from each food type below (g m^{-2} year^{-1}). Numbers adjacent to *arrows* show exact flow values from Table 35.4. In the *I/P* web, *arrows* of different thickness pointing downward indicate ingestion by predator divided by production of prey or resource (theoretical range = 0–1.0). Exact fractions from Table 35.4 are adjacent to *arrows*.

that is converted into biomass and is assumed in this exercise to be 0.50 for all consumers. The ingestion flow from food type *i* to consumer *j* is therefore

$$I_{ij} = \frac{AA_{ij}}{GPE_i} \tag{35.11}$$

where GPE_i = GPE for food type *i*. The flow web shown in Fig. 35.5 thus incorporates all ingestion flows.

The *I/P* web also uses the ingestion flow from food type *i* to consumer *j* (Table 35.4). However in this case, the ingestion flow is divided by production of food type *i* giving the fraction of prey production ingested by the consumer (Fig. 35.5). This is simply calculated as

$$I_{ij}/P_i \tag{35.12}$$

where P_i = the production of the prey species rather than that of the consumer (or predator). Total predation pressure on the prey taxon is therefore the sum of ingestion flows from prey *i* to all predators divided by prey production

$$\sum_{j=1}^{n} I_{ij}/P_i \tag{35.13}$$

The *I/P* web is thus constructed using individual *I/P* values from consumer *j* to food type *i* (Table 35.4). Individual *I/P* values (or links) are sometimes considered to be a measure of "top-down" interaction strength (e.g., Woodward et al., 2005; Hildrew, 2009; Bellmore et al., 2013; Cross et al., 2013). *I/P* values can vary from close to zero (weak link) to one (strong link). Obviously, the sum of *I/P* values from all consumers to a single food type also cannot exceed one, but their cumulative effect could be strong. Note from Table 35.4 and Fig. 35.5 that the sum of *I/P* values from the two consumers of diatoms is >1.0. Because values >1 are impossible, it indicates that the production estimate for diatoms is too low, the production estimate for grazers and/or shredders is too high, or assumptions for ecological efficiencies of one or both consumers are slightly in error.

In contrast to *I/P* webs, the flows in flow webs can be considered as "bottom-up" interaction strengths because they utilize the magnitude of flows across the entire community rather than fractions of "flow mortality" on individual populations. As an extreme example, consider that predator A consumed 1 mg of prey B and 100 mg of prey C, suggesting that the A–C interaction is stronger than the A–B interaction (i.e., bottom-up effects). This would be true even if predator A consumed 50% of B's production and 30% of C's production (top-down effects); i.e., prey C is 100 times more important than prey B in the diet of predator A, even though A has a greater top-down effect on B than C. Interaction strength depends on whether one adopts the perspective of the predator or the prey. Sometimes, *I/P* links and flow links can agree in terms of their interaction strengths and sometimes they do not. In our example (Fig. 35.5), the predator–grazer flow and *I/P* values are both weak in contrast to the predator–shredder in which both are relatively strong. In contrast, the grazer–detritus flow is surprisingly high, whereas the *I/P* value is very weak.

TABLE 35.4 Calculations used in assembling two types of food webs (flow web and *I/P* web) from a hypothetical assemblage of stream invertebrates consisting of two basal food resources, two primary consumers, and an insect predator (Fig. 35.5). Calculations are based on production values for each taxon (column 2), diet proportions of food types in consumer guts (column 3), and assumed ecological efficiencies for each food type (column 4). Assimilation efficiencies (AE = assimilation/ingestion) are assumed to be 0.70 for animal food, 0.40 for diatoms, and 0.10 for detritus. The flow web is based on the amount of food type ingested by each consumer (column 9). The ingestion/production (*I/P*) web is based on the ingestion of a food type by a consumer divided by production of the food type (column 10). Net production efficiencies (NPE = production/assimilation) are assumed to be a constant 0.50 for all consumers. Row of abbreviations (e.g., P_i, G_{ij} etc.) corresponds to equations used in each column found in the text. See also Online Worksheet 35.4.

Consumer Food Type	Production (mg m^{-2} year^{-1}) P_i	Diet Proportion of Food Type i G_{ij}		Assimilation Efficiency of Food Type i AE_i		Relative Amount of Food Type i to Production RA_{ij}	Fractional Amount of Food Type i to Production FA_{ij}	Actual Amount of Food Type i to Production (mg m^{-2} year^{-1}) AA_{ij}	Gross Production Efficiency of Food Type i (AE × NPE) GPE_{ij}		Amount of Food Type i Ingested (mg m^{-2} year^{-1}) I_{ij}		Ingestion/ Production I_{ij}/P_i
Grazer	2,000												
diatoms	10,000	0.5	×	0.4	=	0.2	0.800	1,600	0.20	÷	8,000	=	0.80
detritus	150,000	0.5	×	0.1	=	0.05	0.200	400	0.05	÷	8,000	=	0.05
Shredder	8,000										—		
diatoms	10,000	0.05	×	0.4	=	0.02	0.174	1,391	0.20	÷	6,957	=	0.70
detritus	150,000	0.95	×	0.1	=	0.095	0.826	6,609	0.05	÷	132,174	=	0.88
Predator	2,000												
grazer	10,000	0.1	×	0.7	=	0.07	0.100	200	0.35	÷	571	=	0.06
shredder	8,000	0.9	×	0.7	=	0.63	0.900	1,800	0.35	÷	5,143	=	0.64

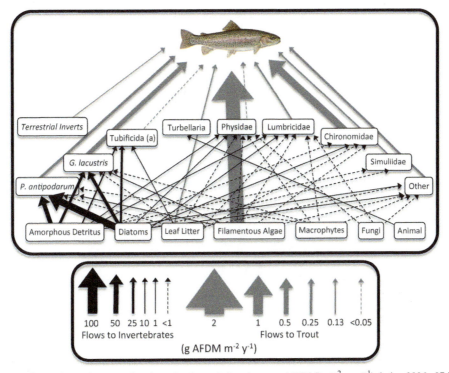

FIGURE 35.6 Flow web of annual organic matter (i.e., ingestion in *g* ash-free dry mass (AFDM) m^{-2} year^{-1}) during 2006−07 in the Glen Canyon of the Colorado River, USA. *Black arrows* are flows from basal resources to invertebrate taxa; *gray arrows* are flows from resources to rainbow trout. The sampling site is downstream of a large dam and the major flows involve nonnative species. *From Cross et al. (2011). Published with permission from Ecological Society of America.*

Real examples of a flow web (Cross et al., 2011) and an *I/P* web (Woodward et al., 2005) are shown in Figs. 35.6 and 35.7, respectively. There are now many other examples of taxon-specific or functional group flow webs, such as Hall et al. (2000), Benke et al. (2001), Rosi-Marshall and Wallace (2002), Frauendorf et al. (2013), Ledger et al. (2013), and Cross et al. (2013). Cross et al. (2013) constructed flow webs, assimilation webs, and *I/P* webs for the same animal community in the Colorado River. Although *I/P* webs are still relatively scarce in stream ecology, it should be possible to construct them given only flow web and production data. Such *I/P* webs provide far more top-down detail in food webs than is possible with experimental removal studies.

35.4 QUESTIONS

1. What is a reasonable value for annual *P/B* if you have a population with two generations per year? For a population with a 5-year life span?
2. When using a cohort table for estimating production (e.g., Table 35.2), do you obtain mean annual biomass by adding or taking the average of values in the biomass column? How about when you use a size−frequency table calculation (e.g., Table 35.3)?
3. When adding the final production column for either the cohort table or the size−frequency table, do you include or exclude negative values from your summation?
4. Approximately what value do you expect to find for the exponent in a power curve that predicts individual dry mass from body length? What does this tell you about a population's growth rate?
5. The "Allen paradox" represents a situation where there does not appear to be enough invertebrate biomass to satisfy the energetic needs of a predator (either a fish or invertebrate predator). How might high values of the *P/B* ratio help resolve this paradox? What other explanations might there be for the paradox?
6. Why might secondary production be a better response variable for comparative or experimental studies than density or biomass?
7. Why is it necessary to use a correction factor (i.e., CPI) in the size−frequency method if the development time is much less than a year? Will you obtain the same annual *P/B* for two univoltine populations, one of which completes its development in 6 months (i.e., 6 months of zero biomass) and one of which completes development in 12 months?
8. From the data in Table 35.2, calculate daily production with the instantaneous growth rate method. How does it compare to the increment-summation estimate?

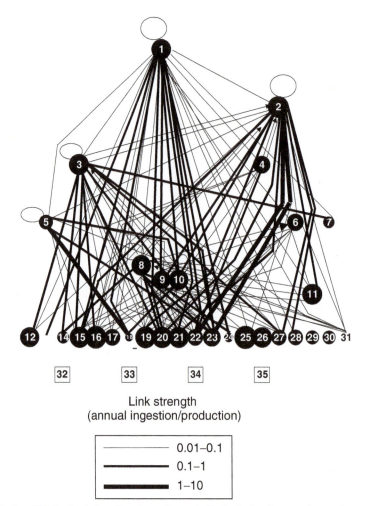

FIGURE 35.7 Ingestion/production (*I/P*) food web from Broadstone Stream in England, showing annual secondary production of 31 invertebrate taxa (production proportional to *circle area*), with the strength of feeding links (*line thickness*) between taxa expressed as annual ingestion by the predator in each link as a proportion of the secondary production of the prey. Numbers 1–31 within circles correspond to invertebrate taxa, with predators at the top and primary consumers at the bottom. Numbers 32–35 within squares correspond to basal resources, but *I/P* links with animals are not shown. *From Woodward et al. (2005). Published with permission from Advances in Ecological Research.*

9. Look at the columns in Table 35.4 and explain how four times more grazer production is attributed to diatoms than detritus, even though grazers consume equal amounts of these food types.
10. What can one say about "top-down" control of a single prey species when multiple predators have low *I/P* values (weak linkage strength), but sum to 0.95?
11. Which food web diagram of Fig. 35.5 is the most convincing that the strongest link in the food web is between shredders and detritus?

35.5 MATERIALS AND SUPPLIES

Field Materials
   Buckets (sturdy, 20 L, with lids)
   Coarse brush
   Forceps
   Plastic bags for temporary storage of sample (size of bag depends on sampler size)
   Preservative (ethanol or formalin) stained with Phloxine B or Rose bengal
   Sampler for quantitative benthic sampling (see Chapter 15)
   Sieves (if samples partially processed in field)
   Growth chambers + microscope for measuring animals

Laboratory Materials
   20-mL scintillation vials or equivalent (for storage of sorted samples)
   100−500 mL jars (for storage of unsorted samples)
   70% Ethanol
   Fine dissecting forceps
   Shallow dishes (Petri dishes)
Laboratory Equipment/Supplies
   Analytical balance (optional)
   Desiccator (optional)
   Dissecting binocular microscope, light source (fiber optic), and ocular micrometer
   Drying oven (optional)
   Sieves (500-μm mesh or smaller)

REFERENCES

Allen, K.R., 1951. The Horokiwi stream: a study of a trout population. New Zealand Department of Fisheries Bulletin 10, 1−238.

Beerstiller, A., Zwick, P., 1995. Biometric studies of some stoneflies and a mayfly (Plecoptera and Ephemeroptera). Hydrobiologia 299, 169−178.

Bellmore, J.R., Baxter, C.V., Martens, K., Connolly, P.J., 2013. The floodplain food web mosaic: a study of its importance to salmon and steelhead with implications for their recovery. Ecological Applications 23, 189−207.

Benke, A.C., 1979. A modification of the Hynes method for estimating secondary production with particular significance for multivoltine populations. Limnology and Oceanography 24, 168−174.

Benke, A.C., 1984. Secondary production of aquatic insects. In: Resh, V.H., Rosenberg, D.M. (Eds.), Ecology of Aquatic Insects. Praeger Scientific, New York, NY, pp. 289−322.

Benke, A.C., 1993. Concepts and patterns of invertebrate production in running waters. Verhandlungen der Internationalen Vereinigung für Theoretische und Angewandte Limnologie 25, 15−38.

Benke, A.C., 1998. Production dynamics of riverine chironomids: extremely high biomass turnover rates of primary consumers. Ecology 79, 899−910.

Benke, A.C., 2011. Secondary production, quantitative food webs, and trophic position. Nature Education Knowledge 3, 26.

Benke, A.C., Huryn, A.D., 2010. Benthic invertebrate production − facilitating answers to ecological riddles in freshwater ecosystems. Journal of the North American Benthological Society 29, 264−285.

Benke, A.C., Huryn, A.D., Smock, L.A., Wallace, J.B., 1999. Length-mass relationships for freshwater macroinvertebrates in North America with particular reference to the southeastern United States. Journal of the North American Benthological Society 18, 308−343.

Benke, A.C., Huryn, A.D., Ward, G.M., 1998. Use of empirical models of stream invertebrate secondary production, as applied to a functional feeding group. Verhandlungen der Internationalen Vereinigung für Theoretische und Angewandte Limnologie 26, 2024−2029.

Benke, A.C., Jacobi, D.I., 1994. Production dynamics and resource utilization of snag-dwelling mayflies in a blackwater river. Ecology 75, 1219−1232.

Benke, A.C., Parsons, K.A., 1990. Modelling black fly production dynamics in blackwater streams. Freshwater Biology 24, 167−180.

Benke, A.C., Wallace, J.B., 1980. Trophic basis of production among net-spinning caddisflies in a southern Appalachian stream. Ecology 61, 108−118.

Benke, A.C., Wallace, J.B., 1997. Trophic basis of production among riverine caddisflies: implications for food web analysis. Ecology 78, 1132−1145.

Benke, A.C., Wallace, J.B., Harrison, J.W., Koebel, J.W., 2001. Food web quantification using secondary production analysis: predacious invertebrates of the snag habitat in a subtropical river. Freshwater Biology 46, 329−346.

Beukema, J.J., Dekker, R., 2013. Evaluation of Brey's production/biomass model on the basis of a long-term data set on a clam population. Marine Ecology Progress Series 489, 163−175.

Brey, T., 1990. Estimating productivity of macrobenthic invertebrates from biomass and mean individual weight. Meeresforschung 32, 329−343.

Burgherr, P., Meyer, E.I., 1997. Regression analysis of linear body dimensions vs. dry mass in stream macroinvertebrates. Archiv für Hydrobiologie 139, 101−112.

Calow, P., 1992. Energy budgets. In: Calow, P., Petts, G.E. (Eds.), The Rivers Handbook: Hydrological and Ecological Principles. Blackwell Scientific, Oxford, UK, pp. 370−378.

Carlisle, D.M., Clements, W.H., 2003. Growth and secondary production of aquatic insects along a gradient of Zn contamination in Rocky Mountain streams. Journal of the North American Benthological Society 22, 582−597.

Cross, W.F., Baxter, C.V., Donner, K.C., Rosi-Marshall, E.J., Kennedy, T.A., Hall Jr., R.O., Wellard Kelly, H.A., Rogers, R.S., 2011. Ecosystem ecology meets adaptive management: food web response to a controlled flood on the Colorado River, Glen Canyon. Ecological Applications 21, 2016−2033.

Cross, W.F., Johnson, B.R., Wallace, J.B., Rosemond, A.D., 2005. Contrasting response of stream detritivores to long-term nutrient enrichment. Limnology and Oceanography 50, 1730−1739.

Cross, W.F., Baxter, C.V., Rosi-Marshall, E.J., Hall Jr., R.O., Kennedy, T.A., Donner, K.C., Wellard Kelly, H.A., Seegert, S.E.Z., Behn, K.E., Yard, M.D., 2013. Food web dynamics in a large river continuum. Ecological Monographs 83, 311−337.

Cross, W.F., Wallace, J.B., Rosemond, A.D., 2007. Nutrient enrichment reduces constraints on material flows in a detritus-based food web. Ecology 88, 2563−2575.

Cross, W.F., Wallace, J.B., Rosemond, A.D., Eggert, S.L., 2006. Whole-system nutrient enrichment increases secondary production in a detritus-based ecosystem. Ecology 87, 1556—1565.

Cummins, K.W., 1973. Trophic relations of aquatic insects. Annual Review of Entomology 18, 183—206.

Dolbeth, M., Cusson, M., Sousa, R., Pardal, M.A., 2012. Secondary production as a tool for better understanding of aquatic ecosystems. Canadian Journal of Fisheries and Aquatic Sciences 69, 1230—1253.

Effron, B., Tibshirani, R., 1993. An Introduction to the Bootstrap. Monographs on Statistics and Applied Probability 57. Chapman and Hall, New York, NY.

Frauendorf, T.C., Colón-Gaud, C., Whiles, M.R., Barnum, T.R., Lips, K.R., Pringle, C.M., Kilham, S.S., 2013. Energy flow and the trophic basis of macroinvertebrate and amphibian production in a neotropical stream food web. Freshwater Biology 58, 1340—1352.

Georgian, T., Wallace, J.B., 1983. Seasonal production dynamics in a guild of periphyton-grazing insects in a southern Appalachian stream. Ecology 64, 1236—1248.

Gillespie, D.M., Benke, A.C., 1979. Methods of calculating cohort production from field data — some relationships. Limnology and Oceanography 24, 171—176.

Gillespie, D.M., Stites, D.L., Benke, A.C., 1985. An inexpensive core sampler for use in sandy substrata. Freshwater Invertebrate Biology 4, 147—151.

Gücker, B., Brauns, M., Solinini, A.G., Voss, M., Walz, N., Pusch, M.T., 2011. Urban stressors alter the trophic basis of secondary production in an agricultural stream. Canadian Journal of Fisheries and Aquatic Sciences 68, 74—88.

Hall Jr., R.O., Taylor, B.W., Flecker, A.S., 2011. Detritivorous fish indirectly reduce insect secondary production in a tropical river. Ecosphere 2, 135.

Hall, R.O., Wallace, J.B., Eggert, S.L., 2000. Organic matter flow in stream food webs with reduced detrital resource base. Ecology 81, 3445—3463.

Hamilton, A.L., 1969. On estimating annual production. Limnology and Oceanography 14, 771—782.

Hannesdóttir, E.R., Gíslason, G.M., Ólafsson, J.S., Ólafsson, Ó.P., O'Gorman, E.J., 2013. Increased stream productivity with warming supports higher trophic levels. Advances in Ecological Research 48, 285—342.

Hauer, F.R., Benke, A.C., 1987. Influence of temperature and river hydrograph on black fly growth rates in a subtropical blackwater river. Journal of the North American Benthological Society 6, 251—261.

Hauer, F.R., Benke, A.C., 1991. Rapid growth of snag-dwelling chironomids in a blackwater river: the influence of temperature and discharge. Journal of the North American Benthological Society 10, 154—164.

Hildrew, A.G., 2009. Sustained research on stream communities: a model system and the comparative approach. Advances in Ecological Research 41, 175—312.

Huryn, A.D., 1996a. Temperature dependent growth and life cycle of *Deleatidium* (Ephemeroptera: Leptophlebiidae) in two high-country streams in New Zealand. Freshwater Biology 36, 351—361.

Huryn, A.D., 1996b. An appraisal of the Allen paradox in a New Zealand trout stream. Limnology and Oceanography 41, 243—252.

Huryn, A.D., 1998. Ecosystem-level evidence for top-down and bottom-up control of production in a grassland stream system. Oecologia 115, 173—183.

Huryn, A.D., 2002. River-floodplain linkage determines production dynamics of detritivorous and predacious mayflies (Ephemeroptera) in a sedge-meadow wetland. Archiv für Hydrobiologie 155, 455—480.

Huryn, A.D., Benke, A.C., Ward, G.M., 1995. Direct and indirect effects of regional geology on the distribution, production and biomass of the freshwater snail Elimia. Journal of the North American Benthological Society 14, 519—534.

Huryn, A.D., Wallace, J.B., 1986. A method for obtaining in situ growth rates of larval Chironomidae (Diptera) and its application to studies of secondary production. Limnology and Oceanography 31, 216—222.

Huryn, A.D., Wallace, J.B., 1987. Local geomorphology as a determinant of macrofaunal production in a mountain stream. Ecology 68, 1932—1942.

Huryn, A.D., Wallace, J.B., 2000. Life history and production of stream insects. Annual Review of Entomology 45, 83—110.

Hynes, H.B.N., Coleman, M.J., 1968. A simple method of assessing the annual production of stream benthos. Limnology and Oceanography 13, 569—573.

Jackson, J.K., Fisher, S.G., 1986. Secondary production, emergence, and export of aquatic insects of a Sonoran Desert stream. Ecology 67, 629—638.

Johnson, B.R., Tarter, D.C., Hutchens Jr., J.J., 2000. Life history and trophic basis of production of the mayfly *Callibaetis fluctuans* (Walsh) (Ephemeroptera: Baetidae) in a mitigated wetland, West Virginia, USA. Wetlands 20, 397—405.

Johnson, B.R., Fritz, K.M., Price, R., 2013a. Estimating benthic secondary production from aquatic insect emergence in streams affected by mountaintop removal coal mining, West Virginia, USA. Fundamental and Applied Limnology 182, 191—204.

Johnson, R.C., Jin, W.-S., Carreiro, M.M., Jack, J.D., 2013b. Macroinvertebrate community structure, secondary production and trophic-level dynamics in urban streams affected by non-point-source pollution. Freshwater Biology 58, 843—857.

Johnston, J.A., Cunjak, R.A., 1999. Dry mass-length relationships for benthic insects: a review with new data from Catamaran Brook, New Brunswick, Canada. Freshwater Biology 41, 653—674.

Keppel, G., 1982. Design & Analysis: A Researcher's Handbook, second ed. Prentice-Hall, Inc., Englewood Cliffs, NJ.

Krueger, C.C., Martin, F.B., 1980. Computation of confidence intervals for the size-frequency (Hynes) method of estimating secondary production. Limnology and Oceanography 25, 773—777.

Ledger, M.E., Brown, L.E., Edwards, F.K., Hudson, L.N., Milner, A.M., Woodward, G., 2013. Extreme climatic events alter aquatic food webs: a synthesis of evidence from a mesocosm drought experiment. Advances in Ecological Research 48, 343—395.

Ledger, M.E., Edwards, F.K., Brown, L.E., Milner, A.M., Woodward, G., 2011. Impact of simulated drought on ecosystem biomass production: an experimental test in stream mesocosms. Global Change Biology 17, 2288—2297.

Lugthart, G.J., Wallace, J.B., 1992. Effects of disturbance on benthic functional structure and production in mountain streams. Journal of the North American Benthological Society 11, 138—164.

Manly, B.F.J., 1991. Randomization and Monte Carlo Methods in Biology. Chapman and Hall, London, UK.

Marchant, R., 1986. Estimates of annual production for some aquatic insects from the La Trobe River, Victoria. Australian Journal of Marine and Freshwater Research 37, 113–120.

Merritt, R.W., Cummins, K.W., Resh, V.H., Batzer, D.P., 2008. Sampling aquatic insects: collection devices, statistical considerations, and rearing procedures. In: Merritt, R.W., Cummins, K.W., Berg, M.B. (Eds.), An Introduction to the Aquatic Insects of North America, fourth ed. Kendall/Hunt Publishing Company, Dubuque, Iowa, pp. 15–37.

Meyer, J.S., Ingersoll, C.G., McDonald, L.L., Boyce, M.S., 1986. Estimating uncertainty in population growth rates: jackknife vs. bootstrap techniques. Ecology 67, 1156–1166.

Meyer, E., 1989. The relationship between body length parameters and dry mass in running water invertebrates. Archiv für Hydrobiologie 117, 191–203.

Miyasaka, H., Genkai-Kato, M., Miyake, Y., Kishi, D., Katano, I., Doi, H., Ohba, S-y., Kuhara, N., 2008. Relationship between length and weight of freshwater macroinvertebrates in Japan. Limnology 9, 75–80.

Morin, A., Bourassa, N., 1992. Empirical models of annual production and productivity-biomass ratios of benthic invertebrates in running waters. Canadian Journal of Fisheries and Aquatic Sciences 49, 532–539.

Morin, A., Dumont, P., 1994. A simple model to estimate growth rate of lotic insect larvae and its value for estimating population and community production. Journal of the North American Benthological Society 13, 357–367.

Morin, A., Mousseau, T.A., Roff, D.A., 1987. Accuracy and precision of secondary production estimates. Limnology and Oceanography 32, 1342–1352.

Newman, R.M., Martin, F.B., 1983. Estimation of fish production rates and associated variances. Canadian Journal of Fisheries and Aquatic Sciences 40, 1729–1736.

Peterson, B.J., et al., 1993. Biological responses of a tundra river to fertilization. Ecology 74, 653–672.

Plante, C., Downing, J.A., 1989. Production of freshwater invertebrate populations in lakes. Canadian Journal of Fisheries and Aquatic Sciences 46, 1489–1498.

Poepperl, R., 2000. Benthic secondary production and biomass of insects emerging from a northern German temperate stream. Freshwater Biology 44, 199–211.

Rabeni, C.F., Collier, K.J., Parkyn, S.M., Hicks, B.J., 1997. Evaluating techniques for sampling stream crayfish (*Paranephrops planifrons*). New Zealand Journal of Marine and Freshwater Research 31, 693–700.

Ramírez, A., Pringle, C.M., 2006. Fast growth and turnover of chironomid assemblages in response to stream phosphorus levels in a tropical lowland landscape. Limnology and Oceanography 51, 189–196.

Rosi-Marshall, E.J., Wallace, J.B., 2002. Invertebrate food webs along a stream resource gradient. Freshwater Biology 47, 129–141.

Ross, D.H., Wallace, J.B., 1981. Production of *Brachycentrus spinae* Ross (Trichoptera: Brachycentridae) and its role in seston dynamics of a southern Appalachian stream (USA). Environmental Entomology 10, 240–246.

Ross, D.H., Wallace, J.B., 1983. Longitudinal patterns of production, food consumption, and seston utilization by net-spinning caddisflies (Trichoptera) in a southern Appalachian stream (USA). Holarctic Ecology 6, 270–284.

Sallenave, R.M., Day, K.E., 1991. Secondary production of benthic stream invertebrates in agricultural watersheds with different land management-practices. Chemosphere 23, 57–76.

Shieh, S.-H., Ward, J.V., Kondratieff, B.C., 2002. Energy flow through macroinvertebrates in a polluted plains stream. Journal of the North American Benthological Society 21, 660–675.

Shieh, S.-H., Ward, J.V., Kondratieff, B.C., 2003. Longitudinal changes in macroinvertebrate production in a stream affected by urban and agricultural activities. Archiv für Hydrobiologie 157, 483–503.

Short, R.A., Stanley, E.H., Harrison, J.W., Epperson, C.R., 1987. Production of *Corydalus cornutus* (Megaloptera) in four streams differing in size, flow, and temperature. Journal of the North American Benthological Society 6, 105–114.

Singer, G.A., Battin, T.J., 2007. Anthropogenic subsidies alter stream consumer–resource stoichiometry, biodiversity, and food chains. Ecological Applications 17, 376–389.

Smock, L.A., 1980. Relationships between body size and biomass of aquatic insects. Freshwater Biology 10, 375–383.

Smock, L.A., Gladden, J.E., Riekenberg, J.L., Smith, L.C., Black, C.R., 1992. Lotic macroinvertebrate production in three dimensions: channel surface, hyporheic and floodplain environments. Ecology 73, 876–886.

Statzner, B., Resh, V.H., 1993. Multiple-site and -year analyses of stream insect emergence: a test of ecological theory. Oecologia 96, 65–79.

Tod, S.P., Schmidt-Araya, J.M., 2009. Meiofauna versus macrofauna: secondary production of invertebrates in a lowland chalk stream. Limnology and Oceanography 54, 450–456.

Towers, D.J., Henderson, I.M., Veltman, C.J., 1994. Predicting dry weight of New Zealand aquatic macroinvertebrates from linear dimensions. New Zealand Journal of Marine and Freshwater Research 28, 159–166.

Tumbiolo, M.L., Downing, J.A., 1994. An empirical-model for the prediction of secondary production in marine benthic invertebrate populations. Marine Ecology Progress Series 114, 165–174.

Venarsky, M.P., Huntsman, B.M., Huryn, A.D., Benstead, J.P., Kuhajda, B.R., 2014. Quantitative food web analysis supports the energy-limitation hypothesis in cave stream ecosystems. Oecologia 176, 859–869.

Wallace, J.B., Eggert, S.L., Meyer, J.L., Webster, J.R., 2015. Stream invertebrate productivity linked to forest subsidies: 37 stream–years of reference and experimental data. Ecology 96, 1213–1228.

Wallace, J.B., Gurtz, M.E., 1986. Response of *Baetis* mayflies (Ephemeroptera) to catchment logging. American Midland Naturalist 115, 25–41.

Wallace, J.B., O'Hop, J., 1985. Life on a fast pad: waterlily leaf beetle impact on water lilies. Ecology 66, 1534–1544.

Waters, T.F., 1969. The turnover ratio in production ecology of freshwater invertebrates. The American Naturalist 103, 173–185.

Waters, T.F., 1977. Secondary production in inland waters. Advances in Ecological Research 10, 91–164.

Waters, T.F., Doherty, M.T., Krueger, C.C., 1990. Annual production and production: biomass ratios for three species of stream trout in Lake Superior tributaries. Transactions of the American Fisheries Society 119, 470–474.

Webster, J.R., Wallace, J.B., Benfield, E.F., 1995. Organic processes in streams of the eastern United States. In: Cushing, C.E., Cummins, K.W., Minshall, G.W. (Eds.), River and Stream Ecosystems. Elsevier, New York, NY, pp. 117–187.

Whitmore, N., Huryn, A.D., 1999. Life history and production of *Paranephrops zealandicus* in a forest stream, with comments about the sustainable harvest of a freshwater crayfish. Freshwater Biology 42, 1–11.

Woodcock, T.S., Huryn, A.D., 2008. The effect of an interstate highway on macroinvertebrate production of headwater streams in Maine (U.S.A.). Fundamental and Applied Limnology 171, 199–218.

Woodcock, T.S., Huryn, A.D., 2007. The response of macroinvertebrate production to a heavy metal pollution gradient in a headwater stream. Freshwater Biology 52, 177–196.

Woodward, G., Speirs, D.C., Hildrew, A.G., 2005. Quantification and resolution of a complex, size-structured food web. Advances in Ecological Research 36, 85–135.

Chapter 36

Elemental Content of Stream Biota

Jonathan P. Benstead[1], Michelle A. Evans-White[2], Catherine A. Gibson[3] and James M. Hood[4]

[1]Department of Biological Sciences, University of Alabama; [2]Department of Biological Sciences, University of Arkansas; [3]The Nature Conservancy; [4]Aquatic Ecology Laboratory, Department of Evolution, Ecology, and Organismal Biology, The Ohio State University

36.1 INTRODUCTION

All organisms are ultimately composed of varying proportions of chemical elements. Whether the ratio of these elements is tightly constrained or relatively flexible, mismatches between organismal demand for elements and their rate of supply from the environment have fundamental consequences, from the level of single cells to the entire biosphere (Sterner and Elser, 2002). Using elements and their relative ratios as common currencies in biological processes is therefore a powerful approach to linking levels of the biological hierarchy, from biochemistry to global ecology. It is this unifying framework that is the great strength of ecological stoichiometry (ES), the body of theory that focuses on the balance of elements in ecological interactions (Sterner and Elser, 2002). First refined in lake ecosystems, ES theory has increasingly been applied by stream ecologists to ask questions related to controls on biomass production, carbon flow, and nutrient cycling (Frost et al., 2002; Cross et al., 2005; McIntyre and Flecker, 2010). Whatever the question, data on the elemental content of diverse types of organic matter are a prerequisite to testing hypotheses. This chapter focuses on the methods required to obtain such data in stream ecosystems.

Although many elements are biologically important (Kaspari and Powers, 2016), carbon (C), nitrogen (N), and phosphorus (P) are of primary focus in ES because they are essential building blocks in the biological polymers that dominate biomass and so their cycling is strongly influenced by biological processes (Sterner and Elser, 2002). The essential biochemical pathways shared by all organisms constrain biomass C:N:P ratios within certain bounds, but the variation around these ratios has important ecological causes and consequences (Reiners, 1986; Sterner and Elser, 2002). Bacteria are comparatively nutrient rich and thought to be relatively homeostatic in their elemental composition, but variation in both nutrient content and growth strategies can occur (Makino et al., 2003; Persson et al., 2010; Scott et al., 2012; Goodwin and Cotner, 2015). Limited data suggest that nutrient content of aquatic fungi is also high but relatively variable, with elemental homeostasis being stronger for N than for P content (Danger et al., 2016; Gulis et al., in review). Algae are also relatively nutrient rich, but their physiology allows for considerable variation in their elemental composition (Hill et al., 2011). Terrestrial detritus is typically nutrient poor, so characterized by relatively high C:N and C:P ratios that can differ among leaf species (Ostrofsky, 1997; Hładyz et al., 2009).

Compared to plants, animals are relatively less flexible in their elemental content (Sterner and Elser, 2002). Stream invertebrates still differ in C:N:P ratios, however, with elemental composition varying not only across phyla, subphyla, and orders, but also across macroinvertebrate functional feeding groups and with insect ontogeny (Cross et al., 2003; Evans-White et al., 2005; Back and King, 2013). Finally, vertebrate taxa also display interspecific variation, as well as intraspecific plasticity that is driven in part by the role that skeletal tissue plays in storing and buffering body P content (Hendrixson et al., 2007; Benstead et al., 2014). Whatever the taxon, imbalances will typically exist between its elemental demand and what is contained in available food resources. Studies of these imbalances have resulted in new insights into population, community, and ecosystem dynamics in streams (Frost et al., 2002, 2005; Cross et al., 2005).

An elemental imbalance between a consumer and its food resource occurs when food resource elemental ratios do not equally match those required for growth and maintenance of the consumer (Sterner and Elser, 2002). Sufficiently large elemental imbalances between primary consumer requirements and their autotroph and detrital food resources result in

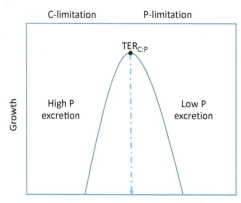

FIGURE 36.1   A conceptual illustration of the relationship between organismal growth and food resource C:P. Maximal growth and minimal loss of C and P should coincide at the threshold element ratio ($TER_{C:P}$). C-limitation of growth and a greater P excretion rate should be observed at food C:P < $TER_{C:P}$ and P-limitation of growth and lower P excretion should be observed at a food C:P > $TER_{C:P}$.

limitation of growth (Elser et al., 2000; Frost and Elser, 2002; Stelzer and Lamberti, 2002; Sterner and Schulz, 1998). In contrast, elemental imbalances between predators and their animal prey are often lower than those between primary consumers and their food resources. Given the existence of significant elemental imbalances that lead to limitation of consumer growth, it follows that threshold elemental ratios (TERs) exist at which the growth-limiting element in the diet switches from one to another. Under a given food quantity and set of environmental conditions, maximal growth and minimal loss of excess elements should occur when the elemental ratio of ingested material matches the TER exactly (see Fig. 36.1). Considerable taxonomic variation in TERs exists (Frost et al., 2006; Benstead et al., 2014). Therefore, quantifying species TERs allows prediction of resource elemental ratios at which individual and population growth limitation by particular elements occurs. Depending on severity, elemental imbalances can have important consequences for population and community dynamics.

Elemental imbalances between animals and their diets also feed back to affect ecosystem processes. Stoichiometric theory makes two predictions about the relationship between diet and nutrient release rates by animals (Sterner, 1990; Elser and Urabe, 1999). These predictions assume that animals maintain strict body C:N:P homeostasis in response to variation in diet C:N or C:P. While this assumption has not held up to close scrutiny (Persson et al., 2010), it is not clear whether the observed deviation from strict homeostasis has serious predictive consequences (Wang et al., 2012). The first prediction states that animals consuming nutrient-rich diets will release (excrete and/or egest) nutrients ingested in excess to maintain elemental homeostasis; nutrient-starved animals will retain limiting nutrients. The second prediction states that N:P release ratios decrease with body N:P ratio, all else being equal, due to differences in N:P demand. Thus, elemental imbalances between animals and their diets influence both the quantity and ratio of nutrients released.

Under some circumstances of high animal biomass, low ambient nutrient availability, and/or large stoichiometric imbalances, nutrient regeneration by animal populations can have a substantial impact on nutrient cycling and, therefore, the growth and composition of autotroph and heterotroph communities (Vanni, 2002; Atkinson et al., 2017). For instance, changes in zooplankton community composition from P-rich cladoceran species to relatively P-poor calanoid copepod species shifted lake algal communities from P-limitation to N-limitation (Elser et al., 1988). Animals have also been shown to make a substantial contribution to nutrient cycles in stream ecosystems. Nutrient release by macroinvertebrates and fish contributed up to 70% of algal N demand in an N-limited desert stream (Grimm, 1988a,b), while invasive snails contributed up to 66% of NH_4-N demand in an N-limited mountain stream (Hall et al., 2003). Similarly, fish contributed up to 49% of N demand in an N-limited neotropical stream and created spatial hot spots of nutrient regeneration through habitat choice and movement patterns (Vanni et al., 2002; McIntyre et al., 2008). Future studies will hopefully provide a more complete understanding of when and where nutrient regeneration by animals is important at ecosystem scales, improving our understanding of stream nutrient cycles and the role that animal biodiversity plays in shaping ecosystem function (Benstead et al., 2010; Capps et al., 2015; Atkinson et al., 2017).

The goal of this chapter is to introduce both basic and more specialized techniques for investigating the role of elemental concentrations and ratios in controlling pattern and process in stream ecology. We first cover the fundamental approaches to the various applications of ES theory in stream habitats, including the measurement of elemental imbalances and excretion rates. In two *basic* methods, we provide detailed protocols for testing ES theory in streams, namely measuring elemental content in (1) food web compartments and (2) waste products. This coverage of basic techniques is followed by protocols for four *advanced* methods: (1) simultaneous measurement of C, N, and P content in single samples; (2) assessing the ecosystem-level importance of excretion; (3) accounting for stress and fasting effects on excretion; and (4) estimation of TERs. Measuring elemental content of bacteria and fungi requires specialized techniques and is not covered here. We encourage those interested in the stoichiometry of heterotrophic microbes to refer to the methodological approaches used in relevant recent publications (e.g., Goodwin and Cotner, 2015; Gulis et al., in review).

36.2 GENERAL DESIGN

36.2.1 Site Selection

Choice of site obviously depends on the motivation for the study, but collection of food web compartments is easiest in wadeable, first- to fourth-order streams with hard substrata (gravel and cobble). Streams with very divergent dissolved nutrient ratios (i.e., very low or high N:P ratios) provide natural contrasts that are highly suited to exploratory analyses of elemental imbalances. It may also be of interest to collect from two or more streams that differ with respect to nutrient or organic matter inputs (e.g., forest vs. urban, nutrient-poor detritus vs. nutrient-rich algal communities). Finally, collection of samples for analysis of elemental content can be combined profitably with other techniques described in this volume, including food web analysis (Chapter 23), solute dynamics (Chapter 30), nutrient limitation (Chapter 31), and nutrient uptake and transformations (Chapters 32 and 33).

36.2.2 Elemental Composition of Food Web Compartments

Identification and documentation of elemental differences in food web compartments is a cornerstone of the ES framework. Elemental imbalances between resources and consumers can affect a diverse suite of ecological processes, including population dynamics, community structure, trophic interactions, and nutrient cycling. Previous work has explored the consequences of large elemental imbalances in low-nutrient streams (Cross et al., 2003; Bowman et al., 2005; Hładyz et al., 2009). Conversely, several studies have examined the effects of increased nutrients on consumer-resource stoichiometry and generally reported reductions in elemental imbalances (Cross et al., 2003; Singer and Battin, 2007; Small and Pringle, 2010; Lauridsen et al., 2012; Morse et al., 2012).

Stream ecosystems can have a diverse suite of basal food resources and consumers that vary in C:N:P. Sampling the elemental composition of food web compartments is initially similar to that used to describe food webs (see Chapter 23). Multiple samples of each food web compartment should be collected. Basal food resources including leaf litter, periphyton, macrophytes, and fine particulate organic matter should be sampled throughout the study reach, whereas seston can be sampled at the end of the undisturbed reach. Leaf litter should be collected as grab samples from multiple locations in the study reach (Chapter 26). Procedures for sampling periphyton, macrophytes, fine benthic organic matter (FBOM) and seston are presented in detail in Chapters 10–13 and 25. Similarly, collection of consumers including insects, crustaceans, amphibians, and fish should follow standard protocols (Chapters 14–17).

36.2.3 Measuring Nutrient Release Rates of Stream Biota

Nutrient release rates of animals have been measured to gain understanding of both the elemental requirements of animals and the role of animals in nutrient cycles (Fig. 36.2). A number of exercises could be designed to explore either of these avenues of research. For example, to explore the influence of diet on elemental imbalances, one could feed individuals diets (leaf litter or algae) spanning a stoichiometric gradient (e.g., Balseiro and Albariño, 2006) and then measure nutrient release rates or, alternatively, measure the release rates of species or age classes with different body stoichiometry (e.g., Vanni et al., 2002). To explore the role of animals in nutrient cycles, one could calculate areal release rates for the dominant animals in a reach and evaluate the relative importance of nutrient cycling by those individuals following the protocols described in Advanced Method 2 (see also McIntyre et al., 2008).

Nutrient release rates can be estimated either indirectly using bioenergetics or mass balance models (Schindler and Eby, 1997; Hood et al., 2005) or directly by placing organisms in a container of filtered water and tracking changes in nutrient concentration (Schaus et al., 1997; McIntyre et al., 2008; Griffiths and Hill, 2014). It is often necessary to modify the container method for specific organisms or systems (Fig. 36.3). Here, we outline a basic approach for small (2–4 cm) tadpoles (following Whiles et al., 2009). In Advanced Method 3, we describe a more complex protocol that accounts for the effects of handling stress and fasting on excretion rates.

FIGURE 36.2 Stoichiometric information for the case-building caddisfly *Lepidostoma* sp. (A) in the South Fork Eel River (CA, USA). *Lepidostoma* larvae were collected from the South Fork Eel River (in the Angelo Coast Range Reserve) for determination of body C, N, and P content (J. M. Hood, C. McNeely, and J. C. Finlay, *unpublished data*). We also measured the N and P excretion rate of *Lepidostoma* in small plastic containers, containing 60 mL of filtered stream water, which were incubated in a shallow part of the stream (B). Body C:N (C) and $\log_{10}$ NH$_4$-N excretion rates (D) of South Fork Eel River *Lepidostoma* increased linearly with $\log_{10}$ dry mass, likely indicating that N demand decreases through ontogeny. *Photo credits: J.M. Hood.*

36.2.4 Estimating Threshold Elemental Ratios for Stream Invertebrates

The TER represents an organism's optimal balance of two elements for growth at a particular food consumption rate and under a given set of environmental conditions. Two approaches are commonly used to quantify TERs: bioenergetics models (Urabe and Watanabe, 1992; Frost et al., 2006) and direct measurements of growth responses across a gradient of diet stoichiometry (Benstead et al., 2014; Bullejos et al., 2014; Halvorson et al., 2015). Bioenergetics models use estimates of growth, physiological efficiencies, and body elemental content to estimate TERs. These models make the simplifying assumptions that (1) element-specific growth efficiencies are constant across stoichiometrically diverse diets and that (2) animals are strictly homeostatic in elemental content. Both assumptions have been challenged and it is still unclear to what degree violation of these assumptions influences TER estimates (Halvorson et al., 2015). Examination of growth rates across a resource ratio gradient may provide more accurate TER estimates (Halvorson et al., 2015) and are no more difficult than generating the appropriate bioenergetics data for models, although designing experiments that manipulate only one element can be extremely challenging. Another drawback to looking only at growth patterns is that one learns less about organismal physiological strategies for dealing with variability in food resource ratios. Future studies of organismal TERs may benefit from the integration of both approaches.

In Advanced Method 4, we outline one method for quantifying the TER$_{C:P}$ (i.e., the TER for a food carbon:phosphorus ratio) of a herbivorous stream invertebrate. We provide a detailed procedure for a feeding experiment that examines grazer growth across a periphyton C:P gradient. This procedure has three main components—incubation of a periphyton food resource, a growth experiment, and data analysis. The incubation component manipulates the C:P of periphyton on tile substrata by allowing algae to grow in stream water receiving different inorganic phosphate amendments. The incubation generates periphyton with a minimum of four different C:P ratios that can be fed to a macroinvertebrate grazer in the growth experiment. The grazer growth data are then analyzed using a regression approach.

FIGURE 36.3 Examples of containers used to measure nutrient release rates for various aquatic organisms. Undergraduate researcher Rachel Moore used plastic jars to measure the nutrient release rates of crayfish (A). Jon Benstead and colleagues used recirculating chambers to measure the nutrient release rates of mussels (B). In this experimental setup, three-way Luer-lock valves allowed the collection of a time series of samples from the hoses, avoiding disturbance of the study organisms. Jim Hood and colleagues used plastic storage bags to measure the nutrient release rates of the armored catfish *Ancistrus triradiatus* (C) in the Río Las Marias (near Guanare, Venezuela) and *Lepidostoma* sp. (D) in Jack of Hearts Creek, a tributary of the South Fork Eel River (CA, USA). *Photo credits: (A) H. Halvorson, (B) J.P. Benstead, (C) J.M. Hood, and (D) J.M. Hood.*

36.3 SPECIFIC METHODS

36.3.1 Basic Method 1: Estimating C, N, and P Content of Food Web Compartments

In this section, we describe the basic steps to analyzing commonly sampled compartments in stream food webs for C, N, and P content. We assume that initial collection of live material will take place under field conditions, but the same steps apply to laboratory-based studies.

36.3.1.1 Protocol for Field Collection

1. The field equipment and methods required for sample collection are similar to those described in other chapters covering periphyton (Chapter 12), macroinvertebrates (Chapter 15), and fish (Chapter 16). We will therefore restrict our descriptions of sampling here to issues particular to sample collection for elemental analysis.
2. Samples need to be as clean of inorganic material and other potential contaminants (e.g., detritus) as possible. Using trays, sort samples from extraneous matter and rinse them with a wash bottle of deionized water when possible. Try to collect as clean and homogenous a sample of each compartment as possible.
3. Periphyton is a heterogeneous mixture of algae, microbes, and detritus, which poses special challenges related to sample purification. Slurries are typically filtered onto preweighed (±0.0001 g) 0.7-μm glass fiber filters after removal of macroscopic contaminants. However, techniques for separation of major components are available (e.g., separation of algae and detritus by centrifugation in colloidal silica; Hamilton et al., 2005).

4. Fine organic material suspended in the water column (i.e., seston) can be collected onto 0.7-μm glass fiber filters using a manifold and side-arm flask, although an in-line filter coupled to a battery-powered peristaltic pump (e.g., Geopump) is much faster.

5. Collected macroinvertebrates should preferably be live-sorted into size classes (1-mm for most taxa), either in the field using a field microscope or in the laboratory. Published length—mass equations (e.g., Benke et al., 1999) are useful for ensuring that sufficient dry mass of invertebrate material has been collected. If possible, keep large predatory taxa separate from smaller invertebrates until they can be sorted.

6. Most samples that are not filtered can be placed in small plastic vials (cryotubes are ideal, preferably prelabeled).

7. Sample and euthanize fish and amphibians using appropriate sampling gear and protocols (see Chapters 16 and 17). Fish and other vertebrates should only be collected under the necessary permits and Institutional Animal Care and Use Committee approval. Whole-body samples are required for the majority of studies, so place large animals in labeled sample bags.

8. Place all sorted samples on ice immediately, and minimize time to laboratory processing whenever possible.

36.3.1.2 Protocol for Laboratory Preparation

1. Presorted material in vials or on filters can be placed immediately either in a drying oven (uncapped vials, 60°C for 2 days for algae and invertebrates) or freezer in preparation for freeze-drying. Generally, freeze-drying is preferable, especially if prefreezing of samples is unavoidable for logistical reasons (thawing before drying in an oven can lead to loss of leaked cell contents).

2. Dried plant material can usually be ground, either by hand using a pestle and mortar or in a ball mill. Very fibrous material is best homogenized using a cutting mill (e.g., Wiley mill).

3. Removal of gut contents from animals should generally be avoided, as it will lead to loss of tissue and hemolymph or blood. Significant effects of stomach contents on whole-body elemental content should be explored in taxa that have proportionally large guts. Waiting for guts to clear is one common approach, but note that many stream insects increase gut passage time when fasting.

4. Small samples (<2 mg dry mass) can be analyzed whole for carbon and nitrogen content, but homogenization may be required for large samples, or when being prepared for P analysis. Loss of material during homogenization is to be avoided and thorough mixing before subsampling is essential. Suitable homogenization methods for small samples include both grinding within the sample vial with a pestle (e.g., disposable pestle or steel rod) and grinding with a miniature ball mill (e.g., Wig-L-Bug). Very small invertebrates (<0.5 mg dry mass) can be analyzed for C and N as composite samples of multiple individuals, but this precludes any estimation of variance in individual body content.

5. Most fish and other whole-body samples of relatively large vertebrates will require homogenization and subsampling. Autoclave large vertebrates in appropriate containers (121°C for 1 h), followed by homogenization in a blender or laboratory homogenizer and thorough mixing in the same container. The resulting material can then be subsampled, dried or freeze-dried, and ground with a ball mill until reduced to a fine homogenous powder.

36.3.1.3 Sample Analysis[1]

1. Dried samples can be prepared for C and N analysis by accurate weighing on a microbalance (± 1 μg) and packing into tin capsules for analysis on a CHN analyzer. Sample weight will depend on C and N content, but a suitable target is 500 μg C and 100 μg N (i.e., 1 mg total sample weight for samples that are 50% C and 10% N). Dried and weighed filters may be subsampled for C, N, and P analysis by cutting out and weighing representative segments. Use the filtrate dry mass (mass of filter + filtrate − original filter mass) to estimate the subsample mass required to obtain a final elemental mass within the desired range. This initial estimate will depend on the elemental content of the filtrate, but a rough estimate for periphyton is 30% C, 1.5% N, and 0.1% P (all based on dry mass).

2. Do not analyze irreplaceable samples until you are confident in your protocol. Always attempt to estimate concentrations of the element of interest in your samples, to optimize the quantification process (i.e., ensure a good match between standards and unknowns). In general, analyzing a representative subsample is preferable to running an entire sample followed by a dilution step (see below), if only because the remaining sample material can be reanalyzed, if necessary.

---

1. Wear personal protective equipment (PPE), including lab coat, eyewear, and gloves, whenever conducting laboratory procedures.

3. P analysis requires wet chemistry methods that involve at least digestion of the solid samples followed by analysis. Here we focus on traditional analytical methods available in most laboratories, but readers should be aware of alternatives, such as the use of inductively coupled plasma mass spectrometry (ICP-MS) or optical emission spectrometry (ICP-OES, sometimes referred to as atomic emission spectrometry). Use of these sophisticated instruments has the advantage of providing data on >20 elements, including P, allowing ionomics approaches to ES. Mass spectrometry is also particularly sensitive, enabling measurement of very low P concentrations. The disadvantages include cost and the inability of these instruments to measure C and N.

4. Here, we describe a method for analysis of organismal P content that involves digesting the sample (incineration followed by acid hydrolysis) and then measuring P content colorimetrically using the ammonium molybdate method, following Murphy and Riley (1962) and Boros and Mozsár (2015). Begin by donning appropriate protective gear (goggles, lab coat, and gloves) and making the following reagents, using glassware that has been scrupulously cleaned, acid-washed, rinsed three times with Milli-Q water and air-dried:

    a. 10 N H_2SO_4: Add ~200 mL Milli-Q water to a clean 500-mL volumetric flask, place in an ice bath, add 138.9 mL concentrated H_2SO_4, swirl to mix, allow the solution to cool, and then fill to 500 mL with Milli-Q water. Cover with Parafilm and invert 20 times to mix. Store in a glass bottle. This solution is stable for several months.

    b. Molybdate reagent: Add ~500 mL Milli-Q water to a clean 1-L volumetric flask, dissolve 0.208 g antimony potassium tartrate, 9.6 g ammonium heptamolybdate 4-hydrate, stir to mix, and then fill to 1 L with Milli-Q water. Cover with Parafilm and invert 20 times to mix. This solution, stored in a glass amber bottle, is stable for several months.

    c. Ascorbic acid: Add ~50 mL Milli-Q water to a clean 100-mL volumetric flask, add 2 g ascorbic acid, stir to dissolve, and then fill to 100 mL with Milli-Q water. Cover with Parafilm and invert 20 times to mix. Store in an amber bottle. Make this solution fresh daily.

    d. Dilution reagent: Add ~100 mL Milli-Q to a clean 250-mL volumetric flask, add 18.5 mL 10 N H_2SO_4, swirl to mix, and fill to 250 mL with Milli-Q. Cover with Parafilm and invert 20 times to mix. Store in a glass bottle.

5. An example data sheet is provided in Online Worksheet 36.1. Sample weights will depend on type, but final concentrations of diluted sample digestates for P analysis using traditional methods (ashing/digestion followed by ammonium molybdate colorimetric analysis) should not exceed 30 μM P if using a 1-cm cell for spectrophotometry (see below and Online Worksheet 36.2). Weigh samples on a microbalance (±1 μg) as above, using tared weigh boats fashioned from tin capsules or aluminum foil. Tap samples carefully into the bottom of acid-washed, dried, and weighed (±0.0001 g) 25-mL digestion tubes (screw-cap borosilicate glass vials) and cap the vials with foil. Reweigh and record the weight of the weigh boat, and discard it if any residual sample remains. Prepare a standard curve by weighing five to six samples containing suitable organic standard (e.g., bovine muscle for animal tissue, wheat, or spinach for plant material, dried at 60°C). Also prepare two to three sample blanks.

6. Ash samples for P analysis at 550°C, while wearing personal protective equipment (PPE) and donning oven mitts when handling hot vessels. Ramp temperature slowly, if possible. A recent study suggests that relatively long ashing times are necessary for complete P extraction (Boros and Mozsár, 2015), so ashing overnight (8 h) is recommended.

7. After cooling, don PPE, and add 10 mL of Milli-Q water followed by 0.8 mL of the 10 N H_2SO_4, tightly cap the digestion tube, vortex, and place in an oven at 105°C for 1 h. Allow tubes to cool.

8. Reweigh uncapped tubes to obtain sample volume by difference.

9. Samples that may result in P concentration greater that 30 μM in the digestion volume (>600 μg dry mass) will require a modified protocol that includes a dilution step. In this case, prepare a second standard curve (which will be diluted) that includes the ranges of P mass expected in the samples. After step 8 above, put 2 mL of the digestate from each sample (and from each of the tubes in the second standard curve) into a new clean tube, add 8 mL of the dilution reagent and vortex. The dilution scheme can be adapted depending on sample weight and approximate P content (see Online Worksheet 36.2). If you dilute samples, obtain weights at each step to calculate volumes and check the results of the undiluted and diluted standard curves, which should be identical.

10. Add 2 mL molybdate reagent, 0.8 mL ascorbic acid solution, and then 6.4 mL of Milli-Q. Vortex well.

11. After 10 min, but within 4 h, measure absorbance of the samples in a 1-cm cuvette at 880 nm on a spectrophotometer.

36.3.1.4 Data Analysis and Assessment of Consumer–Resource Imbalances

1. Check the standard curve used for estimation of P content (the R^2 should be >0.999 with good technique).

2. Calculate the P content of the unknown samples (see Online Worksheet 36.3).

3. Express C, N, and P contents as percentage dry mass of sample. Convert to molar ratios (C:N, C:P, and N:P) by first dividing percent content by the atomic mass of each element (12.01, 14.01, and 30.97 for C, N, and P, respectively) before calculating the ratio of the two elements.

4. Comparison of the molar ratio (e.g., C:P) of a consumer and its principal basal resource(s) is the most basic metric of elemental imbalance in trophic relationships. However, this simple comparison does not account for the inefficiencies of assimilation or the outlay of C required by respiration. More accurate estimations of elemental imbalance involve comparisons of resource elemental ratios with consumer TERs, which account for elemental routing more fully (Frost et al., 2006). See Advanced Method 4 for one method of estimating TERs. Finally, a caveat to calculating elemental resource—consumer imbalances always needs to be stressed, particularly in the case of macroinvertebrates: measurement of bulk material does not reflect what is actually ingested by most animals (Dodds et al., 2014; Hood et al., 2014). Selective feeding by herbivores and detritivores is the rule rather than the exception; this fact should always temper simple comparisons between calculated consumer demand and the elemental concentration in resource samples that are collected and analyzed as bulk material.

36.3.2 Basic Method 2: Measuring Organismal Nutrient Release Rates

In this method, we describe basic steps for measuring N and P excretion and egestion rates of animals. This protocol will work for a frog tadpole between 20 and 40 mm in total length, but could be adapted for other animals by changing the ratio of animal wet mass to filtered water. Published excretion rate—body mass relationships are useful for guiding this aspect of study design (e.g., Hall et al., 2007; Benstead et al., 2010; Vanni and McIntyre, 2016), but bear in mind that considerable variation exists for excretion rates at a given body mass (a very large published data set is available for detailed exploration; Vanni et al., 2017). Although the protocol below measures excreted N as NH_4-N, note that most aquatic organisms (which are typically ammonotelic) do not excrete N purely in the form of ammonia (although ammonia may represent a very high proportion; Wright, 1995). To measure total N excretion, a digestion similar to that described in Basic Method 1 is required, followed by measurement of total dissolved nitrogen (see Benstead et al., 2010).

36.3.2.1 Field Protocol

1. Filter some stream water, and maintain at a constant temperature. The fastest and simplest approach is to use a battery-powered peristaltic pump (e.g., Geopump) with an in-line 0.45-μm filter. If such a pump is not available, use an acid-washed side-arm flask with a 47-mm filter tower. Stream water should ideally be filtered the same day the experiment is conducted. Keep the filtered water in the stream to maintain constant temperature.
2. Measure out 100 mL of filtered water into 17 125-mL wide-mouth Nalgene bottles. Incubate the bottles in the stream or in a cooler with stream water to maintain ambient temperature. Ten of these bottles will receive similarly sized tadpoles, four will be used as container controls (i.e., just water), and three will be used to monitor oxygen availability (also containing a tadpole).
3. Collect and identify the individual tadpoles (see Chapter 17), and put one individual into each of the 10 treatment bottles. Incubations should last approximately 2 h.
4. It is prudent to monitor dissolved oxygen (DO) concentrations in the bottles, since tadpoles will consume oxygen during the incubation. Monitoring oxygen in the sample bottles will likely disturb the animals and contaminate the water; for this reason, create three sham bottles containing filtered water and a tadpole. Throughout the incubation, monitor oxygen concentrations in these sham bottles with a DO probe.
5. Place an ashed and preweighed 25-mm GF/F filter in an acid-washed 25-mm filter cassette. After the 2-h incubation, remove 40 mL of water from each bottle and use the filtered water to rinse four 20-mL glass tubes (two replicates for NH_4 and SRP analysis); then fill all four tubes with 10 mL of filtered sample water. Pull the remaining water from the vial, and push it through the syringe, discarding this filtrate. Retain the filter to estimate egestion rates and stoichiometry. Measure the length and wet weight of each animal, and return it to the stream or, if dry mass is required, develop a length—mass relationship following Benke et al. (1999).

36.3.2.2 Laboratory Protocol: Excretion

1. Analyze the water samples for phosphorus content following the SRP analysis protocol in Chapter 33.
2. Analyze the samples for NH_4-N following Holmes et al. (1999), as modified by Taylor et al. (2007). Make up the reagents as follows in clean, acid-washed, and triple Milli-Q rinsed glassware:
   a. Sodium sulfite solution: Add 0.75 g of sodium sulfite (Sigma S-4672) to ~50 mL of Milli-Q water in a 100-mL volumetric flask. Let dissolve and then top off to 100 mL with Milli-Q. Make a new sodium sulfite solution for each batch of working reagent.

b. Borate buffer: Add 80 g of sodium tetraborate (Sigma S-9640) to ~1.5 L of Milli-Q water in a 2-L volumetric flask. Add a stir bar to the solution, and place on a stirring plate for several hours. Do not heat the solution. Once dissolved, top off the volumetric flask to 2 L with Milli-Q water.

c. OPA solution: Add 4 g of OPA (Sigma P-1378) to ~70 mL of 100% ethanol (high-grade required) in a 100-mL volumetric flask (preferably amber), dissolve, and then fill to 100 mL with the ethanol. OPA is light sensitive, so this reagent should be mixed in dark or dim light and stored in an amber glass bottle.

d. Working reagent: In a 2-L brown polyethylene bottle, mix 2 L of borate buffer solution, 10 mL of sodium sulfite (not 100 mL, a common mistake), and 100 mL of OPA solution. This reagent should also be mixed in dark or dim light. The OPA working reagent is stable for 3 months when stored in the dark at room temperature (Holmes et al., 1999).

e. NH_4-N stock solution: Prepare a ~100 mg NH_4-N L^{-1} stock solution by adding 0.48 g $(NH_4)_2SO_4$ to 1 L of Milli-Q water in a 1-L volumetric flask. Prepare a 0.5 mg NH_4-N L^{-1} intermediate stock solution by adding 4.97 mL of stock solution to 1 L of Milli-Q water in a 1-L volumetric flask. For the greatest precision, measure the standards by mass on a 4-place balance.

f. NH_4-N standards: Minerals and salts in animal excreta create matrix effects in fluorometric analyses, which can lead to over- or underestimates of excretion rates (Whiles et al., 2009; Hood et al., 2014). Standard curves should always be created using sample water. Spike four of the water-only samples with the 0.5 mg NH_4-N L^{-1} intermediate stock to achieve an addition of $5-30\ \mu g\ N\ L^{-1}$ ($5\ \mu g\ N\ L^{-1} = 99.0\ \mu L$; $10\ \mu g\ N\ L^{-1} = 196.1\ \mu L$; $20\ \mu g\ N\ L^{-1} = 384.6\ \mu L$; and $30\ \mu g\ NL^{-1} = 566.0\ \mu L$). Again, measure by mass on a 4-place balance. See Online Worksheet 36.4 for detailed instructions for standard curves.

g. Add 3 mL of the OPA working reagent to each 10-mL water sample, and mix completely (e.g., vortex or cap and invert several times). Incubate the samples for 4 h in a dark place, although incubation times can range from 2 to 20 h with no loss of fluorescence (Holmes et al., 1999).

h. Background fluorescence: Add 3 mL of the OPA working reagent to each unique 10-mL water sample. Immediately measure fluorescence on a suitable fluorometer. Do not measure background fluorescence on spiked samples. *Note*: These samples are very easy to contaminate. Rinse the cuvette multiple times before each measurement.

i. Following the incubation period, measure the fluorescence on a fluorometer.

j. Calculate $\mu g\ NH_4$-N L^{-1} for each sample following Online Worksheet 36.5. Briefly, the NH_4-N concentration of an unknown sample is the product of the background-corrected fluorescence (sample fluorescence minus background fluorescence) and the slope of the relationship between added NH_4-N L^{-1} and the sample fluorescence.

36.3.2.3 Laboratory Protocol: Egestion

This approach for measuring C, N, and P egestion assumes that fecal material is distributed evenly across the filter. If the fecal material is not distributed equally, use the one-vial method described in Advanced Method 1. Dry the filters for particulate C, N, and P at 60°C for 2 days. Weigh each filter on 4-place balance to calculate the dry mass of the filtrate, then cut in half and weigh each half to calculate the proportion of filtrate on each half. Analyze one half for particulate P and the other half for C and N following the protocol in Basic Method 1.

36.3.2.4 Calculations

1. Individual excretion rates can be calculated as:

$$E_I^X = (X_a - X_c) \times V \times T^{-1} \times I^{-1} \tag{36.1}$$

where E_I^X = individual excretion rate of nutrient X (either N or P, mg X individual^{-1} h^{-1}), X_a and X_c are, respectively, the concentrations of nutrient X in the containers with animals or the controls (mg $X\ L^{-1}$), V = volume of water in the container during the incubation (L), T = incubation duration (h), and I = number of individuals used in the incubation. The mass-specific excretion rate (E_M^x, mg X dry mass^{-1} h^{-1}) is the quotient of the individual excretion rate and the mean dry mass of the animals used in the incubation.

2. The individual egestion rate (F_I^X, mg X individual^{-1} h^{-1}) can be calculated as:

$$F_I^x = (F_{x,a} - F_{x,c}) \times T^{-1} \times I^{-1} \tag{36.2}$$

where $F_{x,a}$ and $F_{x,c}$ are, respectively, the mass of nutrient X on the filters from the containers with animals or with filtered water, corrected for the proportion of filtrate analyzed. The mass-specific egestion rate (F_M^x, mg X dry mass^{-1} h^{-1}) is the quotient of the individual egestion rate and the mean mass of individuals in the container.

36.3.3 Advanced Method 1: Measuring C, N, and P Content in a Single Sample With the "One-Vial Technique"

The goal of this method is to determine amounts of C, N, and P (and their ratios) simultaneously in a single sample of particulate organic matter in which the presence of recalcitrant carbon is reasonably low (see Gibson et al., 2015 for additional details). The one-vial method is useful because it allows analysis of all three major elements in a single, low-mass sample. It also enables estimation of true among-sample variance in C:N:P ratios, which is otherwise impossible when elemental concentrations are derived from different samples. However, the standard digestion it employs is not optimized for some biomolecules, including chitin in insect exoskeletons, limiting its utility to certain sample types. Remember to wear your PPE during all phases of this procedure.

36.3.3.1 Preparation

1. Wearing PPE, acid-wash 20-mL borosilicate glass serum vials. Cleaning 27 vials allows one to run a reasonable number of samples in a single batch: 5 standards, 17 samples, 3 check standards, and 2 postdigestion pH checks.
2. Combust vials at 500°C for 2 h.
3. Dry samples at 60°C.

36.3.3.2 Recrystallize the Potassium Persulfate (24 h Before Analysis)

The recrystallization procedure removes impurities present in most potassium persulfate.

1. Wearing PPE, heat approximately 60 mL of double-deionized water to 60°C in a 1000-mL Erlenmeyer flask and stir continuously with a 2.5-cm stir bar. Add a thermometer.
2. While stirring, add 100 g of potassium persulfate. Turn off heat, but keep stirring until the persulfate is completely dissolved.
3. Filter the solution rapidly through a sintered glass funnel. The flask will be hot, so wear appropriate insulated gloves and other PPE.
4. Save filtrate, and rinse the original flask and filter.
5. Pour filtrate back into the original flask.
6. Cool the filtrate to 4°C by placing the Erlenmeyer flask in an ice bath. Swirl continuously to prevent solution from freezing.
7. Filter the 4°C solution through a sintered glass funnel, and rinse the crystals with one or two squeezes of ice-cold double-deionized water. Save the solid crystals.
8. Discard the filtrate, and rinse side-arm flask, the Erlenmeyer flask, and the glass filter with double-deionized water.
9. Fill the Erlenmeyer flask with 400–500 mL of double-deionized water and heat to 60°C, and stir continuously with a stir bar.
10. While stirring, add the solid from step 7 and mix until dissolved.
11. Repeat steps 6 and 7.
12. Dry crystals under vacuum. Store in a desiccator.

36.3.3.3 Digestion

1. Combine 0.15 M low-N NaOH and 3% recrystallized potassium persulfate in a volumetric flask to make digestate. Shake vigorously to dissolve.
2. Make standards. Dilute stock solutions to prepare five standard curve solutions. Stock solutions: carbon standard: 0.5 M sodium carbonate; nitrogen standard: 0.00071 M nitrogen as nitrate; phosphorus standard: 0.0016 M phosphorus as phosphate. Each standard curve solution should contain a mix of C, N, and P (Table 36.1).
3. Make organic check standards for quality assurance and quality control. Stock for check standards: carbon: 0.4 M glucose; phosphorus: 0.4 M tripolyphosphate; nitrogen: dried serine added gravimetrically using a microbalance. Each method standard should contain a mix of C, N, and P, and the ratio of C:N:P should vary and one check standard should be included for every 10 samples.
4. Add standard, check standard, or sample to 20-mL borosilicate glass serum vials.

TABLE 36.1 Suggested mass and stock volumes for creating C, N, and P standards with variation in C:N:P for Advanced Method 1.

Standard	μmol C	μL C Standard	μmol N	μL N Standard	μmol P	μL P Standard	C:N:P
1	5	10	0.5	70	0.05	31.25	100:10:1
2	10	20	0.75	105	0.1	62.5	100:7.5:1
3	30	60	1.5	210	0.5	312.5	60:3:1
4	70	140	4	560	1	625	70:4:1
5	100	200	10	1401	2	1250	50:5:1

5. Add 10 mL of digestate to vial.
6. Flush vial with argon gas for about 30 s. Immediately cap the vial with a rubber septum and crimp closed with an aluminum ring. Label rubber septum to identify vial.
7. Autoclave samples for 2.5 h at 121°C and 15 psi with a slow ramp up and ramp down procedure to minimize leaking.

36.3.3.4 Carbon Analysis

1. The pH of the samples should be ~ 2 after the digestion. At this pH, carbon will be driven out of solution as carbon dioxide gas in equilibrium with the partial pressure within the sealed vial. Carbon is analyzed as CO_2 on a gas chromatograph–mass spectrometer (GC-MS) and quantified relative to the argon.
2. The column for the GC-MS should be a 15-m open tubular column, 0.25-mm inside diameter with 0.25-μm film thickness set to 40°C. Set the injection port to 50:1 split injection at 250°C and 1.0 mL/min He mobile phase flow rate. Set mass selective detector to scan from 10.0 to 50.0 m/z with zero solvent delay.
3. Use a gas-tight syringe to inject 1 μL of headspace gas into the GC-MS. (1) Rinse syringe by inserting syringe directly through the rubber stopper of the sample vial and draw up headspace gas. Remove the syringe from the vial and release gas into the room. (2) Reinsert the syringe into the vial and draw up about 1 μL of headspace gas. (3) Immediately remove the syringe and inject the sample into the GC-MS. Carbon is analyzed relative to argon, so volume of injection does not have to be exactly 1 μL.
4. Determine the relative abundances of mass 44 (CO_2) and 40 (Ar) from the mass spectrum taken at the peak of the chromatogram.

36.3.3.5 Nitrogen and Phosphorus Analysis

1. After digestion, nitrogen and phosphorus liberated by the process are in the form of nitrate and phosphorus, respectively. Prior to analysis, all samples should be diluted by a factor of 10.
2. Analyze nitrogen on an ion chromatograph. If possible, use an eluent and column mixture that allows for the nitrate peak to emerge prior to the sulfate peak and for maximum separation between those two peaks (see Gibson et al., 2015).
3. Analyze phosphorus colorimetrically on a UV-Vis spectrophotometer using standard methods (see Basic Method 1).

36.3.4 Advanced Method 2: Evaluating the Relative Importance of Animal Excretion at the Ecosystem Scale

Evaluating the importance of nutrient release by animals at the ecosystem scale places release measurements in context and provides an understanding of how animal populations and communities influence ecosystem properties. The importance of nutrient excretion by animals is a function of a number of factors, including the absolute magnitude of population or community excretion rates, the size of other sources of nutrient regeneration or supply, autotroph and heterotroph nutrient demand, and the location and timing of nutrient release by animals (Vanni, 2002; McIntyre et al., 2008; Benstead et al., 2010; Griffiths and Hill, 2014; Atkinson et al., 2017). A variety of approaches have been used to evaluate the relative importance of nutrient release by animals. All approaches begin with estimating population or community excretion rates. These rates are then compared to ambient nutrient fluxes, nutrient uptake rates, or autotroph nutrient demand. An alternative approach, which we do not discuss here, is to experimentally separate the direct effects of animals on biofilms (e.g., grazing) from the indirect effect of animal excretion using a series of nested cages in a closed system (e.g., Knoll et al., 2009).

36.3.4.1 Metrics of Relative Ecosystem-Level Importance of Animal Excretion

1. The population or community excretion rate (μg nutrient m^{-2} d^{-1}) is the product of individual excretion rates (μg nutrient individual^{-1} d^{-1}) and density (individuals/m^2). Excretion rates vary with body size (Hall et al., 2007; Vanni and McIntyre, 2016), so when populations contain multiple size classes, the population excretion rate should be calculated as the sum of the excretion rates of each size class. Animal densities and the densities of each size class can be measured following the protocols in Chapters 14–17.

2. Volumetric excretion rates describe the average addition of nutrients to the water column along a focal reach, assuming perfect mixing and no uptake (McIntyre et al., 2008). Volumetric excretion (E_V) can be calculated as:

$$E_V = (E_A \times A \times T)/V \tag{36.3}$$

   where E_A = areal population excretion rate, A = reach area (length $\times$ width), T = travel time through the reach (reach length/mean water velocity), and V = volume (length $\times$ width $\times$ depth).

3. Excretion turnover distance is the distance required for excretion to turn over the ambient nutrient pool and is calculated as:

$$E_T = N \times L/E_V \tag{36.4}$$

   where N = ambient nutrient concentration, and L = reach length (McIntyre et al., 2008).

4. The proportion of nutrient demand supplied by excretion is the quotient of population excretion rate and ambient nutrient uptake rate. Ambient nutrient uptake rates can be measured following the protocols in Chapter 31.

5. The proportion of autotroph nutrient demand supplied by excretion (PADE; Hall et al., 2003) can be calculated as:

$$\text{PADE} = E_A/(\text{GPP} \times B_{XC} \times 0.5) \tag{36.5}$$

   where GPP = gross primary production (μg C m^{-2} d^{-1}), $B_{X:C}$ = nutrient:C ratio of biofilms in the focal stream, and 0.5 reflects the assumption that autotrophic respiration is 50% of GPP (Hall and Tank, 2003). Protocols for measuring biofilm $X:C$ can be found in Basic Method 1 of this chapter, whereas protocols for measuring GPP can be found in Chapter 34.

36.3.5 Advanced Method 3: Accounting for the Effects of Fasting and Stress on Organismal Excretion Rates

In Basic Method 2, we outlined simple procedures for measuring nutrient release rates by animals. Although such approaches are commonly used, they are known to be subject to bias due to the effects of stress (which elevate excretion rates) and cessation of feeding (i.e., fasting, which can depress excretion rates) over the duration of incubation. Refined methods have been developed to correct for these sources of bias (see Whiles et al., 2009), which we now describe.

36.3.5.1 Field Methods

1. Create a holding container out of a 19-L plastic bucket. Position hardware cloth $\sim$10 cm above the bottom to separate the animals from their feces and so prevent coprophagy. Fill the containers with stream water or filtered stream water if the water is particle rich.
2. Filter incubation water as in Basic Method 2.
3. Measure out 100 mL of filtered water into 76 125-mL wide-mouth Nalgene bottles. Following Whiles et al. (2009), we will create four fasting groups (0, 1, 2, and 6 h without food prior to measurement), with four incubation times (30, 60, 120, and 180 min) within each fasting group. Each treatment will be replicated four times. Three replicate container controls will be used for each incubation time.
4. Add $\sim$64 similarly sized individuals to the holding container.
5. At the termination of each fasting period, remove 16 individuals from the holding tanks and randomly distribute each individual to 1 of 16 bottles (four per incubation period).
6. To monitor oxygen consumption, create sham bottles for each incubation period as described in the basic method.
7. At the termination of the incubation period, remove 40 mL of water from each bottle with a 60-mL syringe. Use the filtered water to rinse four 20-mL glass tubes, and then fill each with 10 mL of filtered water. Pull the remaining water from the bottle, and push it through the syringe. Retain the filter to estimate egestion rates and stoichiometry (see Basic Method 2). Measure the length and wet weight of each animal, and return to the stream or, if dry mass is required, develop a length–mass relationship following Benke et al. (1999).
8. Follow the laboratory protocol described in Basic Method 2.

36.3.5.1 Calculations

1. Calculate mass-specific excretion and egestion rates as described in Basic Method 2.
2. Assuming that mass-specific excretion rates decline log-linearly with incubation time, use an ANCOVA to evaluate the relationship between natural log-transformed mass-specific excretion rates and the two predictor variables (incubation time and fasting duration). If the slopes are homogeneous, evaluate differences in intercepts among fasting times. A significant difference among intercepts indicates that there was an effect of fasting time.
3. When fasting effects are not evident (no difference among intercepts), estimate the baseline excretion rate with the nonlinear model:

$$Y = \alpha + \beta e^{-\gamma t} + \varepsilon \tag{36.6}$$

where Y = mass-specific excretion rate at time t, α = baseline excretion rate, β = increase in excretion rate caused by handling stress, γ = rate at which the handling stress signal decays over time, and ε is a normally distributed error term. Use nonlinear regression to fit the above model to the pooled data.
4. If fasting effects are evident, estimate the baseline excretion rate with the nonlinear model:

$$Y = \alpha - \delta\left(1 - e^{-\theta(t+t_s)}\right) + \beta e^{-\gamma t} + \varepsilon \tag{36.7}$$

where x = rate at which the fasting effect increases over time and t_s = fasting time. Use nonlinear regression to fit the above model to the pooled data.

36.3.6 Advanced Method 4: Estimating Threshold Elemental Ratios

36.3.6.1 Periphyton Incubation

1. The goal of the incubation is to provide grazers with a periphyton food source that has a similar community structure but a wide range of C:P. Periphyton community structure changes over time after colonization, so periphyton grown on clay tiles and fed to grazers over the 2-week to month-long feeding experiment will need to be incubated for similar amounts of time.
2. Place clay tiles (~ 10 cm^2) in a low-nutrient stream for approximately 2–6 weeks to allow for periphyton colonization. Place sufficient new tiles in the stream approximately weekly to generate enough algae-colonized tiles to complete the feeding experiment.
3. Use plastic totes (120 L) or similar containers to serve as a nutrient incubator chamber for periphyton. A minimum of four incubation chambers are needed; one chamber per phosphate amendment. Fill each chamber with a known volume (~ 60 L) of stream water, preferably originating from a stream or lake with low dissolved P concentrations (<20 µg P/L). Tap water can be dechlorinated and amended with P if stream water collection is not feasible.
4. Add an appropriate volume of stock solution of Na_2HPO_4 (e.g., 1000 mg P/L) to each incubation chamber to result in an ambient stream water concentration (no P-amendment), and 50, 100, and 500 µg P/L concentrations. Utilize the soluble reactive phosphorus analysis and standard curve protocol described in Chapter 33 to arrive at the appropriate amounts of stock solution to add to each incubation chamber and to verify chamber concentrations over time. Add ammonium or nitrate–nitrogen salt homogenously across treatments at a target concentration of 1 mg dissolved inorganic nitrogen (DIN)/L to encourage P-limitation of periphyton growth. Change the incubation water if floating algae begin to grow in the incubation containers.
5. Place the incubation chambers in a greenhouse where they can receive sufficient light. Promote water circulation and aeration in the chambers using an air pump. It would be ideal to use recirculating or flow-through mesocosms as incubation chambers, if they are available.
6. Periphyton-colonized tiles can be placed in the incubation chambers for 9–18 days or for whatever length of time is necessary to achieve differences in periphyton C:P prior to presenting them to grazers (Stelzer and Lamberti, 2001; Evans-White and Lamberti, 2006).

36.3.6.2 Growth Experiment

Preparation

1. Construct growth chambers from paired 250-mL plastic specimen cups. The bottom of one cup should be cut using a hacksaw or a similar tool. Fasten Nitex mesh or bridal tulle to the bottom using a rubber band. Place the mesh cup inside an unaltered specimen cup. The stacked cup design allows larval egesta to drop out to reduce coprophagy (see Fig. 36.4A). Construct larger growth chambers if necessitated by the size of the study organism. We suggest

FIGURE 36.4 Example of an experimental growth chamber composed of two stacked specimen cups separated by mesh through which feces can fall (A). Typical setup for a growth experiment with air bubblers supplying each chamber (B). Possible relationship between algal P:C and dissolved inorganic phosphorus concentrations in incubation chambers (C). Example of relationship between instantaneous growth rates (IGRs) and food C:P (D). *Photo credits: (A, B) Halvor Halvorson.*

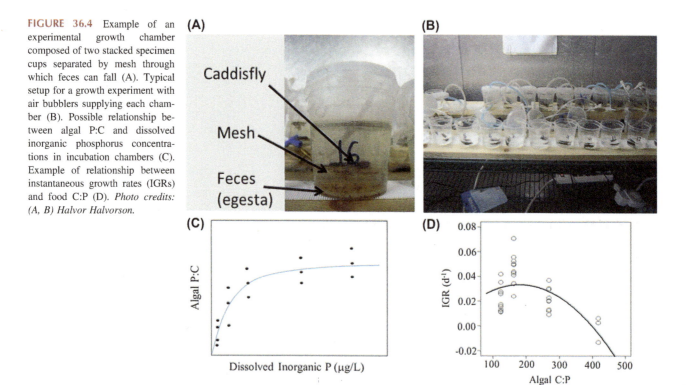

constructing at least 10 replicate growth chambers per food resource ratio level; the total number of experimental replicates in this example growth experiment is 40.

2. Fill chambers with a known volume of filtered stream water (~150 mL), and place the chambers in an environmental chamber at a constant, appropriate temperature and light regime that represents that of the study stream (see Fig. 36.4B). Aerate each chamber. Purchase splitters that allow a single aeration pump to aerate several chambers, if needed.

3. A dominant grazer species should be carefully collected from your study stream and placed in stream water with an aerator for transport back to the laboratory. You will need to collect at least 60—70 similarly sized individuals, even if you plan to stock only one individual per experimental chamber; a subset (10—15 individuals) will be needed for step 4 below.

4. Measure the total length of a subset (10—15 individuals) of your grazer species using a dissecting scope and place individuals in separate preweighed aluminum tins for freeze-drying. After drying (24—48 h), weigh the sample to the nearest 0.01 mg. The average mass of this subsample of the population will represent mass at time 0 to be placed into the instantaneous growth rate (IGR) equation (Chapter 35; Eq. 35.5; W_t). Process the dried study organisms for C, N, and P content as described in Basic Method 1 or Advanced Method 1 to represent initial elemental composition.

5. Measure length of the remaining living individuals, if possible, before carefully placing them in their respective experimental chamber. This measurement step allows estimation of the mean and variance of the initial size of experimental individuals.

36.3.6.3 Experimental Procedure

1. Distribute periphyton-colonized tiles representing each P incubation level randomly across the appropriate growth chambers. Scrape a subset of tiles, and process the resulting algal slurry for C, N, and P content as described in Basic Method 1 in this chapter, so that the resource ratio gradient can be quantified (Fig. 36.4C).

2. Place individuals of your grazer species carefully into their respective experimental chamber. It may be appropriate to place two to three individuals in each cup if the chosen species is small and intraspecific competition can be discounted.

3. The study organisms should never run out of food. Check tiles daily for visible periphyton presence, and replace as needed. Take a sample as described above for periphyton C, N, and P on each feeding event so that changes in periphyton C:N:P can be assessed over the duration of the experiment.

4. Check DO frequently in the feeding chambers with a portable meter to insure that appropriate levels (>5 mg/L) are maintained given the small volume.
5. Note any deaths or emergence of individuals. At the conclusion of the experiment, measure the remaining individuals and place them in tins to freeze-dry. The experiment should run long enough for measurable growth to be detected. Depending on taxon and environmental conditions, this may require a minimum of 2 weeks to a month.
6. After drying (24−48 h), weigh the study organism and tin to the nearest 0.01 mg. Subtract the initial tin mass from the total sample mass to obtain the dry mass of each individual. The average mass of this subsample of the population will represent mass at the end of the growth period to be placed into the IGR equation (Chapter 35; Eq. 35.5; $W_{t+\Delta t}$).
7. Process the dried study organisms for C, N, and P content as described in Basic Method 1 or Advanced Method 1 to represent final elemental composition.

36.3.6.4 Data Analysis

1. IGR based on mass will be calculated using Eq. (35.5) in Chapter 35. Elemental masses at the start and the finish of the experiment can be inserted instead of the initial and final mass to estimate element-specific growth rates.
2. Estimate the relationship between IGR and algal C:P using a quadratic regression (e.g., Fig. 36.4D) as follows,

$$\text{IGR} = cx^2 + bx + a \tag{36.8}$$

where x = food resource ratio of interest, c = quadratic coefficient, b = linear coefficient, and a = y-intercept (Bullejos et al., 2014; Benstead et al., 2014). If a statistically significant quadratic relationship is found, the food resource ratio at which optimal growth occurs (i.e., the TER) can be estimated using the following parameters from the quadratic equation:

$$\text{TER} = \frac{-b}{2c} \tag{36.9}$$

3. Quadratic regressions can be performed in Microsoft Excel or SigmaPlot if access to statistical software is not possible.

36.4 QUESTIONS

1. How do organismal life history traits or characteristics influence a consumer's elemental requirements?
2. Can a consumer's elemental content change predictably through ontogeny? If so, how and why?
3. How does selective feeding affect the estimation of consumer−resource imbalances?
4. If your species has a case or a shell, which mass (whole body or soft tissue) should be used in the threshold elemental ratio growth estimate?
5. Why might two organisms with similar C and P body content differ in their $\text{TER}_{C:P}$?
6. Can excess elements reduce growth rates? If so, how?

36.5 MATERIALS AND SUPPLIES

Laboratory Materials (B# and A# refer to Basic and Advanced Methods)
Sorting trays (B1)
Wash bottles (B1, A1)
In-line 0.45-μm filter (Geotech 73050004) and Geopump (B1, B2, A3)
Forceps (B1)
2-mL cryotubes (B1)
Sample bags (B1)
Ball mill (B1)
Pestle and mortar (B1)
Wiley Mill (B1)
Homogenizer (B1)
Side-arm flask (B1, A1)
NIST standards (bovine muscle, spinach) (B1, A1, A4)
Vortexer (B1, B2, A3, A4)

H_2SO_4 (B1)
Antimony potassium tartrate (B1)
Ammonium heptamolybdate 4-hydrate (B1)
500-mL volumetric flask (B1)
1-L volumetric flask (B1)
1-L amber volumetric flask (B1)
100-mL amber volumetric (B1)
Ascorbic acid (B1)
25-mL borosilicate glass digestion tubes (B1)
Freeze dryer (B1, A4)
Desiccator (B1, A1)
Drying oven (B1, A1)
Muffle furnace (B1, A1)
Ashed and weighed 25-mm GF/F filters (B1, B2, A3)
20-mL glass tubes (B1, B2, A3)
125-mL wide-mouth Nalgene bottles (B2, A3)
Oxygen meter (B2, A3)
60-mL syringes (B2, A3)
25-mm filter cassette for syringes (B2, A3)
Ruler (B2, A3)
Scale (B2, A3)
Sodium sulfite (Sigma S-4672) (B2, A3)
Phthaldialdehyde (OPA, Sigma P-1378) (B2, A3)
Sodium tetraborate (Sigma S-9640) (B2, A3)
100% ethanol (high-grade required) (B2, A3)
100-mL volumetric flask (B2, A3)
2-L volumetric flask (B2, A3)
2-L brown polyethylene bottle (B2, A3)
$(NH_4)_2SO_4$ (B2, A3)
Fluorometer (B2, A3)
Parafilm (A1)
20-mL borosilicate glass serum vials (A1)
Butyl septa (A1)
Aluminum rings (A1)
Crimper (A1)
Decrimper or needle nose pliers (A1)
Pipettes (A1)
Carbonate standard—(Dionex 037162) (A1)
Nitrate standard—(Ricca 5457-16) (A1)
Potassium phosphate (Fisher S80146-1) (A1)
Glucose (Sigma G5767) (A1)
Serine (Sigma 84959 dried at 60°C for 24 h) (A1)
Tripolyphosphate (Sigma 238503) (A1)
Low-N potassium persulfate (J.T. Baker 3239-01) (A1)
Low-N NaOH (Sigma 221465) (A1)
Argon gas (A1)
Erlenmeyer flask (A1)
Sintered glass funnel (A1)
Filtration pump (A1)
Ice bath (A1)
Gas chromatograph—mass spectrophotometer (A1)
Gas-tight syringe (A1)
Ion chromatograph (A1)
UV-Vis spectrophotometer (A1)

Hot plate with stirrer (A1)
Stir bars (A1)
250-mL plastic specimen cups (A4)
500-µm Nitex mesh or bridal tulle (A4)
Rubber bands (A4)
Hacksaw (A4)
Scissors (A4)
Aquarium pump tubing and splitters (A4)
Aquarium pumps (A4)
Environmental chamber (A4)
Aluminum tins (A4)
Dissecting scope with a calibrated ocular micrometer (A4)
Four to five plastic totes for periphyton incubation (A4)
Unglazed clay tiles (A4)
Scrub brush (A4)

ACKNOWLEDGMENTS

We thank Alex Huryn and Carla Atkinson for comments and suggestions that improved this chapter.

REFERENCES

Atkinson, C.L., Capps, K.A., Rugenski, A.T., Vanni, M.J., 2017. Consumer-driven nutrient dynamics in freshwater ecosystems: from individuals to ecosystems. Biological Reviews. http://dx.doi.org/10.1111/brv.12318.

Back, J.A., King, R.S., 2013. Sex and size matter: ontogenetic patterns of nutrient content of aquatic insects. Freshwater Science 32, 837—848.

Balseiro, E., Albariño, R., 2006. C—N mismatch in the leaf litter-shredder relationship of an Andean Patagonian stream detritivore. Journal of the North American Benthological Society 25, 607—615.

Benke, A.C., Huryn, A.D., Smock, L.A., Wallace, J.B., 1999. Length-mass relationships for freshwater macroinvertebrates in North America with particular reference to the southeastern United States. Journal of the North American Benthological Society 18, 308—343.

Benstead, J.P., Cross, W.F., March, J.G., McDowell, W.H., Ramírez, A., Covich, A.P., 2010. Biotic and abiotic controls on the ecosystem significance of consumer excretion in two contrasting tropical streams. Freshwater Biology 55, 2047—2061.

Benstead, J.P., Hood, J.M., Whelan, N.V., Kendrick, M.R., Nelson, D., Hanninen, A.F., Demi, L.M., 2014. Coupling of dietary phosphorus and growth across diverse fish taxa: a meta-analysis of experimental aquaculture studies. Ecology 95, 2768—2777.

Boros, G., Mozsár, A., 2015. Comparison of different methods used for phosphorus determination in aquatic organisms. Hydrobiologia 758, 235—242.

Bowman, M.F., Chambers, P.A., Schindler, D.W., 2005. Changes in stoichiometric constraints on epilithon and benthic macroinvertebrates in response to slight nutrient enrichment of mountain rivers. Freshwater Biology 50, 1836—1852.

Bullejos, F.J., Carillo, P., Gorokhova, E., Medina-Sánchez, J.M., Balseiro, E.G., Villar-Argaiz, M., 2014. Shifts in food quality for herbivorous consumer growth: multiple golden means in the life history. Ecology 95, 1272—1284.

Capps, K.A., Atkinson, C.L., Rugenski, A.T., 2015. Consumer-driven nutrient dynamics in freshwater ecosystems: an introduction. Freshwater Biology 60, 439—442.

Cross, W.F., Benstead, J.P., Frost, P.C., Thomas, S.A., 2005. Ecological stoichiometry in freshwater benthic systems: recent progress and perspectives. Freshwater Biology 50, 1895—1912.

Cross, W.F., Benstead, J.P., Rosemond, A.D., Wallace, J.B., 2003. Consumer-resource stoichiometry in detritus-based streams. Ecology Letters 6, 721—732.

Danger, M., Gessner, M.O., Bärlocher, F., 2016. Ecological stoichiometry of aquatic fungi: current knowledge and perspectives. Fungal Ecology 19, 100—111.

Dodds, W.K., et al., 2014. You are not always what we think you eat: selective assimilation across multiple whole-stream isotopic tracer studies. Ecology 95, 2757—2767.

Elser, J.J., Elser, M.M., MacKay, N.A., Carpenter, S.R., 1988. Zooplankton-mediated transitions between N- and P-limited algal growth. Limnology and Oceanography 33, 1—14.

Elser, J.J., Fagan, W.F., Denno, R.F., Dobberfuhl, D.R., Folarin, A., Huberty, A., Interlandi, S., Kilham, S.S., McCauley, E., Schulz, K.L., Siemann, E.H., Sterner, R.W., 2000. Nutritional constraints in terrestrial and freshwater food webs. Nature 408, 578—580.

Elser, J.J., Urabe, J., 1999. The stoichiometry of consumer-driven nutrient recycling: theory, observations, and consequences. Ecology 80, 735—751.

Evans-White, M.A., Lamberti, G.A., 2006. Stoichiometry of consumer-driven nutrient recycling across nutrient regimes in streams. Ecology Letters 9, 1186—1197.

Evans-White, M.A., Stelzer, R.S., Lamberti, G.A., 2005. Taxonomic and regional patterns in benthic macroinvertebrate elemental composition in streams. Freshwater Biology 50, 1786—1799.

Frost, P.C., Benstead, J.P., Cross, W.F., Hillebrand, H., Larson, J.H., Xenopoulos, M.A., Yoshida, T., 2006. Threshold elemental ratios of carbon and phosphorus in aquatic consumers. Ecology Letters 9, 774−779.

Frost, P.C., Elser, J.J., 2002. Growth responses of littoral mayflies to the phosphorus content of their food. Ecology Letters 5, 232−240.

Frost, P.C., Evans-White, M.A., Finkel, Z.V., Jensen, T.C., Matzek, V., 2005. Are you what you eat? Physiological constraints on organismal stoichiometry in an elementally imbalanced world. Oikos 109, 18−28.

Frost, P.C., Stelzer, R.S., Lamberti, G.A., Elser, J.J., 2002. Ecological stoichiometry of trophic interactions in the benthos: understanding the role of C:N:P ratios in lentic and lotic habitats. Journal of the North American Benthological Society 21, 515−528.

Gibson, C.A., O'Reilly, C.M., Conine, A.L., Jobs, W., Belli, S., 2015. Organic matter carbon, nitrogen, and phosphorus from a single persulfate digestion. Limnology and Oceanography, Methods 13, 202−211.

Goodwin, C.M., Cotner, J.B., 2015. Stoichiometric flexibility in diverse aquatic heterotrophic bacteria is coupled to differences in cellular phosphorus quotas. Frontiers in Microbiology 6, 159.

Griffiths, N.A., Hill, W.R., 2014. Temporal variation in the importance of a dominant consumer to stream nutrient cycling. Ecosystems 17, 1169−1185.

Grimm, N.B., 1988a. Feeding dynamics, nitrogen budgets, and ecosystem role of a desert stream omnivore, *Agosia chrysogaster* (Pisces: Cyprinidae). Environmental Biology of Fishes 21, 143−152.

Grimm, N.B., 1988b. Role of macroinvertebrates in nitrogen dynamics of a desert stream. Ecology 69, 1884−1893.

Gulis, V., Kuehn, K.A., Schoettle, L.N., Leach, D., Benstead, J.P., Rosemond, A.D. Changes in nutrient stoichiometry, elemental homeostasis and growth rate of aquatic litter-associated fungi in response to inorganic nutrient supply. ISME Journal (in review).

Hall, R.O., Koch, B.J., Marshall, M.C., Taylor, B.W., Tronstad, L.M., 2007. How body size mediates the role of animals in nutrient cycling in aquatic ecosystems. In: Hildrew, A.G., Raffaelli, G.G., Edmonds-Brown, R. (Eds.), Body-Size: The Structure and Function of Aquatic Ecosystems. Cambridge University Press, U.K., pp. 286−305

Hall, R.O., Tank, J.L., 2003. Ecosystem metabolism controls nitrogen uptake in streams in Grand Teton National Park, Wyoming. Limnology and Oceanography 48, 1120−1128.

Hall, R.O., Tank, J.L., Dybdahl, M.F., 2003. Exotic snails dominate nitrogen and carbon cycling in a highly productive stream. Frontiers in Ecology 1, 407−411.

Halvorson, H.M., Scott, J.T., Sanders, A.J., Evans-White, M.A., 2015. A stream insect detritivore violates common assumptions of threshold elemental ratio bioenergetics models. Freshwater Science 34, 508−518.

Hamilton, S.K., Sippel, S.J., Bunn, S.E., 2005. Separation of algae from detritus for stable isotope or ecological stoichiometry studies using density fractionation in colloidal silica. Limnology and Oceanography: Methods 3, 149−157.

Hendrixson, H.A., Sterner, R.W., Kay, A.D., 2007. Elemental stoichiometry of freshwater fishes in relation to phylogeny, allometry and ecology. Journal of Fish Biology 70, 121−140.

Hill, W.R., Rinchard, J., Czesny, S., 2011. Light, nutrients, and fatty acid composition of stream periphyton. Freshwater Biology 56, 1825−1836.

Hładyz, S., Gessner, M.O., Giller, P.S., Pozo, J., Woodward, G., 2009. Resource quality and stoichiometric constraints on stream ecosystem functioning. Freshwater Biology 54, 957−970.

Holmes, R.M., Aminot, A., Kerouel, R., Hooker, B.A., Peterson, B.J., 1999. A simple and precise method for measuring ammonium in marine and freshwater ecosystems. Canadian Journal of Fisheries and Aquatic Sciences 56, 1801−1808.

Hood, J.M., McNeely, C., Finlay, J.C., Sterner, R.W., 2014. Selective feeding determines patterns of nutrient release by stream invertebrates. Freshwater Science 33, 1093−1107.

Hood, J.M., Vanni, M.J., Flecker, A.S., 2005. Nutrient recycling by two phosphorus rich grazing catfish: the potential for phosphorus-limitation of fish growth. Oecologia 146, 247−257.

Kaspari, M., Powers, J.S., 2016. Biogeochemistry and geographical ecology: embracing all twenty-five elements required to build organisms. American Naturalist 188, S62−S73.

Knoll, L.B., McIntyre, P.B., Vanni, M.J., Flecker, A.S., 2009. Feedbacks on consumer nutrient recycling on producer biomass and stoichiometry: separating direct and indirect effects. Oikos 118, 1732−1742.

Lauridsen, R.B., Edwards, F.K., Bowes, M.J., Woodward, G., Hildrew, A.G., Ibbotson, A.T., Jones, J.I., 2012. Consumer-resource elemental imbalances in a nutrient-rich stream. Freshwater Science 31, 408−422.

Makino, W., Cotner, J.B., Sterner, R.W., Elser, J.J., 2003. Are bacteria more like plants or animals? Growth rate and resource dependence of bacterial C:N:P stoichiometry. Functional Ecology 17, 121−130.

McIntyre, P.B., Flecker, A.S., 2010. Ecological stoichiometry as an integrative framework in stream fish ecology. American Fisheries Society Symposium 73, 539−558.

McIntyre, P.B., Flecker, A.S., Vanni, M.J., Hood, J.M., Taylor, B.W., Thomas, S.A., 2008. Fish distributions and nutrient recycling in a Neotropical stream: can fish create biogeochemical hotspots? Ecology 89, 2335−2346.

Morse, N.B., Wollheim, W.M., Benstead, J.P., McDowell, W.H., 2012. Effects of suburbanization on foodweb stoichiometry of detritus-based streams. Freshwater Science 31, 1202−1213.

Murphy, J., Riley, J.P., 1962. A modified single solution method for the determination of phosphate in natural waters. Analytica Chimica Acta 27, 31−36.

Ostrofsky, M.L., 1997. Relationship between chemical characteristics of autumn-shed leaves and aquatic processing rates. Journal of the North American Benthological Society 16, 750−759.

Persson, J., Fink, P., Goto, A., Hood, J.M., Jonas, J., Kato, S., 2010. To be or not to be what you eat: regulation of stoichiometric homeostasis among autotrophs and heterotrophs. Oikos 119, 741−751.

Reiners, W.A., 1986. Complementary models for ecosystems. The American Naturalist 127, 59—73.

Schaus, M.H., Vanni, M.J., Wissing, T.E., Bremigan, M.T., Garvey, J.E., Stein, R.A., 1997. Nitrogen and phosphorus excretion by detritivorous gizzard shad in a reservoir ecosystem. Limnology and Oceanography 42, 1386—1397.

Schindler, D.E., Eby, L.A., 1997. Stoichiometry of fishes and their prey: implications for nutrient recycling. Ecology 78, 1816—1831.

Scott, J.T., Cotner, J.B., LaPara, T.M., 2012. Variable stoichiometry and homeostatic regulation of bacterial biomass elemental composition. Frontiers in Microbiology 3, 1—7.

Singer, G.A., Battin, T.J., 2007. Anthropogenic subsidies alter stream consumer-resource stoichiometry, biodiversity, and food chains. Ecological Applications 17, 376—389.

Small, G.E., Pringle, C.M., 2010. Deviation from strict homeostasis across multiple trophic levels in an invertebrate consumer assemblage exposed to high chronic phosphorus enrichment in a Neotropical stream. Oecologia 162, 581—590.

Stelzer, R.S., Lamberti, G.A., 2001. Effects of N:P ratio and total nutrient concentration on stream periphyton community structure, biomass, and elemental composition. Limnology and Oceanography 46, 356—357.

Stelzer, R.S., Lamberti, G.A., 2002. Ecological stoichiometry in running waters: periphyton chemical composition and snail growth. Ecology 83, 1039—1051.

Sterner, R.W., 1990. The ratio of nitrogen to phosphorus resupplied by herbivores: zooplankton and the algal competitive arena. American Naturalist 136, 209—229.

Sterner, R.W., Elser, J.J., 2002. Ecological Stoichiometry: The Biology of Elements from Molecules to the Biosphere. Princeton University Press, Princeton, New Jersey.

Sterner, R.W., Schulz, K.L., 1998. Zooplankton nutrition: recent progress and a reality check. Aquatic Ecology 32, 261—279.

Taylor, B.W., Keep, C.F., Hall, R.O., Koch, B.J., Tronstad, L.M., Flecker, A.S., Ulseth, A.J., 2007. Improving the fluorometric ammonium method: matrix effects, background fluorescence, and standard additions. Journal of the North American Benthological Society 26, 167—177.

Urabe, J., Watanabe, Y., 1992. Possibility of N or P limitation for planktonic cladocerans: an experimental test. Limnology and Oceanography 37, 244—251.

Vanni, M.J., 2002. Nutrient cycling by animals in freshwater ecosystems. Annual Review of Ecology and Systematics 33, 341—370.

Vanni, M.J., Flecker, A.S., Hood, J.M., Headworth, J.L., 2002. Stoichiometry of nutrient recycling by vertebrates in a tropical stream: linking biodiversity and ecosystem function. Ecology Letters 5, 285—293.

Vanni, M.J., McIntyre, P.B., 2016. Predicting nutrient excretion of aquatic animals with metabolic ecology and ecological stoichiometry: a global synthesis. Ecology 97, 3460—3471.

Vanni, M.J., et al., 2017. A global database of nitrogen and phosphorus excretion rates of aquatic animals. Ecology. http://dx.doi.org/10.1002/ecy.1792.

Wang, H., Sterner, R.W., Elser, J.J., 2012. On the "strict homeostasis" assumption in ecological stoichiometry. Ecological Modelling 243, 81—88.

Whiles, M.R., Huryn, A.D., Taylor, B.W., Reeve, J.D., 2009. Influence of handling stress and fasting on estimates of ammonium excretion by tadpoles and fish: recommendations for designing excretion experiments. Limnology and Oceanography, Methods 7, 1—7.

Wright, P.A., 1995. Nitrogen excretion: three end products, many physiological roles. Journal of Experimental Biology 198, 273—281.

Ecosystem Assessment

F. Richard Hauer and Gary A. Lamberti

Stream ecologists observed over a century ago that stream and river organisms respond in very predictable ways to changes in environmental conditions. Since those early observations, the science of ecosystem assessment has been founded on ecological theory and the development of sophisticated methodologies to measure ecological departure from naturally occurring conditions. During the 1970s and 1980s, environmental laws such as the US Clean Water Act and similar laws in Europe, Canada, and Australia created a societal need to develop theory, protocols, and reference data supporting water quality and environmental standards. In many countries, environmental assessment and environmental impact assessment now have specific meanings based on ecological practice and are codified into regulatory law. Assessment refers to an evaluation of environmental conditions such as a proposed plan, policy, or program that could have a measurable effect on the environment. In this section, chapters address the theory and methods behind using biota as indicators of natural, impacted, and impaired stream systems. Chapter 37 provides a basic framework for conducting ecological assessment using the ubiquitous benthic algae; methods include basic rapid bioassessment protocols as well as development of metrics to evaluate ecological integrity. In Chapter 38, methods to measure and evaluate stream macroinvertebrates, which are known to respond dramatically and with taxon specificity to a wide variety of pollution types, are described and include advanced approaches to metric development and application. Chapter 39 provides the basis for criteria and stressors affecting multiple measures of fish assemblage structure and function and "brings to life" the utility of stream fishes in the Index of Biotic Integrity. Chapter 40 focuses on the integration of laboratory, field, and experimental studies and the identification of cause-and-effect relationships that permit the development of "weight-of-evidence" in ecological risk assessments for streams and rivers. These four concluding chapters in *Methods in Stream Ecology* illustrate the breadth of application and knowledge needed across the multiple subdisciplines of stream structure (Volume 1) and stream function (Volume 2) to develop comprehensive and tested methods for assessment of environmental health and ecological integrity of stream ecosystems.

Chapter 37

Ecological Assessment With Benthic Algae

R. Jan Stevenson[1] and Scott L. Rollins[2]

[1]Department of Integrative Biology, Michigan State University; [2]Department of Life Sciences, Spokane Falls Community College

37.1 INTRODUCTION

Integrated ecological assessments provide the information needed to answer a series of questions for managing ecosystems (Stevenson, 2014). What are the conditions of valued attributes in the ecosystem? Are they good enough or are they so degraded that we should plan and implement restoration efforts? If conditions are degraded badly, what pollutants need to be reduced and what human activities are generating the pollutants? Thus, ecological assessments can involve much more than measuring biological conditions. They can also characterize and diagnose pollutants and human activities that impair ecosystem goods and services (see review by Stevenson, 2014).

Excessive algal growth and changes in species composition have long been used to assess ecological conditions in streams. Early assessments inferring water quality based on occurrence of pollution sensitive or tolerant species were common in Europe (Kolkwitz and Marsson, 1908). As the second half of the 20th century started, Patrick et al. (1954) introduced new concepts on the use of species richness and evenness of diatom assemblages as indicators of biological condition. Diatoms and other algae have since been used as indicators of ecological conditions in rivers (Kelly et al., 2008), lakes (Lambert et al., 2008), wetlands (Lane and Brown, 2007; Gaiser, 2009), and coastal zones (Bauer et al., 2007) around the world (Stevenson and Smol, 2015).

Algae are important in ecological assessments because they have valued ecological attributes, are sources of problems, and serve as good indicators. Algae are an important base of food webs in most aquatic ecosystems (Minshall, 1978). Excessive accrual of algal biomass can cause problems by depleting dissolved oxygen supplies, altering habitat structure for aquatic invertebrates and fish, generating taste and odor problems in drinking water supplies, and producing toxic substances with potential health impacts (Palmer, 1962; Holomuzki and Short, 1988; Carmichael, 1997; Stevenson et al., 2012). Algal assessments can more precisely characterize some environmental conditions (e.g., nutrient concentrations, Stevenson, 2001; Stevenson et al., 2010) in aquatic ecosystems than one-time sampling because (1) species are particularly sensitive to some environmental conditions, (2) different species have different sensitivities to contaminants, and (3) the development of species composition of algal assemblages takes long enough that algal indices of environmental conditions vary less than physical and chemical conditions that vary diurnally and with weather-related events (e.g., runoff and floods).

Many characteristics of algal assemblages are used in ecological assessments (see review by Stevenson and Smol, 2015). Biomass of algae can be estimated by direct visual assessments [e.g., Secchi disk and rapid periphyton surveys (RPS)] and by sampling algae from known areas of substratum or volumes of water with subsequent assays of ash-free dry mass (AFDM), chlorophyll a, cell densities, and cell volumes (Humphrey and Stevenson, 1992; Stevenson et al., 2006). Nutrient content, nutrient biomass ratios, and pigment ratios have been used to predict nutrient limitation and health of algal assemblages in streams (Humphrey and Stevenson, 1992; Peterson and Stevenson, 1992). Species composition of algae in samples and species' environmental sensitivities, optima, and tolerances have been used in weighted average indicators of biotic condition and contaminants (Potapova et al., 2004). Historically, diatoms have been used more in assessments than cyanobacteria, green algae, and other types of algae because they are relatively easy to identify to species level, and most algal species in streams are diatoms. However, recent work has shown that using all types of algae and cyanobacteria in assessments may improve the range of conditions that can be detected (Leland and Porter, 2000; Fetscher et al., 2014).

As implied above, algae have been used to assess valued attributes, such as the biotic condition of ecological systems and ecosystem services, as well as infer the contaminants that are causing problems in streams. Multivariate ordination

Methods in Stream Ecology. http://dx.doi.org/10.1016/B978-0-12-813047-6.00015-2

approaches have been used to relate changes in algal species composition to physical and chemical factors altered by humans (Pan et al., 1996, 2000). Multimetric approaches have also been used in which scores of different indicators are combined to reflect the many kinds of changes that can occur in algal assemblages in response to different types of human disturbance (Hill et al., 2000; Fore, 2002). Many algal metrics are, in one form or another, weighted average metrics and are calculated with a slightly modified version of a formula originally used by Zelinka and Marvan (1961):

$$X = \sum_{i=1}^{S} \theta_i p_i \left/ \sum_{i=1}^{S} p_i \right. \tag{37.1}$$

where X is the indicator, θ_i is an indicator value for the ith species, S is the number of species observed in the assemblage, and p_i is the frequency of occurrence of the ith species in the sample and with an indicator value. If taxa do not have indicators values, they are not included in the calculation. To understand this equation, imagine that we characterized the autecologies of species as oligotrophic, mesotrophic, or eutrophic indicating their nutrient requirements. We could assign numeric values for the trophic autecological characterizations to each species of 1, 2, and 3, respectively. If all organisms in a sample from a site were characterized with the same trophic status, then the value of the indicator would equal 1.0, 2.0, or 3.0. If, for example, half of the organisms in the sample were characterized as oligotrophic (1) and the other half was mesotrophic (2), then the indicator value would equal 1.5. Thus, the indicator value is the average of indicator values of species after weighting for the relative (proportional) abundance of species with different autecological categories.

This chapter will introduce several techniques used in ecological assessments of stream algae. In addition, it will provide an example of an integrated assessment of ecological condition of streams in which biological and pollution criteria are established, biological condition is assessed, and potential pollutants are diagnosed. Information from integrated assessments is important for managing streams because such assessments help guide protection and restoration strategies depending on the biological condition and pollutants in streams. More details on these subjects can be found elsewhere in this book (see Chapters 38–40). In addition, an excellent review of ecological assessment of streams was compiled by Barbour et al. (2004). Two chapters in the book review design and implementation of assessments (Stevenson et al., 2004a,b) and provided background on how elements of the algal assessment exercise are used in ecological assessments. Details about algal assessments in aquatic habitats and streams specifically can be found in Lowe and Pan (1996), Stevenson et al. (2010), Stevenson (2014), and Stevenson and Smol (2015).

37.2 GENERAL DESIGN

37.2.1 Ecological Assessment

Ecological assessments can be delineated into three stages (Stevenson et al., 2004a). The first is the *design stage*, in which the objectives, important attributes, and likely pollutants should be clearly defined. In addition, a conceptual model should be developed, which sets logical hypotheses relating cause–effect relationships between valued attributes, pollutants, and human activities. Finally, the study design for the assessment is established based on the objectives, conceptual model, and economic constraints on the study. In the *characterization stage* of assessment, the observed conditions at the assessed site are compared to criteria or expectations for valued attributes and pollutants to determine whether conditions meet or fail expectations (Stevenson et al., 2004b). The third stage of ecological assessment is *diagnosis of pollutants* that are most likely causing system impairment or threatening valued ecosystem attributes. Valued ecosystem attributes are now more commonly referred to as ecosystem goods and services.

Many exercises in the different chapters of this book could be modified to accomplish goals of an integrated assessment, as described in this chapter. In particular, Chapters 11, 12, and 31 could be integrated because they involve quantifying benthic algal species composition, biomass, and nutrient limitation. In addition, other exercises involving measurement of physical and chemical habitat characteristics and macroinvertebrate assessment could be incorporated into the study design of this chapter (see especially Volume 1 of this book). Some of these activities will be suggested during the following discussion of the project plan. At the least, Basic Exercise 1 (rapid periphyton assessment) and Basic Exercise 2 (genus-level taxonomic assessment) should be completed because they will provide sufficient information about algal biomass and taxonomic composition to conduct an integrated assessment. Sample collection can easily be integrated with fieldwork associated with other chapter exercises. Advanced Exercises 1–3 are recommended because they provide additional information and experience about algal division biomass and taxonomic composition. Advanced Exercise 4 is a data analysis exercise designed to demonstrate the calculation of environmental optima and their application to infer environmental conditions.

37.2.2 Project Plan

Algal assessments will be used in this chapter for the following three objectives: (1) characterize expected condition and criteria for assessing algal production and biodiversity; (2) determine whether pollutants are affecting the productivity and algal biodiversity of streams; and (3) diagnose which pollutants are important. Sediments and nutrients are two of the most common pollutants that affect productivity and biodiversity in streams. Sediments reduce productivity in streams by shading, burying, or coating benthic algae. Sediments alter biodiversity by inhibiting species that cannot move vertically through sediments. Nutrients stimulate productivity by enabling accrual of higher biomass. Nutrients alter biodiversity by enabling invasion of taxa that require high nutrients concentrations for survival. Therefore, algal biomass and taxonomic composition, nutrients, and sediments will be important to measure.

37.2.3 Sampling Plan

37.2.3.1 Site Selection

Develop criteria for conditions expected in the absence of human disturbance by assessing physical, chemical, and algal conditions in at least three streams in the region with low human disturbance in their watersheds. These *reference streams* should be selected randomly from the set of all streams that meet qualifications of having low human disturbance in watersheds. The condition of three or more test sites will be evaluated by assessing the same conditions in other streams of the region. Refine the goal of the assessment to either targeted assessment of streams with particular interest or a characterization of regional streams. In targeted assessments of specific streams, streams can be compared to criteria for expected condition to determine whether the targeted stream meets expectations or not. In a regional assessment, streams must be selected randomly from the set of all possible streams in the region. Random sampling enables generalizing results to all streams that were included in the original set of possible sites. The precision of regional assessments will increase with the number of streams sampled. Therefore, increase both the number of reference and test streams sampled corresponding to the number of students in the class.

37.2.3.2 Field Sampling and Laboratory Assays

Algal biomass should be measured by using RPS as described in Basic Exercise 1. Algal biomass can also be measured with chlorophyll *a*, AFDM, or algal biovolume as described elsewhere in this book (see Chapters 11 and 12, and Advanced Exercises 1 and 2 in this chapter). Taxonomic composition of algae should be measured using methods described in Basic Exercise 2.

Relative levels of nutrients and sediments can be inferred based on taxonomic composition of the diatom assemblages characterized in Basic Exercise 2 and ecological preferences and tolerances provided for some genera. Nutrients can be measured using techniques described in Wetzel and Likens (2000) and in Chapters 31−33. Sediments can be measured using techniques described in Chapter 5 or for biofilm as the difference in dry mass and AFDM of benthic algal samples (see Chapter 11).

37.2.4 Data Analysis Plan

37.2.4.1 Objective A

Expected conditions of physical, chemical, and biological characteristics of ecosystems can be determined in many ways (Stevenson et al., 2004b). We could expect conditions to be as similar to natural as possible, such as the "physical, chemical, and biological integrity" prescribed in the US Clean Water Act of 1972. Alternatively, we could expect high productivity and many big fish. Thus, reaching expectations for different stream management goals of streams may require compromises and different selection criteria for reference sites for each of these management goals. For Advanced Exercises 1−3, expected condition will be based on central tendency and variation in conditions at reference sites—i.e., those sites with best attainable conditions in the region. Often, either the 25th or 75th percentiles of conditions at reference sites are used as criteria for expected condition (Fig. 37.1). For ease of calculation, we recommend using the standard deviation of observations, which roughly estimates the 16th and 84th percentiles. These criteria can be calculated for all attributes, but especially for the important project attributes describing algal biomass, taxonomic composition, nutrients, and sediments. For positive attributes (e.g., biodiversity), use the mean of reference conditions minus the standard deviation (i.e., the 16th percentile) to provide a criterion that would be the lowest value that would meet management expectations. For negative attributes (e.g., biomass), use the mean of reference conditions minus plus the standard deviation (i.e., the 84th percentile) to provide a criterion that would be the highest value that would meet management expectations.

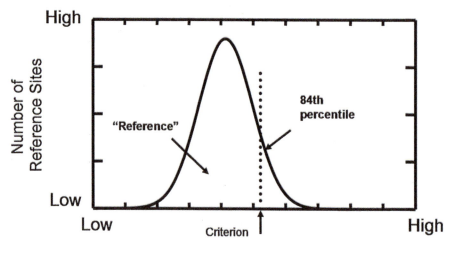

FIGURE 37.1 Criterion for a pollutant based on the 84th percentile of conditions at a reference site. The number of sites with specific concentrations of that pollutant is plotted as a function of the concentrations of that pollutant. Sites with pollution concentrations above the criterion would be characterized as impaired for that pollutant. Sites with pollution concentrations below the criterion would be characterized as meeting expectations.

37.2.4.2 Objective B

Assessment of conditions at test sites will be done by comparing them with criteria developed from reference sites. If observed conditions at test sites fall within the range delineated by criteria, then we lack evidence to conclude that the test site is impaired. If observed conditions do not meet criteria, then the test site is impaired.

37.2.4.3 Objective C

Diagnosing the relative importance of stressors, nutrients, and sediments in this case, can be estimated as the ratio between the observed condition and the criterion. The greater this ratio, the more likely the stressor is a major cause of problems in impaired streams or threatening sustainability in unimpaired streams. If this ratio is greater than 1, then the stressor is likely having an effect on valued attributes, but is not necessarily the only stressor having an effect.

37.3 SPECIFIC EXERCISES

37.3.1 Basic Exercise 1: Rapid Periphyton Survey

Circumstances may prevent the collection, proper preservation or analysis of algal biomass using chlorophyll *a* or AFDM. In such cases, *rapid periphyton surveys* can be used to estimate algal biomass. RPS can also be used to separate sources of benthic primary production into rough functional categories, which cannot be done using standard chlorophyll *a* and AFDM methods. In this exercise, you will evaluate whether macrophyte and algal growth in test sites differs from that in reference sites using RPS to measure macrophyte and algal biomass. The RPS method presented is a modification of the method described by Stevenson and Bahls (1999) and eliminates the requirement of a viewing bucket.

37.3.1.1 Field Assessment

1. The RPS will require two people—one "sampler" to evaluate algal biomass and the other to record information on the datasheet provided in Table 37.1. Copies of Table 37.1 (see Online Worksheet 37.1 for downloadable excel spreadsheet) should be made for each site that will be sampled.
2. Within the stream reach, establish five transects that cross the stream perpendicular to the direction of stream flow. Transects should be spaced relatively evenly and far enough apart to span the full reach. Assessments of reaches should be restricted to riffles or runs if they are available.
3. Along each transect, sample 10 evenly spaced points, beginning with the downstream transect. For narrow streams, you may wish to use more transects, each containing fewer sampling points. Reach down and touch the substratum with your index finger. Do this without looking.
4. For each of these points, image a circle with the radius of your hand, from fingertip to wrist. Estimate percent macrophyte cover, macroalgal cover, and microalgal biofilm thickness (see Chapter 13 for additional description of stream

TABLE 37.1 Field data sheet for rapid periphyton survey and algae sample collection. *Trns*, Transect number; *Sz*, size (check to indicate substratum >2 cm). See also Online Worksheet 37.1.

Stream: _____ Date: _____ Sampler: _____ Recorder: _____

Point	Trns	Macrophyte	Macro	Micro	Sz	Point	Trns	Macrophyte	Macro	Micro	Sz
1						26					
2						27					
3						28					
4						29					
5						30					
6						31					
7						32					
8						33					
9						34					
10						35					
11						36					
12						37					
13						38					
14						39					
15						40					
16						41					
17						42					
18						43					
19						44					
20						45					
21						46					
22						47					
23						48					
24						49					
25						50					

Total Algae Sample Volume = _____
Identification Subsample Volume = _____
Chlorophyll Subsample Volume = _____
AFDM Subsample Volume = _____

Total Surface Area Sampled = _____
Substrata Sampled (Circle): rock/wood/plant/sand/silt/other

macrophytes). In addition, determine suitability of substrata for algal accrual by putting a check in the Sz column of the table if the substratum is greater than 2 cm in diameter in its longest dimension. Shout your estimates of macrophyte cover, macroalgal cover, microalgal biofilm thickness, and substratum suitability to the data recorder. The recorder should repeat the list of estimates, and enter them on the datasheet using macrophyte and macroalgal cover classes and microalgal thickness classes in Table 37.2. If the information being repeated back to the sampler is incorrect, the numbers should be corrected before moving to the next sampling point. If the sampler is unable to make these evaluations for a given point, the recorder should mark "NA" in the appropriate columns on the datasheet. In particular, thin films of microalgae can be difficult to evaluate if the substratum is less than 2 cm in diameter.

5. Calculate RPS metrics for algal biomass using Table 37.3.

   a. Transfer information from Table 37.2 to Table 37.3 by recording the number of points with each cover and thickness class for macrophytes, macroalgae, and microalgae. Do not transfer information if Sz < 2 cm, which is information that can be used in other calculations that we will not do in this chapter.

TABLE 37.2 Algae and macrophyte cover and thickness class descriptions.

Macrophyte and Macroalgae Cover Classes

Class	0	1	2	3	4
Cover	0%	<5%	5%–25%	25%–50%	>50%

Microalgal Thickness Class

Class	0	1	2	3	4	5
Thickness	0 mm	<0.5 mm	Visible to 1 mm	1–5 mm	5–20 mm	>20 mm
Characteristics	Rough	Slimy, visible evidence of biofilm absent	Biofilm visible; may require scraping rock surface to distinguish			

TABLE 37.3 Calculations for extent and magnitude of macrophyte and benthic algal cover. Shaded areas do not have appropriate records for calculations. See also Online Worksheet 37.3.

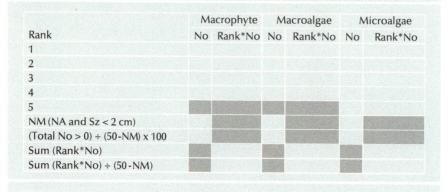

Rank	Macrophyte No	Rank*No	Macroalgae No	Rank*No	Microalgae No	Rank*No
1						
2						
3						
4						
5						
NM (NA and Sz < 2 cm)						
(Total No > 0) ÷ (50-NM) x 100						
Sum (Rank*No)						
Sum (Rank*No) ÷ (50-NM)						

b. Determine extent of stream bottom covered by macrophytes, macroalgae, and microalgae as the percent of sampling points having these elements with the following steps:

   i. Count the number of measurements that were not recorded (NM) as the number of points with NA and substratum less than 2 cm, which will be the same for macrophytes, macroalgae, and microalgae and enter that number into Table 37.3—row NM.

   ii. Determine the number of points where macrophyte, macroalgal, and microalgal cover classes were greater than 0 (Total No > 0), divide it by the number of all points with RPS measurements (50 − NM), and multiply that by 100; and then enter that in the row labeled (Total No > 0) ÷ (50 − NM) × 100.

c. Determine the magnitude of macrophyte and algal biomass on the stream bottom with the following steps:

   i. Multiply the number of points with a cover or thickness class by the rank of that cover or thickness class, and enter it into Table 37.3 in the column labeled Rank*No.

   ii. Sum the products Rank*No for each cover or thickness class and enter it into Table 37.3 in the row labeled Sum (Rank*No).

   iii. Divide Sum (Rank*No) by the number of points with NA or substratum less than 2 cm (50 − NM) and enter it into the last row of Table 37.3.

37.3.2 Basic Exercise 2: Genus-Level Periphyton Assays and Index of Biotic Condition

37.3.2.1 Collecting Periphyton Samples

1. Survey the reach to determine the best habitat for sampling. The habitat should be shallow enough to have light that supports benthic algal growth. You should be able to see algae on substrata. Prioritize the habitat selected for sampling using the following three criteria: (1) select rock, wood, or plant (firm) substrata in current velocities greater than

15 cm/s, but not too high (e.g., 50 cm/s), to minimize silt in the sample and effects of current on species composition; (2) if you cannot find firm substrata in fast current, sample them in slow current; (3) if sufficient firm substrata are not present in the reach, then sample sediments (ranging from sand, to fine detrital material, to silt) in slow current habitats (often along the stream margins) where sediments have been stable and periphyton accrual is evident or at least likely.

2. *Sample periphyton on firm substrata* by randomly selecting and removing at least five substrata from the stream and placing them in a pan for processing onshore.

   a. Scrape periphyton into another white pan from all areas of the substratum that were exposed to light and not buried in sediments. Scrape thick accumulations with a knife or spoon to remove most of the algae and then brush with a toothbrush to remove more tightly attached individuals. Rinse the sampled surface and sampling tools into the pan with distilled water from a squirt bottle. Cut long filaments of algae into short pieces with scissors.

   b. Rinse all algae from the pan into a 500-mL sampling bottle marked with volumetric graduations. Fill the subsampling bottle with distilled water to the next volumetric graduation. Record the total volume of the sample on the field data sheet (Table 37.1).

   c. Measure length and width of the scraped areas with a ruler and calculate the total area of sampled surfaces of substrata. Alternatively, use a preformed template to scrape within a known area or define a known rock area with a pencil (see Chapter 12).

   d. Record the total area sampled on the field data sheet.

3. If firm substrata are not available, sample periphyton on sediments from at least five representative locations of the targeted habitat in the stream using a Petri dish and spatula.

   a. Place the open side of a dish onto the sediment, embed it gently into the sediment, and then slide the spatula under the dish to capture a short core of sediments.

   b. Remove the sediments from the stream, invert the core, and rinse the core from the dish into a 1-L beaker. If the sediment is organic and fine, and algae are not expected to occur deep within the sediment, the bottom of the core could be removed to reduce silt in the final sample.

   c. Depending on substratum size, use different methods to transfer the sample from the 1-L beaker to a sample bottle.

      i. *If the sample is not sandy*, pour the sample into a 500-mL sample bottle marked with volumetric graduations. Rinse the 1-L beaker with distilled water to get all sediments into the sample bottle. Fill the sampling bottle with distilled water to the next volumetric graduation. Record the total volume of the sample on the field data sheet (Table 37.1).

      ii. *If the sediment is sandy or coarser*, a swirl-and-pour technique should be employed to remove algae from the coarse inorganic substrata before subsampling the collected periphyton (Stevenson and Stoermer, 1981). The swirl-and-pour technique removes algae from coarse substrata by repeatedly adding small amounts of water to the sediment sample, swirling the sample to tumble sediments and thereby scouring algae from the fine substrata. Gently pour the suspended algae from the swirled sample into a 500-mL sample bottle marked with volumetric graduations. This step should be repeated for 5–10 times or until the poured water appears relatively clean. Fill the sample bottle with distilled water to the next volumetric graduation.

   d. Record the total sample volume and the area sampled on the field data sheet (Table 37.1).

37.3.2.2 Subsampling for Different Assays

1. Subsample separately for algal assays that require preserved and unpreserved sample. Algal identification, cell counting, and AFDM assessments require preservation. Pigment analysis requires subsampling, storage in the field on ice, and freezing in the laboratory until assay. Subsampling should be done by removing two or more aliquots with a turkey baster or large pipette and placing them in a sample bottle.

2. Record the volume of the subsample in the sample bottle on the field data sheet (Table 37.1).

3. Subsample a relatively large proportion of the total sample for microscopic and AFDM analysis.

4. Preserve these samples with 3%–5% buffered formaldehyde or glutaraldehyde[1].

5. Subsample the algal suspension for chlorophyll *a* and pigment analysis, and place this sample on ice in the dark until returned to the lab. Freeze the pigment samples in the lab if they will not be analyzed immediately.

6. Check for correct labels on the sample bottles before leaving the sample site.

7. Unused sample remaining after subsampling may be discarded.

---

1. Note that formaldehyde and glutaraldehyde are known carcinogens and thus exposure should be minimized by using Personal Protective Equipment (PPE) including gloves, goggles, and lab coat.

37.3.2.3 Identifying Diatom Genera and Counting Cells

Assessing taxonomic composition of algal assemblages usually calls for identifications to the species level of taxonomy; however, recent papers have shown utility of genus-level assays (Hill et al., 2001; Wang et al., 2005). Genus-level assays may not be as precise as species-level assays and assessment of biological condition, but they do have value and can be conducted with less technical expertise. In this exercise, we assume that genus-level metrics will transfer from the regions and stream types in which they were developed to study streams selected for your project.

1. Clean diatoms and mount them in a high-resolution mounting medium as described in Chapter 11.
2. Using the genus key provided in that chapter, identify the diatom genus of each valve and count 300 valves of diatoms in random fields around the coverglass. Record the genera and number of valves observed on Table 37.4 (see also Online Worksheet 37.4). (*Option*: keep track of the sample volume put on the coverglass and number of fields counted to determine the numbers of diatoms per area of substratum as detailed in Advanced Exercise 1). Be careful to distinguish whether one or two valves are present in the diatom cell wall (frustule) and whether one or more cells are aggregated together.
3. If time or resources are a constraint, SimRiver (http://www.u-gakugei.ac.jp/~diatom/en/simriver) provides an online simulation alternative that demonstrates the application of diatoms in ecological assessment. The simulation allows users to modify watershed land use characteristics and to select sampling locations. After modifying the landscape, users are provided a simulated microscopic field of diatoms to identify and count.

37.3.2.4 Calculating Diatom Metrics

1. Using the diatom bench sheet (Table 37.4), tally the number of valves observed for each genus of diatom and divide by 2 to get the number of cells of each genus (n_i).
2. Sum the n_i to determine the total number of cells observed (N) in the count.
3. Calculate the proportional relative abundance of each ith genus (p_i) of diatom as $p_i = n_i/N$.
4. Calculate the following metrics to assess biotic condition of diatom assemblages: % *Achnanthidium*, % *Cymbella* plus *Encyonema*, % *Epithemia* plus *Rhopalodia*, % *Navicula*, and % *Nitzschia*. Contrary to suggestions in Stevenson and Bahls (1999), high relative abundances of *Achnanthidium* are characteristic of low nutrient reference streams in many regions of the country, particularly *Achnanthidium minutissima* and *A. deflexum* (Wang et al., 2005; Stevenson, unpublished data). *Cymbella* and *Encyonema* are also common in references streams, while *Navicula* are more common in disturbed streams (Wang et al., 2005; Stevenson, unpublished data). *Epithemia* and *Rhopalodia* have nitrogen-fixing endosymbiotic cyanobacteria. In regions where nitrate pollution is present, these genera tend to decrease in abundance (Stancheva et al., 2013; Rollins, unpublished data). *Nitzschia* are pollution-tolerant (Hill et al., 2001). Record these in the assessment table (Table 37.5; see also Online Worksheet 37.5).
5. Calculate the following metrics to infer relative pollution conditions: % acidobiontic, % eutraphentic, and % motile. According to Hill et al. (2001), the average environmental preference of species in the genera *Eunotia*, *Frustulia*, and *Tabellaria* would classify them as *acidobiontic*, meaning taxa with tolerances for pH < 5.5 (Lowe, 1974; Van Dam et al., 1994). On average, species in the genera *Amphora*, *Cocconeis*, *Diatoma*, *Gyrosigma*, *Meridion*, *Nitzschia*, and *Synedra* are *eutraphentic*, meaning taxa with requirements for nutrient-enriched waters (Van Dam et al., 1994). *Navicula*, *Nitzschia*, *Surirella*, *Cymatopleura*, and *Gyrosigma* are relatively common genera that are motile and commonly found in sediments. Relatively rare and planktonic genera with acidobiontic or eutraphentic environmental preferences are not listed above.
6. Record metrics in the assessment table (Table 37.5).

37.3.3 Advanced Exercise 1: Identification and Counting of All Algae (Optional)

1. Prepare a Palmer Counting Chamber with 0.1 mL of algal suspension from a known proportion of the original sample.
2. Identify the size and division of algae for 300 cells in random fields around the counting chamber (see Chapter 11). Do not include diatoms without protoplasm in the frustules in your count of 300 cells, as these are considered dead.
3. Record the number of cells of green algae, cyanobacteria, diatoms, and other algae by size category in Table 37.6 (see also Online Worksheet 37.6).
4. Usually, biovolume of each species is assessed (see Chapter 11). However, when only identifying to division, use the size categories provided to group algae with relatively similar biovolumes.

TABLE 37.4 Bench sheet example for cleaned diatom counts. PRA is proportional relative abundance (p_i). See also Online Worksheet 37.4.

Stream: _____ Date: _____ Counter: _____ Fields Counted: _____

Genus	Valve Count (Tick Marks)	Cell Count (Valves ÷ 2)	PRA (p_i)
Sum (N)			

Total Sample Volume = _____mL
Volume Cleaned = _____mL
Volume on Coverglass = _____mL

37.3.3.1 Calculating Algal Metrics

1. Calculate cell densities and biovolumes of all algae, cyanobacteria (blue-green algae), and diatoms using equations at the bottom of Table 37.6.
2. Calculate the percent of biovolume of cyanobacteria and diatoms. Usually, high biovolumes and percent biovolumes of cyanobacteria indicate nutrient and organic enrichment by human activities, and high percent diatoms is considered more natural (Hill et al., 2000). However, any deviation from reference conditions should be considered a decrease in biological condition.

TABLE 37.5 Assessment calculation table. MT refers to metric type where BC is a metric of biological condition and IPC is a metric that infers pollution condition. See also Online Worksheet 37.5.

Metric	MT	Reference Mean	Lower Criterion (e.g., 16%)	Upper Criterion (e.g., 84%)	Site Value	Site Value ÷ Ref. Mean	Meets Criterion? (Yes/No)
% *Achnanthidium*	BC						
% *Cymbella* and *Encyonema*	BC						
% *Epithemia* and *Rhopalodia*	IPC						
% *Navicula*	BC						
% *Nitzschia*	IPC						
% Acidobiontic genera	IPC						
% Eutraphentic genera	IPC						
% Motile genera	IPC						
Macrophyte cover extent	BC						
Macrophyte cover magnitude	BC						
Macroalgae cover extent	BC						
Macroalgae cover magnitude	BC						
Microalgae cover extent	BC						
Microalgae cover magnitude	BC						
Chlorophyll *a*	BC						
AFDM	BC						
Cyanobacteria density	BC						
% Cyanobacteria	BC						
Cyanobacteria biovolume	BC						
% Biovolume cyanobacteria	BC						
Green algae density	BC						
% Green algae	BC						
Green algae biovolume	BC						
% Biovolume green algae	BC						
Diatom density	BC						
% Diatoms	BC						
Diatom biovolume	BC						
% Biovolume diatoms	BC						

37.3.4 Advanced Exercise 2: Species Autecologies and Inferring Environmental Condition (Optional)

Weighted average models are one of the simplest and most tested techniques for inferring environmental conditions based on species composition and their optimal environmental conditions. This technique has been used to characterize historic conditions in lakes using species composition in sediment records and to infer nutrient concentrations in streams where it can be so variable (Stevenson, 2001; Potapova et al., 2004; Stevenson and Smol, 2015). Developing weighted average models requires two steps: (1) characterizing species environmental optima and (2) testing and calibrating the model. In this exercise, you will first calculate the optimum environmental condition for one species, assume model testing and

TABLE 37.6 Bench sheet for all algae counts. See also Online Worksheet 37.6.

Stream: _____ Date: _____ Counter: _____ Fields Counted: _____

Taxonomic Size Class	Biovolume Weight	Cell Count
Diatoms		
Width < 5 μm	4	
5 μm ≤ width < 12 μm	6	
Width ≥ 12 μm or >25 μm long	8	
Dead diatoms	0	
Green algae		
Width < 5 μm, length ≤ 5 μm	3	
Width ≥ 5 μm, length ≤ 5 μm	5	
Width < 5 μm, length > 5 μm	4	
Width ≥ 5 μm, length > 5 μm	8	
Cyanobacteria		
Width < 3 μm, length ≤ 5 μm	1	
Width ≥ 3 μm, length ≤ 5 μm	3.5	
Width < 3 μm, length > 5 μm	3	
Width ≥ 3 μm, length > 5 μm	5.5	
Other		

Palmer Chamber Dilution Factor = _____
   (Volume of Deionized Water ÷ Volume of Sample)
Total Algae Sample Volume = _____mL
Stream Surface Area Sampled = _____cm^2
Microscope Field-of-view Area = _____μm^2
Palmer Chamber Depth = _____μm
Proportion of Palmer Chamber Counted = _____
   (# Fields × Field-of-view Area × Palmer Chamber Depth) ÷ 100 μL Palmer Vol.
Volume of Original Sample Counted = _____mL
   (Proportion of Palmer Counted × 0.1 mL Palmer Vol.) ÷ Dilution Factor
Stream Area Counted = _____cm^2
   (Original Vol. Counted ÷ Total Algae Sample Vol.) × Stream Area Sampled
Cell Density = _____cells/cm^2
   (Number of Cells Counted ÷ Stream Area Counted)

calibration were successful, and then you will apply the model to infer environmental conditions in a stream based on species composition of diatoms in a sample and the species' environmental optima.

37.3.4.1 Calculating an Environmental Optimum for a Species

Species environmental optima are calculated using a calibration dataset and a weighted average model. The calibration dataset is composed of species abundances and environmental characteristics at sites throughout a region. The weighted average model for a species' environmental optimum is as follows:

$$\theta_i = \sum_{j=1}^{N} E_j p_{ij} \Big/ \sum_{j=1}^{N} p_{ij} \qquad (37.2)$$

where the environmental optimum of the *i*th species (θ_i) is the sum of the products of *i*th species abundance at site *j* (p_{ij}) and measured environmental conditions at site *j* (E_j) for all *N* sites, divided by the sum of the *i*th species' abundance for all sites.

Using data provided in Table 37.7, calculate the total phosphorus optimum for *Navicula cryptocephala* Kützing.

TABLE 37.7 Proportional relative abundances (PRA) of *Navicula cryptocephala* and total phosphorus (TP) concentrations at 17 stream sites. See also Online Worksheet 37.7.

Taxon	Site#	PRA (p_i)	TP (E_i)	PRA*TP ($p_i{*}E_i$)
Navicula cryptocephala	1	0.05	55	
Navicula cryptocephala	2	0.15	25	
Navicula cryptocephala	3	0.10	10	
Navicula cryptocephala	4	0.10	45	
Navicula cryptocephala	6	0.20	35	
Navicula cryptocephala	8	0.10	5	
Navicula cryptocephala	9	0.10	20	
Navicula cryptocephala	10	0.40	30	
Navicula cryptocephala	12	0.05	15	
Navicula cryptocephala	13	0.15	50	
Navicula cryptocephala	14	0.05	5	
Navicula cryptocephala	15	0.10	60	
Navicula cryptocephala	16	0.35	30	
Navicula cryptocephala	17	0.10	40	
Sums				
Optimum				

1. Determine the products of species abundance (p_{ij}) and measured environmental conditions (E_j) for all sites.
2. Independently sum the species abundances and products of species abundances and environmental conditions.
3. For this example, determine the total phosphorus optimum for *N. cryptocephala* by dividing the sum of the products of species relative abundances and total phosphorus concentrations by the sum of the species relative abundances.

37.3.4.2 Calculate Inferred Total Phosphorus Concentration for a Stream

Here you will apply a weighted average model to infer an environmental condition, total phosphorus, in a stream based on species composition of diatoms in a sample and the species' environmental optima. Using Eq. (37.1) and data in Table 37.8, calculate *X*, which in this case will be the inferred total phosphorus concentration for the sampled stream.

1. Record relative abundances of species for which the total phosphorus optimum is known (p_{ij}).
2. Calculate the products of the species relative abundances and their total phosphorus optima.
3. Sum the proportional relative abundances of species for which total phosphorus is known and the products of species relative abundances and total phosphorus optima.
4. Calculate inferred total phosphorus concentrations for the stream by dividing the sum of the products of species relative abundances and total phosphorus by the sum of the species relative abundances for which total phosphorus optima are known.

37.3.5 Advanced Exercise 3: Biomass Assays (Optional)

Use the subsamples saved during sampling for AFDM and chlorophyll *a* analysis. Follow instructions in Chapter 12 to determine AFDM and chlorophyll *a* in the subsamples. Calculate the proportions of total samples represented by AFDM and chlorophyll *a* subsamples by dividing the volume used in each analysis by the total sample volume recorded on the field data sheets. Calculate the area of substrate from which the subsamples were collected by multiplying these proportions by the area sampled. Calculate AFDM and chlorophyll *a* per unit area by dividing AFDM and chlorophyll *a* in samples by the area of substrate from which the subsamples were collected.

TABLE 37.8 Proportional relative abundances (PRA) and total phosphorus optima (TP$_{Opt}$) of diatom species in a benthic algal sample from a stream. See also Online Worksheet 37.8.

Taxon	PRA	PRA With TP$_{Opt}$ (p_i)	TP$_{Opt}$	PRA*TP$_{Opt}$ (θ_i)
Navicula reichardtiana	0.2		34	
Nitzchia frustulum	0.2		71	
Amphora perpusilla	0.1		67	
Hippodonta capitata	0.1		44	
Navicula cryptocephala	0.05		31	
Gyrosigma acuminatum	0.1		59	
Gomphonema angustatum	0.05		66	
Gomphonema parvulum	0.05		50	
Achnanthidium minutissimum	0.05		33	
Staurosirella lapponica	0.1		-	
Sums				
Inferred TP				

37.3.6 Advanced Exercise 4: Analysis and Interpretation of Data

37.3.6.1 Determine Expected Conditions

List all the attributes of algal biomass and taxonomic condition, algal-inferred stressor conditions, and measured nutrient concentrations and sediments in the assessment table (Table 37.5). Include algal division biovolumes and percent biovolumes if determined. Predict whether human activities would increase or decrease each attribute and mark it in the assessment table. Calculate and record the mean and standard deviation of all attributes at reference sites. Calculate and record the 16th and 84th percentiles of each attribute by, respectively, subtracting and adding the standard deviation from the mean.

37.3.6.2 Assess Stream Conditions for Targeted Streams

Compare values of all metrics of biological condition (algal biomass and taxonomic composition) to respective criteria for reference conditions in each targeted stream. Determine the number of metrics for biological condition and pollutants that meet criteria for each targeted stream.

37.3.6.3 Assess Stream Conditions for the Region

Compare values of all metrics of biological condition (algal biomass and taxonomic composition) to respective criteria for reference conditions. Determine the proportion of test streams in the region that are impaired for each biological condition. Since these test streams were randomly selected from the set of all regional streams (although perhaps a limited size and landscape type), we can assume that this proportion is a reasonable estimate of the proportion of all regional streams that are impaired for these conditions.

37.3.6.4 Diagnose Likely Causes of Impairment

Calculate ratios of all biologically inferred and measured pollutants (acidification, nutrient enrichment, and sedimentation) in each test stream with criteria for reference conditions. For each impaired stream, determine whether pollution ratios are greater or less than 1 and, therefore, which pollutants are likely impairing stream conditions. For targeted streams, this information can be used to restore streams. With randomly selected streams, the proportion of regional streams impaired and threatened by different stressors can be determined.

37.4 QUESTIONS

1. Researchers have long suggested that species composition should be more sensitive to environmental change than aggregate community properties such as biomass (chlorophyll *a* and AFDM) and primary production. Why might this be true? Do your data support these assertions?

2. Did you have any difficulty finding good reference sites to assess your test sites? In regions where this is a problem, how might water quality criteria be developed?

3. The nature of many environmental problems prevents traditional experimental approaches. Thus, environmental assessments often take a survey approach similar to those used in epidemiology. How do survey approaches differ from more traditional scientific experiments? What are the pros and cons of each approach? How might inferences drawn from surveys be strengthened? Can hypotheses be tested using both approaches?

4. For your impaired streams, which stressors are most important? For unimpaired streams, which stressors most threaten impairment? How might these affect other organisms in these streams? How might these affect various uses of these streams by humans?

5. Are there alternative explanations for the patterns you have observed? How would you design a study to determine which explanations are most probable?

6. What management actions would you recommend to improve or maintain the water quality of these streams?

7. Estimates of standard deviation are influenced by sample size. How could this influence water quality criteria if percentiles are estimated as above? How would an increased number of test sites influence the probability of declaring the region impaired? How might this influence monitoring decisions?

37.5 MATERIALS AND SUPPLIES

Field
    Spoon or knife
    Toothbrush
    Scissors
    Pans, 2 (white plastic dish tubs works well)
    1-L beaker
    Squirt bottle filled with deionized water
    500-mL sample bottles with marked volumetric graduations
    Ruler
    Petri dish (one half)
    Kitchen spatula (without openings)
    Graduated sample bottles (plastic screw-cap centrifuge tubes that work well)
    Turkey baster
    Formaldehyde or glutaraldehyde[1]
    Pipette for dispensing known volume of preservative
    Labels
    Permanent markers
    Data sheets (waterproof)
    Pencils
Laboratory
    Microscope
    Bench sheets
    Slides
    Coverglass
    Mounting medium
    Palmer counting chamber
    10–100 μL adjustable pipette
    100–1000 μL Adjustable Pipette
    Optional supplies and equipment for measuring chlorophyll *a* and AFDM (see Chapter 12)

REFERENCES

Barbour, M., Norton, S., Preston, R., Thornton, K. (Eds.), 2004. Ecological Assessment of Aquatic Resources: Linking Science to Decision-Making. Society of Environmental Toxicology and Contamination Publication, Pensacola, FL.

Bauer, D.E., Gomez, N., Hualde, P.R., 2007. Biofilms coating *Schoenoplectus californicus* as indicators of water quality in the Rio de la Plata Estuary (Argentina). Environmental Monitoring and Assessment 133, 309–320.

Carmichael, W.W., 1997. The cyanotoxins. Advances in Botany Research 27, 211–256.

Fetscher, A.E., Stancheva, R., Kociolek, J.P., Sheath, R.G., Stein, E.D., Mazor, R.D., Ode, P.R., Busse, L.B., 2014. Development and comparison of stream indices of biotic integrity using diatoms vs. non-diatom algae vs. a combination. Journal of Applied Phycology 26, 433–450.

Fore, L.S., 2002. Response of diatom assemblages to human disturbance: development and testing of a multimetric index for the Mid-Atlantic Region (USA). In: Simon, T.P. (Ed.), Biological Response Signatures: Indicator Patterns Using Aquatic Communities. CRC Press, Boca Raton, pp. 445–480.

Gaiser, E.E., 2009. Periphyton as an indicator of restoration in the Florida Everglades. Ecological Indicators 9, 37–45.

Hill, B.H., Herlihy, A.T., Kaufmann, P.R., Stevenson, R.J., McCormick, F.H., 2000. The use of periphyton assemblage data as an index of biotic integrity. Journal of the North American Benthological Society 19, 50–67.

Hill, B.H., Stevenson, R.J., Pan, Y.D., Herlihy, A.T., Kaufmann, P.R., Johnson, C.B., 2001. Comparison of correlations between environmental characteristics and stream diatom assemblages characterized at genus and species levels. Journal of the North American Benthological Society 20, 299–310.

Holomuzki, J.R., Short, T.M., 1988. Habitat use and fish avoidance behaviors by the stream-dwelling isopod, *Lirceus fontinalis*. Oikos 52, 79–86.

Humphrey, K.P., Stevenson, R.J., 1992. Responses of benthic algae to pulses in current and nutrients during simulations of subscouring spates. Journal of the North American Benthological Society 11, 37–48.

Kelly, M., Juggins, S., Guthrie, R., Pritchard, S., Jamieson, J., Rippey, B., Hirst, H., Yallop, M., 2008. Assessment of ecological status in UK rivers using diatoms. Freshwater Biology 53, 403–422.

Kolkwitz, R., Marsson, M., 1908. Ökologie der pflanzliche Saprobien. Berichte der Deutschen Botanischen Gesellschaft 26, 505–519.

Lambert, D., Cattaneo, A., Carignan, R., 2008. Periphyton as an early indicator or perturbation in recreational lakes. Canadian Journal of Fisheries and Aquatic Sciences 65, 258–265.

Lane, C.R., Brown, M.T., 2007. Diatoms as indicators of isolated herbaceous wetland condition in Florida, USA. Ecological Indicators 7, 521–540.

Leland, H.V., Porter, S.D., 2000. Distribution of benthic algae in the upper Illinois River basin in relation to geology and land use. Freshwater Biology 44, 279–301.

Lowe, R.L., 1974. Environmental Requirements and Pollution Tolerance of Freshwater Diatoms. EPA-670/4-74-005. US Environmental Protection Agency, Cincinnati, Ohio, USA.

Lowe, R.L., Pan, Y., 1996. Benthic algal communities and biological monitors. In: Stevenson, R.J., Bothwell, M., Lowe, R.L. (Eds.), Algal Ecology: Freshwater Benthic Ecosystems. Academic Press, San Diego, California, USA, pp. 705–739.

Minshall, G.W., 1978. Autotrophy in stream ecosystems. BioScience 28, 767–771.

Palmer, C.M., 1962. Algae in Water Supplies. Publication 657. U.S. Public Health Service, Washington, DC.

Pan, Y., Stevenson, R.J., Hill, B.H., Herlihy, A.T., 2000. Ecoregions and benthic diatom assemblages in the Mid-Atlantic Highland streams, USA. Journal of the North American Benthological Society 19, 518–540.

Pan, Y., Stevenson, R.J., Hill, B.H., Herlihy, A.T., Collins, G.B., 1996. Using diatoms as indicators of ecological conditions in lotic systems: a regional assessment. Journal of the North American Benthological Society 15, 481–495.

Patrick, R., Hohn, M.H., Wallace, J.H., 1954. A new method for determining the pattern of the diatom flora. Notulae Naturae 259, 1–12.

Peterson, C.G., Stevenson, R.J., 1992. Resistance and recovery of lotic algal communities: importance of disturbance timing, disturbance history, and current. Ecology 73, 1445–1461.

Potapova, M.G., Charles, D.F., Ponader, K.C., Winter, D.M., 2004. Quantifying species indicator values for trophic diatom indices: a comparison of approaches. Hydrobiologia 517, 25–41.

Stancheva, R., Sheath, R.G., Read, B.A., McArthur, K.D., Schroepfer, C., Kociolek, J.P., Fetscher, A.E., 2013. Nitrogen-fixing cyanobacteria (free-living and diatom endosymbionts): their use in southern California stream bioassessment. Hydrobiologia 720, 111–127.

Stevenson, R.J., 2001. Using algae to assess wetlands with multivariate statistics, multimetric indices, and an ecological risk assessment framework. In: Rader, R.R., Batzger, D.P., Wissinger, S.A. (Eds.), Biomonitoring and Management of North American Freshwater Wetlands. John Wiley & Sons, Inc., New York, NY, pp. 113–140.

Stevenson, R.J., 2014. Ecological assessment with algae: a review and synthesis. Journal of Phycology 50, 437–461.

Stevenson, R.J., Bahls, L.L., 1999. Periphyton protocols. In: Barbour, M.T., Gerritsen, J., Snyder, B.D. (Eds.), Bioassessment Protocols for Use in Wadeable Streams and Rivers: Periphyton, Benthic Macroinvertebrates, and Fish, second ed. U.S. Environmental Protection Agency, Washington, DC. Pages 6, 1–23.

Stevenson, R.J., Bailey, R.C., Harass, M.C., Hawkins, C.P., Alba-Tercedor, J., Couch, C., Dyer, S., Fulk, F.A., Harrington, J.M., Hunsaker, C.T., Johnson, R.K., 2004a. Designing data collection for ecological assessments. In: Barbour, M.T., Norton, S.B., Preston, H.R., Thornton, K.W. (Eds.), Ecological Assessment of Aquatic Resources: Linking Science to Decision-Making. Society of Environmental Toxicology and Contamination Publication, Pensacola, Florida, pp. 55–84.

Stevenson, R.J., Bailey, R.C., Harass, M.C., Hawkins, C.P., Alba-Tercedor, J., Couch, C., Dyer, S., Fulk, F.A., Harrington, J.M., Hunsaker, C.T., Johnson, R.K., 2004b. Interpreting results of ecological assessments. In: Barbour, M.T., Norton, S.B., Preston, H.R., Thornton, K.W. (Eds.),

Ecological Assessment of Aquatic Resources: Linking Science to Decision-Making. Society of Environmental Toxicology and Contamination Publication, Pensacola, Florida, pp. 85–111.

Stevenson, R.J., Bennett, B.J., Jordan, D.N., French, R.D., 2012. Phosphorus regulates stream injury by filamentous algae, DO, and pH with thresholds in responses. Hydrobiologia 695, 25–42.

Stevenson, R.J., Pan, Y., van Dam, H., 2010. Assessing ecological conditions in rivers and streams with diatoms. In: Smol, J.P., Stoermer, E.F. (Eds.), The Diatoms: Applications to the Environmental and Earth Sciences. Cambridge University Press, Cambridge, UK, pp. 57–85.

Stevenson, R.J., Rier, S.T., Riseng, C.M., Schultz, R.E., Wiley, M.J., 2006. Comparing effects of nutrients on algal biomass in streams in two regions with different disturbance regimes and with applications for developing nutrient criteria. Hydrobiologia 561, 140–165.

Stevenson, R.J., Smol, J.P., 2015. Use of algae in environmental assessments. In: Wehr, J.D., Sheath, R.G., Kociolek, J.P. (Eds.), Freshwater Algae in North America: Classification and Ecology. Academic Press, San Diego, pp. 921–962.

Stevenson, R.J., Stoermer, E.F., 1981. Quantitative differences between benthic algal communities along a depth gradient in Lake Michigan. Journal of Phycology 17, 29–36.

Van Dam, H., Mertenes, A., Sinkeldam, J., 1994. A coded checklist and ecological indicator values of freshwater diatoms from the Netherlands. Netherlands Journal of Aquatic Ecology 28, 117–133.

Wang, Y.K., Stevenson, R.J., Metzmeier, L., 2005. Development and evaluation of a diatom-based index of biotic integrity for the Interior Plateau Ecoregion. Journal of the North American Benthological Society 24, 990–1008.

Wetzel, R.G., Likens, G.E., 2000. Limnological Analyses, third ed. Springer, New York, NY.

Zelinka, M., Marvan, P., 1961. Zur Prazisierung der biologischen Klassifikation des Reinheit fliessender Gewässer. Archiv für Hydrobiologie 57, 389–407.

Chapter 38

Macroinvertebrates as Biotic Indicators of Environmental Quality

James L. Carter[1], Vincent H. Resh[2] and Morgan J. Hannaford[3]

[1]U.S. Geological Survey, Menlo Park, CA; [2]Department of Environmental Science, Policy & Management, University of California, Berkeley; [3]Science, Industry and Natural Resources, Shasta College

38.1 INTRODUCTION

The use of aquatic organisms to assess water quality is a century-old approach (Kolkwitz and Marsson, 1909; Cairns and Pratt, 1993), and biomonitoring of streams and rivers is now well established throughout most of the developed world. Monitoring programs in North America relied mainly on chemical and physical monitoring until the 1970s. One problem in relying solely on chemical and physical measurements to evaluate water quality is that they provide data that primarily reflect conditions that exist when the sample is taken. In essence, a physicochemical approach provides a "snapshot" of water quality conditions. In contrast, biological monitoring provides a "moving picture" of past and present conditions, and hence, a more spatially and temporally integrated measure of ecosystem health.

Benthic macroinvertebrates (mainly consisting of aquatic insects, mites, molluscs, crustaceans, and annelids; see also Chapters 15 and 20) are most often used when monitoring lotic systems (Hellawell, 1986; Bonada et al., 2006; Carter et al., 2006). All states in the United States use macroinvertebrates in water quality monitoring (Carter and Resh, 2013), whereas only about two-thirds of the programs use fish and only one-third use algae (USEPA, 2002). In this chapter we describe the many ways macroinvertebrates are used as indicators of environmental quality, from the molecular-through the community-level of biological organization. We highlight more recent advances in the use of DNA bar coding and species traits and then describe in detail the most commonly used macroinvertebrate-based methods for assessing the quality of streams and rivers. Even though there are many advantages to using macroinvertebrates in water-quality monitoring, as with all methods of environmental assessment, the disadvantages must also be considered and ways to overcome them be developed (see Table 38.1).

38.1.1 Evaluating Stressors

Macroinvertebrates have been used to evaluate the effects of various anthropogenic stressors at all levels of biological organization, from the molecular level to the ecosystem (Rosenberg and Resh, 1993). At the molecular level, the effects of pesticides have been examined by measuring depressions in acetylcholinesterase levels (Pestana et al., 2014). Likewise, changes in levels of mixed-function oxidases, enzyme activities, metallothioneins, and the extent of DNA damage have been shown to be useful in identifying the effects of metals and synthetic organic compounds (Brix et al., 2011; Cain et al., 2011; Damásio et al., 2011; Martínez-Paz et al., 2013; Nair et al., 2013). Collectively, these biochemical changes are referred to as biomarkers (Johnson et al., 1993). Studies of the effects of stressors on macroinvertebrates at the molecular level are beginning to take into account the influences of phylogeny (Buchwalter et al., 2008) and species traits (Rubach et al., 2011; Poteat et al., 2015). The latter will be further discussed later.

At the organism level, the changes in growth and reproduction and rates of morphological deformities have been evaluated as responses to increased pollution (Martin et al., 2007; Di Veroli et al., 2014). Likewise, various physiological

TABLE 38.1 Advantages and difficulties to consider in using benthic macroinvertebrates for biological monitoring.

	Advantages		Difficulties to Consider
(1)	Being ubiquitous, they are affected by perturbations in all types of waters and habitats	(1)	Quantitative sampling requires large numbers of samples, which can be costly
(2)	Large numbers of species offer a spectrum of responses to perturbations	(2)	Factors other than water quality can affect their distribution and abundance
(3)	The sedentary nature of many species allows spatial analysis of disturbance effects	(3)	Propensity of some macroinvertebrates to drift may offset the advantage gained by the sedentary nature of many species; Seasonal variation abundance may complicate interpretations or comparisons
(4)	Long life cycles allow effects to be examined temporally	(4)	Seasonal variation abundance may complicate interpretations or comparisons
(5)	Qualitative sampling and analysis are well developed and can be done using simple, inexpensive equipment	(5)	Certain groups are not well known taxonomically
(6)	Taxonomy of many groups is well known and identification keys are available	(6)	Identification of individuals to species in the larval form can be difficult to impossible
(7)	Modern genomic methods are available for precise identification of species	(7)	May not be sensitive to some perturbations, such as human pathogens and trace amounts of some pollutants
(8)	Many methods of data analysis have been developed for macroinvertebrate assemblages	(8)	Poorly established relationships between specific stressors and most commonly used metrics
(9)	Responses of many common species to different types of pollution have been established	(9)	All useable traits are not available for all species
(10)	Well-established species traits databases exist		
(11)	Macroinvertebrates are well suited to experimental studies of perturbation		
(12)	Biochemical and physiological measures of the response of individual organisms to perturbations are being developed		

Summarized from Rosenberg and Resh (1993), who also discuss how to overcome the difficulties mentioned.

responses, such as changes in respiration, metabolism, and bioenergetics have been examined in terms of their response to specific pollutants (Buchwalter and Luoma, 2005; Runck, 2007). While many of these organism-based processes have been evaluated in the field, most often they have been evaluated in a laboratory setting, usually by performing bioassays. The most realistic estimates of the effects of contaminants on populations are through full life-cycle and multiple generation testing (Diepens et al., 2014).

Most commonly, the population and community (= assemblage) levels are evaluated when the effects of pollution and habitat degradation are examined in nature. The abundance of populations and the abundance, taxonomic richness, and evenness of macroinvertebrate assemblages have been routinely examined in water-quality studies for decades. Numerous metrics, such as total richness (i.e., the total number of species, genera, families or more commonly, a combination of taxonomic levels identified from a sample) or the richness of specific groups of taxa, especially Ephemeroptera, Plecoptera, and Trichoptera (known in biomonitoring as EPT) have been found to be diagnostic of various environmental impacts (Lenat, 1988; Carter and Resh, 2013). Metrics based on richness are by far the most commonly used measures of the benthic assemblage by US state biomonitoring programs (Carter and Resh, 2013). In addition to richness metrics, the percentage composition of various tolerant (e.g., Chironomidae) and intolerant (EPT) taxonomic groups have been used to evaluate both natural and anthropogenic factors. These types of metrics have often been combined to form a multimetric index (MMI).

One problem associated with taxonomically-based metrics is that they often are most useful when based on species identifications. However, identifying macroinvertebrates to the species level can be challenging because immature and/ or damaged specimens may lack necessary diagnostic characteristics (Pfrender et al., 2010). Moreover, practically all aquatic insect species descriptions are based on the adult stage, whereas bioassessments principally collect the immature stage. Last, few identification keys for use at the species-level have been created, although keys to the generic-level among the aquatic insects in North America are fairly well developed (Merritt et al., 2008).

38.1.2 Advanced Approaches to the Identification of Macroinvertebrates

Over the past several decades, two new approaches for assessing macroinvertebrate assemblages have developed. One differentiates species based on their genome compared to the more traditional method of differentiating species based on their morphological characteristics (Hebert et al., 2003). The second method evaluates the traits of species and assemblages (Poff, 1997). Traits are measurable attributes that species possess and are based in part on physiology, morphology, behavior, and life history characteristics.

38.1.2.1 Genetic Approach

There have been many recent advances in delineating the species of freshwater macroinvertebrates present in a habitat. Over the last several decades, many molecular genomic methods have been developed and used in freshwater ecological and bioassessment studies (Pauls et al., 2014). The method typically used to differentiate species is known as deoxyribonucleic acid (DNA) bar coding (Hebert et al., 2003) and typically enhances the taxonomic resolution well beyond the ability of using only morphology. The portion of the genome that is most often used is the cytochrome c oxidase subunit I (COI), which is extracted from the mitochondrial DNA. Identifying species based on DNA bar coding involves acquiring tissue from an organism, amplifying the DNA associated with the COI portion of the mitochondrial DNA, determining the sequence of nucleotides (base pairs), and comparing the observed sequence to a database of sequences representing known species (Cold Spring Harbor Learning Center, 2014). Although there are many databases of sequences available for comparison with the material collected from the specimens, two that are often used are the BarCode of Life—http://www.boldsystems.org/ and GenBank—http://www.ncbi.nlm.nih.gov/genbank/.

The increase in the number of species identified from a macroinvertebrate sample using bar coding compared to traditional morphological characteristics is dramatic. For example, Sweeney et al. (2011) studied two sites on White Clay Creek, in Pennsylvania, United States, to test the influence of the level of taxonomic resolution on identifying impairment. They showed a 70% increase in the number of species identified using DNA bar coding compared to "expert genus-species level" identification using traditional morphological techniques. In a study of five streams in southern California, Jackson et al. (2014) also discovered there were far more (200 vs. 96) putative species identified using DNA bar coding than would be identified morphologically. This difference in the number of species detected allowed the identification of the impact of channel armoring on the benthic assemblage that was not observable using morphological techniques (Stein et al., 2014). Pilgrim et al. (2011) identified species, in part, as molecular operational taxonomic units (mOTUs) and found within just the EPT collected as part of Maryland's Biological Stream Survey, an approximate threefold increase in the number of species using DNA bar coding.

The use of genomic methods in biomonitoring clearly will continue its rapid development (Pauls et al., 2014); however, there are numerous issues that need to be addressed for its full potential to be realized (Pfrender et al., 2010). First, more research is needed on the methods used for determining species delimitations (White et al., 2014). Second, databases of DNA sequences need to be sufficiently populated with the sequences of known and verified species, so sequences derived from biomonitoring collections can be identified. Third, as Sweeney et al. (2011) noted, current information on the tolerance of many species is unknown. Therefore, determining some commonly used metrics such as biotic indices (see the first exercise below) is problematic.

In addition to COI, there are many other portions of the genome that can be used for species identifications (Pauls et al., 2014). In time, entire genomes will be compared as next-generation sequencing techniques become more common. Last, the development of eDNA techniques, which is DNA collected from environmental media (such as water samples) instead of DNA collected from specific tissue (Deiner et al., 2015; Thomsen and Willerslev, 2015), may obviate the need in the future for even physically sampling macroinvertebrates to address some environmental questions.

38.1.2.2 Species Traits Approach

In addition to structural characteristics of macroinvertebrate assemblages such as richness and composition, biological and ecological characteristics that are related to the physiology, morphology, behavior, and life history of species, populations, and higher taxa are used in environmental monitoring. Collectively these characteristics are known as species traits (Menezes et al., 2010). One of the limitations to the use of the taxonomy of macroinvertebrate assemblages in environmental monitoring is that species (or higher taxa) distributions are geographically and temporally specific

(Verberk et al., 2013). In contrast, functional aspects of species tend to be habitat specific, in contrast to being geographically specific (Poff, 1997; Statzner and Bêche, 2010). Therefore, comparing the response of macroinvertebrate assemblages among locations, or assessing the effects of environmental perturbations over large geographic areas is more fruitful when species traits are used vs. the species themselves (Culp et al., 2011). A further advantage of using species traits is that they have the potential to provide a mechanistic understanding for species distributions.

Although the founding concept of species traits dates back to the mid- to late 19th century (Statzner et al., 2001), the more recent beginnings stem from Southwood (1988) and the concept of the habitat template. This approach was expanded upon by Poff (1997) when he suggested that the existence of a species in a location is a function of whether it can pass through a hierarchical series (from large scale to small scale) of habitat filters. One of the first applications of the use of species traits was by Cummins and Klug (1979) who established the usefulness of categorizing macroinvertebrates by their mode of feeding (functional feeding groups). Since then, there have been numerous studies demonstrating the importance of evaluating macroinvertebrate assemblages based on their traits in contrast to their taxonomy (Resh et al., 1994; Statzner et al., 1994). The use of species traits in environmental monitoring has been aided by the establishment of several, easily accessible databases listing the traits of macroinvertebrates (Vieira et al., 2006; USEPA, 2012; Schmidt-Kloiber and Hering, 2015).

Although, using species traits in environmental monitoring will help generalize the response of macroinvertebrate assemblages across geographies and provide a mechanistic understanding for species distributions, there are several aspects that need further attention before species traits can be universally used. First, there is a need to develop a common nomenclature among databases containing species traits information and among researchers (Schmera et al., 2015). Second, and similar to DNA databases, databases on species traits must be more completely populated, which will involve substantially more research on the life histories of individual macroinvertebrate species (Menezes et al., 2010). These studies also should include information on the presence of ontogenetic shifts in traits within species if it exists (Poff et al., 2006). Third, common coding methods of scoring different traits within trait groups should be developed among trait databases. Fourth, the effects of nonindependence among traits because of phylogeny (Poff et al., 2006) and trade-offs (Resh et al., 1994) need to be addressed. Last, the methods used for the analysis of species traits need to be further developed because of the complexity of data types among traits (nominal through continuous) and the simultaneous analysis of several data matrices representing environmental characteristics, traits, and species abundances (Poff et al., 2006; Menezes et al., 2010; Verberk et al., 2013; Schmera et al., 2015).

38.1.3 Volunteer (Citizen-Based) Assessments

The simplicity and low cost of macroinvertebrate collecting and the ease with which water quality evaluations can be made has led to considerable development of volunteer monitoring programs in the United States (Ely, 2005). Conservation groups such as the Isaac Walton League of America (IWLA) popularized simplified field assessments for use by concerned citizens with the Save Our Streams (SOS) program (Firehock and West, 1995). Early USEPA Rapid Bioassessment Protocols (RBP; Plafkin et al., 1989) also described a cursory, or "RBP I," approach that was generally accepted as suitable for nonprofessionals given it was based on the IWLA SOS protocol (Firehock and West, 1995). The quality of data obtained by volunteers using good equipment (e.g., microscopes, undamaged nets) and adhering to accepted protocols can be very similar to data obtained by professionals when the same techniques are followed (Fore et al., 2001). However, the level of training received by volunteers has a significant effect on the quality of laboratory processing and identification. The involvement of professionals and the constancy of personnel in volunteer programs contribute positively to data quality (Ely, 2005). Because the taxonomic resolution achieved by volunteer monitoring groups is often not as detailed as professional assessments, the number and types of indices and analyses that can be used for a stream assessment are somewhat limited. However, detailed taxonomy is not necessary for deriving many commonly used metrics. Some volunteer programs base their assessments on the concept of indicator organisms (Connecticut Department of Energy and Environmental Protection, 2013). The presence of a short list of easily identified macroinvertebrates that indicate different environmental qualities simplifies the need for detailed identifications and therefore has the potential to reduce both errors and effort by volunteers. Last, in the future it even may be possible to use eDNA methods along with large-scale volunteer programs to greatly expand our knowledge of the distribution of macroinvertebrates. These methods were recently applied in United Kingdom by Biggs et al. (2015) in a study of the great crested newt (*Triturus cristatus*).

Regardless of whether the measures (metrics) evaluated are structural or functional attributes, assessment proceeds by comparing these values between unimpaired (=reference) sites and putatively (presumed) impacted (=test) sites. Although bioassessments based on metrics dominate United States programs, many sophisticated multivariate statistical procedures are used to evaluate stream impairment using macroinvertebrate assemblages as well. It is important to recognize that changes in the use of benthic macroinvertebrates in biomonitoring will continue to occur. For example, Bonada et al. (2006) evaluated a range of approaches in terms of how they met preestablished criteria of an ideal biomonitoring tool based on their underlying rationale, implementation, and performance. They found that many newer applications, performed far better than the oldest, widely used *Saprobian* approach (Niemi and McDonald, 2004). Furthermore, debates today on which organisms should be used for aquatic bioassessments (such as fish, macroinvertebrates, or diatoms), taxonomic levels needed (family, genus, or species), and which analytical techniques (summary statistics, univariate approaches, or multivariate approaches) seem far from being resolved. However, large-scale state and federal programs desiring increased comparability may ultimately resolve many of these issues.

Benthic macroinvertebrates represent an integral part of lotic systems by processing organic matter and providing energy to higher trophic levels; therefore, an understanding of the effects of anthropogenic, as well as natural stressors, on their distribution and abundance is critical for comprehensive impact assessment of streams and rivers. In this part of the chapter we describe the fundamental processes and considerations necessary for using macroinvertebrate assemblages for environmental assessment. We provide two procedures—a basic method that requires effort similar to that used in a student assignment or a volunteer-based site-specific project, and an advanced procedure that involves effort similar to that used in a graduate student or larger-scale project. An evaluation of stream habitat is also presented because of the importance of habitat in the distribution of macroinvertebrates and because anthropogenic effects on habitat are often the impact of concern.

38.2 GENERAL DESIGN

The basic principle behind assessing impairment by evaluating the structure and function of macroinvertebrate assemblages is the comparison of putatively (or presumed) impaired sites to unimpaired sites (see also Chapter 40). Unimpaired sites are known as control or reference sites and putatively impaired sites are known as test sites. The phrase "control site" is rarely used today however, and has been replaced by the concepts of reference site or reference condition (see below). Also, it is generally accepted that pristine reference conditions are rarely available in most study areas, so comparisons are normally made between putatively impaired sites and least impaired sites (Stoddard et al., 2006).

A fundamental consideration before making comparisons between reference and test sites of attributes derived from the analysis of macroinvertebrate assemblages is that both sets of sites have a similar biological potential in the absence of impact (Stoddard et al., 2006; Carter et al., 2009). Therefore, studies are most often restricted to areas that have similar gross physiography. Ecoregions, subecoregions, type of land cover, stream size, and elevation are just a few of the criteria to consider when restricting the range of physical variables that could confound comparisons of macroinvertebrate assemblages among sites in impact assessment (Hawkins et al., 2010a,b; Carter and Resh, 2013). The importance of these variables to the design of a bioassessment is often a function of the scale of the question being addressed.

Spatial and temporal considerations are critical in all assessment designs. In fact, geographic scale often dictates whether a study will follow a point source or regional assessment design. Small-scale, point source studies frequently use one of many BACI-type (before-after-control-impact) designs (Stewart-Oaten et al., 1986). In the simplest case with these designs, comparisons are made before and after an impact occurs at both control (reference) and impacted sites. The comparisons are based on analysis of variance (ANOVA) approaches that assess impact by appropriately partitioning variability (Downes et al., 2002). A very readable account of basic experimental design in ecology is presented by Underwood (1997); a more thorough treatment of the many complicating aspects associated with proper impact study design in streams is presented by Downes et al. (2002).

Large-scale regional assessments are the foundation of most state and national bioassessment programs. Although, the variety of designs used in these assessments is extensive (Carter and Resh, 2013), the basic principles are similar. In most studies, putatively impaired sites are compared to a reference condition. The use of reference condition rather than a single specific site has become increasing popular and thought necessary (Stoddard et al., 2006; Hawkins et al., 2010a,b) because establishing a reference condition aids in accounting for some of the variability among reference sites that is inherent in all large-scale studies. When reference conditions are unknown or unknowable, gradient-type assessments can be performed and reference conditions predicted (Carter et al., 2009).

38.2.1 Analytical Approaches

The process of analyzing macroinvertebrate assemblage data for bioassessments is often divided into two approaches—multimetric and multivariate. A principal distinction between the two approaches lies in how variables are defined. Both use the same raw species by sample data matrix (Hawkins et al., 2010b); however, in the multimetric approach, the variables (metrics) analyzed are derived by estimating certain summary characteristics from the species by sample data on a per sample basis. These characteristics include estimates of richness, percentage composition of certain taxa or traits (e.g., those that represent the trait group—feeding mode), measures of species diversity and evenness, and biotic indices based on tolerance scores. Once metrics are estimated, the value of each metric (or a multimetric comprised of a combination of metrics) is compared between the reference condition and the test site.

Alternatively, in the multivariate approach metrics are not estimated as they are in the multimetric approach. Rather, samples are compared by their position in species space by using the presence (or a measure of abundance) of each taxon in each sample as input to a classification and/or ordination procedure. Also, multivariate in design is the analysis of taxonomic completeness (Hawkins, 2006). A wide range of differences exists among programs in the application of these two data analysis approaches; nevertheless, the data necessary for performing bioassessments based on both multimetric and multivariate methods are practically identical (Fig. 38.1).

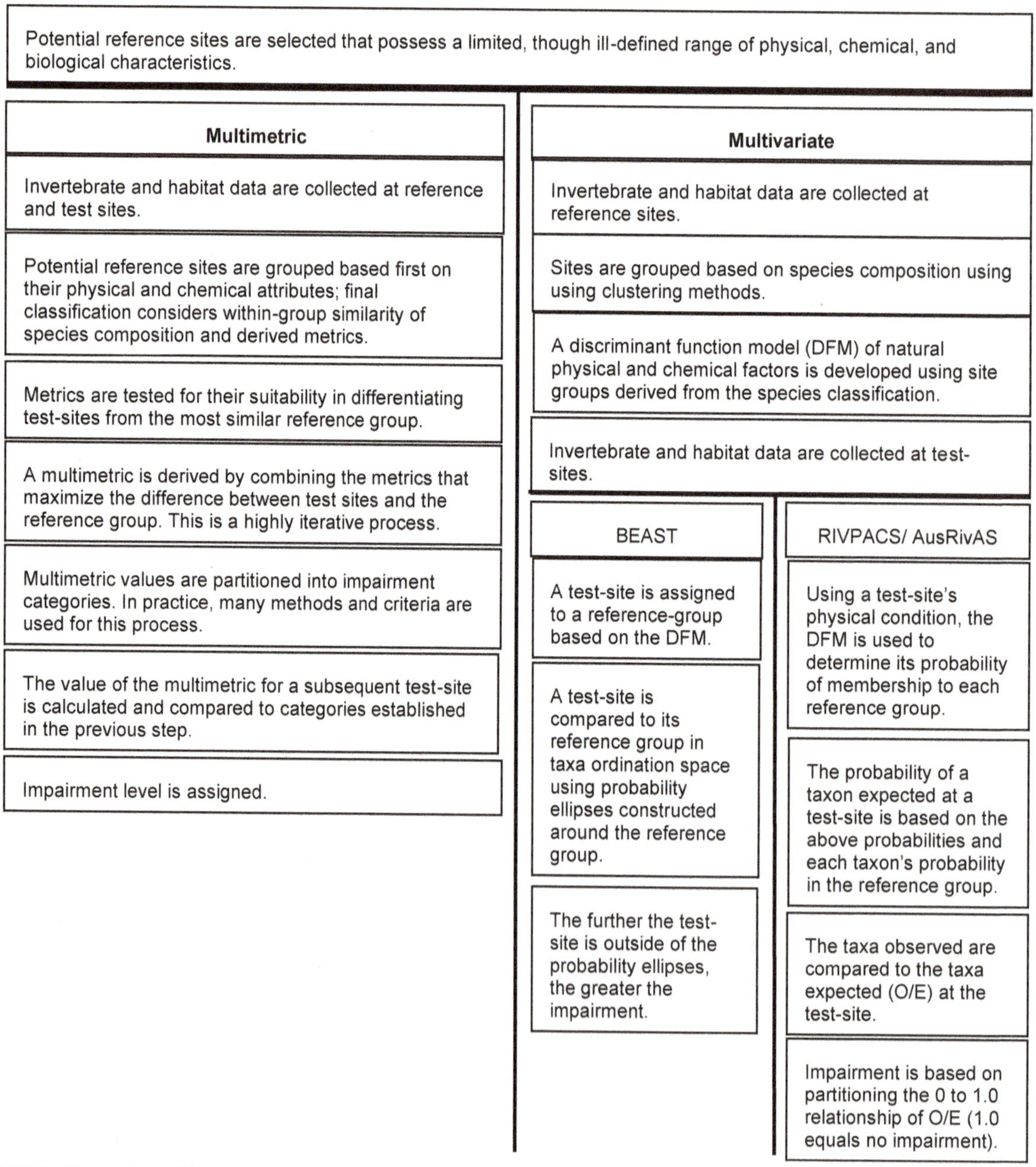

FIGURE 38.1 Comparison of the steps in multimetric and multivariate bioassessments; BEAST, RIVPACS, AUSRIVAS are three of the most widely used multivariate models. *Modified from Reynoldson et al. (1997) and Barbour et al. (1999).*

TABLE 38.2 Eleven metrics that represent 50% of all metrics used by US state biomonitoring programs.

Category	Metric
Richness	Total
Richness	EPT
Biotic index	Hilsenhoff biotic index (HBI)
Percentage	EPT
Richness	Mayflies
Percentage	% Dominant taxon
Diversity	Shannon
Percentage	Mayflies
Percentage	Chironomidae
Percentage	Scrapers
Richness	Caddisflies

Listed in the order of most frequently used to least frequently used.
From Carter and Resh (2013).

38.2.1.1 Multimetric Approach

The multimetric approach is based on the premise that certain measures of the benthic assemblage can be used to indicate its ecological condition and, by extension, the condition of the stream ecosystem. Many well established metrics have been used in stream assessments. Carter and Resh (2013) identified 112 metrics used by US state biomonitoring programs. Of these metrics, none were universally used by all programs; however, 11 of these metrics represented 50% of all metrics in use (Table 38.2). Although an underlying assumption has been that most metrics are firmly based on accepted ecological theory, little testing of this premise has been done.

Most contemporary survey approaches rely on multiple measures of community structure and function. These measures (metrics) can be grouped into several categories such as (1) taxa richness (e.g., family level, generic level, species level) of either the entire benthic assemblage or specific components of that assemblage viewed to be tolerant (e.g., Chironomidae) or intolerant (e.g., EPT) to pollution; (2) enumerations (e.g., number of all macroinvertebrates collected) or proportions of selected orders such as the EPT; (3) diversity indices which generally reflect dominance (e.g., Shannon's index); (4) percentage composition of the assemblage that possess a given trait (e.g., percentage of "shredders" within the trait group "feeding mode"; see also Chapter 20); and (5) biotic indices.

An expected change in species richness accompanying impairment is based on the premise that a loss of species occurs with increased impairment. A change in the number of individuals within a certain taxon is based on the notion that, with some types of pollution, more intolerant individuals may be lost [e.g., the decrease in EPT as a result of high trace metals (Clements et al., 2013)], while the numbers of tolerant individuals may rise (e.g., certain species of Chironomidae). Diversity and evenness indices summarize the distribution of the number of individuals among the number of species present in a sample into a single number. The functional feeding group concept assumes that organisms that have evolved specific traits (e.g., shredders) to use a certain food source through the shape of their mouthparts should be present when those resources (e.g., packs of leaves) are present. When stressors alter these resources, these functional traits are hypothesized to change within the benthic assemblage as species composition changes (Masese et al., 2014). Conversely, the possession of a specific functional feeding group trait may lead a species to be particularly susceptible to a specific stressor [e.g., toxic metal uptake by algae would affect grazers (Cain et al., 2013)]. This type of evaluation provides a mechanistic understanding that relates macroinvertebrate composition to stream processes via the traits they possess.

Biotic indices, used in one of the exercises below, are based on the premise that pollution tolerance differs among various benthic organisms. Tolerance values for each taxon are intended for a single type of pollution, typically for organic (nutrient) pollution, but, tolerance scores for temperature (Dallas and Rivers-Moore, 2012), sediments (Relyea et al., 2012), metals (Clements et al., 1992) and acidification (Chessman and McEvoy, 2012) have been developed. In most biotic indices, the taxon-specific tolerance value and the abundance (actual, proportional, or categorical) of each taxon in the assemblage are used to calculate a single score using a weighted average approach (formula 38.1). Various biotic indices and their formulae are discussed by Metcalfe (1989) and Resh and Jackson (1993).

Tolerance values for individual taxa are derived in a number of ways. One approach is to use collections of benthic invertebrates from streams of varying water quality and relate the presence or abundance of individual taxa to these conditions. These types of surveys have been published for Wisconsin (Hilsenhoff, 1988), the southeastern United States (Lenat, 1993), and for a number of other regions (Barbour et al., 1999). Yuan (2004) used general linear models to derive tolerance scores for genera collected in the Mid-Atlantic region of the United States. This method was designed to control for covarying natural gradients. However, the more common method of assigning tolerance scores in different regions uses "expert opinion." Carter and Resh (2013) found that the use of "local expertise" was the most often cited method used for establishing tolerance values by US state biomonitoring programs. Unfortunately, there is circularity in the use of tolerance values in that they are based on where the organisms are found and then applied in an assessment based on an organism's distribution. Additionally, tolerances values based on correlative studies are not mechanistic (Poff, 1997). An inferentially stronger approach is to base tolerances values on empirically derived laboratory and field testing (Buchwalter and Luoma, 2005; Clements, 2000). Nevertheless, currently used tolerance values tend to have broad application and often summarize the effects of multiple stressors; however, better documentation of how tolerance scores are determined is desirable (Chang et al., 2014).

Community similarity indices (e.g., percentage similarity, Bray—Curtis similarity) represent another method for comparing macroinvertebrate composition between sites and are used in both multimetric and multivariate approaches. Most similarity indices compare the composition of two samples on a taxon-by-taxon basis. These indices also may form the basis for cluster and classification analyses as well as advanced multivariate techniques.

The examples of metrics mentioned above are just a small suite of the measurements that can be evaluated when assessing water quality with structural and functional aspects of macroinvertebrate assemblages. An expanded list is presented in Barbour et al. (1999) (http://www.epa.gov/owow/monitoring/rbp/ch07b.html), and Carter and Resh (2013) list the 112 metrics currently used by state biomonitoring programs. Many other metrics [e.g., the proportions of individuals with morphological deformities (Di Veroli et al., 2014), fluctuating asymmetry, changes in behavior] have shown promise but are not yet widely used in bioassessments. It should be emphasized that biomonitoring procedures are not static; they continue to evolve as new knowledge becomes available.

38.2.1.2 Multivariate Approaches

Multivariate approaches consider each taxon (or trait) to be a variable and the presence or abundance of each taxon as an attribute of a site or a point in time (Norris and Georges, 1993). In contrast to the multimetric approach, the value associated with any given site is a function of that site's composition in relationship to the composition of all other sites in the analysis, or in some analyses, a reference condition. Multivariate approaches are used more often in large-scale assessments than in point source studies; although, there are good reasons to use them in point source studies as well (Carter and Resh, 2013). A variety of procedures are used including many different types of clustering and ordination techniques. Results of clustering and ordination analyses are often combined with other multivariate techniques such as multiple linear regression and discriminant function analysis when developing multivariate methods and relating biological patterns to environmental variables (Reynoldson et al., 2014).

Almost concurrently with the onset of the multimetric approach in the United States, the use of multivariate models in biomonitoring began in the United Kingdom (Wright et al., 1984). These techniques formed the basis for the development of similar techniques in Canada (Reynoldson et al., 1995) and Australia (Davies, 1991) and have been developed and used for biomonitoring in the United States (Hawkins, 2006). The initial approach by the United Kingdom led to the development of RIVPACS (River InVertebrate Prediction And Classification System), which is the progenitor of most other multivariate-based biomonitoring approaches used for evaluating streams and rivers. RIVPACS is based on the concept of taxonomic completeness (Hawkins, 2006) and is a sequential set of multivariate analyses that provide site-specific predictions of the macroinvertebrate fauna to be expected in the absence of major anthropogenic stressors (Wright et al., 2000). The expected fauna is derived using a database of species presence at reference sites and a suite of environmental site characteristics. The fauna observed at a test site are compared to the model-derived taxa expected at a site (Wright et al., 2000; Hawkins, 2006) hence, they are known as O/E-type models.

Two websites illustrating a multivariate approach that is a derivative of RIVPACS offer detailed presentations about how this approach is being used in wide-scale monitoring programs in Australia (http://ausrivas.ewater.org.au/ausrivas/index.php/introduction) and the United States (http://www.cnr.usu.edu/wmc).

Multivariate and multimetric approaches can be compared schematically (Fig. 38.1). Both approaches require establishing what characteristics (metrics or assemblages) would be typical of unimpaired conditions. Although these approaches differ considerably in the method used for determining whether a test site is equivalent to a reference condition, both methods begin from the same premise and require similar biological data. Models based on metrics and multimetrics generally require greater attention to habitat similarity than some multivariate methods because O/E-type models account for some of the influence of natural habitat variability among sites.

A question that comes up repeatedly in the application of both multimetric and multivariate methods is the level of taxonomy at which macroinvertebrates must be identified for bioassessments. This is a particularly important consideration when evaluating taxon richness (Resh and Unzicker, 1975), but is also important regardless of the approach used (Lenat and Resh, 2001; Bailey et al., 2001). This decision influences many metrics but it is especially critical for richness metrics and the assignment of tolerance values. In some citizens' monitoring programs where volunteer participants depend on "picture keys" to name the organisms collected, identification is usually to a mixture of the order- and family-levels (Firehock and West, 1995); however, in some programs that use a small suite of "indicator organisms," identifications can be to genus and species (Connecticut Department of Energy and Environmental Protection, 2013). Because the tolerance of many benthic macroinvertebrates differs within a family and even within a genus (Lenat and Resh, 2001), the more detailed the taxonomic resolution the more reliable the assignment of tolerance values can be. Most state and federal agencies in the United States involved in water quality monitoring typically use a mixture of generic- and species-level identifications for most, but not all taxa (e.g., Chironomidae and Annelida are generally not identified to the genus-species level) (Carter and Resh, 2001). In most bioassessments, the level of taxonomy necessary is likely a function of the level of impairment one wishes to detect, with more detailed taxonomy capable of detecting smaller effects.

38.2.2 Habitat Assessment

In the past four decades that benthic macroinvertebrates have been widely used in biomonitoring in North America, there has been a shift in how studies have been conducted. Prior to the early 1970s, emphasis was placed on the use of qualitative sampling with subjective comparisons made to evaluate differences between test and reference sites. Emphasis then shifted to more quantitative studies involving replicated, fixed-area sampling and the use of inferential statistical tests. Importantly, both types of studies were confined to relatively small scales, often point source type studies. By 1990, emphasis shifted back toward more qualitative sampling approaches and analyses because of a general change from effluent-based monitoring to ambient monitoring (National Research Center, 2001; but see US Environmental Protection Agency, 2010). Along with this latter shift was the development and promulgation of rapid bioassessment procedures by the USEPA (e.g., Plafkin et al., 1989), which were developed to address these larger-scale, non—point source effects. As the spatial scale of studies has increased, the influence of habitat also has increased and an appreciation for its effects at various geographic scales has developed (Carter et al., 1996).

The rationale for including a habitat assessment in a biomonitoring study is that benthic macroinvertebrates are influenced by habitat quality (e.g., bank stability, fine sediment deposition, temperature) just as they are by water chemistry (e.g., a pollutant present in the water). Habitat evaluations may have multiple functions. They can be a component of the design of an assessment, as when defining reference conditions and identifying test sites for comparison. Additionally, habitat can be a source of impairment, or a confounding factor in assessment interpretation when habitat and water quality impacts occur simultaneously. Although visual estimates of habitat quality do not substitute for rigorously measured habitat characteristics in describing and assessing impairment to stream processes that may affect macroinvertebrates (e.g., Chapters 1—8), in many cases they provide an adequate view of general habitat conditions. For example, over 50% of state programs rely on visual estimates of habitat in their biomonitoring programs (Carter and Resh, 2013).

When considering water pollution, the relationship between biological and physical condition may best be viewed as:

$$\text{Biological condition} = \text{habitat quality} + \text{water quality}$$

In the above equation, *biological condition* represents a metric (or multimetric) or multivariate measures derived from the macroinvertebrate assemblage; *water quality* represents water chemistry, including toxins; and *habitat quality* represents the physical, geomorphological, and biological (e.g., temperature, condition of the riparian zone, in-stream algae, macrophytes, introduced species) conditions at the site.

As important as an analysis of habitat is for confident interpretation of bioassessment results, an overriding consideration in making any physical measurement is to ask: *What is the purpose of measuring this characteristic of the lotic environment?* The answer will often determine how much effort and what method should be used for the measurement of each variable chosen. A great deal of effort is often misspent in bioassessments collecting physical measurements that have little to do with the question(s) being addressed. An intensive geomorphological description of a stream reach can take days to years of effort by a large crew of trained fluvial geomorphologists (Fitzpatrick, 2001; see also Chapters 1, 5, and 28 for application of remote sensing tools). Restricting the physical evaluation to a suite of variables that are predicted to have a potential influence on macroinvertebrate distributions within the region of the assessment can be extremely efficient (Rankin, 1995). Customizing the effort per variable is also useful in reducing total effort (Fend et al., 2005). When choosing a specific habitat protocol, it is important to note that many protocols focus on geomorphological influences on channel processes and fish habitat, and do not adequately or specifically evaluate macroinvertebrate habitat.

Prior to beginning any study, appropriate sampling sites must be chosen. The linear dimensions of a site (i.e., reach length), can be based on the geomorphology of the stream (e.g., one pool–riffle sequence; 40 × channel widths) or a fixed length of stream (Carter and Resh, 2001). When sites are restricted to just a few locations, as in many point source studies, reaches are often chosen above and below a potential source of pollution that is suspected of impairing the biological condition of a single stream. The placement of sites in this manner is a common practice, although it is not an optimal design because upstream and downstream sites could differ in terms of many uncontrolled variables (e.g., discharge could be higher downstream, riparian conditions could differ), and values estimated (mean and variance) using within-site "replicates" would be spatially confounded. A more robust design is to choose sites on multiple (≥3) reference and test streams[1] that have comparable characteristics except for the presumed impairment. The inclusion of multiple streams in the design greatly increases the generality of the conclusions.

The principle that should be remembered when sites are compared is that their physical characteristics should be as similar as possible save the putative chemical- or habitat-based impairment that is being evaluated. These characteristics should include (1) the gradient of the compared reaches be very similar [e.g., comparing a high-gradient (5% slope) reach with a meandering (0.5% slope) reach is not meaningful]; (2) the substrate composition or at least the dominant substratum size of each reach be similar (e.g., a sand-dominated channel will have different macroinvertebrates than a cobble-dominated channel); (3) the streams be of similar order and have similar discharge regimes (see also Chapters 3 and 4); (4) the streams be either permanent *or* intermittent (e.g., although an intermittent stream may appear perennial in winter and spring, its invertebrate fauna will be very different than the fauna of a perennial stream). It should also be remembered that sites have "legacies," which could include historical anthropogenic impacts (e.g., early agriculture or logging) or natural (e.g., floods or droughts) events that can greatly influence the fauna present at the time of sampling. Establishing similarity of these factors among study sites will greatly increase confidence in the resultant assessment. However, it should be remembered that in practice there will always be differences among the sites being compared. The remaining differences are comprised of uncontrolled variables that produce variation that is not considered in the design of the study.

Once the study reaches have been chosen, individual sampling sites within them must be selected. Frequently, riffles are chosen as sites for macroinvertebrate sampling because of the abundance and diversity of organisms often found in them; however, a combination of riffles and pools, sampling all habitats in proportion to their occurrence or sampling systematically (USEPA, 2007) also are options. Regardless of the habitat(s) chosen for collecting macroinvertebrates, standardization of the sampling protocol at all sites is critically important for comparisons to be valid (Cao and Hawkins, 2011; Carter and Resh, 2013).

38.3 SPECIFIC METHODS

The process of conducting a biological assessment, including a physical habitat assessment of a stream and its surrounding riparian area and basin, can range from a quick estimate of present conditions to detailed measurements made over long periods of time (months or even years). Below, we present methods for rapid biological and physical habitat assessments.

The *Basic Method* can be completed during 1 day in the field and/or 1 day in the laboratory. This method is particularly appropriate for a survey or reconnaissance study when little is known about a stream, when a contaminant spill or pollution problem needs to be evaluated quickly, or for comparisons over large geographic areas within similar environmental

---

1. When conducting these procedures in a class setting, if an impacted stream cannot be compared with a reference stream, then researchers could sample macroinvertebrates in distinctly different habitats within one stream (e.g., pools, riffles, vegetated stream margins) or repeatedly sample the same type of habitat within a single stream. Thus, the researchers could examine variability in sampling while still learning the concepts and techniques used in biological assessment.

settings. Additionally, it is practical for volunteer monitoring groups and student projects. With modification, it can also be applied in establishing pollution control programs in newly industrialized and developing countries (Resh, 1995).

The *Advanced Method* builds on the techniques used in the basic method. It represents a level of effort and analytical sophistication that is more similar to state and national biomonitoring programs. The following exercises cannot begin to cover the full range of techniques and analytical tools available for bioassessments (see also Chapter 40); nevertheless, they provide the foundation for more intensive studies. Reference works mentioned throughout this chapter and book should be consulted for more intensive studies. Prior to beginning any study, consultation with an aquatic invertebrate ecologist and statistician to develop an experimental design that will adequately address the questions being asked is highly advised.

38.3.1 Basic Method: Assessment of Two Sites

38.3.1.1 Site Selection

Site selection is an important component of all biomonitoring procedures. Because this *basic exercise* serves more as a demonstration of biomonitoring approaches, strict adherence to ideal experimental design principles is not necessary. For example, sites can be located on two different (but physically similar) streams or simply in two different reaches on the same stream. Keep in mind that the physical factors listed previously should be similar between both sites: try to restrict your study to streams with pools and riffles; select one site that would be considered a reference site and a second site that is likely to be impaired; and collect an equal number of macroinvertebrate samples (≥ 3) from each site (see below; Chapter 15).

38.3.1.2 Physical Habitat Description

A reasonable on-site description of a stream can be generated by a few simple habitat measures. In most stream studies, these habitat characteristics are nearly always measured and recorded to describe the area under study at the time of sampling: stream width and depth, flow velocity, water temperature, and weather conditions (see Chapters 2–7). Along with recording the geographic location of the site (preferably using a global positioning system and recording what datum is used), these observations provide fundamental information about the size of the stream, what types of organisms might be living there (e.g., some organisms typically live only in cold water, others only in warm water), and summarize the conditions at the time of sampling which can influence sampling efficiency and provide possible explanations when data are analyzed. A simple narrative description of the appearance of the reach, in particular noting the presence of in-stream structures, point source inputs of effluents, and other features is extremely important.

Which variables are measured depend on the question(s) being addressed by the study and the general conditions of the reach. For example, an investigator might not need to measure dissolved oxygen in a clean, high-gradient mountain stream because it is likely that the water is saturated with dissolved oxygen. However, in a slow-moving warm stream that may be subject to organic pollution (e.g., sewage), measurements of dissolved oxygen (both during the day and the night) are crucial.

Other characteristics often measured include the gradient (slope), composition of the substratum, median particle size (as determined by pebble counts; see Chapter 5), stability of the stream banks, the percentage of the stream that is shaded by riparian vegetation, the riparian width and composition, the complexity of microhabitats within the stream, and the number of pieces of large wood present (Barbour et al., 1999; see also Chapters 2, 6, 7, and 26). Water quality measurements such as pH, conductivity, the concentration of dissolved oxygen and nutrients, are also often measured (see Chapters 6, 7, 9–11, 24, 30–33).

The first three measurements mentioned below are all related to discharge at the time of sampling (see also Chapter 3). These three measurements should be taken together in a reach having few obstructions and a uniform flow. Record the measurements of physical habitat as they are taken.

1. *Mean stream wetted width*: Measure the width of the stream in meters, from one edge to the other edge and perpendicular to the flow, for three different transects across the stream. This measurement is dependent on the discharge at that time and the form or shape of the channel.
2. *Mean stream depth*: Along the same transects as above, measure the depth (in meters) at one-fourth the distance from the water's edge, again at half the distance (midstream), and at three-fourths of the way across. Add the three values and divide by 4 (divide by 4 to account for the shallow water from the bank edge to the one-fourth distance mark). Record the average depths (in meters) for each transect.
3. *Current velocity*: Follow the protocols described in Chapter 3. Alternatively, lay a tape measure along the edge of the stream (5–10 m is sufficient). Drop a neutrally buoyant float (e.g., orange or lemon) into the water several meters upstream of where

the tape measure begins and measure the amount of time it takes for the float to pass the length of the tape (if the water is very shallow a twig or cork can be used). Repeat five times, record each value, and calculate the average velocity (m/s).

4. *Water temperature*: Using an electronic, liquid or bimetal thermometer, read the temperature while the probe is still in the water and after a reasonable equilibration period. Always record the time of day that temperatures are taken. Relatively inexpensive continuous recording thermometers are now widely available (see Chapter 7).

5. *Water clarity and quality*: Record whether the water is clear, slightly turbid, or muddy (e.g., whether you can or cannot see the bottom of the stream). Try to note any source of sediment (e.g., storm runoff, construction activities). Also, visually evaluate and record whether any oil is apparent and/or whether the water has an unusual odor.

38.3.1.3 Macroinvertebrate Field Collection Option

1. *Sampling benthic invertebrates*: Although many devices and techniques are available for collecting aquatic invertebrates (see Chapter 15 and Merritt et al., 2008), a D-frame net is most commonly used in macroinvertebrate biomonitoring studies (Carter and Resh, 2001). Collect at least three ~0.1 m^2 samples per site using a standard 0.33-m wide D-frame kicknet fitted with a 500-μm mesh net. In cobble-to-gravel bottom streams collect macroinvertebrates from riffles. In very large substrate rivers where the mean cobble diameter is >10 cm, use the Hauer–Stanford kicknet (Hauer and Stanford, 1981; see Chapter 15). Locate each placement of the kicknet in a random fashion; however, begin collecting at the most downstream location and move in an upstream direction. A video by Resh et al. (1990) illustrates how various sampling devices are used.

2. *Field sample processing*: Once the sample is collected, macroinvertebrates can either be "picked" in the field or the entire sample can be taken to the laboratory for more controlled processing (see below). If the samples are to be field picked, place the contents of the net into a large, white-enamel pan with enough water from the stream to cover the invertebrates. Using forceps, an eyedropper, or the "bug spatula" described in Chapter 20, pick out the first 100 invertebrates that you *randomly* encounter and place them in a jar with 70% alcohol. Although the tendency will be to pick out the largest organisms, it is essential that all species and size classes are sampled proportionately. Faster-moving organisms will be harder to catch but every effort should be made to sample all taxa evenly.[2] Be aware of organisms in cryptic cases that resemble pieces of substratum or debris (some insects, such as caddisflies, build cases out of natural materials). If upon field sorting, one D-frame net collection does not provide 100 individuals, take additional collections until the total number of invertebrates picked reaches 100 for each of the 3 samples per site. Remember that organisms that are extremely small (e.g., early instars of insects) will often be difficult to identify but are necessary to include if an unbiased sample is desired.

   Many factors influence the effectiveness and comparability of field sample processing, including mesh size of sampling devices, the area sampled, magnification used in picking, and the time of day and lighting conditions, among others. In general, field picking is far less reliable than laboratory processing.

3. *Subsampling*: If a sample contains far more individuals than needed, subsampling is required. This can be done by marking a grid in the bottom of the pan, using random numbers to select individual squares, and then picking out macroinvertebrates from the selected squares. Systematically pick out individuals until a total of 100 have been separated.[3] The important thing to remember about subsampling is that the procedure should not over- or underrepresent any particular group.

   When subsampling is based on a fixed number of individuals it is essential that the same number of individuals be used for comparisons among sites. This is because metrics based on richness (e.g., total number of species) may be confounded by the nonlinear relationship between the number of species estimated and the number of individuals examined. For example, if a 100-organism subsample is the goal as in this exercise, but a 150 organism subsample is picked, the 150 organism sample will likely overestimate the number of species present compared to the number of species estimated from a 100 organism subsample. Many protocols recommend that the total number of organisms sorted from the sample be within ±10% of the sorting goal (e.g., 90–110 individuals if the goal is 100) (Barbour et al., 1999). However, to our knowledge, the validity of this criterion has not been tested and would make a good follow-up research project.

---

2. Soda water or "club soda" can be added to the pan to anesthetize the animals.

3. A single square will likely have either fewer than 100 or greater than 100 individuals. If all the organisms are sorted from one or more squares it is likely that >100 organisms will be sorted. However, sort only 100 identifiable organisms because in this exercise the same number of organisms should be used to compare richness among samples and an estimate of density is not necessary.

If the sample is to be taken back to the laboratory for picking, place the contents of the sample and drain off water into an appropriate-sized bottle, plastic container, or a sealed plastic bag. Add enough 95% ethanol to cover the contents. The residual water in the sample should produce a final concentration of about 70% ethanol. If the sample is to be kept for any length of time, the initial ethanol should be changed to fresh 70% ethanol within the first week of storage. Label the sample as described below. See Chapter 15 or Moulton et al. (2000) for laboratory subsampling and sorting procedures.

4. *Sample labeling*: Label the sample clearly, giving the date; a clear description of the location including the stream, county, and state; a brief description of the habitat type (e.g., pool, riffle); and the collector's name(s). Write the information on a paper tag using a pencil, and place the tag in the jar with the sample.

5. *Invertebrate identification*: Either while picking the sample or after the sample has been picked, separate the macroinvertebrates into groups of similar-looking organisms (i.e., those you think represent a single species or taxon). Use the general key in Chapter 15 to identify an individual from each group to the family level. Record the information on the data sheet provided (Table 38.3). Good general keys for more detailed identifications are available for all benthic macroinvertebrate groups (e.g., Thorp and Covich, 2014; Smith, 2001; Voshell, 2003), specific groups such as the insects (e.g., Lehmkuhl, 1979; Merritt et al., 2008), macroinvertebrates of specific regions (e.g., Clifford, 1991), and insects of specific regions (e.g., Usinger, 1956; Peckarsky et al., 1990). In addition, a video that demonstrates how to use a dichotomous identification key for benthic macroinvertebrates is available (Merritt, 2002).

38.3.1.4 Macroinvertebrate Laboratory-Only Option

If a demonstration (e.g., to a class or volunteer monitoring group) is required because weather conditions or the size of the group do not allow a field visit, the following activity may be appropriate to illustrate the principles behind biomonitoring with macroinvertebrates.

1. *Laboratory preparation*: This exercise is intended for demonstration only. The instructor must assemble macroinvertebrates from previous collections or make special collections prior to the laboratory session. Plan to provide at least 100 macroinvertebrates per student or group of students. Assemble two "macroinvertebrate soups," one that represents the macroinvertebrate fauna of a reference stream and one that represents that of an impacted stream. To make the "soup," place all the macroinvertebrates that represent the reference site together in a bowl and cover with 70% ethanol. Do the same for the macroinvertebrates from the impacted site.

2. *Sampling*: To take a sample from the invertebrate soup, swirl the "soup" to evenly distribute the organisms. Then using a tea strainer or a small aquarium net, dip into the "soup" to obtain a sample. The purpose of this is to obtain a random sample of 100 invertebrates. The sample is then placed in a petri dish with ethanol, sorted, and identified as outlined in item 5 of the previous section.

38.3.1.5 Data Analyses

Just as errors in sampling, sorting, and identification can lead to inappropriate conclusions, errors in data analysis can lead to misinterpretations as well. A systematic approach to each step, from data entry through statistical analysis is necessary for confident data interpretation.

1. *Family biotic index*: On the worksheet provided (Table 38.3), list the names of the macroinvertebrate families collected and the number of individuals in each family in the sample. Look up the tolerance score (Table 38.4) for each family or higher taxon and write it in the next column. After multiplying the value in the number column by the corresponding tolerance score for that taxon, add the resulting numbers and then divide this sum by the total number of individuals. Eq. (38.1) is used to calculate the *family biotic index* (FBI) (Hilsenhoff, 1988), which is a weighted average:

$$\text{FBI} = \sum_{i=1}^{S} n_i \times t_i \left/ \sum_{i=1}^{S} n_i \right. \tag{38.1}$$

where n_i and t_i = number of individuals and the tolerance, respectively, of the *i*th family and S = number of families included in the analysis.

The information recorded on Table 38.3 can also be used to calculate several other useful metrics (e.g., the total number of families or family richness; percentage of total organisms that are EPT; percentage of total organisms of a particular

TABLE 38.3 Form to record macroinvertebrate data.

DATE: _____

NAME: _____

SITE: _____

	A	B		C		D
	ORDER/FAMILY	# OF ORGANISMS		TOLERANCE SCORE		TOTAL
1	_____	_____	X	_____	=	_____
2	_____	_____	X	_____	=	_____
3	_____	_____	X	_____	=	_____
4	_____	_____	X	_____	=	_____
5	_____	_____	X	_____	=	_____
6	_____	_____	X	_____	=	_____
7	_____	_____	X	_____	=	_____
8	_____	_____	X	_____	=	_____
9	_____	_____	X	_____	=	_____
10	_____	_____	X	_____	=	_____
11	_____	_____	X	_____	=	_____
12	_____	_____	X	_____	=	_____
13	_____	_____	X	_____	=	_____
14	_____	_____	X	_____	=	_____
15	_____	_____	X	_____	=	_____
16	_____	_____	X	_____	=	_____
17	_____	_____	X	_____	=	_____
18	_____	_____	X	_____	=	_____
19	_____	_____	X	_____	=	_____
20	_____	_____	X	_____	=	_____

Family Biotic Index = Total of Column D divided by Total of Column B = _____

% EPT = Total Ephemeroptera, Trichoptera, and Plecoptera divided by total of Column B = _____

Family richness = total number of families = _____

functional feeding group). Additional commonly used metrics can be found in Table 38.2. Barbour et al. (1999) provide many examples of the various metrics that can be calculated, and Carter and Resh (2013) list metrics used by US state biomonitoring programs. These measures can then be analyzed in the same way outlined below for the FBI. Analysis of FBI values are conducted in two steps in this exercise: (1) individual participants or groups evaluate their own data from the habitat assessment, calculate an FBI, and determine the water quality category of the sites selected; (2) within-site and between-site variabilities are examined by using the data from all participants.

2. *Biotic index—analysis of individual data*: Water quality can be evaluated using the FBI by comparing the index value calculated from a benthic sample with a predetermined scale of "biological condition." A scale developed for use in Wisconsin to determine the degree of organic pollution is provided in Table 38.5 (Hilsenhoff, 1988). For example, find Group A in the sample data set (Table 38.6). An index of 4.5 was calculated from a test stream sample, which

TABLE 38.4 Tolerance values for macroinvertebrates.

Plecoptera		Polycentropodidae	6
Capniidae	1	Psychomyiidae	2
Chloroperlidae	1	Rhyacophilidae	0
Leuctridae	0	Sericostomatidae	3
Nemouridae	2	Uenoidae	3
Perlidae	1	**Megaloptera**	
Perlodidae	2	Corydalidae	0
Pteronarcyidae	0	Sialidae	4
Taeniopterygidae	2	**Lepidoptera**	
Ephemeroptera		Pyralidae	5
Baetidae	4	**Coleoptera**	
Baetiscidae	3	Dryopidae	5
Caenidae	7	Elmidae	4
Ephemerellidae	1	Psephenidae	4
Ephemeridae	4	**Diptera**	
Heptageniidae	4	Athericidae	2
Leptophlebiidae	2	Blephariceridae	0
Metretopodidae	2	Ceratopogonidae	6
Oligoneuriidae	2	Blood-red Chironomidae	8
Polymitarcyidae	2	Other Chironomidae	6
Potomanthidae	4	Dolichopodidae	4
Siphlonuridae	7	Empididae	6
Tricorythidae	4	Ephydridae	6
Odonata		Psychodidae	10
Aeshnidae	3	Simuliidae	6
Calopterygidae	5	Muscidae	6
Coenagrionidae	9	Syrphidae	10
Cordulegastridae	3	Tabanidae	6
Corduliidae	5	Tipulidae	3
Gomphidae	1	**Amphipoda**[b]	
Lestidae	9	Gammaridae	4
Libellulidae	9	Talitridae	8
Macromiidae	3	**Isopoda**[b]	
Trichoptera		Asellidae	8
Brachycentridae	1	**Acariformes**[b]	8
Calamoceratidae[a]	3	**Decapoda**[b]	
Glossosomatidae	0	Astacidae	8
Helicopsychidae	3	**Gastropoda**[b]	
Hydropsychidae	4	Lymnaeidae	6
Hydroptilidae	4	Physidae	8
Lepidostomatidae	1	**Pelecypoda**	
Leptoceridae	4	Pisidiidae	8
Limnephilidae	4	**Oligochaeta**[b]	5
Molannidae	6	**Hirudinea**[b]	10
Odontoceridae	0	**Turbellaria**[b]	4
Philopotamidae	3		
Phryganeidae	4		

[a]Adapted from Lenat (1993).
[b]From Barbour et al. (1999).
Remainder from Hilsenhoff (1988).

TABLE 38.5 Water quality based on family biotic index values from Hilsenhoff (1988).

Family Biotic Index	Water Quality
0.00–3.75	Excellent
3.76–4.25	Very good
4.26–5.00	Good
5.01–5.75	Fair
5.76–6.50	Fairly poor
6.51–7.25	Poor
7.26–10.00	Very poor

TABLE 38.6 Sample data set for *t*-test (see Narf et al., 1984).

Group	Reference Site	Test Site
A	3.4	4.5
B	3.2	5.2
C	3.9	6.1
D	5.6	7.9
E	3.1	5.2
F	5.3	5.7
G	4.3	6.5
H	4.3	5.4
I	5.1	6.3
J	3.2	4.7

Summary statistics:

Reference site: $n_1 = 10$; $\bar{x}_1 = 4.1$; $S_1^2 = 0.89$; $\sum x_1 = 41.4$; $\sum (x^2)_1 = 179.3$

Test site: $n_2 = 10$; $\bar{x}_2 = 5.8$; $S_2^2 = 1.00$; $\sum x_2 = 57.5$; $\sum (x^2)_2 = 339.6$

would indicate a "good" water quality rating on a scale of "excellent" to "very poor." Find the water quality rating that describes the FBI scores you calculated.

Alternatively, you can assess water quality by comparing the test site (the assumed impaired site) to the reference condition. The similarity (Table 38.7) between the test site and the reference condition can be calculated, and expressed as a percentage as follows:

$$\text{Similarity (\%)} = (\text{reference FBI/test FBI}) \times 100 \tag{38.2}$$

In the example provided, Group A would show a 76% similarity between sites [=(3.4/4.5) × 100], indicating that the test site is "slightly impaired" relative to the reference. Calculate the percent similarity between your reference and test site using Eq. (38.2), then match this to the water quality thresholds provided in Table 38.7.

3. *Graphical analysis of group data*: Examine the variability of the rapid assessment data. The easiest way to view variability is to graph it. With each point representing a single collection, place each of the sites on the *x*-axis and the calculated measure (e.g., FBI, family richness, % EPT) on the *y*-axis. Mark the predetermined water quality thresholds directly on this graph and observe the range of water quality assessments obtained within the same site. Download and install R (R Core Team, 2015) using the information provided in Appendix 38.1. A R (R Core Teams, 2015; Version 3.2.2) script is provided that plots the test FBI values found in Table 38.6.

TABLE 38.7 Biological condition using percent similarity of family biotic index calculated between test site and reference site samples.

% Similarity	Biological Condition
≥85%	Unimpaired
84−70%	Slightly impaired
69−50%	Moderately impaired
<50%	Severely impaired

Modified from Plafkin et al. (1989).

4. *Measure of variability*: One way to compare the variability of metrics (e.g., FBI) is to express the standard deviation of the sample as a percent of the sample mean. This is the coefficient of variation (CV) and is computed as:

$$CV = (\text{standard deviation/mean}) \times 100 \tag{38.3}$$

The mean is computed as:

$$\bar{x} = \frac{1}{n} \sum_{i=1}^{n} x_i \tag{38.4}$$

where $x_i = i$th value of the metric and $n = $ total number of metric values. The standard deviation is computed as:

$$\text{S.D.} = \left(\frac{\sum_{i=1}^{n} (x_i - \bar{x})^2}{n-1} \right)^{0.5} \quad \text{or} \quad = \left(\frac{1}{n-1} \left(\sum_{i=1}^{n} x^2 - \frac{\left(\sum_{i=1}^{n} x \right)^2}{n} \right) \right)^{0.5} \tag{38.5}$$

The advantage of calculating the CV is that it has no units and the variability of metrics that differ greatly in the magnitude of their means can be compared. Therefore, it can be used to compare the variability of different types of measures (e.g., % EPT, taxa richness). Barbour et al. (1992) provide an example of its use.

38.3.2 Advanced Method: Assessment of Multiple Sites

38.3.2.1 Site Selection

Site selection is an important component of all biomonitoring procedures. In this exercise two groups of sites will be selected—reference sites and putatively impaired sites. Because this exercise focuses on a larger-scale question than the basic exercise, each site is a replicate of its respective group. An example of a possible impairment to investigate could include the effects of land cover or land use (suburban or urban compared to forested lands). Each group should contain the same number of sites (≥5), and sites should be in separate streams if possible. Keep in mind the factors listed on page 10 that should be similar among all sites analyzed. Try to restrict your study to pool−riffle streams. However, if only low gradient, sand-silt bed streams are available, the reach-wide method in USEPA (2007) has been shown to function well regardless of gradient (Flotemersch et al., 2014).

38.3.2.2 Macroinvertebrate Collections

Collect a single, composited macroinvertebrate sample from each site using a standard 0.33-m wide D-frame kicknet fitted with a 500-μm mesh net. Collect from riffles in cobble-to-gravel bottom streams. As in the *Basic Method* (above), if collections are made in very large substrate rivers where the mean cobble diameter is >10 cm, use the Hauer−Stanford kicknet (see Chapter 15). Locate each of the 10 placements of the kicknet either randomly of systematically (USEPA, 2007). Each composited sample should be composed of ten ~0.1 m^2 collections per site. Carefully clean samples of excess debris and water, place in a container (e.g., quart canning jars), insert a standard collecting label, and fix the sample with ethanol so the final concentration is 70−80%. Be certain that macroinvertebrate collections are complete before disturbing the study reach as you conduct the habitat assessment.

38.3.2.3 Habitat Assessment

As mentioned above, the rationale for including a habitat assessment in a biomonitoring study is that benthic macro-invertebrates may be influenced by habitat quality (e.g., bank stability, fine sediment deposition, temperature) just as they are by water chemistry (e.g., a pollutant present in the water). Habitat parameters describe components of the stream channel, the surrounding riparian area, and the basin. Habitat assessments can be visually based, such as when different conditions for parameters are described verbally or pictorially, scores ascribed to the different conditions, and summations made to describe overall habitat conditions (see Habitat Assessment in the General Design section of this chapter). The most widely used of these is that presented in Barbour et al. (1999).

We have prepared a visual habitat assessment useful for benthic macroinvertebrate studies in which characteristics are evaluated from large scale (basin land cover) to in-stream features (Table 38.8). In conducting this assessment, the user judges the condition of these parameters as optimal, suboptimal, marginal, or poor. The variables we included are only a few of many that could be used.

For each parameter, read the description given for each of the four conditions, select the one that most clearly identifies what you see or know about your basin, and circle the number that corresponds to the condition class. When you have finished estimating all the parameters, sum the circled values. Determine the habitat condition of both the reference sites and the test sites. These values will be used in later exercises.

38.3.2.4 Subsampling, Sorting, and Identification

It is likely that most samples will contain far more than the 300 individuals per sample needed for the exercise and therefore subsampling is required. This can be done by marking a grid in the bottom of the pan, using random numbers to select individual squares, and then picking out all the macroinvertebrates from the selected squares until you have a total of 300 individuals. An inexpensive, more efficient method of subsampling can be found in Moulton et al. (2000). Sort all organisms from the subsample(s) using a dissecting microscope set at $7-10\times$ magnification and identify them to the lowest taxonomic level in which you are confident of your determinations. If you want to distinguish invertebrates to the lowest taxonomic level possible, use operational taxonomic units (e.g., species a, species b) if the species is unknown and you can accurately categorize the invertebrates you are identifying. Be certain that your identifications are uniformly done among all samples. Input the species $\times$ sample data into a spreadsheet. Most statistical and ordination computer programs accept data input from commonly used spreadsheet programs.

38.3.2.5 Calculate Metrics

Create a list of metrics that you hypothesize will differ between the putatively impaired sites and the reference sites using the list of metrics in Table 38.2 and/or in Barbour et al., 1999. Calculate the value of each metric for each sample.

38.3.2.6 Comparison of Replicate Samples

We will determine if there is a statistically significant difference in the FBI between the two sets of sites. A t-test can be used to indicate if the two sample means are the same, or if they are significantly different from a statistical point-of-view. To do this test, calculate the mean ($\bar{x}$), variance (S^2), sum of observations ($\sum x_i$), sum of squared observations ($\sum x_i^2$), and determine the number of observations (n) for both the reference and test groups (see Zar, 2010). An example is provided in Table 38.6 (see Narf et al., 1984).

A t-test is used to choose between two hypotheses regarding the two populations that were sampled. The null hypothesis (H_0) is that the two population means are equal. The alternative hypothesis (H_A) is that the two means come from different populations. The t-test actually tells us the probability (p) that the null hypothesis (H_0) is true (i.e., the probability that mean$_1$ = mean$_2$). By convention, when the probability is less than 1 in 20 (i.e., $p < .05$) the two means are considered to be significantly different. To calculate the t-statistic, take the difference between the sample means and divide by the standard error of this difference, as shown by the following equation:

$$t = \frac{\bar{x}_1 - \bar{x}_2}{\sqrt{S^2_{n_1+n_2}\left(\frac{1}{n_1}+\frac{1}{n_2}\right)}} \tag{38.6}$$

TABLE 38.8 Form to record the physical habitat assessment.

Habitat Parameter	Optimal	Suboptimal	Marginal	Poor	Estimated Value
Basin land cover	Near 100% natural	50–75% natural	25–50% natural	0–25% natural	
Score	20	15	10	5	
Riparian width[a]	>18 m	12–18 m	6–12 m	<6 m	
Score	20	15	10	5	
Riparian structure and composition	Natural structure with predominantly native vegetation	Natural structure but a high percentage of nonnative vegetation	Natural vegetation structure modified and nonnative plants predominate	Riparian structure and composition highly disrupted	
Score	20	15	10	5	
Shading[b]	>75% of the water surface of sample reach is shaded	50% shaded in reach	20–50% shaded in reach	<20% shaded in reach	
Score	20	15	10	5	
Channel alteration	Channelization absent or minimal; stream with normal pattern	Slight localized channelization or evidence of historic channelization	40–80% of stream reach channelized	>80% of stream reach channelized	
Score	20	15	10	5	
Embeddedness	Boulder, cobble, and gravel particles with obvious open interstices	Boulder, cobble, and gravel particles ≤25% embedded predominately by sand; negligible silt	Boulder, cobble, and gravel particles 25–50% embedded by a mixture of sand and silt	Boulder, cobble, and gravel particles >50% embedded by sand and silt	
Score	20	15	10	5	
Benthic silt cover	No obvious surface deposited silt in the reach	Silt deposits common along stream margins	Interstitial silt extensive in midchannel	Top surface of substratum silt-covered	
Score	20	15	10	5	
Water appearance	Clear and odorless with no oil sheen	Slightly turbid; bottom visible in pools	Turbid; bottom visible in shallow riffles	Very turbid, odiferous, or oily	
Score	20	15	10	5	
				Total score	_____

Numbers (20, 15, 10, and 5) refer to the scores to beapplied for each condition selected.
[a]*Natural riparian conditions may lack trees.*
[b]*If the stream is located in a region lacking shaded streams or the stream is >50 m wide, disregard this factor.*
Based on Petersen (1992), USDA (1998), Barbour et al. (1999), and Fend et al. (2005).

where the means and sample sizes are taken directly from the summary statistics. The pooled variance $S^2_{n_1+n_2}$ is calculated as follows:

$$S^2_{n_1+n_2} = \frac{\left(\sum_{i=1}^{n_1} x_i^2 - \bar{x}_1 \sum_{i=1}^{n_1} x_i\right)_1 + \left(\sum_{i=1}^{n_2} x_i^2 - \bar{x}_2 \sum_{i=1}^{n_2} x_i\right)_2}{n_1 + n_2 - 2} \tag{38.7}$$

A computed t-test using the sample data in Table 38.6 is:

$$t = \frac{5.75 - 4.14}{\sqrt{0.94\left(\frac{1}{10} + \frac{1}{10}\right)}} = \frac{1.61}{0.43} = 3.74 \tag{38.8}$$

The critical value for t (t_{crit}) is then looked up in a t-table. The t_{crit} depends on the number of samples used to calculate t. The data above has a total of 20 samples, and therefore has 18 degrees of freedom (df); where df $= n_1 + n_2 - 2$. For 18 df, the $t_{crit\,(0.05)} = 2.10$ (Table 38.9). Our calculated t is greater than this critical t-value, and therefore we conclude that the two means are significantly different ($p < .05$). (If the calculated t is negative, use the absolute value.) A published t-table should be consulted for critical values (e.g., Zar, 2010; Sokal and Rohlf, 2012) but Table 38.9 contains some 2-tailed $t_{crit\,(0.05)}$ values for sample sizes ($N = n_1 + n_2$) ranging from 7 to 37 (df $= 5-35$). Determine if the test and reference groups that you sampled are significantly different using the FBI.

The above procedure can be done with many different types of measures (e.g., those listed earlier). Note that the t-test assumes that both samples come from a normally distributed population and that the variances (S^2) of both samples are equal. When sample sizes are equal, these assumptions may be "relaxed." However, it is always a good idea to know if these assumptions are being met. A quick rule-of-thumb to check for equal variances is to make sure the ratio of the sample variances is less than 2. If this ratio is greater than 2, then you may need to transform the data or use a test that does not assume equal variance. Refer to a statistical text (e.g., Zar, 2010; Sokal and Rohlf, 2012) for further information on these alternatives. Always be confident that the various assumptions for parametric statistics are met before drawing a conclusion based on the test.

Download and install R (R Core Team, 2015) using the information provided in Appendix 38.1. There you will find R script that sequentially calculates the values necessary for performing the above t-test. There is an example of a t-test found in R.

38.3.2.7 Generate Ordination Scores

Using one of the ecology-based statistical packages such as PC-ORD (McCune and Mefford, 1999), ordinate your species by sample data using the indirect gradient analysis method of Detrended Correspondence Analysis (DCA) choosing default program settings. We suggest using DCA instead of Nonmetric Multidimensional Scaling (NMDS) because in our

TABLE 38.9 List of 2-tailed critical t-values ($p < .05$) for various degrees of freedom (df). Use the next smaller N if your sample size is between the values provided or consult a more extensive t-table (e.g., Zar, 2010).

N	df	Critical t
7	5	2.57
12	10	2.23
17	15	2.13
22	20	2.09
27	25	2.06
32	30	2.04
42	40	2.02

experience single physical variables are usually more highly correlated with the site scores of the first axis of a DCA than with NMDS axes. Save the site scores from at least the first axis for use in the next step.

If ecological type software is not available, download and install R (R Core Team, 2015) using the information provided in Appendix 38.1. Locate the R script that demonstrates performing a DCA and follow the instructions for downloading and installing the vegan package (Oksanen et al., 2015). The provided R script performs a DCA using both raw and log_{10} transformed data. Two plots are created. The first is the default vegan plot that includes both the taxa and the sites. The second is a plot of the site scores with the sites labeled by a grouping variable.

38.3.2.8 Correlations

Using an available statistical program, determine whether there is a correlation between the above generated site scores from axis 1 and 2 of the DCA and the per site index of habitat condition calculated under *Habitat Assessment* above. To further explain these relationships calculate the correlations between the site score and selected habitat variables, plot each correlation.

38.4 QUESTIONS

38.4.1 General Questions for Both Basic and Advanced Methods

1. Choose two physical parameters and describe what the optimal condition would be and what the poor condition would be in reference to the abundance and distribution of macroinvertebrates. Would these optimal values be the same for other stream organisms such as fish, amphibians, or algae?
2. You are planning to conduct a biomonitoring comparison between two streams, one presumed to be pristine and the other impaired. How would you avoid the possibility of "uncontrolled variables" influencing your results? That is, account for factors in your experimental design that would indicate impairment when no impairment has actually occurred.
3. You have been placed in charge of a team to monitor urban streams. What type of impacts might be present and how would you design a biomonitoring study that would detect these impacts?
4. Oftentimes, only a single measurement of a physical (e.g., temperature) or a water chemistry (e.g., conductivity) variable is taken at a site; however, we often take replicate biological measures because we assume that intersample variability will occur. After collecting replicate biological, physical, and chemical samples, test whether the biological variability was greater than or less than the variability in physical and water chemistry parameters you measured?
5. How do you think that physical, chemical, and biological measurements would vary in streams in your region throughout the year? How would numerical values of benthic macroinvertebrate metrics vary throughout the year?
6. What types of legacy effects (i.e., effects that occurred at your site(s) or in the basin prior to you sampling) could influence the results of a biomonitoring survey?

38.4.2 Questions for Basic Method

7. In Table 38.5, FBI scores are ranked into category ranges that denote specific water quality judgments (e.g., excellent, good, poor). Are there any caveats you would warn the public about in using these judgments based on FBI scores that you calculated from your samples?
8. If several samples were taken from each stream, how similar were the FBI scores within a stream? Do you think you would get different FBI scores if you looked in different habitats (e.g., riffles, pools, stream edges, vegetation)? How could the variability among samples within a stream been reduced?
9. If you sampled a reference stream and an impaired stream, was the variability among samples the same in each stream? Do you think the variability among samples should be higher in a reference stream or an impaired stream?

38.4.3 Questions for Advanced Method

10. What are the strengths and weaknesses of visually based habitat assessment approaches?
11. In reviewing quantitative approaches to physical habitat assessments in streams (Chapters 1−7 in this book), which methods also have potential applications in interpreting results of biomonitoring programs?
12. Metrics are presumed to be based on accepted ecological theory. Does the expectation of a decline in species richness with increased impairment reflect the predictions of the intermediate disturbance hypothesis (Townsend and Scarsbrook, 1997), the ecological theory on which the predicted response of this metric is based?
13. Your results were based on identifying macroinvertebrates to the lowest taxonomic level possible. Do you think your conclusions would be different if identification was only to the family level? Why or why not?

14. Were metrics statistically different between the two groups of sites? Could you sample fewer sites and still detect a difference for these metrics?
15. How did the ordination scores change when the example data were $\log_{10}$ transformed? Why do you suppose the change occurred?
16. Were there significant correlations between the physical variables and the site scores? If so, how would you incorporate this information into designing a larger-scale biomonitoring program?

38.5 MATERIALS AND SUPPLIES

Maps showing collecting site locations and land cover
Global positioning system
Calculator
Computer software (e.g., spreadsheet, ecological)
Dissecting microscopes
Enamel pans
Equipment for collection of macroinvertebrates (see Chapter 15)
Equipment to measure current velocity (see Chapter 3)
Ethanol (95% and 70%)
Forceps (fine-tipped)
Keys to identify macroinvertebrates (e.g., Appendix 20.1; Merritt et al. 2008)
Petri dishes to sort samples
Thermometer
Vials, jars, or sealable plastic bags

ACKNOWLEDGMENTS

We thank Marilyn Meyers for all the effort she put in to the first several editions of this chapter. We also thank Steve Fend and Terry Short for their fine and most helpful reviews. Last, our editors added immensely to the quality of this chapter.

REFERENCES

Bailey, R.C., Norris, R.H., Reynoldson, T.B., 2001. Taxonomic resolution of benthic macroinvertebrate communities in bioassessments. Journal of the North American Benthological Society 20, 280−286.

Barbour, M.T., Gerritsen, J., Snyder, B.D., Stribling, J.B., 1999. Rapid Bioassessment Protocols for Use in Streams and Wadeable Rivers: Periphyton, Benthic Macroinvertebrates and Fish. EPA 841-B-99−002. Office of Water, US Environmental Protection Agency, Washington, DC.

Barbour, M.T., Graves, C.G., Plafkin, J.L., Wisseman, R.W., Bradley, B.P., 1992. Evaluation of EPA's rapid bioassessment benthic metrics: metric redundancy and variability among reference stream sites. Environmental Toxicology and Chemistry 11, 437−449.

Biggs, J., Ewald, N., Valentini, A., Gaboriaud, C., Dejean, T., Griffiths, R.A., Foster, J., Wilkinson, J.W., Arnell, A., Brotherton, P., Williams, P., Dunn, F., 2015. Using eDNA to develop a national citizen science-based monitoring programme for the great crested newt (*Triturus cristatus*). Biological Conservation 183, 19−28.

Bonada, N., Prat, N., Resh, V.H., Statzner, B., 2006. Developments in aquatic insect biomonitoring: a comparative analysis of recent approaches. Annual Review of Entomology 51, 495−524.

Brix, K.V., DeForest, D.K., Adams, W.J., 2011. The sensitivity of aquatic insects to divalent metals: a comparative analysis of laboratory and field data. Science of the Total Environment 409, 4187−4197.

Buchwalter, D.B., Cain, D.J., Martin, C.A., Xie, L., Luoma, S.N., Garland, T., 2008. Aquatic insect ecophysiological traits reveal phylogenetically based differences in dissolved cadmium susceptibility. Proceedings of the National Academy of Sciences 105, 8321−8326.

Buchwalter, D.B., Luoma, S.N., 2005. Differences in dissolved cadmium and zinc uptake among stream insects: mechanistic explanations. Environmental Science and Technology 39, 498−504.

Cain, D.J., Croteau, M.-N., Fuller, C., 2013. Dietary bioavailability of Cu absorbed to colloida hydrous ferric oxide. Environmental Science and Technology 47, 2869−2876.

Cain, D.J., Croteau, M.-N., Luoma, S.N., 2011. Bioaccumulation dynamics and exposure routes of Cd and Cu among species of aquatic mayflies. Environmental Toxicology and Chemistry 30, 2532−2541.

Cairns Jr., J., Pratt, J.R., 1993. A history of biological monitoring using benthic macroinvertebrates. In: Rosenberg, D.M., Resh, V.H. (Eds.), Freshwater Biomonitoring and Benthic Macroinvertebrates. Chapman & Hall, New York, NY, USA, pp. 10−27.

Cao, Y., Hawkins, C.P., 2011. The comparability of bioassessments: a review of conceptual and methodological issues. Journal of the North American Benthological Society 30, 680−701.

Carter, J.L., Fend, S.V., Kennelly, S.S., 1996. The relationships among three habitat scales and stream benthic invertebrate community structure. Freshwater Biology 35, 109–124.

Carter, J.L., Purcell, A.H., Fend, S.V., Resh, V.H., 2009. Development of a local-scale urban assessment method using benthic macroinvertebrates: an example from the Santa Clara Basin, California. Journal of the North American Benthological Society 28, 1007–1021.

Carter, J.L., Resh, V.H., 2001. After site selection and before data analysis: sampling, sorting, and laboratory procedures used in stream benthic macroinvertebrate monitoring programs by USA state agencies. Journal of the North American Benthological Society 20, 658–682.

Carter, J.L., Resh, V.H., 2013. Analytical Approaches Used in Stream Benthic Macroinvertebrate Biomonitoring Programs of State Agencies in the United States. Open-File Report 2013-1129. U.S. Geological Survey, Reston, Virginia, USA.

Carter, J.L., Resh, V.H., Rosenberg, D.M., Reynoldson, T.B., 2006. Biomonitoring in North American rivers: a comparison of methods used for benthic macroinvertebrates in Canada and the United States. In: Ziglio, G., Siligardi, M., Flaim, G. (Eds.), Biological Monitoring of Rivers: Applications and Perspectives. John Wiley & Sons, NY, pp. 203–228.

Chang, F.-H., Lawrence, J., Rios-Touma, B., Resh, V.H., 2014. Tolerance values of benthic macroinvertebrates for stream biomonitoring: assessment of assumptions underlying scoring systems worldwide. Environmental Monitoring and Assessment 186, 2135–2149.

Chessman, B.C., McEvoy, P.K., 2012. Insights into human impacts on streams from tolerance profiles of macroinvertebrate assemblages. Water, Air, and Soil Pollution 223, 1343–1352.

Clements, W.H., Cherry, D.S., Van Hassel, J.H., 1992. Assessment of the impact of heavy metals on benthic communities at the Clinch River (Virginia) – evaluation of an index of community sensitivity. Canadian Journal of Fisheries and Aquatic Sciences 49, 1686–1694.

Clements, W.H., 2000. Integrating effects of contaminants across levels of biological organization: an overview. Aquatic Ecosystem Stress Recovery 7, 113–116.

Clements, W.H., Cadmus, P., Brinkman, S.F., 2013. Responses of aquatic insects to Cu and Zn in stream microcosms: understanding differences between single species tests and field responses. Environmental Science and Technology 47, 7506–7513.

Clifford, H.F., 1991. Aquatic Invertebrates of Alberta. The University of Alberta Press, Edmonton, Alberta, Canada.

Cold Spring Harbor Learning Center, 2014. Using DNA Barcodes to Identify and Classify Living Things. Cold Spring Harbor Learning Center.

Connecticut Department of Energy and Environmental Protection, 2013. A Tiered Approach for Volunteer Monitoring of Wadeable Streams and Rivers. Connecticut Department of Energy and Environmental Protection, Hartford, CT 06106.

Culp, J.M., Armanini, D.G., Dunbar, M.J., Orlofske, J.M., Poff, N.L., Pollard, A.I., Yates, A.G., Hose, G.C., 2011. Incorporating traits in aquatic biomonitoring to enhance causal diagnosis and prediction. Integrated Environmental Assessment and Management 7, 187–197.

Cummins, K.W., Klug, M.J., 1979. Feeding ecology of stream invertebrates. Annual Review of Ecology and Systematics 10, 147–172.

Dallas, H.F., Rivers-Moore, N.A., 2012. Critical thermal maxima of aquatic macroinvertebrates: toward identifying bioindicators of thermal alteration. Hydrobiologia 679, 61–76.

Damásio, J., Barceló, D., Brix, R., Postigo, C., Gros, M., Petrovic, M., Sabater, S., Guasch, H., de Alda, M.L., Barata, C., 2011. Are pharmaceuticals more harmful than other pollutants to aquatic invertebrate species: a hypothesis tested using multi-biomarker and multi-species responses in field collected and transplanted organisms. Chemosphere 85, 1548–1554.

Davies, P.E., 1991. Development of a national river bioassessment system (AUSRIVAS) in Australia. In: Wright, J.F., Sutcliffe, D.W., Furse, M.T. (Eds.), Assessing the Biological Quality of Fresh Waters: RIVPACS and Other Techniques. Freshwater Biological Association, Ambleside, Cumbria, UK, pp. 113–124.

Deiner, K., Walser, J.C., Mächler, E., Altermatt, F., 2015. Choice of capture and extraction methods affect detection of freshwater biodiversity from environmental DNA. Biological Conservation 183, 53–63.

Diepens, N.J., Arts, G.H.P., Brock, T.C.M., Smidt, H., Van Den Brink, P.J., Van Den Heuvel-Greve, M.J., Koelmans, A.A., 2014. Sediment toxicity testing of organic chemicals in the context of prospective risk assessment: a review. Critical Reviews in Environmental Science and Technology 44, 255–302.

Di Veroli, A., Santoro, F., Pallottini, M., Selvaggi, R., Scardazza, F., Cappelletti, D., Goretti, E., 2014. Deformities of chironomid larvae and heavy metal pollution: from laboratory to field studies. Chemosphere 112, 9–17.

Downes, B.J., Barmuta, L.A., Fairweather, P.G., Faith, D.P., Keough, M.J., Lake, P.S., Mapstone, B.D., Quinn, G.P., 2002. Monitoring Ecological Impacts: Concepts and Practice in Flowing Waters. Cambridge University Press, United Kingdom.

Ely, E., 2005. Volunteer macroinvertebrate monitoring: a panoramic view. The National newsletter of Volunteer Watershed Monitoring 17, 1.

Fend, S.V., Carter, J.L., Kearns, F.R., 2005. Relationships of field habitat measurements, visual habitat indices, and land cover to benthic macroinvertebrates in urbanized streams of the Santa Clara Valley, California. In: Brown, L.R., Gray, R.H., Hughes, R.M., Meador, M.R. (Eds.), Effects of Urbanization on Stream Ecosystems. American Fisheries Society, Symposium 47, Bethesda, MD, pp. 193–212.

Firehock, K., West, J., 1995. A brief history of volunteer biological water monitoring using macroinvertebrates. Journal of the North American Benthological Society 14, 197–202.

Fitzpatrick, F.A., 2001. A comparison of multi-disciplinary methods for measuring physical conditions of streams. In: Dorava, J.B., Montgomery, D.R., Palcsak, B., Fitzpatrick, F.A. (Eds.), Geomorphic Processes and Riverine Habitat. American Geophysical Union Monograph, Washington, DC, pp. 7–18.

Flotemersch, J.E., North, S., Blocksom, K.A., 2014. Evaluation of an alternate method for sampling benthic macroinvertebrates in low-gradient streams sampled as part of the National Rivers and Streams Assessment. Environmental Monitoring and Assessment 186, 949–959.

Fore, L.S., Paulsen, K., O'Laughlin, K., 2001. Assessing the performance of volunteers monitoring streams. Freshwater Biology 46, 109–123.

Hauer, F.R., Stanford, J.A., 1981. Larval specialization and phenotypic variation in *Arctopsyche grandis* (Trichoptera: Hydropsychidae). Ecology 62, 645–653.

Hawkins, C.P., 2006. Quantifying biological integrity by taxonomic completeness: its utility in regional and global assessments. Ecological Applications 16, 1277–1294.

Hawkins, C.P., Cao, Y., Roper, B., 2010a. Method of predicting reference condition biota affects the performance and interpretation of ecological indices. Freshwater Biology 55, 1066–1085.

Hawkins, C.P., Olson, J.R., Hill, R.A., 2010b. The reference condition: predicting benchmarks for ecological and water-quality assessments. Journal of the North American Benthological Society 29, 312–343.

Hebert, P.D.N., Ratnasingham, S., deWaard, J.R., 2003. Barcoding animal life: cytochrome c oxidase subunit 1 divergences among closely related species. Proceedings of the Royal Society B: Biological Sciences 270, S96–S99.

Hellawell, J.M., 1986. Biological Indicators of Freshwater Pollution and Environmental Management. Elsevier, New York, NY.

Hilsenhoff, W.L., 1988. Rapid field assessment of organic pollution with a family-level biotic index. Journal of the North American Benthological Society 7, 65–68.

Jackson, J.K., Battle, J.M., White, B.P., Pilgrim, E.M., Stein, E.D., Miller, P.E., Sweeney, B.W., 2014. Cryptic biodiversity in streams: a comparison of macroinvertebrate communities based on morphological and DNA barcode identifications. Freshwater Science 33, 312–324.

Johnson, R.K., Wiederholm, T., Rosenberg, D.M., 1993. Freshwater biomonitoring using individual organisms, populations, and species assemblages of benthic macroinvertebrates. In: Rosenberg, D.M., Resh, V.H. (Eds.), Freshwater Biomonitoring and Benthic Macroinvertebrates. Chapman & Hall, New York, pp. 40–158.

Kolkwitz, R., Marsson, M., 1909. Ökologie der tierischen Saprobien. Beiträge zur Lehre von des biologischen Gewasserbeurteilung. Internationale Revue der gesamten Hydrobiologie und Hydrographie 2, 126–152.

Lehmkuhl, D.M., 1979. How to Know the Aquatic Insects. W.M.C. Brown Company, Dubuque, IA.

Lenat, D.R., 1988. Water quality assessment of streams using a qualitative collection method for benthic macroinvertebrates. Journal of the North American Benthological Society 7, 222–233.

Lenat, D.R., Resh, V.H., 2001. Taxonomy and stream ecology—the benefits of genus- and species-level identifications. Journal of the North American Benthological Society 20, 287–298.

Lenat, D.R., 1993. A biotic index for the southeastern United States: derivation and list of tolerance values, with criteria for assigning water-quality ratings. Journal of the North American Benthological Society 12, 279–290.

Martin, C.A., Luoma, S.N., Cain, D.J., Buchwalter, D.B., 2007. Cadmium ecophysiology in seven stonefly (Plecoptera) species: delineating sources and estimating susceptibility. Environmental Science and Technology 41, 7171–7177.

Martínez-Paz, P., Morales, M., Martínez-Guitarte, J.L., Morcillo, G., 2013. Genotoxic effects of environmental endocrine disruptors on the aquatic insect *Chironomus riparius* evaluated using the comet assay. Mutation Research – Genetic Toxicology and Environmental Mutagenesis 758, 41–47.

Masese, F.O., Kitaka, N., Kipkemboi, J., Gettel, G.M., Irvine, K., McClain, M.E., 2014. Macroinvertebrate functional feeding groups in Kenyan highland streams: evidence for a diverse shredder guild. Freshwater Science 33, 435–450.

McCune, B., Mefford, M.J., 1999. Multivariate Analysis of Ecological Data. Version 2.20. MjM Software, Gleneden Beach, OR.

Menezes, S., Baird, D.J., Soares, A.M.V.M., 2010. Beyond taxonomy: a review of macroinvertebrate trait-based community descriptors as tools for freshwater biomonitoring. Journal of Applied Ecology 47, 711–719.

Merritt, R.W., 2002. How to Use a Dichotomous Key in Identifying Aquatic Insects. Video. Kendall/Hunt, Dubuque, IA.

Merritt, R.W., Cummins, K.W., Berg, M.B. (Eds.), 2008. An Introduction to the Aquatic Insects of North America, fourth ed. Kendall/Hunt Publishing Co., Dubuque, Iowa. 52002.

Metcalfe, J.L., 1989. Biological water quality assessment of running waters based on macroinvertebrate communities: history and present states in Europe. Environmental Pollution 60, 101–139.

Moulton II, S.R., Carter, J.L., Grotheer, S.A., Cuffney, T.F., Short, T.M., 2000. Methods of Analysis by the U.S. Geological Survey National Water Quality Laboratory – Processing, Taxonomy, and Quality Control of Benthic Macroinvertebrate Samples. U.S. Geological Survery, Denver, Colorado. Open-File Report 00–212.

Nair, P.M.G., Park, S.Y., Choi, J., 2013. Characterization and expression of cytochrome p450 cDNA (CYP9AT2) in *Chironomus riparius* fourth instar larvae exposed to multiple xenobiotics. Environmental Toxicology and Pharmacology 36, 1133–1140.

Narf, R.P., Lange, E.L., Wildman, R.C., 1984. Statistical procedures for applying Hilsenhoff's biotic index. Journal of Freshwater Ecology 2, 441–448.

Niemi, G.J., McDonald, M.E., 2004. Application of ecological indicators. Annual Review of Ecology and Systematics 35, 89–111.

Norris, R.H., Georges, A., 1993. Analysis and interpretation of benthic macroinvertebrate surveys. In: Rosenberg, D.M., Resh, V.H. (Eds.), Freshwater Biomonitoring and Benthic Macroinvertebrates. Chapman & Hall, New York, pp. 234–286.

NRC (National Research Council), 2001. Assessing the TMDL Approach to Water Quality Management. National Research Council. Committee to Assess the Scientific Basis of the Total Maximum Daily Load Approach to Pollution Reduction. National Academy Press, Washington, DC.

Oksanen, J., Blanchet, F.G., Kindt, R., Legendre, P., Minchin, P.R., O'hara, R.B., Simpson, G.L., Solymos, P., Stevens, M.H.H., Wagner, H., 2015. vegan: Community Ecology Package. R Package Version 2.3-0.

Pauls, S.U., Alp, M., Bálint, M., Bernabò, P., Čiampor, F., Čiamporová-Zaťovičová, Z., Finn, D.S., Kohout, J., Leese, F., Lencioni, V., Paz-Vinas, I., Monaghan, M.T., 2014. Integrating molecular tools into freshwater ecology: developments and opportunities. Freshwater Biology 59, 1559–1576.

Peckarsky, R.L., Fraissinet, P.R., Penton, M.A., Conklin, D.J., 1990. Freshwater Macroinvertebates of Northeastern North America. Cornell University Press, Ithaca, NY.

Pestana, J.L.T., Novais, S.C., Lemos, M.F.L., Soares, A.M.V.M., 2014. Cholinesterase activity in the caddisfly *Sericostoma vittatum*: biochemical enzyme characterization and in vitro effects of insecticides and psychiatric drugs. Ecotoxicology and Environmental Safety 104, 263–268.

Petersen Jr., R.C., 1992. The RCE: a riparian, channel, and environmental inventory for small streams in the agricultural landscape. Freshwater Biology 27, 295–306.

Pfrender, M.E., Ferrington Jr., L.C., Hawkins, C.P., Hartzell, P.L., Bagley, M., Jackson, S., Courtney, G.W., Larsen, D.P., Creutzburg, B.R., Lévesque, C.A., Epler, J.H., Morse, J.C., Fend, S., Petersen, M.J., Ruiter, D., Schindel, D., Whiting, M., 2010. Assessing macroinvertebrate biodiversity in freshwater ecosystems: advances and challenges in DNA-based approaches. Quarterly Review of Biology 85, 319–340.

Pilgrim, E.M., Jackson, S.A., Swenson, S., Turcsanyi, I., Friedman, E., Weigt, L., Bagley, M.J., 2011. Incorporation of DNA barcoding into a large-scale biomonitoring program: opportunities and pitfalls. Journal of the North American Benthological Society 30, 217–231.

Plafkin, J.L., Barbour, M.T., Porter, K.D., Gross, S.K., Hughes, R.M., 1989. Rapid Bioassessment Protocols for Use in Streams and Rivers: Benthic Macroinvertebrates and Fish. Report No. 444/4-89-001. U.S. Environmental Protection Agency, Washington, DC.

Poff, N., Olden, J., Vieira, N., Finn, D., Simmons, M., Kondratieff, B., 2006. Functional trait niches of North American lotic insects: traits-based ecological applications in light of phylogenetic relationships. Journal of the North American Benthological Society 25, 730–755.

Poff, N.L., 1997. Landscape filters and species traits: towards mechanistic understanding and prediction in stream ecology. Journal of the North American Benthological Society 391–409.

Poteat, M.D., Jacobus, L.M., Buchwalter, D.B., 2015. The importance of retaining a phylogenetic perspective in traits-based community analyses. Freshwater Biology 60, 1330–1339.

Rankin, E.T., 1995. Habitat indices in water resource quality assessments. In: Davis, W.S., Simon, T.P. (Eds.), Biological Assessment and Criteria: Tools for Water Resource Planning and Decision Making. Lewis Publisher, Boca Raton, FL, pp. 181–208.

R Core Team, 2015. R: A Language and Environment for Statistical Computing. R Foundation for Statistical Computing, Vienna, Austria.

Relyea, C.D., Minshall, G.W., Danehy, R.J., 2012. Development and validation of an aquatic fine sediment biotic index. Environmental Management 49, 242–252.

Resh, V.H., 1995. Freshwater benthic macroinvertebrates and rapid assessment procedures for water quality monitoring in developing and newly industrialized countries. In: Davis, W.S., Simon, T.P. (Eds.), Biological Assessment and Criteria. Lewis Publishers, Boca Raton, FL, pp. 167–177.

Resh, V.H., Feminella, J.W., McElravy, E.P., 1990. Sampling Aquatic Insects. 38 minutes Video tape, Office of Media Services, University of California, Berkeley.

Resh, V.H., Hildrew, A.G., Statzner, B., Townsend, C.R., 1994. Theoretical habitat templets, species traits, and species richness: a synthesis of long-term ecological research on the Upper Rhône River in the context of concurrently developed ecological theory. Freshwater Biology 31, 539–554.

Resh, V.H., Jackson, J.K., 1993. Rapid assessment approaches to biomonitoring using benthic macroinvertebrates. In: Rosenberg, D.M., Resh, V.H. (Eds.), Freshwater Biomonitoring and Benthic Macroinvertebrates. Chapman & Hall, New York, pp. 195–223.

Resh, V.H., Unzicker, J.D., 1975. Water quality monitoring and aquatic organisms: the importance of species identification. Water Quality Monitoring 47, 9–19.

Reynoldson, T.B., Bailey, R.C., Day, K.E., Norris, R.H., 1995. Biological guidelines for freshwater sediment based on BEnthic Assessment of SedimenT (the BEAST) using a multivariate approach for predicting biological state. Australian Journal of Ecology 20, 198–219.

Reynoldson, T.B., Norris, R.H., Resh, V.H., Day, K.E., Rosenberg, D.M., 1997. The reference condition: a comparison of multimetric and multivariate approaches to assess water-quality impairment using benthic macroinvertebrates. Journal of the North American Benthological Society 16, 833–852.

Reynoldson, T.B., Strachan, S., Bailey, J.L., 2014. A tiered method for discriminant function analysis models for the reference condition approach: model performance and assessment. Freshwater Science 33, 1238–1248.

Rosenberg, D.M., Resh, V.H., 1993. Freshwater Biomonitoring and Benthic Macroinvertebrates. Chapman & Hall, New York, NY.

Rubach, M.N., Ashauer, R., Buchwalter, D.B., De Lange, H.J., Hamer, M., Preuss, T.G., Töpke, K., Maund, S.J., 2011. Framework for traits-based assessment in ecotoxicology. Integrated Environmental Assessment and Management 7, 172–186.

Runck, C., 2007. Macroinvertebrate production and food web energetics in an industrially contaminated stream. Ecological Applications 17, 740–753.

Schmera, D., Podani, J., Heino, J., Erős, T., Poff, N.L., 2015. A proposed unified terminology of species traits in stream ecology. Freshwater Science 34, 823–830.

Schmidt-Kloiber, A., Hering, D., 2015. www.freshwaterecology.info — an online tool that unifies, standardises and codifies more than 20,000 European freshwater organisms and their ecological preferences. Ecological Indicators 53, 271–282.

Smith, D.G., 2001. Pennak's Freshwater Invertebrates of the United States: Porifera to Crustecea, fourth ed. John Wiley & Sons, Inc., New York, NY.

Sokal, R.R., Rohlf, F.J., 2012. Biometry, fourth ed. W.H. Freeman.

Southwood, T.R.E., 1988. Tactics, strategies and templets. Oikos 52, 3–18.

Statzner, B., Bêche, L.A., 2010. Can biological invertebrate traits resolve effects of multiple stressors on running water ecosystems? Freshwater Biology 55, 80–119.

Statzner, B., Hildrew, A.G., Resh, V.H., 2001. Species traits and environmental constraints: entomological research and the history of ecological theory. Annual Review of Entomology 46, 291–316.

Statzner, B., Resh, V.H., Roux, A.L., 1994. The synthesis of long-term ecological research in the context of concurrently developed ecological theory: design of a research strategy for the Upper Rhône River and its floodplain. Freshwater Biology 31, 253–263.

Stein, E.D., White, B.P., Mazor, R.D., Jackson, J.K., Battle, J.M., Miller, P.E., Pilgrim, E.M., Sweeney, B.W., 2014. Does DNA barcoding improve performance of traditional stream bioassessment metrics? Freshwater Science 33, 302–311.

Stewart-Oaten, A., Murdoch, W.W., Parker, K.R., 1986. Environmental impact assessment: "Pseudoreplication" in time? Ecology 67, 929–940.

Stoddard, J., Larsen, D., Hawkins, C., Johnson, R., Norris, R., 2006. Setting expectations for the ecological condition of streams: the concept of reference condition. Ecological Applications 16, 1267–1276.

Sweeney, B.W., Battle, J.M., Jackson, J.K., Dapkey, T., 2011. Can DNA barcodes of stream macroinvertebrates improve descriptions of community structure and water quality? Journal of the North American Benthological Society 30, 195–216.

Thomsen, P.F., Willerslev, E., 2015. Environmental DNA – an emerging tool in conservation for monitoring past and present biodiversity. Biological Conservation 183, 4–18.

Thorp, J.H., Covich, A.P., 2014. Ecology and Classification of North American Freshwater Invertebrates, fourth ed. Academic Press, Inc., San Diego, CA.

Townsend, C.R., Scarsbrook, M.R., 1997. The intermediate disturbance hypothesis, refugia, and biodiversity in streams. Limnology and Oceanography 42, 938–949.

Underwood, A.J., 1997. Experiments in Ecology: Their Logical Design and Interpretation Using Analysis of Variance. Cambridge University Press, Cambridge, UK.

USDA (United States Department of Agriculture), 1998. Stream Visual Assessment Protocol. Technical Note 99-1. National Water and Climate Center.

USEPA (United States Environmental Protection Agency), 2002. Summary of Biological Assessment Programs and Biocriteria Development for States, Tribes, Territories, and Interstate Commissions: Streams and Wadeable Rivers. EPA 822-R-02–048. Washington, DC.

USEPA (United States Environmental Protection Agency), 2007. National Rivers and Streams Assessment: Field Operations Manual. EPA-841-B-07–009. Washington, DC.

USEPA (United States Environmental Protection Agency), 2010. MS4 Improvement Guide. EPA 833-R-10–001. Office of Water, Washington DC.

USEPA (United States Environmental Protection Agency), 2012. Freshwater Traits Database. EPA/600/R-11/038F. Global Change Research Program. National Center for Environmental Assessment, Washington, DC. Available from the National Technical Information Service, Springfield, VA, and online at: http://www.epa.gov/ncea.

Usinger, R.L. (Ed.), 1956. Aquatic Insects of California. University of California Press, Berkeley, CA.

Verberk, W.C.E.P., van Noordwijk, C.G.E., Hildrew, A.G., 2013. Delivering on a promise: integrating species traits to transform descriptive community ecology into a predictive science. Freshwater Science 32, 531–547.

Vieira, N.K.M., Poff, N.L., Carlisle, D.M., Moulton Ii, S.R., Koshi, M.L., Kondratieff, B.C., 2006. A Database of Lotic Invertebrate Traits for North America. Data Series 187. U.S. Geological Survey, Washington, DC.

Voshell, J.R., 2003. A Guide to Common Freshwater Invertebrates of North America. The McDonald & Woodward Publishing Company, Blacksburg, VA.

White, B.P., Pilgrim, E.M., Boykin, L.M., Stein, E.D., Mazor, R.D., 2014. Comparison of four species-delimitation methods applied to a DNA barcode data set of insect larvae for use in routine bioassessment. Freshwater Science 33, 338–348.

Wright, J.F., Moss, D., Armitage, P.D., Furse, M.T., 1984. A preliminary classification of running water sites in Great Britain based on macroinvertebrate species and prediction of community type using environmental data. Freshwater Biology 14, 221–256.

Wright, J.F., Sutcliffe, D.W., Furse, M.T. (Eds.), 2000. Assessing the Biological Quality of Fresh Water: RIVPACS and Other Techniques. Freshwater Biological Association, Ambleside, UK.

Yuan, L.L., 2004. Assigning macroinvertebrate tolerance classifications using generalised additive models. Freshwater Biology 49, 662–677.

Zar, J.H., 2010. Biostatistical Analysis, fifth ed. Prentice Hall, Upper Saddle River, NJ.

Chapter 39

Environmental Quality Assessment Using Stream Fishes

Thomas P. Simon[1,a] and Nathan T. Evans[2]

[1]*School of Public and Environmental Affairs, Indiana University;* [2]*Southeast Environmental Research Center, Florida International University*

39.1 INTRODUCTION

The science of environmental assessment has changed dramatically over the last century (Hermoso and Clavero, 2013; Ruaro and Gubiani, 2013), with improvements in methods, theory, and detection. Unfortunately, over this same time period we have seen rampant aquatic habitat degradation as a result of declining water quality (Karr, 1993, 1995). Increases in complex effluent and sediment toxicity, pesticide toxicity, and the synergistic, additive, as well as antagonistic effects of these complex contaminants, often confound the ability of scientists to determine the cause and source of impacts (Simon, 2003; see also Chapter 40). Contemporary bioassessment techniques that seek to diagnosis of cause and effect must make use of data patterns from biological databases, use of dose–response curves, implementing traveling zones that replace the conventional upstream versus downstream approaches, and the increasing use of models that evaluate predicted versus observed patterns in heterogeneous landscapes (Emery and Thomas, 2003). Therefore, contemporary bioassessment approaches are frequently rooted in elucidating patterns in multimetric indices or determining signal-to-noise (S:N) ratios for select impacts. Likewise, contemporary bioassessment approaches often employ the use of logical response patterns, multivariate analysis, and analysis of specific metric patterns to increase discriminatory power when evaluating potential water quality impacts.

This chapter is focused on the methods necessary to implement biological criteria based on its three pillars of (1) regionalization, (2) reference standard development, and (3) use of multimetric indices (Simon, 2000). This chapter further describes the index of biotic integrity (IBI) using stream fishes as the focal organisms. It is important to note, however, that IBIs exist for various groups of aquatic organisms, including algae and macroinvertebrates (see Chapters 37 and 38). The purpose of this fish-based IBI is to determine whether stream reaches are meeting designated uses for aquatic life.

39.1.1 A Brief History of the Index of Biotic Integrity

The multimetric index concept and the IBI was first developed for use in small warmwater streams (i.e., too warm to support salmonids) in central Illinois and Indiana, USA (Karr, 1981), in response to the Clean Water Act of 1972, and as amended in 1977, which mandated restoration and maintenance of the biological integrity of the United States' surface waters. The original version included 12 "metrics" or attributes that reflected fish species richness and composition, number and abundance of indicator species, trophic organization and function, reproductive behavior, fish abundance, and condition of individual fish. Each metric receives a score either based on the original 5-point system (Karr, 1981) or a modified 10-point system (Lyons, 1992). In the original system, a metric received a score of five points if it had a value similar to that expected for a fish community characteristic of a system with little human influence, a score of one point if it

---

a *In Memoriam*—Tom Simon passed away on July 16, 2016, during the production of this chapter. Tom was a superb ichthyologist and a champion of preserving fish biodiversity and stream health in his writing and with his actions. Tom will be sorely missed by his family, friends, and colleagues. This chapter, one of Tom's final projects, is dedicated to his memory.

Methods in Stream Ecology. http://dx.doi.org/10.1016/B978-0-12-813047-6.00017-6

had a value similar to that expected for a fish community that departs significantly from the reference condition, and a score of three points if it had an intermediate value. This system is also reflected in the modified scoring criteria (Lyons, 1992) with higher point values associated with reaches most similar to the regional expectations.

Despite initial skepticism by the scientific community, which complained that the information content provided by an IBI approach would be less valuable than the full-scale annual measurement of a single reach, and that the assessment based on a rapid approach would not provide sufficient information to assess biological integrity, the IBI quickly became popular and was used by many investigators to assess streams throughout the central United States (e.g., Berkman et al., 1986; Angermeier and Schlosser, 1987; Hite, 1988; Osborne et al., 1992). Eventually, most criticisms were addressed with simple designations of index periods, sampling effort, and standard operating procedures. Moreover, early success of the IBI capitalized on the recognition of fish as important indicators that were recognizable to the public (Table 39.1).

As the IBI has become more widely used, different versions have been developed for different regions and ecosystems. Refinements have emerged for various water thermal regimes including coldwater (e.g., Lyons et al., 1996; Hughes et al., 2004) and coolwater streams (e.g., Simon, 1991; Lyons, 2012). Further applications also exist for different water body types including intermittent headwater streams (Lyons, 2006; Megan et al., 2007), wetlands (Simon and Stewart, 1998; Simon et al., 2000), lakes and reservoirs (Lenhardt et al., 2009; Ivasauskas and Bettoli, 2014), coastal wetlands of the Great Lakes (Simon and Stewart, 2006), large and great rivers (Simon and Emery, 1995; Simon, 2006), and coastal estuaries (Daniel et al., 2014; Fisch et al., 2016). Moreover, reference condition—based multimetric approaches, analogous to the IBI, have been developed for and frequently used to assess water quality and ecosystem health in Canada (Steedman, 1988; Minns et al., 1994), Australia (Davies et al., 2010; Hallett et al., 2012), South Africa

TABLE 39.1 Attributes of fishes that make them desirable components of biological assessment and monitoring programs.

Goal/Quality	Attributes
Accurate assessment of environmental health	Fish populations and individuals generally remain in the same area during summer seasons
	Communities are persistent and recover rapidly from natural disturbances
	Comparable results can be expected from an unperturbed site at various times
	Fish have large ranges and are less affected by natural microhabitat differences than smaller organisms; this makes fish extremely useful for assessing regional and macrohabitat differences
	Most fish species have long life spans (2—10+ years) and can reflect both long-term and current water resource quality
	Fish continually inhabit the receiving water and integrate the chemical, physical, and biological histories of the waters
	Fish represent a broad spectrum of community tolerances from very sensitive to highly tolerant and respond to chemical, physical, and biological degradation in characteristic response patterns
Public visibility	Fish are highly visible and valuable components of the aquatic community to the public
	Aquatic life uses and regulatory language are generally characterized in terms of fish (i.e., fishable and swimmable goal of the US Clean Water Act of 1972)
Ease of use and interpretation	The sampling frequency for trend assessment is less than for short-lived organisms
	Taxonomy of fishes is well established, enabling professional biologists the ability to reduce laboratory time by identifying many specimens in the field
	Distribution, life histories, trophic feeding dynamics, reproductive guilds, and tolerance to environmental stresses of many species of freshwater fish are documented in the literature

Adapted from Karr (1981), Karr et al. (1986), and Simon (1991).

(Harrison and Whitfield, 2006; Republic of South Africa Department of Water and Sanitation, 2016), and throughout Europe (Hering et al., 2006; Pont et al., 2007). For example, reference condition—based multimetric indices have been applied extensively to assess the ecological condition of streams, rivers, lakes, and wetlands as part of the European Water Framework Directive of achieving "good ecological status" in all European water bodies (Pont et al., 2006; Uriarte and Borja, 2009) including development of a continental-scale multimetric fish-based index of river biotic condition (Pont et al., 2007). Likewise, IBI-like multimetric indices are at the foundation of Australia's Sustainable Rivers Audit (Davies et al., 2010) and the South Africa's River Eco-status Monitoring Programme (Republic of South Africa Department of Water and Sanitation, 2016).

39.1.2 Patterns in "Noise" Versus "Signal"

The IBI has proven to be responsive to a wide variety of disturbances that affect fish assemblage stability and function (Karr et al., 1986; Simon, 2003). The use of multimetric indices to assess human disturbance to aquatic systems has often shown that repeat visits to the same site generate a range of index scores that were perceived to be too variable; however, the range of this "noise" is often a signal in itself that can provide information on the health and status of the community. For example, Yoder and Rankin (1995a) found that greater "noise" in the IBI was found at sites with low biological integrity compared to high biological integrity for streams in Ohio, USA. Similar results were observed on the Ohio River where repeated sampling over 11 weeks showed that sites with hard, stable substrate produced less noise in the IBI than soft, unstable substrate (Simon and Sanders, 1999). Similarly, Karr et al. (1985) showed how changes in chlorine and ammonia levels from wastewater treatment facilities caused changes in IBI scores.

39.1.3 Motivation and Regionalization

The ability to protect biological resources is dependent on our ability to detect differences between natural and human-induced variation in biological condition (Karr and Chu, 1999). To determine changes as a result of human disturbance, sampling and analyses should concentrate on multiple sites within the same environmental setting across a range of conditions from "least impacted" to severely disturbed as a result of human activity (Emery and Thomas, 2003).

The development of a properly calibrated IBI is dependent on sampling a variety of disturbance intensities from only a single region; thus, a changing biological response is similar to a "dose—response" curve. While it is difficult to successfully locate and sample such a disturbance gradient, this approach can produce biological response patterns for that particular activity (Karr et al., 1986; Yoder and Rankin, 1995b) or predominant land-use stressors within a region. Knowledge of such biological response patterns can give researchers a diagnostic tool for watersheds influenced by unknown or multiple human activities. However, the complex nature of chemical contaminants from outfalls and divergent land-use practices makes it virtually impossible to separate stressors into single human actions. Although it is often desirable to diagnosis specific contaminants causing impacts, it is often not feasible or not necessary since the biological indicators can be used to classify impacts into specific classes of human disturbance. For example, nonpoint source impacts from logging and agriculture can cause different response signatures. Logging impacts may include a warming of the stream, removal of riparian vegetation, and increased sedimentation, while agriculture may result in increased nutrient runoff to the stream, increased channelization, and loss of instream cover. Sedimentation may have the biggest impacts on logged streams with the loss of sensitive benthic habitat specialists, decline in the percentage of specialized insectivores, and decline in species abundance. In agricultural streams, changes in fish assemblage structure and function would be detected as increased percentage of tolerant species, loss of sensitive species, and decline of simple lithophilic spawning species that require clean gravel and cobble to lay their eggs. Both of these human disturbance types would be different from contaminants from point source discharge (Table 39.2).

Diverse human activities generally interact to affect watersheds, which may enable sites to be grouped and placed on a gradient according to activities and their effects (Rossano, 1995; Karr and Chu, 1999). For example, since industrial effluents are more toxic than domestic effluents and both are more serious than low-head dams, weirs, or levees, a dichotomous flowchart of human disturbance threats could be produced that group sites into categories of biological conditions across a gradient of human disturbance. Sometimes a single variable can capture and integrate multiple sources of influence. For example, relatively simple descriptors that act as surrogates for human disturbance can explain these biological differences, such as percent impervious area. Alternatively, sites can be grouped in qualitative disturbance categories. For example, Patterson (1996) classified Rocky Mountain, USA, stream sites into four categories of human activity: (1) little or no human influence on the watershed, (2) light recreational use (e.g., hiking, backpacking), (3) heavy recreational use (e.g., major trailheads, camping areas), and (4) urbanization, grazing, agriculture, or wastewater discharge.

TABLE 39.2 Characteristics of impact types including sources, characteristics, and aspects of multimetric indices affected. This list of nine impact types from Yoder and Rankin (1995a) is intended to describe biological response in Ohio, USA, rivers and streams.

Type	Major Source	Characteristics	Biocriteria Effects
Complex toxic	Major municipal WWTP; industrial point sources	These facilities comprise a significant portion of the summer base flow of the receiving stream and generally have one of the following characteristics: (1) serious instream chemical water quality impairments involving toxics; (2) recurrent whole effluent toxicity, fish kills, or severe sediment contamination involving toxics; or (3) this may include areas that have CSOs and/or urban areas located upstream from the point source	Lowest quality for IBI, darter species, percent round-bodied suckers, sensitive species, percent DELT anomalies, intolerant species, and density (less tolerant species)
Conventional municipal/industrial	Municipal WWTPs that discharge conventional substances	These facilities may or may not dominate stream flows and no serious or recurrent whole effluent toxicity is evident or small industrial discharges that may be toxic, but do not comprise a significant fraction of the summer base flow; other influences, i.e., CSOs and urban runoff may be present upstream from the point sources	High incidence of extreme outliers for darter species, number of species, percent carnivores, percent simple lithophils, density, (minus tolerants) and biomass. Extreme range for DELT anomalies >10% observed within or close proximity to WWTP mixing zones
Combined sewer overflows/urban	Impacts from CSOs and urban runoff within cities and metropolitan areas that are in direct proximity to sampling sites	Areas include both free-flowing and impounded areas upstream from the major WWTP discharges. Minor point sources may also be present in some areas	Moderate decline in IBI, loss of darter, intolerant, and sensitive species, decline in percent round-bodied suckers, number of species and few DELT anomalies
Channelization	Areas impacted by extensive, large-scale channel modification projects	Little or no habitat recovery has occurred, and some minor point source influences may be present	Low or even lower metric values for percent round-bodied suckers, intolerant species, sunfish species(lowest), percent top carnivores, percent simple lithophils, and biomass. Exhibits the highest maximums and outliers for density (including tolerants), however, did not indicate toxic impacts (e.g., DELT anomalies very low)
Agricultural nonpoint	Areas that are principally impacted from row crop agriculture	Dominant land use in the Corn Belt of the United States. Some minor point source and localized habitat influences may be present	Metric and index values indicative of good and exceptional performance. May also show fair and poor scores under extended low flows due to water withdrawal, higher effluent loads, and more intensive land use and riparian impacts
Flow alteration	Controlled releases	Sites affected by flow alteration include controlled releases downstream from major reservoirs or areas affected by water withdrawals as the predominant impact	Good to exceptional performance of IBI, little effect on number of sensitive, darter, and number of species
Impoundment	Navigation dams, low-head dams, flood control and water supply reservoirs	River segments that have been artificially impounded by low-head dams or flood control and water supply reservoirs	Good performance of IBI, loss of sensitive and intolerant species, including darters and round-bodied suckers, decline in number of species. Low numbers of omnivores and tolerant species
CSO/urban with toxics	Same as CSO/urban conventional	A significant presence of toxics usually associated with municipal CSO systems with significant pretreatment programs and sources of industrial contributions to the sewer system	Metric values consistently show lowest quality for IBI, darter species, percent round-bodied suckers, sensitive species, percent DELT anomalies, intolerant species, percent tolerant species, and density (less tolerant species)
Livestock access	Sites directly impacted by livestock operations	Animals have unrestricted access to adjacent streams	Declines in sensitive benthic species, increase in percent tolerant taxa

CSO, combined sewer overflow; DELT, deformities, eroded fins, lesions, and tumor; IBI, index of biotic integrity; WWTP, wastewater treatment plant.

Data collected over a number of years at the same site can also reveal biological responses as human activities change during that period. Regardless of how a range of human influences are selected among study sites, sampling at sites with different intensities and types of human activity is essential to detect and understand biological responses to human influence.

39.2 GENERAL DESIGN

39.2.1 Family of Indices

With many different versions of the IBI (and analogous multimetric indices) now in existence, the IBI is best thought of as a family of related indices rather than a single index (Simon and Lyons, 1995; Simon, 2000). The IBI includes attributes of the biota that range from individual health to population, community, and ecosystem levels (Karr, 1981; Karr et al., 1986). The IBI is broadly defined as any index that is based on the sum or ratings for several different measures, termed *metrics*, of fish structure and function in this case, with the rating for each metric based on quantitative expectations of what comprises high biological integrity (Karr, 1991). For some metrics, expectations will vary depending on ecosystem size and location. The IBI is not a community analysis but rather is an analysis of several hierarchical levels of biology that uses a sample of the assemblage.

39.2.2 Regional Calibration

Each metric is calibrated based on a reference or "least impacted" stream of similar size from the same region (Hughes et al., 1986). These "least impacted" areas represent the best attainable condition possible for a watershed within a region (Hughes et al., 1986).

Multiple published approaches exist to determine the reference standard (Stoddard et al., 2006). Hughes et al. (1986) and Ohio EPA (1987) were among the first to define reference standards as the best quality sites that would be able to define Karr et al.'s (1986) classification. The problem is that few of these areas remain, especially in anthropogenically disturbed landscapes. For example, in the Eastern Corn Belt Plain Ecoregion of North America, Simon and Morris (2014) found that less than 5% of the ecoregion met the criteria for reference conditions as defined by Hughes et al. (1986). Therefore, it is often not possible to calibrate an IBI using multiple reference sites that represent the full species area curve spectrum of stream sizes present in a region. Consequently, the reference standard is sometimes "modeled" using data collected via stratified random sampling of remaining subtle differences across a collection of sites that are representative of a region (Simon, 1991). This "reference condition" approach evaluates patterns in the data and avoids the need for a defined "least impacted" reference standard as it does not anticipate that any single site would be expected from the region to score the highest for all metrics (Simon, 1991). Instead, the reference condition is a composite comprised of the cumulative pattern of best attainable attributes obtained from multiple sampling sites of comparable condition or from historical or repeated sampling of the same and random stations (Simon and Stewart, 2006).

39.2.3 Major Assumptions

Substantial research has been conducted to determine the most reliable and reproducible methods for assessing fish assemblage structure in wadeable streams. This research includes studies focused on determining: minimum sampling distances and intensities, relationships between stream order and surface area, efficiency of single-pass and multiple-pass methods, and effects of personnel training and equipment bias (see Chapter 16). Complementary research has been conducted to evaluate the responses of calibrated IBIs to assessing fish assemblage integrity in wadeable streams. Simon and Morris (2014) evaluated the influence of sampling distance on metrics used to calculate an IBI in warmwater, headwater streams, and found that few of the metrics showed any significant difference based on over- or under-sampling, rare species influence, or due to scoring differences. In contrast, Dauwalter and Pert (2003) found that both IBI scores and species richness metric scores increased as electrofishing effort (measured as reach length in mean stream widths) increased for Arkansas, USA, streams located in the Ozark highland ecoregion. Similar results were reported by Angermeier and Karr (1986), who found that IBI scores increased as the stream length sampled increased when studying Jordan Creek, Illinois, USA. These results emphasize the need to calibrate regional IBIs to specific stream conditions.

39.2.4 Establishment of Baseline Considerations

A five-step process in IBI development, validation, and application has been the key to the development of index and baseline conditions. The approach was modified by Lyons et al. (2001) after recommendations by Hughes et al. (1998) and Karr and Chu (1999). *First*, an appropriate sampling methodology must be identified and tested. *Second*, this methodology is used to collect fish assemblage data in a standardized manner from reaches across the specific region. Some reaches will have minimal human impact, while others will have varying amounts and types of impacts from point and nonpoint source pollution. *Third*, fish assemblage data are used to evaluate potential metrics and develop the candidate metrics for an IBI (see Appendix 39.1). Metric replacement should follow the same intent as the original metric, but can be modified for appropriate regional application (Appendix 39.1). Data from these least impacted sites are used to characterize relatively high-quality fish communities and to investigate the influence of natural factors on community attributes. These reference conditions are contrasted with data from least impacted sites and the most degraded sites to quantify the metric range and sensitivity to human impacts. The final metrics and their scoring criteria are then selected and incorporated into the IBI. *Fourth*, the final IBI is validated with a new set of independent field data that was not used in the development phase. *Fifth*, compared IBI scores and ratings among river reaches are grouped by type of human impact to assess the relative effect of each impact on biotic integrity.

39.2.5 Metric Selection and Replacement

The following methods are general guidelines that can be used to narrow candidate metrics to appropriately responsive metrics for the final index development (Krause et al., 2013).

39.2.5.1 Range Test

First, metrics with zero values at >33% of sites should be eliminated from consideration. Second, richness metrics with a range of less than three species should be eliminated from consideration. Third, any metric for which more than 75% of the values are identical should be eliminated from consideration. Lastly, for sites with repeat visits, the most recent sample data should be used.

39.2.5.2 Signal-to-Noise

The ratio of variance among sites (signal) to the variance of repeated visits at the same site (noise) can be calculated as an F-value in a one-way analysis of variance (ANOVA) for parametric metrics or as an H-value in a Kruskal–Wallis test for nonparametric metrics (Zar, 2010). Only metrics with a high ratio (e.g., >3) should be retained.

39.2.5.3 Correlation With Natural Gradients

Metrics should not be strongly correlated with a natural gradient of watershed area. This type of correlation can be inferred by regressing metric values from least disturbed sites against watershed area using a general linear regression model for parametric data or a rank-based regression for nonparametric data. If the two lines drawn between the upper and lower 95% confidence bands at the limits of the watershed area data have overlapping y-values, then correlation with the natural gradient should be considered strong and the metric should be corrected or eliminated.

39.2.5.4 Responsiveness Test

Candidate metrics should be screened for responsiveness to various anthropogenic disturbances by running a Kruskal–Wallis test on the raw metrics to test the ability of metrics to distinguish between the least and most disturbed sites. The metrics with the highest significant Kruskal–Wallis H-statistic in each class should be retained.

39.2.5.5 Redundancy Test

Redundancy in metrics can be tested via a Spearman rank correlation (Zar, 2010). Metrics should be considered redundant if one metric explains greater than 70% of the variability in another metric ($r_s > 0.70$). If a pair of redundant metrics is found, the metric with the highest Kruskal–Wallis statistic should be retained. Only the least disturbed sites should be used to assess redundancy so as not to inadvertently exclude metrics that covaried as a result of anthropogenic disturbance (Stoddard et al., 2008).

39.2.5.6 Range Test for Metric Values

Data from least disturbed and most disturbed sites should return the minimum and maximum observed values for each metric. Box plots can be used to analyze for overlap in the distribution of the data. If more than 50% of the values were similar, then the metric should be rejected. Only metrics with less than half of the total distribution shared between responses from least disturbed and most disturbed sites should be retained.

39.2.5.7 Metric Scoring

Metrics can be scored on Karr's (1981) original trisected scale of 1, 3, and 5 or on a continuous scale from 0 to 10. Floor and ceiling values are typically set at the 5th and 95th percentile of metric values across all sites (Fig. 39.1). Positive metric values less than the 5th percentile are given a score of zero, and those above the 95th percentile are given a score of 10. Metric scores between the 5th and 95th percentiles are interpolated linearly. Negative metrics are scored similarly, but the floor and ceiling values are reversed. The scored metrics are then summed and the index is often scaled to a range of 0−60 or 0−100 (Table 39.3).

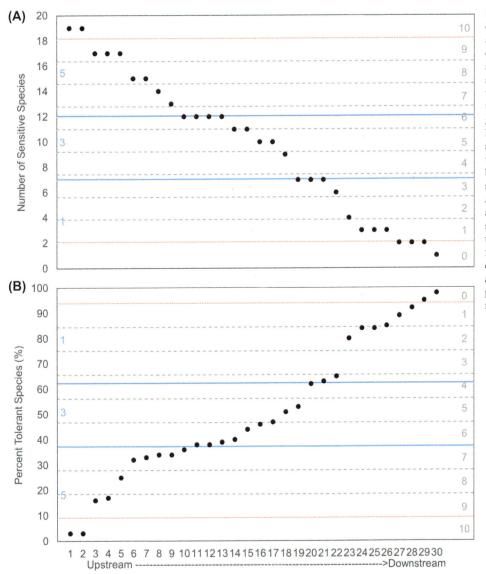

FIGURE 39.1 Examples of index of biotic integrity scoring criteria for a positive scoring metric (A; number of sensitive species) and negative scoring metric (B; percent tolerant species). Data for 30 sampling sites were simulated along a hypothetical upstream to downstream disturbance gradient where biotic integrity was lowest at the downstream sites. For simulation purposes both metrics were highly sensitive to the disturbance. Karr's (1981) trisected scoring criteria are illustrated by the *solid blue lines and blue numbers*. In accordance with a 0 to 10 scoring scale, the *red dotted lines* represent the 5th and 95th percentiles that function as the scoring floor and ceiling. *Gray dashed lines and gray numbers* represent the linearly interpolated scoring criteria for each metric.

TABLE 39.3 Example calculation of an index of biotic integrity (IBI) calibrated for the Wabash River, Indiana, USA. Scoria criteria follow the 60-point scale proposed by Karr (1981). The example IBI score of 46 suggests that the condition of the stream is good to fair.

Metric	Scoring Criteria and Rating (Points)			
	Poor (1)	Fair (3)	Good (5)	Example
Native species (total)	<10	10–20	>20	15 spp. = 3
Number of centrachid species	≤2	3–4	≥5	3 sp. = 3
Number of round-bodied suckers	<2	2–4	≥5	6 spp. = 5
Number of sensitive species	≤3	4–7	≥8	4 spp. = 3
% Tolerant species	>71.6%	43.3–71.6%	<43.3%	65% = 3
% Omnivores	>68.3%	36.7–68.3%	<36.7%	20% = 5
% Insectivores	<25.0%	25.0–50.0%	>50.0%	40% = 3
% Carnivores	<10% or >40%	10–20% and 30–40%	>20–30%	28% = 5
% Large-river species	<28.3%	28.3–56.6%	>56.6%	44% = 3
% Lithophilic species	<15%	15–30%	>30%	50% = 5
% Deformities, eroded fins, lesions, and tumor	>1.3%	0.1–1.3%	<0.1%	0.08% = 5
Catch per unit effort	<600	600–1200	>1200	900 = 3
				$\sum$ Score = 46

From Simon (2006).

39.2.6 Additional Considerations

Although the IBI enables for the assessment of anthropogenic impact at the index or metric levels, selection of certain metrics may not always be diagnostic for select disturbances. Therefore, metric selection during IBI calibration should consider the ability of the individual metrics to indicate causal relationships with specific stressors. For this task, the United States Environmental Protection Agency (USEPA) has developed an iterative approach that resembles risk management methods and enables for determining cause and effect of specific impacts (USEPA, 2000). The method is based on a strength-of-evidence approach, which is evaluated using Koch's postulate. Koch's postulate combines different lines of evidence in a manner to provide compelling evidence for causation. The approach was originally developed for pathogen-induced diseases (Yerushalmy and Palmer, 1959; Hackney and Kinn, 1979) and then adapted for ecological effects (Adams, 1963; Woodman and Cowling, 1987) and ecological risk assessment (USEPA, 1998). Koch's postulate infers that (1) the injury, dysfunction, or other effect of the stressor must be regularly associated with exposure to the stressor in association with any contributing causal factors; (2) the stressor or a specific indicator of exposure must be found in the affected organism; (3) the effects must be seen when healthy organisms are exposed to the stressor under controlled conditions, and any contributory factors should contribute in the same way during the controlled exposures; and (4) the stressor or a specific indicator of exposure must be found in the experimentally affected organism.

The power of Koch's postulate is how the four types of evidence are combined. For example, the requirement of regular association cannot be determined in the field because it usually cannot be controlled in such a manner as to establish whether the stressor works alone or in combination with other correlated causes. In addition, field associations cannot document the temporal sequence of cause and effect. The second and fourth postulates suggest that field observations must correspond to experimental exposures, and therefore exposure and field correspondence are unlikely to be coincidental. Thus, each cause can be evaluated separately or in combination with other stressors. Yoder and DeShon (2002) illustrated how a biological stressor gradient can be used to identify the cause of environmental impacts using such a logical argument progression.

39.3 SPECIFIC METHODS

39.3.1 Basic Method: Application of the Index of Biotic Integrity to a Reference Site and an Impacted Site

In this exercise, you will identify a high-quality (reference) site and a low-quality (impacted) site within a stream network and then obtain a fish sample from both stream reaches (see Chapter 16 for fish sampling approaches). Sampling duration should range from 15 to 90 min, depending on stream size and complexity. The objective of the sampling is to collect a "representative sample" of the fish assemblage using methods designed to collect all except very rare species and provide an unbiased measure of the proportional abundances of species. You will then calculate IBI scores for the impacted site based on the condition of the reference site using the metrics and scoring criteria from the original IBI designed by Karr (1981) and summarized in Appendix 39.1. The simplicity of the original IBI provides flexibility for modifying the IBI to other stream types and regions as illustrated in Appendix 39.1.

39.3.1.1 Field Methods

1. Before proceeding with field work, obtain appropriate fish collection approvals (institutional IACUC), collection permits (generally state permits in the United States), and any land entry permissions. Notify land owners prior to accessing the site.
2. Obtain a sample of fish from the designated reach following procedures in Chapter 16. The sampling reach should be at least 15 mean wetted-channel widths (MWW) in length and rounded up to the nearest 50 m (e.g., $2.4 \, \text{MWW} \times 15 = 36 \, \text{m}$, which rounds up to 50 m). If the water level appears to be substantially (>0.15 m) above normal, sampling should not occur. Reaches should not contain permanent tributaries or hydraulic controls (e.g., dams, old bridge abutments).
3. For all fish collected, identify to species, record length and mass measurements, examine for and record any deformities, eroded fins, lesions, and tumor (DELT) anomalies on a waterproof data sheet similar to **Catch Summary** provided in Online Supplement 39.1.
4. For each species, measure total length (TL) and mass for all individual fish ≥ 25 mm in TL. Fish specimens <25 mm TL should not be counted in the number of specimens but recorded separately unless the individual species only reaches a maximum length of 25 mm TL.
5. Fish should be handled carefully to minimize mortality, including during collection and afterward during processing. All individuals should be maintained in a live well or aerated container.
6. Following measurement, either return fishes to the water or preserve as voucher specimens if identification is uncertain. Store any voucher specimens (two to three individuals) in a designated museum collection. Voucher specimens should be tagged with labels similar to those provided in Online Supplement 39.2. Rare or imperiled species should be photographed and released. Nonnative fish species should never be returned to the water, but instead sacrificed using MS222 (see Chapter 16).
7. Fill out a habitat data sheet similar to the **Station Summary** provided in Online Supplement 39.1.

39.3.1.2 Data Analysis

1. Compare the species richness and abundances from the reference and impacted reaches.
2. Based on the data collected at the reference reach, score each metric in Appendix 39.1 as either:
   a. Minus ($-$): The fish community observed in the impacted reach deviates significantly from the reference reach (i.e., impacted reach is highly disturbed compared to the reference reach)
   b. Plus ($+$): The fish community observed in the impacted reach is similar to that of the reference reach
   c. Zero (0): The fish community observed in the impacted reach is intermediate between the reference reach and a highly disturbed reach (i.e., impacted reach is moderately disturbed)
3. Quantify each metric by converting the qualitative scores as: ($-$) = 1, (0) = 3, and ($+$) = 5
4. Sum all 12 quantified metrics
5. Scale the summed IBI score between 0 and 100 by multiplying by 1.66
6. Classify the condition of the impacted reach relative to the reference reach using Table 39.4.

TABLE 39.4 Classification boundaries suggested by Karr (1981) for classifying stream condition based on index of biotic integrity scores. Karr's original 0 to 60-point scale has been rescaled from 0 to 100 by multiplying by 1.66.

Stream Condition	60-Point Scale	100-Point Scale
Excellent (E)	57–60	95–100
E–G	53–56	88–94
Good (G)	48–52	80–87
G–F	45–47	75–79
Fair (F)	39–44	65–74
F–P	36–38	60–64
Poor (P)	28–35	47–59
P–VP	24–27	40–46
Very poor (VP)	≤23	≤39

39.3.2 Advanced Method: Calibration and Testing a Stream Fish Index of Biotic Integrity

In this exercise, you will develop a fish IBI for a selected region using the framework developed by Whittier et al. (2007) and expounded upon by Krause et al. (2013). You will then test the developed IBI against known "least impacted" reference sites.

39.3.2.1 Field Methods

1. To enable for regression analyses, select at least 14 known-impact reference sites (7 least impacted and 7 highly impacted) and 14 random condition sites. Both least impacted and highly impacted reference condition sites are needed to qualify the effect of disturbance on each metric.
2. Using appropriate electrofishing methods for a multiple-pass depletion survey (see Chapter 16), sample the fish assemblages of each of the selected sites.
3. For all fish collected, identify to species, record length and mass measurements, examine for and record any deformities, DELT anomalies on a waterproof data sheet similar to **Catch Summary** provided in Online Supplement 39.1.
4. For each species, measure or calculate batch mass t for all fish ≥25 mm in TL as well as minimum and maximum length. Fish specimens <25 mm TL should not be counted in the number of specimens but recorded separately unless the individual species only reaches a maximum length of 25 mm TL.
5. Fish should be handled carefully to minimize mortality, including during collection and afterward during processing. All individuals should be maintained in a live well or aerated container.
6. Following measurement, either return fishes to the water or preserve as voucher specimens if identification is uncertain. Store any voucher specimens (two to three individuals) in a designated museum collection. Voucher specimens should be tagged with labels similar to those provided in Online Supplement 39.2. Rare or imperiled species should be photographed and released. Nonnative fish species should never be returned to the water, but instead sacrificed using MS222.
7. Fill out a habitat data sheet similar to the **Station Summary** provided in Online Supplement 39.1.

39.3.2.2 Data Analysis

1. Classify species characteristics based on habitat preferences, tolerance levels, trophic position, reproductive strategy, life history traits, and native status.
2. Calculate scores for as many potential metrics as possible (an in-depth compilation of candidate metrics is available in Appendix A2 of Krause et al., 2013).
3. When possible, correct metrics for normality and heteroscedasticity using appropriate transformations.
4. Test each metric for range, S:N, correlation with natural gradients, responsiveness, redundancy, and range test for values (see above Section 39.2.5).
5. Sum all retained metrics and scale the IBI to a range of 0–100.
6. Validate IBI scores against additional site information that was not used in the original calibration phase to ensure that the approach is capable of differentiating between least disturbed and most impacted sites in the region.

39.4 QUESTIONS

39.4.1 General Questions

1. The use of multimetric indices is one of the three pillars for biocriteria. What are the other two? How do the three pillars facilitate the completion of an assessment using biological information? Discuss how the three pillars might need to be adjusted for different applications.
2. What effect would land use and regionalization have on developing the reference expectations for the index of biotic integrity?
3. Contrast the impact that a decision to use a modeled "reference condition" approach compared to a "reference site" approach would have on the biological expectations for the index of biotic integrity development? In IBI development, what might be the potential pitfalls for using an entirely random approach versus targeted or stratified random designs to select reference sites?

39.4.2 Questions for Basic Method

4. How might the stream length sampled affect the metric scores for the total index of biotic integrity score?
5. Effort can be defined based on units of intensity, distance, or time. How might each of these definitions of effort change the sampling protocol and scope of inference for an index of biotic integrity?
6. What additional chemical and physical data might one want to collect along with the biological data for calibrating and validating the IBI?
7. How would one determine the amount of effort necessary to sample (with the two extremes being a census versus a "representative" sampling approach)? Under what conditions might the decision be preferring one extreme compared to the other?
8. When selecting an appropriate sampling gear, what might be the deciding factors that would determine the criteria for different regions?

39.4.3 Questions for Advanced Method

9. One of the key premises of the IBI is that it will respond to changes in anthropogenic disturbance. How can we ensure that the index is responsive to predominant anthropogenic disturbances characteristic of the region?
10. Candidate metric and projected response should be considered a priori to sampling and testing of candidate metrics. What is the purpose for this a priori determination? Why is this critical in the evaluation of candidate metrics?
11. The IBI is considered a family of indices due to the recalibration and development of alternative versions for different regions. Karr et al. (1986) suggested that the 12 original metrics were selected to evaluate a wide range of species composition based on habitat considerations, tolerance and sensitivity, trophic dynamics, relative abundance, and disease. Why would failure to account for metric replacement cause issues in the proper performance of the metrics and ultimately the diagnosis of water quality conditions?
12. Karr and Chu (1999) advocated against a statistically generated IBI. Part of the reason for this is that the IBI was never intended to answer every question. If statistical validation was not the original intention of the author, why should the final metric decision be based on statistical approaches?

39.5 MATERIALS AND SUPPLIES

Items needed before implementing these methods

IACUC approval for fish handling

Collection permits from local authorities

Site access permissions from public or private land owners

Latitude and longitude (for random statistically based sites only)

County plat maps; state-specific atlas and gazetteers (DeLorme)

Aerial photographs (if available)

GPS receiver, battery, and antenna (if necessary)

Digital camera for taking fish photographs

Fish identification key for region

Items needed for field procedures

Pencils

Permanent/alcohol-proof markers

Labeling tape

Flagging for designating sampling reaches

Block nets (1-cm mesh lead-line seines)

Fish sample identification labels

Waterproof data sheets

Chest waders and rain gear

Jars or bottles, in which the sample can be preserved

Box or crate to store sample bottles

Cooler with ice for voucher specimens

Backpack electrofishing unit with appropriate cathode and anode (see Chapter 16)

Livewell with aerators or instream live car to maintain fish safely

Dip nets, 5-mm mesh

Buckets and miscellaneous sorting chambers to keep fish individuals alive

Fish measuring board (metric units)

Digital weighing scale (grams) plus hanging scales for larger individuals

REFERENCES

Adams, D.F., 1963. Recognition of the effects of fluorides on vegetation. Journal of Air Pollution Control Association 13, 360–362.

Angermeier, P.L., Karr, J.R., 1986. Applying an index of biotic integrity based on stream-fish communities: considerations in sampling and interpretation. North American Journal of Fisheries Management 6, 418–429.

Angermeier, P.L., Schlosser, I.J., 1987. Assessing biotic integrity of the fish community in a small Illinois stream. North American Journal of Fisheries Management 7, 331–338.

Berkman, H.F., Rabeni, C.F., Boyle, T.P., 1986. Biomonitors of stream quality in agriculture areas: fish vs. invertebrates. Environmental Management 10, 413–419.

Daniel, W.M., Brown, K.M., Kaller, M.D., 2014. A tiered aquatic life unit bioassessment model for Gulf of Mexico coastal streams. Fisheries Management and Ecology 21, 491–502.

Dauwalter, D.C., Pert, E.J., 2003. Effect of electrofishing effort on an index of biotic integrity. North American Journal of Fisheries Management 23, 1247–1252.

Davies, P.E., Harris, J.H., Hillman, T.J., Walker, K.F., 2010. The Sustainable Rivers Audit: assessing river ecosystem health in the Murray-Darling Basin, Australia. Marine and Freshwater Research 61, 764–777.

Emery, E.B., Thomas, J.A., 2003. A method for assessing outfall effects on Great River fish populations: the traveling zone approach. In: Simon, T.P. (Ed.), Biological Response Signatures: Patterns in Biological Integrity for Assessment of Freshwater Aquatic Assemblages. CRC Press, Boca Raton, FL, USA, pp. 157–165.

Fisch, F., Branco, J.O., de Menezes, J.T., 2016. Ichthyofauna as indicator of biotic integrity of an estuarine area. Acta Biologica Colombiana 21, 27–38.

Hackney, J.D., Kinn, W.S., 1979. Koch's postulates updated: a potentially useful application to laboratory research and policy analysis in environmental toxicology. American Review in Respiration and Disease 119, 849–852.

Hallett, C.S., Valesini, F.J., Clarke, K.R., Hesp, S.A., Hoeksema, S.D., 2012. Development and validation of fish-based, multimetric indices for assessing the ecological health of Western Australian estuaries. Estuarine, Coastal and Shelf Science 104–105, 102–113.

Harrison, T.D., Whitfield, A.K., 2006. Application of a multimetric fish index to assess the environmental condition of South African estuaries. Estuaries and Coasts 29, 1108–1120.

Hermoso, V., Clavero, M., 2013. Revisiting ecological integrity 30 years later: non-native species and the misdiagnosis of freshwater ecosystem health. Fish and Fisheries 14, 416–423.

Hering, D., Feld, C.K., Moog, O., Ofenböck, T., 2006. Cook book for the development of a Multimetric Index for biological condition of aquatic ecosystems: experiences from the European AQEM and STAR projects and related initiatives. Hydrobiologia 566, 311–324.

Hite, R.L., 1988. Overview of stream quality assessments and stream classification in Illinois. In: Simon, T.P., Holst, L.L., Shepard, L.J. (Eds.), Proceedings of the First National Workshop on Biological Criteria, 1987, Lincolnwood, Illinois. United States Environmental Protection Agency, Chicago, IL, USA., pp. 99–144. EPA 905-9-89-003.

Hughes, R.M., Larsen, D.P., Omernik, J.M., 1986. Regional reference sites: a method for assessing stream potentials. Environmental Management 10, 629–635.

Hughes, R.M., Kaufmann, P.R., Herlihy, A.T., Kincaid, T.M., Reynolds, L., Larsen, D.P., 1998. A process for developing and evaluating indices of fish assemblage integrity. Canadian Journal of Fisheries and Aquatic Sciences 55, 1618–1631.

Hughes, R.M., Howlin, S., Kaufmann, P.R., 2004. A biointegrity index (IBI) for coldwater streams of Western Oregon and Washington. Transactions of the American Fisheries Society 133, 1497–1515.

Ivasauskas, T.J., Bettoli, P.W., 2014. Development of a multimetric index of fish assemblages in a cold tailwater in Tennessee. Transactions of the American Fisheries Society 143, 495–507.

Karr, J.R., 1981. Assessment of biological integrity using fish communities. Fisheries 6, 21–27.

Karr, J.R., 1991. Biological integrity: a long-neglected aspect of water resource management. Ecological Applications 1, 66–84.

Karr, J.R., 1993. Defining and assessing ecological integrity: beyond water quality. Environmental Toxicology and Chemistry 12, 1521–1531.

Karr, J.R., 1995. Protecting aquatic ecosystems: clean water is not enough. In: Davis, W.S., Simon, T.P. (Eds.), Biological Assessment and Criteria: Tools for Water Resource Planning and Decision Making. Lewis Publishers, Boca Raton, FL, USA, pp. 7–14.

Karr, J.R., Chu, E.W., 1999. Restoring Life in Running Waters: Better Biological Monitoring. Island Press, Washington, DC, USA.

Karr, J.R., Toth, L.A., Dudley, D.R., 1985. Fish communities of midwestern rivers: a history of degradation. BioScience 35, 90–95.

Karr, J.R., Fausch, K.D., Angermeier, P.L., Yant, P.R., Schlosser, I.J., 1986. Assessing the Biological Integrity in Running Waters: A Method and Its Rationale, vol. 5. Illinois Natural History Survey, Special Publication, Champaign, IL, USA.

Krause, J.R., Bertrand, K.N., Kafle, A., Troelstrup Jr., N.H., 2013. A fish index of biotic integrity for South Dakota's Northern Glaciated Plains Ecoregion. Ecological Indicators 34, 313–322.

Lenhardt, M., Markovic, G., Gacic, Z., 2009. Decline in the index of biotic integrity of the fish assemblage as a response to reservoir aging. Water Resources Management 23, 1713–1723.

Lyons, J., 1992. Using the Index of Biotic Integrity (IBI) to Measure Environmental Quality in Warmwater Streams in Wisconsin. General Technical Report NC-149. North Central Forest Experiment Station, U.S. Department of Agriculture, St. Paul, MN, USA.

Lyons, J., 2006. A fish-based index of biotic integrity to assess intermittent headwater streams in Wisconsin, USA. Environmental Monitoring and Assessment 122, 239–258.

Lyons, J., 2012. Development and validation of two fish-based indices of biotic integrity for assessing perennial coolwater streams in Wisconsin, USA. Ecological Indicators 23, 402–412.

Lyons, J., Wang, L., Simonson, T.D., 1996. Development and validation of an index of biotic integrity for coldwater streams in Wisconsin. North American Journal of Fisheries Management 16, 241–256.

Lyons, J., Piette, R.R., Niermeyer, K.W., 2001. Development, validation, and application of a fish-based index of biotic integrity for Wisconsin's large warmwater rivers. Transactions of the American Fisheries Society 130, 1077–1094.

Megan, M.H., Nash, M.S., Neale, A.C., Pitchford, A.M., 2007. Biological integrity in mid-atlantic coastal plains headwater streams. Environmental Monitoring and Assessment 124, 141–156.

Minns, C.K., Cairns, V.W., Randall, R.G., Moore, J.E., 1994. An index of biotic integrity (IBI) for fish assemblages in the littoral zone of Great Lakes' areas of concern. Canadian Journal of Fisheries and Aquatic Sciences 51, 1804–1822.

Ohio Environmental Protection Agency (OEPA), 1987. Biological Criteria for the Protection of Aquatic Life, vol. I. OEPA, Columbus, OH, USA.

Osborne, L.L., Kohler, S.L., Bayley, P.B., Day, D.M., Bertran, W.A., Wiley, M.J., Sauer, R., 1992. Influence of stream location in a drainage network on the index of biotic integrity. Transactions of the American Fisheries Society 121, 635–643.

Patterson, A.J., 1996. The Effect of Recreation on Biotic Integrity of Small Streams in Grand Teton National Park (M.S. thesis). University of Washington, Seattle, WA, USA.

Pont, D., Hugueny, B., Rogers, C., 2007. Development of a fish-based index for the assessment of river health in Europe: the European Fish Index. Fisheries Management and Ecology 14, 427–439.

Pont, D., Hugueny, B., Beier, U., Goffaux, D., Melcher, A., Noble, R., Rogers, C., Roset, N., Schmutz, S., 2006. Assessing river biotic condition at a continental scale: a European approach using functional metrics and fish assemblages. Journal of Applied Ecology 43, 70–80.

Republic of South Africa Department of Water and Sanitation, 2016. River Eco-Status Monitoring Programme. Available: https://www.dwa.gov.za/IWQS/rhp/default.aspx.

Rossano, E.M., 1995. Development of an Index of Biological Integrity for Japanese Streams (IBI-J) (M.S. thesis). University of Washington, Seattle, WA, USA.

Ruaro, R., Gubiani, E.A., 2013. A scientometric assessment of 30 years of the Index of Biotic Integrity in aquatic ecosystems: applications and main flaws. Ecological Indicators 29, 105–110.

Simon, T.P., 1991. Development of Index of Biotic Integrity Expectations for the Ecoregions of Indiana. I. Central Corn Belt Plain. EPA 905/9–91/025. U.S. Environmental Protection Agency, Chicago, IL, USA.

Simon, T.P., 2000. The use of biological criteria as a tool for water resource management. Environmental Science and Policy 3, S43–S49.

Simon, T.P., 2003. Biological Response Signatures: Indicator Patterns Using Aquatic Communities. CRC Press, Boca Raton, FL, USA.

Simon, T.P., 2006. Development, calibration, and validation of an index of biotic integrity for the Wabash River. Proceedings of the Indiana Academy of Science 115, 170–186.

Simon, T.P., Emery, E.B., 1995. Modification and assessment of an index of biotic integrity to quantify water resource quality in great rivers. Regulated Rivers: Research and Management 11, 283–298.

Simon, T.P., Lyons, J., 1995. Application of the index of biotic integrity to evaluate water resource integrity in freshwater ecosystems. In: Davis, W.S., Simon, T.P. (Eds.), Biological Assessment and Criteria: Tools for Water Resource Planning and Decision Making. Lewis Publishers, Boca Raton, FL, USA, pp. 245–262.

Simon, T.P., Morris, C.C., 2014. Relationships among varying sampling distance and the IBI in warmwater, headwater streams of the Eastern Corn Belt Plain. Journal of Monitoring and Assessment 186, 6537–6551.

Simon, T.P., Sanders, R.E., 1999. Applying an index of biotic integrity based on Great River fish communities: considerations in sampling and interpretations. In: Simon, T.P. (Ed.), Assessing the Sustainability and Biological Integrity of Water Resources Using Fish Communities. CRC Press, Boca Raton, FL, USA, pp. 475—505.

Simon, T.P., Stewart, P.M., 1998. Validation of an index of biotic integrity for evaluating dunal palustrine wetlands with emphasis on the Grand Calumet Lagoons. Aquatic Ecosystem Health and Management 1, 71—82.

Simon, T.P., Stewart, P.M., 2006. Coastal Wetlands of the Laurentian Great Lakes: Health, Habitat, and Indicators. Authorhouse Press, Bloomington, IN, USA.

Simon, T.P., Jankowski, R., Morris, C., 2000. Modification of an index of biotic integrity for assessing vernal ponds and small palustrine wetlands using fish, crayfish, and amphibian assemblages along southern Lake Michigan. Aquatic Ecosystem Health & Management 3, 407—418.

Steedman, R.J., 1988. Modification and assessment of an index of biotic integrity to quantify stream quality in southern Ontario. Canadian Journal of Fisheries and Aquatic Sciences 45, 492—501.

Stoddard, J.L., Herlihy, A.T., Peck, D.V., Hughes, R.M., Whittier, T.R., Tarquinio, E., 2008. A process for creating multimetric indices for large-scale aquatic surveys. Journal of the North American Benthological Society 27, 878—891.

Stoddard, J.L., Larsen, D.P., Hawkins, C.P., Johnson, R.K., Norris, R.H., 2006. Setting expectations for the ecological condition of streams: the concept of reference condition. Ecological Applications 16, 1267—1276.

Uriarte, A., Borja, A., 2009. Assessing fish quality status in transitional waters, within the European Water Framework Directive: setting boundary classes and responding to anthropogenic pressures. Estuarine, Coastal and Shelf Science 82, 214—224.

U.S. Environmental Protection Agency (USEPA), 1998. Guidelines for Ecological Risk Assessment. EPA 822-R-98—008. USEPA, Office of Research and Development. Risk Assessment Forum, Washington, DC, USA.

U.S. Environmental Protection Agency (USEPA), 2000. Stressor Identification Guidance Document. EPA 822/B-00/025. USEPA, Office of Research and Development, Washington, DC, USA.

Whittier, T.R., Hughes, R.M., Stoddard, J.L., Lomnicky, G.A., Peck, D.V., Herlihy, A.T., 2007. A structured approach for developing indices of biotic integrity: three examples from streams and rivers in the western USA. Transactions of the American Fisheries Society 136, 718—735.

Woodman, J.N., Cowling, E.B., 1987. Airborne chemicals and forest health. Environmental Science 21, 120—126.

Yerushalmy, J., Palmer, C.E., 1959. On the methodology of investigations of etiologic factors in chronic disease. Journal of Chronic Disease 10, 27—40.

Yoder, C.O., DeShon, J.E., 2002. Using biological response signatures within a framework of multiple indicators to assess and diagnose causes and sources of impairments to aquatic assemblages in selected Ohio rivers and streams. In: Simon, T.P. (Ed.), Biological Response Signatures: Patterns in Biological Integrity for Assessment of Freshwater Aquatic Assemblages. CRC Press, Boca Raton, FL, USA, pp. 23—81.

Yoder, C.O., Rankin, E.T., 1995a. Biological criteria program development and implementation in Ohio. In: Davis, W.S., Simon, T.P. (Eds.), Biological Assessment and Criteria: Tools for Water Resource Planning and Decision Making. Lewis Publishers, Boca Raton, FL, USA, pp. 109—144.

Yoder, C.O., Rankin, E.T., 1995b. Biological response signatures and the area of degradation value: new tools for interpreting multimetric data. In: Davis, W.S., Simon, T.P. (Eds.), Biological Assessment and Criteria: Tools for Water Resource Planning and Decision Making. Lewis Publishers, Boca Raton, FL, USA, pp. 263—286.

Zar, J.H., 2010. Biostatistical Analysis, fifth ed. Prentice-Hall, Upper Saddle River, NJ, USA.

APPENDIX 39.1

List of original index of biotic integrity metrics (Arabic number) proposed by Karr (1981) for streams in the central United States, followed by replacement or modifications (capital letters) proposed by subsequent authors (citations in parentheses; see Online Supplement 39.3) for streams in other regions or for different water body types in North America. While the table lists references for North American Indices of Biotic Integrity (IBIs), the principles behind metric selection are applicable to other fish-based multimetric indices used in other geographic areas.

Metric (Reference Number)

Species Richness and Composition Metrics

1. Total number of fish species (1—7, 11, 13—15, 18, 19, 22, 23, 25—29, 31)
   A. Number of native fish species (8—10, 12, 16, 17, 20, 22—24, 30, 33)
   B. Number of fish species, excluding Salmonidae (13)
   C. Number of amphibian species (3, 13)
   D. Number of salmonid age classes (23)
   E. Number of resident lotic species—coldwater (26)

2. Number of darter species (Percidae genera *Crystallaria*, *Etheostoma*, *Percina*, and *Ammocrypt*) (1, 2, 4, 5, 8, 9, 12, 15—17, 19, 22, 24, 31)
   A. Number of darter and Cottidae species (9, 10, 23)
   B. Number of darter, Cottidae, and *Noturus* (Ictaluridae) species (15, 16, 19)
   C. Number of darter, Cottidae, and round-bodied sucker species (17)
   D. Number or percent of Cottidae species (6, 13, 23, 25)
   E. Abundance of Cottidae individuals (3)
   F. Number of benthic species (11, 18, 28—30)

    G. Percent of individuals that are native benthic species (11)

    H. Number of benthic insectivore species—coldwater (7, 23, 26, 27)

    I. Number of darter species, excluding "tolerant darter species" (headwater sites) (21)

    J. Percent cyprinids with subterminal mouths (22, 27)

    K. Number of salmonid yearlings (individuals) (23)

    L. Number of sculpin individuals (23)

    M. Number or native species (25)

    N. Number of coldwater species (28)

    O. Percent individuals as coolwater species (28, 32)

    P. This metric deleted from IBI (4)

3. Number of sunfish species (Centrarchidae, excluding *Micropterus*) (1, 4, 5, 14–17, 20, 24, 29, 30, 31)

    A. Number of native sunfish species (9, 12)

    B. Number of sunfish and Salmonidae species (9,12)

    C. Number of sunfish species and *Perca flavescens* (Percidae) (16)

    D. Number of headwater (restricted to small streams) species (9, 15, 22, 27)

    E. Number of water column (nonbenthic) species (7, 17, 18)

    F. Number of water column cyprinid species (17, 25, 26)

    G. Number of sunfish species including *Micropterus* (19, 22, 32, 33)

    H. Number of sunfish and trout species (23)

    I. Number of salmonid species (23)

    J. Number of headwater species (23)

    K. Number of introduced species (25)

    L. This metric deleted from IBI (11)

4. Number of Catostomidae species (1, 4–6, 8, 9, 12, 15, 17, 19, 20, 26)

    A. Percent of individuals that are Catostomidae (9)

    B. Percent of individuals that are round-bodied suckers (genera *Cycleptus, Hypentelium, Minytrema,* and *Moxostoma*) (9, 17, 19, 27)

    C. Number of Catostomidae and Ictaluridae species (10, 23)

    D. Number of Catostomidae and Cyprinidae species (17)

    E. Number of benthic insectivorous species (7, 11, 17, 22)

    F. Number of laterally compressed minnow species (21)

    G. Number of minnow species (4, 6, 9, 14, 15, 17, 22, 23, 27, 28, 30, 33)

    H. Number of adult trout species (23)

    I. Number of native minnow species (24, 31)

    J. Number of phytophilic species (29, 30, 33)

    K. This metric deleted from IBI (2, 3, 13, 14)

Indicator Species Metrics

5. Number of intolerant or sensitive species (1–4, 6–8, 10–12, 14, 15, 17–20, 22, 23, 25–31)

    A. Number of Salmonidae species (3, 15, 25)

    B. Percent individuals that are Salmonidae (11, 25)

    C. Juvenile Salmonidae presence or abundance (3, 18)

    D. Large (>15–20 cm) or adult Salmonidae presence or abundance (3, 6, 18)

    E. Abundance or biomass of all sizes of Salmonidae (3, 13)

    F. Mean length or weight of Salmonidae (13)

    G. Percent of individuals that are anadromous *Oncorhynchus mykiss* (Salmonidae) older than age 1 (3)

    H. Percent individuals as *Salvelinus fontinalis* (Salmonidae) (10, 23, 28)

    I. Presence of juvenile or large *Esox lucius* (Esocidae) (18)

    J. Number of large-river species (19, 22)

    K. Percent of species that are native species (3)

    L. Percent of individuals that are native species (3)

    M. Number of amphibian species (23)

    N. Percent lake associates species (29, 33)

    O. Percent dominance (8)

    P. Percent individuals as nontolerant native invertivores (32)

    Q. This metric deleted from IBI (2, 14, 24)

6. Percent of individuals that are *Lepomis cyanellus* (Centrarchidae) (1, 17, 23)

    A. Percent of individuals that are *Lepomis megalotis* (Centrarchidae) (5)

    B. Percent of individuals that are *Cyprinus carpio* (Cyprinidae) (6, 23)

    C. Percent of individuals that are *Semotilus atromaculatus* (Cyprinidae) (2, 23, 26)

    D. Percent of individuals that are *Rutilus rutilus* (Cyprinidae) (18)

    E. Percent of individuals that are *Rhinichthys* species (Cyprinidae)—coldwater (10, 23, 26)

    F. Percent of individuals that are *Catostomus commersonii* (Catostomidae) (4, 7, 11, 23, 26, 28)

   G. Percent of individuals that are tolerant species (8, 9, 12, 14—17, 19, 22—24, 27—29)
   H. Percent of individuals that are "pioneering species" (9, 15, 22, 27)
   I. Percent of individuals that are introduced species (4, 6, 12, 14, 24, 25)
   J. Number of introduced species (12, 13, 25, 26, 29)
   K. Evenness (22, 27)
   L. This metric deleted from IBI (3)

Trophic Function Metrics

7. Percent of individuals that are omnivores (1—4, 6—8, 10, 12, 14—20, 22, 26, 29—31)
   A. Percent of individuals that are omnivorous Cyprinidae species (10)
   B. Percent of individuals that are *Luxilus cornutus* or *Cyprinella spiloptera* (Cyprinidae) facultative omnivores (9)
   C. Percent of individuals that are generalized feeders that eat a wide range of animal material but limited plant material (2, 9, 11, 24, 26)
   D. Percent biomass of omnivores (22)
   E. Percent individuals as yearling salmonids (23, 24)
   F. This metric deleted from IBI (3, 13)

8. Percent of individuals that are insectivorous Cyprinidae (1, 17, 23, 26, 31)
   A. Percent of individuals that are insectivores/invertivores (5—7, 9, 12,14—19, 24)
   B. Percent of individuals that are specialized insectivores (2, 4, 20, 23)
   C. Percent of individuals that are specialized insectivorous minnows and darters (8)
   D. Percent biomass of insectivorous cyprinids (22, 26, 27)
   E. Number of juvenile trout (23)
   F. This metric deleted from IBI (3, 10, 13)

9. Percent of individuals that are top carnivores or piscivores (1, 5, 7—9, 11, 12, 15—20, 26, 28—31)
   A. Percent of individuals that are large (>20 cm) piscivores (1, 5, 7—9, 11, 12, 15—20)
   B. Percent biomass of top carnivores (22, 27)
   C. Percent catchable salmonids (23)
   D. Percent catchable trout (23)
   E. Percent pioneering species (23)
   F. Density of catchable wild trout (23)
   G. Proportion as specialized carnivores (24)
   H. Percent individuals as lentic piscivores—coldwater (26)
   I. This metric deleted from IBI (2—4, 5, 13—15)

Abundance and Condition Metrics

10. Abundance or catch per unit effort of fish (1—8, 10—11, 14—15, 18, 19, 22, 26, 27, 29, 31)
   A. Catch per unit effort of fish excluding tolerant species (9, 16, 20)
   B. Biomass of fish (6, 13, 22)
   C. Biomass of amphibians (13)
   D. Density of macroinvertebrates (13)
   E. Density of individuals (23)
   F. Number individuals as lotic residents (26)
   G. Number of individuals as lentic fishes (26)
   H. Number of warmwater individuals (28)
   I. This metric deleted from IBI (17)

11. Percent of individuals that are hybrids (1, 7, 8, 13, 26)
   A. Percent of individuals that are simple lithophilic species: spawn on gravel, no nest, no parental care (9, 15—17, 19, 20, 22—24, 27, 30, 31)
   B. Percent of individuals that are gravel spawners (18)
   C. Ratio of broadcast spawning to nest building cyprinids (22, 24)
   D. Percent introduced species (23)
   E. Number of simple lithophilic species (23)
   F. Number of native species (23)
   G. Percentage of native wild individuals (23)
   H. Percent ration of benthic to water column insectivores—coldwater (26)
   I. Number of coldwater individuals (28)
   J. Percent of species as lithophilic spawners (32)
   K. This metric deleted from IBI (2—6, 10—12, 14)

12. Percent of individuals that are diseased, deformed, or have eroded fins, lesions, or tumors (1—9, 11, 12, 14—16, 18, 19, 20, 22, 23, 26, 27, 29, 31)
   A. Percent of individuals with heavy infestation of cysts of the parasite *Neascus* (10)
   B. This metric deleted from the IBI (13, 17, 24, 28)

Chapter 40

Establishing Cause–Effect Relationships in Multistressor Environments

Joseph M. Culp[1], Adam G. Yates[2], David G. Armanini[2,3] and Donald J. Baird[1]

[1]*Environment and Climate Change Canada and Canadian Rivers Institute, Department of Biology, University of New Brunswick;* [2]*Department of Geography, Western University and Canadian Rivers Institute;* [3]*Prothea*

40.1 INTRODUCTION

Adverse pollution effects on riverine environments are invariably the result of a combination of stressors. This outcome is inevitable, as rivers receive multiple, interacting effluent discharges from municipalities and industries, and diffuse inputs from nonpoint sources (e.g., agriculture) and poses a challenge in establishing cause–effect relationships through standard field biomonitoring of rivers. This challenge is compounded by the fact that the strength and duration of effluent exposure is often poorly described or unknown. Adequate replication of stressor intensity along pollution gradients (e.g., effluent plumes) can also be difficult to achieve, due to the confounding effects of spatial heterogeneity within river habitats (Glozier et al., 2002). Expert reviews, such as those of the Canadian Environmental Effects Monitoring (EEM) programs, have concluded that unsuccessful field assessments can often be linked to the presence of multiple effluent discharges, interaction of contemporary effluent stressors with legacy effects of past pollution, and uncertainties regarding effluent exposure (Megraw et al., 1997). Moreover, Dafforn et al. (2016) highlight the importance of considering temporal (i.e., past, current, and future stressor effects, and their duration) and spatial scale (e.g., the ability to aggregate and integrate monitoring and other observational data from site to regional scales) in determining appropriateness of study design. Ultimately, approaches for ecological *causal assessment* should seek to identify the specific causes of undesirable ecological effects, and this is best supported by robust scientific evidence interpreted in the context of a well-posed questions generated by the prevailing assessment process (Cormier et al., 2015).

Stressor effects can be cumulative, in that repeated pulse or continuous press exposures to individual stressors will continuously degrade the health of organisms within river communities, altering ecosystem structure (e.g., through species loss) and function (e.g., productivity, nutrient processing, and retention), contributing to system-wide degradation. Culp et al. (2000a) identify three different categories of impacts that can retard the establishment of causal relationships in river ecosystems. *Incremental impacts* represent the additive effects of similar stressor events whose combined effect exceeds a critical ecological threshold. *Multiple source impacts* occur when sources of stressors and their effects overlap spatially. *Multiple stressor impacts* include scenarios where different classes of stressors interact in a synergistic or antagonistic fashion preventing a priori prediction of biotic responses. In this chapter, we focus on a combination of methodologies for investigating multiple stressor effects of effluent discharges.

Adams (2003) and, more recently, Norton et al. (2015a) evaluated the wide variety of approaches that researchers have implemented to establish cause-and-effect relationships between environmental stressors and biological response variables (i.e., response endpoints). The broad categories identified include laboratory toxicity tests, field bioassessments, field experiments, simulation modeling, and hybrid methodologies that combine aspects of two or more approaches. The integration of laboratory, field, and experimental manipulation studies is particularly useful for establishing causality and has been employed to assess the impact of heavy metals (Clements and Kiffney, 1994) and pulp mill effluents (Alexander et al., 2015) on the benthos of rivers. These integrated approaches have the advantage of using field observations to focus hypothesis generation for experimental studies that identify cause-and-effect relationships (Culp et al., 2015). Historically, data for such ecological causal assessment have been limited by the spatial scale and comparability of field assessments

(Buss et al., 2015). However, contemporary availability of large regional data sets and establishment of biological reference condition (Gerritsen et al., 2015), as well as technological advancements such as satellite earth observation, are poised to overcome such constraints and allow more comprehensive ecosystem-level assessments (Dafforn et al., 2016).

During the last decade, the number of publications reporting the results of field biomonitoring studies aimed at linking cause and effect in single and multiple stressor environments has increased substantially (e.g., Webb et al., 2015; Armanini et al., 2014; Kellar et al., 2014). A systematic analysis of existing literature (Norris et al., 2011), as well as consideration of observed local environmental stressors and biotic effects (Norton et al., 2015b), can provide an efficient approach to identifying plausible causal linkages. Additionally, Webb et al. (2015) have developed a software package to support the integration of available literature on causal effects linked to flow alteration using the ecoepidemiological approaches mentioned above. Such approaches provide a useful platform for developing the list of candidate causes of ecological effects.

Establishment of strong causal linkages between stressors and biological responses can be facilitated by integrating this diverse information through *weight-of-evidence* methodologies (Cormier et al., 2015; Suter et al., 2015). This concept incorporates an ecoepidemiological approach that evaluates the strength of the causal relationship by using a formalized set of criteria developed previously in the field of epidemiology. Several authors (Fox, 1991; Suter, 1993; Gilbertson, 1997; Beyers, 1998; Lowell et al., 2000) forwarded weight-of-evidence postulates that provide logical guidelines for establishing causation in *ecological risk assessment*. These postulates were incorporated by the USEPA into a stressor identification protocol (USEPA, 2000; Norton et al., 2015b). The multiple criteria proposed by various authors are listed by Adams (2003) as seven assembly rules that can be consistently applied in studies of ecological risk assessment. These criteria are outlined in Table 40.1 and include (1) strength of association; (2) consistency of association; (3) specificity of association; (4) time order or temporality; (5) biological gradient; (6) experimental evidence; and (7) biological plausibility. Essentially, this method requires that biological effects be associated with stressor exposure, plausible mechanisms that link cause and effect, and experimental verification of causality that is in concordance with available field evidence. Studies that include the methods described in this chapter will produce the baseline information needed for application of this ecoepidemiological approach to assess ecological risk of an effluent discharge.

40.1.1 Linking Field Biomonitoring to In Situ Bioassay Experiments

Field biomonitoring is an important component of contemporary impact assessment because field surveys can identify the biological responses to pollution. Nevertheless, field surveys alone cannot easily link cause and effect (Adams, 2003), which is why we combine biological assessment with field experiments. Most industrial and municipal effluents contain an array of compounds, and their effects on aquatic organisms can be stimulatory as a result of nutrient enrichment, or inhibitory because of contaminant toxicity. *In situ bioassays* are useful tools in this regard as their application can help researchers tease apart the ecological effects of nutrients from those of contaminants. An important consideration for bioassay application is the importance of selecting test species because the primary aim is to demonstrate a connection

TABLE 40.1 Formalized set of causal criteria forming part of a weight-of-evidence approach for ecological risk assessment. Causal criteria are modified from Adams (2003).

Causal Criterion	Support for Criterion
Strength of association	Cause and effect are observed to coincide. Many individuals/species are affected in the exposure relative to the reference area
Consistency of association	An association between a particular stressor or stressors and an effect has been observed by other investigators in similar studies at other times and places
Specificity of association	The observed effect is diagnostic of exposure
Time order or temporality	The cause (i.e., the stressor agent) precedes the observed effect, *and* the effect is observed to decrease when the stressor agent is observed to decrease or disappear
Stressor gradient	An incremental change in the strength of an observed effect is observed to coincide with an incremental increase or decrease in a specific stressor or stressors either spatially or temporally within the system
Experimental evidence	Experimental studies support the proposed cause-and-effect relationship
Plausibility	Credible ecological, physiological, or toxicological basis for the hypothesized mechanism linking the proposed cause and effect

between stressor exposure and biological effect under field conditions (Liber et al., 2007). This selection process may be aided by considering biological traits (i.e., phenotypic or ecological characters of an organism) that identify cause–effect linkages between environmental stressor and ecological effect (Rubach et al., 2010; Statzner and Bêche 2010; Culp et al., 2011). The incorporation of bioassays into field biomonitoring thus has the potential to improve the ability to separate multiple stressor effects and identify the cause of biological responses (Crane et al., 2007).

Nutrient diffusing substrates (NDS) are an ideal method for establishing the effects of effluents on nutrient limitation (Tank and Dodds, 2003; see also Chapter 31). Chambers et al. (2000) used NDS experiments to demonstrate that periphyton biomass was maintained at low levels by insufficient P upstream of point source discharges, while effluent loading from pulp mill and sewage inputs alleviated nutrient limitation downstream of major discharges. Others have deployed NDS bioassays throughout multiple river basins to draw broad conclusions about the *cumulative effects* of effluent discharges on algal standing crop (Scrimgeour and Chambers, 2000). Similarly, by performing NDS experiments in the autumn and winter, Dubé et al. (1997) demonstrated that the effects of nutrient additions from effluents varied seasonally. In these studies, the researchers developed their experimental hypotheses based on patterns observed initially in field surveys.

Field-based, toxicity bioassays are not a new approach—for centuries, miners used caged birds to detect the presence of carbon monoxide in tunnels. However, the use of caged organisms to detect aquatic pollution is a more recent development, and these methods are more sophisticated, involving the measurement of sublethal endpoints such as feeding behavior and growth, in addition to mortality (Burton et al., 2001; McWilliam and Baird, 2002; Crane et al., 2007; Agostinho et al., 2012). These in situ bioassays employ a wide variety of organisms and can provide useful insight into the ability of animals to perform their functional role within the ecosystem. Similarly, *body burdens* and biomarkers are useful endpoints for causal assessment because they provide evidence of contaminant exposure and effects (Suter, 2015). Body burdens of contaminants (e.g., metals, synthetic organic compounds) are widely used to assess effluent impacts (e.g., Gagnon et al., 2006; Allert et al., 2009; Smith et al., 2016). Application of biomarkers is currently limited by a lack of studies linking individual biomarkers to specific causal agents. Ongoing research to develop toxicity profiles for a wider array of contaminants and organisms will enhance the utility of biomarkers for causal assessment in the future (e.g., Hamers et al., 2013). In this chapter, we describe an in situ bioassay that integrates mortality, growth rate, and body burdens to evaluate the toxicity response of biota to effluent discharge.

40.1.2 Artificial Stream Approaches

In contrast to field biomonitoring or in situ bioassays, artificial stream experiments can control relevant variables and help isolate potential agents, such as nutrients or contaminants that cause the biological response. This research tool incorporates greater ecological complexity than is possible to include in laboratory toxicity tests and can generate important information on the chronic effects of pollutants on riverine communities. *Stream mesocosms* (i.e., artificial streams) vary widely in design, from simple laboratory systems to elaborate outdoor complexes (Lamberti and Steinman, 1993), and have been used to examine many levels of biological organization, ranging from single species to multispecies tests (Guckert, 1993). The use and application of stream mesocosms in ecotoxicology has been reviewed by many authors over the last several decades (Shriner and Gregory, 1984; Kosinski, 1989; Guckert, 1993; Pontasch, 1995; Culp et al., 2000d, 2015). Integration of stream mesocosm studies with field biomonitoring is a particularly beneficial approach for retrospective ecological risk assessments (Suter, 1993) and has been used to generate weight-of-evidence risk assessments for large rivers (Culp et al., 2000b).

In this chapter, we combine artificial stream use with field biomonitoring and in situ bioassays to better understand the relationship between exposure to specific pollutant sources and the resulting ecological consequences. The specific objectives of this chapter are to (1) outline a basic methodology for assessing the biological responses to an effluent discharge (i.e., a point source); (2) demonstrate how NDS bioassays can be used to evaluate the effect of effluents on nutrient limitation of algal biomass (Advanced Method 1); (3) illustrate the usefulness of field-based toxicity bioassays to determine contaminant effects of effluents (Advanced Method 2); and (4) introduce the use of stream mesocosms as an approach to establish causality between stressors and biological effects (Advanced Method 3). Together, the different methods provide a process through which key stressors are identified, ecological effects are measured, and subsequent investigations into the cause of effects are conducted.

40.2 GENERAL DESIGN

This chapter describes methods and a conceptual framework for evaluating the effects on benthic communities of a point source effluent discharge that contains both nutrients and contaminants. Environmental assessment proceeds sequentially from basic to advanced approaches such that studies that complete the sequence will be able to use weight-of-evidence criteria to draw conclusions about the cause of the measured biological effects. The tiered approach follows a decision tree of logic similar to the Canadian EEM program for metal mining (Glozier et al., 2002) and methods of causal analysis outlined by Burton et al. (2001).

Under this approach, the questions posed within each assessment tier become progressively more focused with the ultimate goal of identifying the cause of the measured biological effects. In the first tier, the existing information for the effluent exposure site is summarized including the identification of potential reference and exposure areas, review of historical data on water quality and effluent composition, and consideration of existing benthic invertebrate data. The effluent constituents are characterized, key stressors are identified, and effluent dilution in the exposure area is estimated. The second tier determines if the benthic invertebrate community is affected within the immediate vicinity of the effluent discharge. The most likely cause of the ecological effects is determined by comparing the ecological effects data with the list of major stressors in the effluent. Within this assessment tier, the research can also identify the spatial extent of the effect by locating additional exposure sites downstream. The third tier aims to produce mechanistic understanding of the responses of the benthic invertebrate community to the key stressors by conducting one or more field experiments chosen from a wide variety of available assessment tools. The application of explicit weight-of-evidence criteria to the information collected within each tier will strengthen cause-and-effect understanding of stressor and ecological effects relationships.

40.2.1 Site Selection

Third- to fourth-order streams, which have cobble and pebble substrate and are wadeable, are ideal for the application of these methods. However, the approach is also appropriate for larger rivers when riffle areas along the shoreline can be safely waded. Reference and exposure habitats must be comparable if the ecological effects of the effluent stressors are to be separated from natural habitat variability. The reference area should be free from effluent exposure. Care must be taken to choose reference and effluent exposure sites with similar habitat features including stream channel geometry (i.e., bankfull width and depth, channel gradient; see Chapter 2), substrate particle size, current velocity, discharge (see Chapter 3), and riparian vegetation (see Chapter 28). Comparability of catchment physiography and land use, as well as stream and catchment size (see Chapter 1), should also be carefully considered when circumstances necessitate the use of reference sites located on streams other than that for which effluent effects are to be assessed.

Your choice to examine the ecological effects on the benthic community of a particular effluent discharge will be based on the assumption that the effluent contains potentially deleterious contaminants or nutrients that may cause unacceptable environmental effects. Your assumption can be strengthened through qualitative observations such as changes below the effluent discharge in plant biomass, the composition of the benthic community, or records of acute events such as fish kills. In the absence of such observations, you can choose to examine the effects of a particular discharge based on public concern that the point source is causing ecosystem impairment. Regardless of the pathway of problem initiation, the objective of the ecological assessment will be to determine if the effluent discharge is having an effect on the benthic invertebrate community (measured as a statistical difference between the reference and exposure areas in a control versus impact study design).

40.2.2 General Procedures

Ecological effects of the effluent discharge for the *Basic Method* will be assessed by changes in algal biomass [measured as chlorophyll *a* (Chl *a*)] and several benthic invertebrate endpoints [i.e., total numerical abundance, taxonomic richness, evenness, Bray–Curtis (BC) index]. Most field sampling techniques and laboratory analyses incorporated in the basic and advanced methods are described in detail elsewhere in this book. Methods for the collection and processing of benthic invertebrate samples are fully described in Chapter 15. Sample collection and laboratory processing of Chl *a* is detailed in Chapter 12, while water collection and analytical methods for nutrient analysis (N, P) are covered in Chapters 32 and 33. Analysis of water samples for general ions and metals, and analysis of the full-strength effluent for nutrients and contaminants (i.e., metals, acids and derivatives, phenols, alcohols, aldehydes and ketones, hydrocarbons) will best be done by a commercial lab following methods outlined in Rice et al. (2012).

The needs of the researcher will determine which of the advanced methods (i.e., 1–3) should be applied. However, by combining two or more of the techniques, you will be able to assign any measured effects to the presence of nutrients or contaminants in the effluent. The NDS technique (Capps et al., 2011; see also Chapter 31) can be used to detect whether the effluent modifies nutrient limitation in the river. We use in situ bioassays to measure sublethal toxicity effects. Artificial stream techniques are excellent methods for establishing cause-and-effect relationships as the researcher can control relevant environmental variables and separate the effects of multiple stressors in the effluent (e.g., nutrient versus contaminant effects).

40.3 SPECIFIC METHODS

40.3.1 Basic Method: Retrospective Ecological Risk Assessment of an Effluent Discharge

This method employs a conceptual framework for assessing environmental effects that is modified from Burton et al. (2001) and Glozier et al. (2002). The approach evaluates existing databases to help focus the assessment and analyzes effluent composition to identify possible nutrient and contaminant stressors of concern. Following this sequence of problem definition, site characterization, and identification of potentially important stressors, the method extends to assessment of ecological effects by comparing benthic communities at reference and exposure sites. Such field assessments can adequately measure environmental impacts on biological communities and, further, can suggest a potential cause of the ecological effect (e.g., excessive nutrient addition). Nevertheless, it is difficult to establish cause-and-effect relationships unequivocally with this type of field bioassessment design because the approach relies on statistical inference (Stewart-Oaten et al., 1992; Cooper and Barmuta, 1993). For example, the limited knowledge of the concentration and duration of stressor exposure and excessive spatial heterogeneity complicates the assignment of causality. The advanced methods below provide further approaches that produce information to link cause with ecological effect.

The primary objective of the ecological assessment will be to assess quantitatively if the effluent discharge is affecting the benthic invertebrate community. This type of effects examination is termed *retrospective ecological risk assessment* because the effluent pollution began in the past and likely has ongoing consequences (Suter, 1993). Effects will be measured as a statistical difference between the reference and exposure areas in a control versus impact study design. You must complete the steps of site characterization, field assessment of ecological effects, and data interpretation to complete the basic retrospective effects assessment.

40.3.1.1 Site Characterization

A principal role of the site characterization step is to gather sufficient information to identify potential exposure and reference areas that have similar habitat characteristics (Table 40.2). The upstream catchment area(s) should also be assessed when data are available to assist in identification of additional pollutant sources, both point and nonpoint, that could mask effluent effects or interact with the effluent to influence ecological responses to the effluent. The six steps below of site characterization require you to collect catchment and habitat information for the sites, assess the quality of previous benthic invertebrate data, estimate the concentration of effluent in the exposure area, examine the effluent constituents, and produce a list of stressors that may cause ecological impairment.

TABLE 40.2 Site characterization information for the retrospective assessment of a point source discharge.

Information Category	Reference and Exposure Site Descriptors
Catchment characteristics	Land use; physiography; catchment area; stream order; tributaries; additional point sources
Physical characteristics	Reach gradient; substrate particle size composition; mean annual discharge and range; mean annual water temperature and range
Effluent treatment	Summary of effluent type and description of the treatment process
Effluent constituents	Summary of the major constituents of the effluent including nutrients, general ions, metals, and contaminants (i.e., acids and derivatives, phenols, alcohols, aldehydes, ketones, hydrocarbons)
Effluent mixing	General description of how the focal effluent mixes with the receiving water with a semiquantitative or quantitative estimate of effluent concentration at the exposure site
Other effluent discharges	Location and description of other effluent discharges (e.g., stormwater outfalls)
Species at risk	Summary of any rare, threatened, or endangered aquatic species in the study area

1. Collect information on the stream catchment (i.e., physiography (e.g., soil types, topography), land use, catchment area, stream size, tributary location) using a geographic information system (GIS) following the protocols in Chapter 1. Identify and describe any pollutant sources (e.g., agricultural lands, sewer outfalls, forest harvesting) located in the stream catchment that could mask effects of the effluent source being assessed. Many of these sources can be identified using publically available GIS layers generated by government sources or by visually analyzing recent, remotely sensed images of the catchment.

2. Collect information on the channel geometry (i.e., bankfull width and depth, channel gradient) and substrate particle size following the protocols in Chapter 2. Measure the current velocity and discharge at the potential sample sites following methods in Chapter 3. Note any other inputs to the exposure or reference areas (e.g., stormwater, sewer outfalls). Identify and describe all other factors, either anthropogenic or natural (e.g., riparian vegetation), that are not related to the effluent under study and that might confound the comparison of observed differences in effects measures between the reference and exposure sites. Much of this information may be available in government reports or other public domain sources that describe earlier environmental assessments of the discharge. Briefly summarize the production processes that contribute to the effluent source (e.g., type of industrial facility), any effluent treatment processes employed by the discharger, and the mean daily amount of effluent discharged to the receiving water. Use the above information to justify your choice of a reference site above the influence of the effluent discharge and an exposure site 500–1000 m below the effluent outfall. In this example you will use a simple *Control* versus *Impact* study design. However, more complicated designs that incorporate additional reference and exposure sites are available (see Glozier et al., 2002).

3. Summarize the available benthic invertebrate community data collected during previous environmental assessments. Determine the adequacy of the historical data set by assessing the quality assurance and quality control methods employed for the field and laboratory procedures. This review will list the sampling device and mesh size used in the earlier study, and determine whether the study design and replication was appropriate. Assess whether the quantity or physicochemical quality of the discharge has changed since the previous study. If the data sets are of good quality and the effluent discharge is unchanged, then these data can be used to focus your study question to one or more of the advanced methods. If historical benthic invertebrate data do not exist or are of poor quality, or the effluent has changed substantially, then you will conduct a field assessment following the protocol described below.

4. Estimate the concentration of effluent at the exposure site area by conducting a conductivity survey in the field using a standard conductivity meter. This survey will measure the conductivity in the full-strength effluent (C_e), at the reference site upstream of the effluent (C_u), and at the downstream exposure site (C_d). Conductivity measurements from the field survey can be converted into relative effluent concentrations (C_r) ranging from 1 (full-strength effluent) to 0 (background) by applying the following formula:

$$C_r = \frac{C_d - C_u}{C_e - C_u} \tag{40.1}$$

where C_e = effluent conductivity (μS/cm), C_u = reference site conductivity (μS/cm), C_d = exposure site conductivity (μS/cm), and C_r = relative concentration (i.e., dilution ratio) of the effluent at the exposure site. If a conductivity survey is not undertaken, an indication of effluent concentration at the point of complete mixing can be estimated by dividing the mean daily discharge of effluent by mean daily stream discharge during the period proposed for the benthic invertebrate survey.

5. Collect a sample of the full-strength effluent, and analyze the sample for major ions, nutrients, and contaminants (i.e., metals, acids and derivatives, phenols, alcohols, aldehydes and ketones, hydrocarbons). Requisite sample volume is typically at least 1 L, and replicate water samples are advised. Effluent chemistry is best done in a commercial lab following methods outlined in Rice et al. (2012).

6. Make a candidate list of the nutrient and contaminant stressors that are most likely to cause impairment in the benthic invertebrate community. Estimate the stressor concentration (S_c) in the exposure environment using the following formula:

$$S_c = C_r \cdot S_i \tag{40.2}$$

where C_r = dilution ratio, S_i = concentration of the nutrient or contaminant stressor in the full-strength effluent, and S_c = estimated concentration of the stressor at the exposure site. Compare the estimates for the environmental concentration of nutrients and contaminants to recommended water quality guidelines (e.g., CCME, 1999; USEPA, 2002). S_c estimates that exceed these guidelines should be considered as contributors to potential causes of impairment. Further discussion on identification of stressors can be found in USEPA (2000).

40.3.1.2 Field Assessment of Ecological Effects

1. Collect five replicate samples of benthic invertebrates from the reference and exposure sites following the quantitative methods outlined in Chapter 15. Transfer each sample from the collection net to a storage container that is clearly labeled (site, replicate number, date, sample collector's name). Preserve the sample in 95% ethanol (final concentration in container of ~70% ethanol). Place a permanent paper label (i.e., written in pencil) inside the sample container. It is best to process samples for retrospective analysis in the laboratory.
2. Collect five replicate periphyton samples from cobble substrate, selected haphazardly, by using a scalpel to scrape the periphyton from within a 10-cm^2 area. Caps from 50-mL centrifuge tubes are ideal templates for delimiting the sample area, either by etching the rock with a pencil or by scraping away the periphyton from outside the cap to reveal the sampling area. Other simple quantitative samplers are described in Chapter 12. Transfer the periphyton sample to a labeled vial or bag, and store the samples on ice in the dark until they can be processed in the laboratory (preferably within 4 h).
3. Collect water samples for measurement of major ions and nutrients, store the samples on ice in the dark, and analyze in the laboratory (nutrients) or ship immediately to an analytical laboratory (major ions, metals, etc.) for processing.

40.3.1.3 Laboratory Methods

1. Follow the methods of Chapter 38 to sort and identify the benthic invertebrates to the taxonomic level of family. Record the abundance of each family for each sample.
2. Determine Chl *a* abundance for each periphyton sample following the protocol in Chapter 12.
3. Process water samples according to methods in Rice et al. (2012).

40.3.1.4 Data Analysis

1. Calculate the total invertebrate density (i.e., total number of individuals in all taxa) for each sample expressed per unit area (numbers/m^2).
2. Calculate the taxonomic richness (i.e., total number of different taxa) for each sample (i.e., family richness/unit area sampled).
3. Calculate evenness (*E*) for each sample as:

$$E = \frac{1}{\sum_{i=1}^{S} (p_i)^2 / S} \tag{40.3}$$

where p_i = proportion of the *i*th taxon in the sample and S = total number of taxa in the sample.
4. Calculate the BC index to estimate the dissimilarity of each sample from the median taxonomic density at the reference site. The BC index is a distance coefficient that reaches a maximum value of 1 for two sites that are entirely different and a minimum value of 0 for two sites that possess identical taxonomic composition. The BC index measures the difference between sites and this is calculated as:

$$\text{BC} = \frac{\sum_{i=1}^{S} |y_{in} - y_{ir}|}{\sum_{i=1}^{S} (y_{in} + y_{ir})} \tag{40.4}$$

where BC = Bray—Curtis distance between sample *n* and the reference median, y_{in} = count for taxon *i* at site *n*, y_{ir} = median count for taxon *i* at the reference site, and S = total number of taxa present at site *n* and the reference sites. Tables 40.3—40.5 illustrate the steps for calculating the BC distance between each sample and the median taxon density at the reference site (note that Excel spreadsheets for Tables 40.3—40.5 are available with the online website for this book). First, determine the median taxon density at the reference site (see Table 40.3A). A similar table is constructed for the exposure stations without the median calculation (Table 40.3B). Calculate the distances between each sample and the reference median following the example illustrated in Table 40.4; repeat this procedure for all reference and exposure site samples. For this approach, the reference median for taxon *i* becomes y_{ir} in Eq. (40.4). To determine whether there is an effect at the exposure site, the mean BC distance between the reference stations and the reference median (0.18 ± 0.06 in Table 40.5) is compared to the mean distance between the exposure stations and the reference median (0.43 ± 0.03 in Table 40.5).
5. Compute the mean and standard error (SE) for each of the four endpoints above. Determine if there is a statistical difference between the reference and exposure sites for each of the four endpoints using a *t*-test to compare the two

TABLE 40.3 Examples of (A) determination of the median taxon density at the reference site from a study design with five replicate samples; and (B) taxon densities at the exposure site without the median calculation. See also Online Worksheet 40.3.

(A)	Taxon Density (numbers/m^2)				
Reference Sample Number	Taxon 1	Taxon 2	Taxon 3	Taxon 4	Taxon 5
Reference 1	2	3	2	3	1
Reference 2	3	5	2	4	3
Reference 3	9	1	1	1	1
Reference 4	4	6	3	4	1
Reference 5	5	4	2	3	2
Reference site median value	4	4	2	3	1

(B)	Taxon Density (numbers/m^2)				
Exposure Sample Number	Taxon 1	Taxon 2	Taxon 3	Taxon 4	Taxon 5
Exposure 1	23	4	2	10	1
Exposure 2	12	2	2	8	3
Exposure 3	14	6	1	6	2
Exposure 4	13	1	3	12	2
Exposure 5	15	3	2	4	1

TABLE 40.4 Example calculation of Bray–Curtis (BC) Index for a reference sample. Values are illustrated for y_{in} of the reference sample 1, y_{ir} of the reference median, and subsequent calculations of $|y_{i1} - y_{ir}|$, $(y_{i1} + y_{ir})$, and BC. See also Online Worksheet 40.4.

Variable	Taxon 1	Taxon 2	Taxon 3	Taxon 4	Taxon 5
Reference 1 (y_{i1})	2	3	2	3	1
Reference median (y_{ir})	4	4	2	3	1
$|y_{i1} - y_{ir}|$ or reference 1 − reference median	2	1	0	0	0
$(y_{i1} + y_{ir})$	6	7	4	6	2

Substituting into Eq. (40.4):

$$BC = \frac{2+1+0+0+0}{6+7+4+6+2} = \frac{3}{25} = 0.12$$

where BC is the dissimilarity value between the taxonomic composition of reference sample 1 and the median composition for the reference site.

sample means (see Zar, 2010 or Chapter 38 for a detailed statistical procedure). Record the mean ($\pm$1SE) values of the four endpoints (total abundance, richness, evenness, BC index) for the reference and exposure sites in a data table.

40.3.2 Advanced Method 1: Determining Nutrient Limitation Using Nutrient Diffusing Substrate Bioassays

Effluent discharges often contain nutrients that can stimulate food web productivity and contaminants that can lead to various levels of toxicity. This exercise employs NDS bioassays to determine whether the effluent discharge modifies nutrient limitation in the receiving waters. This simple and inexpensive technique follows the method outlined in Chapter

TABLE 40.5 Example calculation of the Bray–Curtis (BC) distance from the reference median to each reference or exposure sample. See also Online Worksheet 40.5.

Reference or Exposure Site Sample Number	$\|y_{i1} - y_{ir}\|$	$(y_{i1} + y_{ir})$	Sample BC Distance to Median	Mean (±SE) BC Distance to Median
Reference 1	3	25	0.12	0.18 ± 0.06
Reference 2	5	31	0.16	
Reference 3	11	27	0.41	
Reference 4	4	32	0.13	
Reference 5	2	30	0.07	
Exposure 1	26	54	0.48	0.43 ± 0.03
Exposure 2	17	41	0.41	
Exposure 3	17	43	0.40	
Exposure 4	23	45	0.51	
Exposure 5	13	39	0.33	

31 (also see Capps et al., 2011). NDS should be placed in the stream in uniform conditions of depth, current velocity, and light. NDS units are then retrieved after a 21-day incubation period in the stream and processed for Chl a.

1. Build the NDS following Basic Method 1 described in Chapter 31, and prepare five replicates of each treatment (i.e., control, N, P, N + P; 5 replicates × 4 treatments × 2 sites = 40 containers). Use an NDS preparation sheet as in Chapter 31 (Table 31.2).
2. Place the NDS in the field according to Basic Method 1 in Chapter 31. Repeat the placement procedure until all four replicates are deployed at the reference and exposure sites.
3. Retrieve the NDS after 21 days, and remove the disk from each container as in Basic Method 1 in Chapter 31. Place the disk in a labeled plastic bag, and store the sample on ice in the dark until frozen (within 4 h). Chl a should be processed following the protocol in Chapter 12.
4. Use a two-factor ANOVA (factors of N and P levels) to test whether algal biofilms were significantly affected by the treatments. Possible interpretations of the responses to nutrient treatments are provided in Chapter 31 (Table 31.1).

40.3.3 Advanced Method 2: In Situ Determination of Sublethal Effects of the Effluent

The bioassay described below uses the freshwater crayfish, *Orconectes virilis* (Hagen), obtained from reference streams. If this species is not available, then another crayfish species can be substituted (see below). Using crayfish offers key advantages for bioassay procedures. First, they are large-bodied organisms that are widely distributed in streams around the world and hold an intermediate trophic position where they facilitate key ecological processes. Crayfish are also easily collected in the field or cultured in laboratory settings (or may be purchased from biological supply companies). Second, and perhaps most importantly, their biology and responses to a wide range of toxic substances are well documented (Lodge et al., 1995; Schilderman et al., 1999; Allert et al., 2009; Vioque-Fernandez et al., 2009; Kouba et al., 2010).

This bioassay assumes that the user has access to *O. virilis* at reference streams; however, any well-described crayfish species indigenous to the study region may be substituted. Nonindigenous species should not be used in bioassays to avoid unintended species introductions. Information on the collection and/or culturing of crayfish is available from a wide range of literature sources (e.g., Allert et al., 2009) and is not covered further here. The crayfish bioassay employs the traditional endpoints of mortality and growth, as well as the measurement of contaminant body burden. Using multiple endpoints with varying levels of sensitivity has the advantage of allowing evaluation of a wide range of scenarios from gross pollution to less obviously polluted sites. The experimental design described below requires 16 replicate bioassay cages at the reference and exposure sites.

1. *O. virilis* can be easily collected in reference streams using baited minnow traps. Minnow traps can be effectively baited using commercially available dog food pellets placed in mesh bags. Baited traps should be deployed in areas known to be inhabited by crayfish with the trap resting on the bed of the stream. Light rope or chain can be used to secure the trap to a tree or anchor stake on the stream bank. Deployment of several (5–10) traps is recommended to increase trapping efficiency. Baited

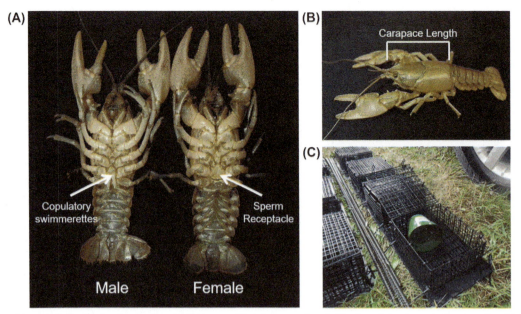

FIGURE 40.1   Diagrams indicating key features for establishing sex (A) and carapace length (B) of crayfish prior to deployment. Crayfish bioassays are conducted in wire cages with a flower pot added for crayfish shelter (C).

traps should be left overnight and retrieved the following morning. Place the retrieved traps in an aerated container or bucket for opening. Remove all crayfish, and return unwanted taxa (e.g., fish and other invertebrates) to the stream.

2. Collected crayfish should be sexed by exposing the underside of the crayfish and observing if a small port (female) or an extra set of swimmerets (males) is present (Fig. 40.1A). To account for inherent differences in the physiological profiles of males and females, only one sex should be retained. Once all animals have been sexed, randomly allocate the retained individuals to the reference site(s) and exposure site. Transport the individuals allocated to the exposure site.

3. At each field site, prepare cages made of 4-mm wire or plastic mesh. Commercially available bird suet feeders (21 cm L × 18 cm W × 10 cm H) wrapped in 1-cm plastic mesh can be used as the structure of the cage (Fig. 40.1C). To each cage, add small cobbles and/or a small flower pot to serve as shelter for the crayfish. Label each cage with a unique identification code.

4. Randomly select an individual crayfish from the transportation container, and weigh and measure the crayfish. Crayfish should be weighed using a conventional field balance (to the nearest 0.01 g) by placing the crayfish in a preweighed plastic container. Determine the carapace length using a pair of vernier calipers. Carapace length is typically measured from the tip of the rostrum to the end of the carapace (Fig. 40.1B). Following weighing and measuring, place the crayfish in a cage, and record the cage identification number being sure to link this number with the weight and length measurements. Place a single individual in each cage to eliminate competitive, conspecific interactions. Close the cages, and secure with cable ties. Repeat the process until all crayfish have been measured and assigned to a cage.

5. Place the cages in the stream at each of the sites. In streams with coarse substrate (gravel or cobble), cages should be partially embedded in the substrate by removing the top layer of substrate. In softer bottom streams (i.e., silt and fine sands), we recommend fixing cages to a sheet of plexiglass to prevent settling into the streambed. In both instances, cages should be secured to the bed using cable ties and rebar (or bricks) and nylon cord. Rebar can be driven into the substrate next to the cage, and cable ties are used to secure the cage to the rebar. In harder substrate a brick can be tied to the base of each cage to anchor cages to the streambed.

6. Animals are exposed in situ for a 2-week period. Check and clean cages regularly during the exposure period to avoid fouling that may impede water flow to limit effects on contaminant absorption, sediment, and waste accumulation, as well as physicochemical conditions (e.g., dissolved oxygen concentration).

7. After 2 weeks, retrieve the cages containing the crayfish. Place the individual cages in a water-filled bucket for opening. Remove the crayfish from the cage, record the cage identification number, and weigh and measure the crayfish as described in step 4. Record any mortality. Freeze the surviving crayfish on site using a portable freezer or dry ice, and transport to the laboratory for body burden analysis.

8. Tissue samples should be analyzed for body burdens in the laboratory using standard techniques for the contaminants of interest (e.g., trace metals and synthetic organic compounds). Because analytical methods vary widely depending on

the contaminant(s) of interest, detailed descriptions of analytical methods are beyond the scope of this chapter. Readers are directed to Crawford and Luoma (1993), Krahn et al. (1988) and Rice et al. (2012), among others, for more information on contaminant analyses.

9. Determine any statistically significant differences in mortality, growth (calculated as difference in lengths and weights at beginning and end of exposure), or contaminant body burden of animals placed at the impacted site and reference site. Compare single sites upstream and downstream of the effluent by conducting a one-tailed t-test (e.g., H_0: growth upstream $\leq$ growth downstream; H_A: growth upstream $>$ growth downstream). Alternatively, multiple sites can be compared using an ANOVA design, with differences between individual "impacted" sites and reference sites tested using a post hoc Bonferroni test (Zar, 2010).

40.3.4 Advanced Method 3: Separating Nutrient and Contaminant Effects on Benthic Food Webs

Experimentation using artificial streams complements biomonitoring and in situ bioassay studies because artificial streams provide control over relevant environmental variables and allow for the separation of multiple stressors contained within complex effluents (Pestana et al., 2009; Piggott et al., 2015). In this example, we describe a general method that can assign cause-and-effect definitively by isolating the effects of effluents on metrics of community structure and ecosystem processes. A detailed description of artificial stream design is beyond the scope of this chapter, and the reader is directed to the vast literature on this topic (Shriner and Gregory, 1984; Kosinski, 1989; Lamberti and Steinman, 1993; Pontasch, 1995; Culp et al., 2000d, 2015). The response variables (i.e., endpoints) can encompass multiple trophic levels (i.e., periphyton and benthic invertebrate assemblages) and ecosystem processes (primary production and decomposition). The example below uses periphyton growth rates (measured as change in Chl a concentration, composition of periphyton and benthic macroinvertebrate communities, and decomposition rate endpoints).

1. Choose an appropriate artificial stream design. The simplest systems use flow-through troughs located alongside the stream to which reference water is continuously delivered from a head tank. Reference water can also be obtained from local groundwater sources or other reference streams provided the water has limited nutrient concentrations and is free of the contaminants of concern. The head tank can be filled by pumps or through piping that siphons water from upstream. A simple, inexpensive, and nonelectrical release apparatus is the Mariotte bottle (Webster and Ehrman, 1996); however, battery-powered metering pumps (e.g., Fluid Metering, Inc. Syosset, NY, United States) are generally more reliable, and pump rates can be easily adjusted to field conditions. Chapter 31 describes pumping procedures in detail.

2. The basic experimental design follows Culp et al. (2000c, 2003) and includes at least four replicates of each of three treatments: (1) raw reference water collected upstream of effluent discharge; (2) an N + P treatment simulating the nutrient concentration at the exposure site; and (3) an effluent treatment simulating the exposure site concentration of effluent (Fig. 40.2). Generally, the exposure concentration for the effluent and nutrient treatments is targeted to simulate the dilution ratio that corresponds to the stream reach where the effluent becomes completely mixed.

3. Effluent for the experiments should be collected from the effluent treatment system just prior to discharge to the river. Strictly follow all necessary safety precautions, including the use of personal protective equipment (PPE), when handling effluent or other chemical solutions. N and P estimates for the full-strength effluent and the dilution ratio determined previously should be used to determine the N and P concentrations in the treatment additions.

4. To create a standardized benthic environment in the artificial streams, it is necessary to first deploy trays of substrate at the reference site for colonization by the natural periphyton and invertebrate communities. This standardized, colonized material will later be transferred to the experimental streams. The substrate used should be consistent with the grain size distribution of the exposure and reference reaches and can be purchased from local aggregate companies. The volume of substrate purchased should be based on the amount required to cover the surface area of the artificial channels. Deploy the substrate at the reference site using common garden trays lined with 250-μm mesh to aid in retaining invertebrates on collection. Anchor the trays to the streambed. Retrieve the trays from the reference site following 3—4 weeks of colonization. Place a kick net immediately downstream of the tray to catch invertebrates that may drift from the trays. Add the colonized substrate to the artificial stream channels.

5. Deploy 26 unglazed ceramic tiles (5×5 cm) in each of the artificial stream channels for algal colonization during the experiments. Tiles should be randomly distributed within the artificial stream. Randomization can be achieved by marking the underside of each tile with a unique identification code using a permanent marker and mapping the position of each tile in the channel. Use a random number generator to randomly select tiles for collection throughout the experiment. Use of tiles simplifies and standardizes collection of periphyton samples (see Lamberti and Resh, 1985).

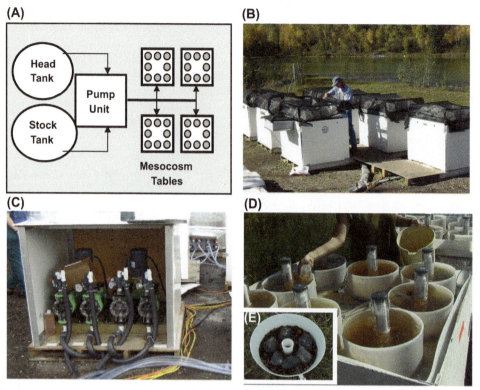

FIGURE 40.2 Mesocosm system used by Culp et al. (2003). (A) Schematic overview of mesocosm system with *arrows* indicating the direction of water flow; (B) photograph of mesocosm system deployed along the Wapiti River, AB, Canada; (C) pumping unit containing four positive displacement, reciprocating pumps (Pulsa Feeder© Series Model 25 H) that transfer water from source tanks to the mesocosm tables; (D) mesocosm table containing eight artificial streams; and (E) an artificial stream with gravel and cobble substrate prior to experiment initiation.

6. Deploy 8 cotton strips prepped as described by Tiegs et al. (2013) in each of the artificial streams. In brief, cut a 2.5 × 8 cm strip of cotton canvas from a bolts of Fredrix-brand, unprimed 12-oz. heavy-weight cotton fabric, Style #548 (Fredrix, Lawrenceville, GA, United States) using a rotary cutter and fray the edges. Gently insert a large pin or small nail between the threads approximately 1 cm from one end of the cotton strip. Feed a thin cable tie through the gap in the threads. Attach a large metal washer to the cable tie to serve as an anchor. Deploy the cotton strips throughout the channel.

7. Allow reference water to flow through the systems for 1 day prior to beginning the nutrient and effluent additions. Monitor the artificial stream system each day throughout the 4-week exposure to ensure that water delivery and flow rates are maintained and that effluent is continuously delivered at the desired concentrations.

8. Following 1 week of exposure, randomly select and collect three tiles from each of the artificial streams. Remove periphyton from each tile using methods described in Chapter 12 and analyze for Chl *a* concentration. Repeat this process every 3 or 4 days until the end of the experiment.

9. At 4 weeks of exposure, collect the remaining eight tiles. Process three tiles for Chl *a* (see Chapter 12) and the remaining five tiles for community composition (see Chapter 11). At the same time, collect the eight cotton strips and clean with a small brush in an ethanol bath following the methods described in Tiegs et al. (2013). After removal of tiles and cotton strips, wash all the substrate particles in buckets partly filled with water to separate the benthic macroinvertebrates from the substrate. Following washing, pour the water through a 250-μm mesh sieve or net and preserve the collected invertebrates as described above. If a smaller substrate has been used, such as fine gravel, samples can be elutriated by placing the substrate in a bucket partly filled with water, gently swirling the bucket, then pouring off the separated organic materials into a 250-μm mesh sieve or net prior to preservation

10. Following processing of the samples, calculate the rate of change in Chl *a* concentration (as a surrogate of growth rate) between each sampling interval. Dry the cotton strips at 40°C, and assess tensile strength using a tensiometer mounted on a motorized test stand as described in Tiegs et al. (2013). Calculate the loss of tensile strength (CTSL, sensu Slocum et al., 2009) using Eq. (40.5) as a metric of decomposition:

$$\mathrm{CTSL} = (1 - (N/C)) \times 100 \tag{40.5}$$

where N = strength of the experimental strip in Newtons and C = mean tensile strength of the reference strips.

11. Enumerate and identify algal cells and benthic macroinvertebrate taxa using protocols described in Chapters 37 and 38, respectively. Use the algal and macroinvertebrate data to calculate community metrics described in Chapters 37 and 38. Use one-way ANOVA to determine if the treatments significantly affected the community and ecosystem process endpoints.

12. Review the set of causal criteria in Table 40.1, and determine the weight-of-evidence you have accumulated from the use of field bioassessment, field bioassays, and artificial stream experiments. Following the approach of Lowell et al. (2000), create an interpretive table that integrates the results of the field survey, field bioassays, and artificial stream experiments (e.g., see Table 40.2 in Culp et al., 2000b). The advantages to applying these formalized causal criteria in this way include helping to tie together diverse assemblages of data on the effects of multiple stressors, thereby providing a weight-of-evidence approach to ecological risk assessment. Using the results of your weight-of-evidence assessment, determine the major effects of the effluent on the benthic food web. What is the most plausible mechanism that explains the changes observed in the benthic food webs exposed to effluent discharge?

40.4 QUESTIONS

40.4.1 Basic Method: Retrospective Ecological Risk Assessment of an Effluent Discharge

1. Why do channel geometry, current velocity, and discharge need to be standardized between the reference and exposure site? Are there other physicochemical variables that should be similar at these sites? Why? Should you proceed if sites are highly divergent in habitat characteristics?

2. How might an online macroinvertebrate database aid you in selecting appropriate reference and exposure sites? If historical data are available, determine whether the effluent quality and quantity have changed substantially since the earlier macroinvertebrate survey. Were the collection methods used in the previous survey acceptable for a quantitative analysis? (Consider factors including mesh size of the sampling net, laboratory quality assurance and control, replication, etc.)

3. Where is the zone of complete effluent mixing at the exposure site? What is the disadvantage of not undertaking a conductivity survey? What assumptions are made when you estimate effluent concentration in the river based on simply calculating the dilution ratio from the ratio of effluent to river discharge volume?

4. Which chemical compounds in the effluent are most likely to affect the biota at the exposure site? Do any of the chemical concentrations exceed water quality guidelines?

5. Which benthic invertebrate response variables are expected to be sensitive to water pollution? What can you conclude about the effect of the effluent on the stream biota? For example, did the effluent affect productivity or biodiversity?

40.4.2 Advanced Method 1: Determining Nutrient Limitation Using Nutrient Diffusing Substrate Bioassays

6. Is algal biomass limited by nutrient availability at the reference site?

7. Examine Table 31.1 in Chapter 31 and determine the possible interpretations of the responses to nutrient and effluent treatments.

8. Was nutrient limitation mitigated or otherwise reduced by the effluent discharge?

40.4.3 Advanced Method 2: In Situ Determination of Sublethal Effects of the Effluent

9. Was crayfish mortality significantly affected by exposure to the effluent? Were there significant effects of effluent exposure on growth? Was there any indication that the animals were exposed to high levels of suspended solids during the field experiment? If high levels of suspended solids were detected, how might this affect your results?

10. Were significant lethal and sublethal effects of effluent exposure related to contaminant body burden of crayfish?

11. Is the information gained from the field bioassays worth the extra effort in terms of improving your ability to interpret field biomonitoring results?

40.4.4 Advanced Method 3: Separating Nutrient and Contaminant Effects on Benthic Food Webs

12. What are the advantages and disadvantages of employing artificial stream designs in retrospective risk assessment? What are the limitations of restricting the experiment to a 28-day period?

13. Is the pattern of ecological effect similar for endpoints related to primary producers, decomposers, and secondary producers? If the patterns of effect are not the same, how do these dissimilarities help you separate the effects of nutrients and contaminants on the benthic food web?

14. What are the advantages to sampling endpoints multiple times throughout the 28-day experimental period? How might data on insect emergence from the artificial streams help with the interpretation of effluent effects?

15. What is the most plausible mechanism that explains the changes observed in the exposure site communities?

40.5 MATERIALS AND SUPPLIES

Basic Method

Site Characterization

Equipment for measuring channel geometry and substrate size (see Chapters 2 and 5)

Equipment for measuring current velocity and discharge (see Chapter 3)

Conductivity meter

1-L polyethylene bottles for collection of effluent

Field Assessment of Ecological Effects

Equipment for collection of water samples (see Chapters 32 and 33)

Equipment for collection of benthic invertebrate samples (see Chapters 15 and 38)

Equipment for collection of algal samples (see Chapters 11 and 12)

Scintillation vials or bags for storage of algal samples

Laboratory Methods

Equipment for processing benthic invertebrate samples (see Chapters 15 and 38)

Equipment for processing algal samples (see Chapters 11 and 12)

Advanced Method 1 (see also Chapter 31)

32 60-mL plastic containers

Agar

KNO_3

NaH_2PO_4

Glass fiber filters

Eight plastic L-bars

Stakes and nylon cord for attaching L-bars to streambed

Advanced Method 2

Crayfish species (e.g., *O. virilis*) collected from reference stream

10 minnow traps

Nylon cord and stakes for securing minnow traps

Dog food for baiting minnow traps

Utility buckets

Field analytical balance

Vernier caliper

32 lidded cages (21 cm L × 18 cm W × 10 cm H)

7.6-cm diameter plastic pots or longitudinally cut PVC pipe for cover in cages

Roll of 1-cm plastic mesh for covering cages

Stakes and cable ties for anchoring cages to streambed

Advanced Method 3

Artificial stream system (e.g., flow-through troughs, such as gutters or large PVC pipe cut longitudinally)

KNO_3 and NaH_2PO_4 for simulating effluent nutrient concentrations

Peristaltic pumps (see Chapter 31) or Mariotte bottles (see Chapter 30)

Unglazed 5 × 5-cm tiles

Cotton strips and large metal washers for weights

5-L plastic tanks for effluent transport (amounts will vary depending on stream design)

Washed gravel and cobble for artificial streams (quantities will vary depending on stream design)

Garden trays for colonization deployments of substrate

Wire cable and rebar for fastening garden trays to streambed

Conductivity meter

Equipment for collection of water samples (see Chapters 32 and 33)

Equipment for collection of benthic invertebrate samples (see Chapters 15 and 38)

Equipment for collection of algal samples (see Chapters 11 and 12)

Scintillation vials or bags for storage of algal samples

ACKNOWLEDGMENTS

Financial support for this work was provided by Natural Sciences and Engineering Research Council Discovery Grants to JMC, AGY, and DJB. Helpful advice during the development of our approach to multiple stressor assessment was provided by Drs. Bob Brua, Max Bothwell, Patricia Chambers, and Alexa Alexander Trusiuk. The technical design skills and assistance of Daryl Halliwell, Dave Hryn, Edward Krynak, Eric Luiker, and Sarah MacKenzie contributed greatly to the design of mesocosm and bioassay sampling techniques.

REFERENCES

Adams, S.M., 2003. Establishing causality between environmental stressors and effects on aquatic ecosystems. Human and Ecological Risk Assessment 9, 17–35.

Agostinho, M., Moreira-Santos, M., Ribeiro, R., 2012. A freshwater amphipod toxicity test based on post-exposure feeding and the population consumption inhibitory concentration. Chemosphere 87, 43–48.

Alexander, A.C., Chambers, P.A., Brua, R.B., Culp, J.M., 2015. Northern rivers basin study and the Athabasca River: the value of experimental approaches in a weight-of-evidence assessment. In: Norton, S.B., Cormier, S.M., Suter II, G.W. (Eds.), Ecological Causal Assessment. CRC Press, New York, NY, pp. 385–396.

Allert, A.L., Fairchild, J.F., DiStefano, R.J., Schmitt, C.J., Brumbaugh, W.G., Besser, J.M., 2009. Ecological effects of lead mining on Ozark streams: in-situ toxicity to woodland crayfish (*Orconectes hylas*). Ecotoxicology and Environmental Safety 72, 1207–1219.

Armanini, D.G., Chaumel, A.I., Monk, W.A., Marty, J., Smokorowski, K., Power, M., Baird, D.J., 2014. Benthic macroinvertebrate flow sensitivity as a tool to assess effects of hydropower related ramping activities in streams in Ontario (Canada). Ecological Indicators 46, 466–476.

Beyers, D.W., 1998. Causal inference in environmental impact studies. Journal of the North American Benthological Society 17, 367–373.

Burton Jr., G.A., Dyer, S.D., Cormier, S.M., Suter II, G.W., Dorward-King, E.J., 2001. Identifying watershed stressors using database evaluations linked with field and laboratory studies: a case example. In: Baird, D.J., Burton Jr., G.A. (Eds.), Ecological Variability: Separating Anthropogenic from Natural Causes of Ecosystem Impairment. SETAC Press, Pensacola, FL, pp. 233–253.

Buss, D.F., Carlisle, D.M., Chon, T.-S., Culp, J.M., Harding, J.S., Keizer-Vlek, H.E., Robinson, W.A., Strachan, S., Thirion, C., Hughes, R.M., 2015. Stream biomonitoring using macroinvertebrates around the globe: a comparison of large-scale programs. Environmental Monitoring and Assessment 187, 1573–2959.

Capps, K.A., Booth, M.T., Collins, S.M., Davison, M.A., Moslemi, J.M., El-Sabaawi, R.W., Simonis, J.L., Flecker, A.S., 2011. Nutrient diffusing substrata: a field comparison of commonly used methods to assess nutrient limitation. Journal of the North American Benthological Society 30, 522–532.

[CCME] Canadian Council of Ministers of the Environment, 1999. Canadian Environmental Quality Guidelines. Canadian Council of Ministers of the Environment, Winnipeg, Manitoba, Canada.

Chambers, P.A., Dale, A.R., Scrimgeour, G.J., Bothwell, M.L., 2000. Nutrient enrichment of northern rivers in response to pulp mill and municipal discharges. Journal of Aquatic Ecosystem Stress and Recovery 8, 53–66.

Clements, W.H., Kiffney, P.M., 1994. An integrated approach for assessing the impact of heavy metals at the Arkansas River, Colorado. Environmental Toxicology and Chemistry 13, 397–404.

Cooper, S.D., Barmuta, L., 1993. Field experiments in biomonitoring. In: Rosenberg, D.M., Resh, V.H. (Eds.), Biomonitoring and Benthic Invertebrates. Chapman and Hall, New York, NY, pp. 399–441.

Cormier, S.M., Norton, S.B., Suter II, G.W., 2015. Our approach for identifying causes. In: Norton, S.B., Cormier, S.M., Suter II, G.W. (Eds.), Ecological Causal Assessment. CRC Press, New York, USA, pp. 79–88.

Crane, M., Burton, G.A., Culp, J.M., Greenberg, M.S., Munkittrick, K.R., Ribeiro, R., Salazar, M.H., St-Jean, S.D., 2007. Review of aquatic in situ approaches for stressor and effect diagnosis. Integrated Environmental Assessment and Management 3, 234–245.

Crawford, J.K., Luoma, S.N., 1993. Guidelines for Studies of Contaminants in Biological Tissues for the National Water-Quality Assessment Program. U.S. Geological Survey Open-File Report, pp. 92–494.

Culp, J.M., Cash, K.J., Wrona, F.J., 2000a. Cumulative effects assessment for the Northern River Basins Study. Journal Aquatic Ecosystem Stress and Recovery 8, 87–94.

Culp, J.M., Lowell, R.B., Cash, K.J., 2000b. Integrating in situ community experiments with field studies to generate weight-of-evidence risk assessments for large rivers. Environmental Toxicology and Chemistry 19, 1167–1173.

Culp, J.M., Podemski, C.L., Cash, K.J., 2000c. Interactive effects of nutrients and contaminants from pulp mill effluents on riverine benthos. Journal of Aquatic Ecosystem Stress and Recovery 8, 67–75.

Culp, J.M., Podemski, C.L., Cash, K.J., Lowell, R.B., 2000d. A research strategy for using stream microcosms in ecotoxicology: integrating single population and community experiments with field data. Journal of Aquatic Ecosystem Stress and Recovery 7, 167–176.

Culp, J.M., Cash, K.J., Glozier, N.E., Brua, R.B., 2003. Effects of pulp mill effluent on benthic assemblages in mesocosms along the Saint John River, Canada. Environmental Toxicology and Chemistry 12, 2916–2925.

Culp, J.M., Armanini, D.G., Dunbar, M.J., Orlofske, J.M., Poff, N.L., Pollard, A.I., Yates, A.G., Hose, G.C., 2011. Incorporating traits in aquatic biomonitoring to enhance causal diagnosis and prediction. Integrated Environmental Assessment and Management 7, 187–197.

Culp, J.M., Alexandra, A.C., Brua, R.B., 2015. Mesocosm studies. In: Norton, S.B., Cormier, S.M., Suter II, G.W. (Eds.), Ecological Causal Assessment. CRC Press, Boca Raton, FL, USA, pp. 225–231.

Dafforn, K., Johnston, E.L., Ferguson, A., Humphrey, C., Monk, W.A., Nichols, S., Simpson, S., Tulbure, M., Baird, D.J., 2016. Big data opportunities and challenges for assessing multiple stressors across scales in aquatic ecosystems. Marine and Freshwater Research 67, 393–413.

Dubé, M.G., Culp, J.M., Scrimgeour, G.J., 1997. Nutrient limitation and herbivory: processes influenced by bleached kraft pulp mill effluent. Canadian Journal of Fisheries and Aquatic Sciences 54, 2584–2595.

Fox, G.A., 1991. Practical causal inference for ecoepidemiologists. Journal of Toxicology and Environment Health 33, 359–379.

Gagnon, C., Gagné, F., Turcotte, P., Saulnier, I., Blaise, C., Salazar, M.H., Salazar, S.M., 2006. Exposure of caged mussels to metals in primary-treated municipal wastewater plume. Chemosphere 62, 998–1010.

Gerritsen, J., Yuan, L.L., Shaw-Allen, P., Farrar, D., 2015. Regional observational studies: assembling and exploring data. In: Norton, S.B., Cormier, S.M., Suter II, G.W. (Eds.), Ecological Causal Assessment. CRC Press, New York, USA, pp. 155–168.

Gilbertson, M., 1997. Advances in forensic toxicology for establishing causality between Great Lakes epizootics and specific persistent toxic chemicals. Environmental Toxicology and Chemistry 16, 1771–1778.

Glozier, N.E., Culp, J.M., Reynoldson, T.B., Bailey, R.C., Lowell, R.B., Trudel, L., 2002. Assessing metal mine effects using benthic invertebrates for Canada's Environmental Effects Program. Water Quality Research Journal of Canada 37, 251–278.

Guckert, J.B., 1993. Artificial streams in ecotoxicology, 350–356. In: Lamberti, G.A., Steinman, A.D. (Eds.), Journal of the North American Benthological Society Research in Artificial Streams: Applications, Uses, and Abuses 12, 313–384.

Hamers, T., Legler, J., Blaha, L., Hylland, K., Mariogomez, I., Schipper, C.A., Segner, H., Vethaak, A.D., Witters, H., de Zwart, D., Leonards, P.E.G., 2013. Expert opinion on toxicity profiling—report from a NORMAN expert group meeting. Integrated Environmental Assessment and Management 9, 185–191.

Kellar, C.R., Hassell, K.L., Long, S.M., Myers, J.H., Golding, L., Rose, G., Kumar, A., Hoffmann, A.A., Pettigrove, V., 2014. Ecological evidence links adverse biological effects to pesticide and metal contamination in an urban Australian watershed. Journal of Applied Ecology 51, 426–439.

Kosinski, R.J., 1989. Artificial streams in ecotoxicological research. In: Boudou, A., Ribeyre, F. (Eds.), Aquatic Ecotoxicology: Fundamental Concepts and Methodologies, vol. I. CRC Press, Boca Raton, FL, pp. 297–316.

Kouba, A., Buric, M., Kozak, P., 2010. Bioaccumulation and effects of heavy metals in crayfish: a review. Water Air and Soil Pollution 211, 5–16.

Krahn, M.M., Wigren, C.A., Pearce, R.W., Moore, L.K., Bogar, R.G., Macleod Jr., W.D., Chan, S., Brown, D.W., 1988. Standard Analytical Procedures of the NOAA National Analytical Facility, 1988: New HPLC Cleanup and Revised Extraction Procedures for Organic Contaminants. National Oceanic and Atmospheric Administration, Springfield, VA.

Lamberti, G.A., Resh, V.H., 1985. Comparability of introduced tiles and natural substrates for sampling lotic bacteria, algae, and macro-invertebrates. Freshwater Biology 15, 21–30.

Research in artificial streams: applications, uses, and abuses. In: Lamberti, G.A., Steinman, A.D. (Eds.), Journal of the North American Benthological Society 12, 313–384.

Liber, K., Goodfellow, W., Green, A., Clements, W., den Besten, P., Galloway, T., Gerhardt, A., Simpson, S., 2007. In-situ-based effects measures: considerations for improving methods and approaches. Integrated Environmental Assessment and Management 3, 246–258.

Lodge, D.M., Kershner, J.W., Aloi, J.E., 1995. Effects of an omnivorous crayfish (*Orconectes rusticus*) on a freshwater littoral food web. Ecology 75, 165–1281.

Lowell, R.B., Culp, J.M., Dubé, M.G., 2000. A weight-of evidence approach for northern river risk assessment: integrating the effects of multiple stressors. Environmental Toxicology and Chemistry 19, 1182–1190.

McWilliam, R.A., Baird, D.J., 2002. Application of post-exposure feeding depression bioassays with *Daphnia magna* for assessment of toxic effluents in rivers. Environmental Toxicology and Chemistry 22, 1462–1468.

Megraw, S., Reynoldson, T., Bailey, R., Burd, B., Corkum, L., Culp, J., Langlois, C., Porter, E., Rosenberg, D., Wildish, D., Wrona, F., 1997. Benthic Invertebrate Community Expert Working Group Final Report: Recommendations from Cycle 1 Review. EEM/1997/7. Ottawa, Canada.

Norris, R.H., Webb, J.A., Nichols, S.J., Stewardson, M.J., Harrison, E.T., 2011. Analyzing cause and effect in environmental assessments: using weighted evidence from the literature. Freshwater Science 31, 5–21.

Norton, S.B., Cormier, S.M., Suter II, G.W. (Eds.), 2015a. Ecological Causal Assessment. CRC Press, Boca Raton, FL.

Norton, S.B., Schofield, K., Suter II, G.W., Cormier, S.M., 2015b. Listing candidate causes. In: Norton, S.B., Cormier, S.M., Suter II, G.W. (Eds.), Ecological Causal Assessment. CRC Press, New York, NY, pp. 101–121.

Pestana, J.L.T., Alexander, A.C., Culp, J.M., Baird, D.J., Cessna, A.J., Soares, A.M.V.M., 2009. Structural and functional responses of benthic invertebrates to imidacloprid in outdoor stream mesocosms. Environmental Pollution 157, 2328–2334.

Piggott, J.J., Salis, R.K., Lear, G., Townsend, C.R., Matthaei, C.D., 2015. Climate warming and agricultural stressors interact to determine stream periphyton community composition. Global Change Biology 21, 206–222.

Pontasch, K.W., 1995. The use of stream microcosms in multispecies testing. In: Cairns, J., Niederlehner, B.R. (Eds.), Ecological Toxicity Testing: Scale, Complexity and Relevance. Lewis Publishers, Boca Raton, FL, pp. 169–191.

Rice, E.W., Baird, R.B., Eaton, A.D., Clesceri, L.S. (Eds.), 2012. Standard Methods for the Examination of Water and Wastewater, nineteenthth ed. American Public Health Association, Washington, DC.

Rubach, M.N., Baird, D.J., Van den Brink, P.J., 2010. A new method for ranking mode-specific sensitivity of freshwater arthropods to insecticides and its relationship to biological traits. Environmental Toxicology and Chemistry 29, 476—487.

Schilderman, P.A.E.L., Moonen, E.J.C., Maas, L.M., Welle, I., Klienjans, J.C.S., 1999. Use of crayfish in biomonitoring studies of environmental pollution of the river Meuse. Ecotoxicology and Environmental Safety 44, 241—252.

Scrimgeour, G.J., Chambers, P.A., 2000. Cumulative effects of pulp mill and municipal effluents on epilithic biomass and nutrient limitation in a large northern river ecosystem. Canadian Journal of Fisheries and Aquatic Sciences 57, 1342—1354.

Shriner, C., Gregory, T., 1984. Use of artificial streams for toxicological research. Critical Reviews in Toxicology 13, 253—281.

Slocum, M.G., Roberts, J., Mendelsohn, I.A., 2009. Artist canvas as a new standard for the cotton-strip assay. Journal of Plant Nutrition and Soil Science 172, 71—74.

Smith, D.L., Cooper, M.J., Kosiara, J.M., Lamberti, G.A., 2016. Body burdens of heavy metals in Lake Michigan wetland turtles. Environmental Monitoring and Assessment 188, 128—142.

Statzner, B., Bêche, L.A., 2010. Can biological invertebrate traits resolve effects of multiple stressors on running water ecosystems? Freshwater Biology 55, 80—119.

Stewart-Oaten, A., Bence, J.R., Osenberg, C.W., 1992. Assessing effects of unreplicated perturbations: no simple solutions. Ecology 67, 929—940.

Suter II, G.W., 1993. Ecological Risk Assessment. Lewis Publishers, Chelsea, MI.

Suter II, G.W., 2015. Symptoms, body burdens and biomarkers. In: Norton, S.B., Cormier, S.M., Suter II, G.W. (Eds.), Ecological Causal Assessment. CRC Press, New York, NY, pp. 233—241.

Suter, G.W., Cormier, S.M., Norton, S.B., 2015. Forming casual conclusions. In: Norton, S.B., Cormier, S.M., Suter II, G.W. (Eds.), Ecological Causal Assessment. CRC Press, New York, NY, pp. 252—270.

Tank, J.L., Dodds, W.K., 2003. Nutrient limitation of epilithic and epixylic biofilms in ten North American streams. Freshwater Biology 48, 1031—1049.

Tiegs, S.D., Clapcott, J.E., Griffiths, N.A., Boulton, A.J., 2013. A standardized cotton-strip assay for measuring organic-matter decomposition in streams. Ecological Indicators 32, 131—139.

[USEPA] U.S. Environmental Protection Agency, 2000. Stressor Identification Guidance Document. USEPA, Washington, DC. EPA/822/B-00/025.

[USEPA] U.S. Environmental Protection Agency, 2002. National Recommended Water Quality Criteria: 2002. Washington, DC: EPA-822-R-02—047.

Vioque-Fernandez, A., de Almeida, E.A., Lopez-Barea, J., 2009. Assessment of Donana National Park contamination in *Procambarus clarkii*: integration of conventional biomarkers and proteomic approaches. Science of the Total Environment 407, 1784—1797.

Webb, J.A., Miller, K.A., Stewardson, M.J., de Little, S.C., Nichols, S.J., Wealands, S.R., 2015. An online database and desktop assessment software to simplify systematic reviews in environmental science. Environmental Modelling and Software 64, 72—79.

Webster, J.R., Ehrman, T.P., 1996. Solute dynamics. In: Hauer, F.R., Lamberti, G.A. (Eds.), Methods in Stream Ecology. Academic Press, San Diego, CA, pp. 145—160.

Zar, J.H., 2010. Biostatistical Analysis, fifth ed. Prentice-Hall, Upper Saddle River, NJ.

Glossary

Acetylene reduction An assay for nitrogen fixation that makes use of the fact that the enzyme that converts N_2 gas to ammonium also converts the gas acetylene to ethylene.

Actual nitrification Nitrification rates under natural conditions.

Allochthonous Originating from a place different from where found.

Allochthonous material Organic matter derived from outside of the stream ecosystem, such as from the riparian zone.

Aquatic bioassessment An integrated assessment that compares biological measures, habitat condition, and water quality against defined reference conditions.

Areal uptake (U) Mass of nutrient taken up per unit stream bed area, representing the amount of processing occurring within a system; heavily influenced by background concentration.

Assimilation A nutrient transformation that results in incorporation of the nutrient into a cell.

Assimilation efficiency The fraction or percentage of organic matter ingested that becomes assimilated for use in growth and respiration.

Autecology The study of an individual organism's interactions with its environment; in bioassessment, the environmental optima and/or tolerance estimates for specific environmental characteristics (e.g., pH or total P) for individual species.

Autochthonous Originating within the place where found.

Autochthonous material Organic matter produced within the stream via the process of primary production, including algae, mosses, and vascular plants.

Autotrophic In the context of an organism, one that uses an external energy source to reduce CO_2 to organic C molecules; in the context of an ecosystem, one that has net ecosystem production (NEP) > 0.

Autotrophic respiration (R_a) Consumption of oxygen by autotrophic organisms for their own metabolism.

Benthic organic matter (BOM) Organic particles deposited on the stream substratum.

Bioassay A system to measure the effects of a substance or physical property (e.g., temperature) on cells, tissues, or whole organisms, usually laboratory based.

Biomass The amount of living tissue of a population or community present at one instant in time (or averaged over several periods of time); units are mass (or energy) per unit area (typically g dry mass or $AFDM/m^2$).

Biotic index A scale developed from the presence or abundance of specific organisms, often weighted by the organisms' tolerance to a stressor.

Body burden The amount of a toxic substance accumulated by an organism over a specified period of time or over part of its life cycle.

Carbon-13 (^{13}C) Natural, stable isotope of carbon with a nucleus containing one more neutron than the more abundant ^{12}C; makes up about 1% of all natural C on Earth.

Carbon spiraling Downstream travel of organic carbon prior to being respired.

Causal assessment The systematic analysis of potential environmental drivers, pressures, or stressors to allow ranking of influence and to gauge statistical significance of any observed pattern.

Chemoautotroph Organisms that obtain energy by the oxidation of electron donors in their environment, in contrast to phototrophs that utilize solar energy.

Coarse particulate organic matter (CPOM) Particles of organic matter larger than 1 mm, such as leaves, twigs, wood fragments, and even entire trees; CPOM is often separated into different size classes and may be distinguished as woody or nonwoody material.

Conditioning The colonization and growth of microbial communities on organic matter such that its nutritive quality to detritivores increases.

Consumer—resource imbalance The difference in stoichiometry, generally C:N:P, between an animal (the consumer) and its food source (the resource).

Cumulative effects The combined effects on an ecosystem of past and current multiple stressors.

Decay (or decomposition) Biological breakdown or loss of mass through both biological decay and physical abrasion.

Denitrification The conversion of nitrate to N_2 gas where nitrate is used as an electron acceptor to oxidize organic carbon in a form of anaerobic respiration.

Detritus Nonliving organic debris usually of plant origin and in a state of decay.

Deuterium A stable isotope of hydrogen with symbol D or ^2H (known as heavy hydrogen) and used as a tracer to aid in partitioning allochthonous and autochthonous energy sources in aquatic food webs and in terrestrial plant species.

Dissimilatory A biologically mediated nutrient transformation that does not result in incorporation of the nutrient into the cell.

Dissimilatory nitrate reduction to ammonium (DNRA) A microbial process that converts nitrate to ammonium for energetic gain.

Dissolved organic carbon (DOC) Usually the largest fraction of the DOM pool, consisting of many different carbon-based molecules.

Dissolved organic matter (DOM) Organic material in solution that passes through a 0.45-μm filter, typically representing >90% of the transported OM in a stream or river.

DNA bar coding A biochemical technique for species identification using a portion of the organism's DNA.

Ecological assessment The process of evaluating environmental conditions using biological endpoints at levels of organization above the individual organism (i.e., population, community, and ecosystem structures and processes).

Ecological efficiency Various measures of the efficiency with which organisms convert matter or energy from one form to another.

Ecological risk assessment The quantitative, probabilistic analysis of threats to ecosystems, in terms of likely ecological responses of preidentified organisms or groups of organisms.

Ecosystem respiration (ER) Total consumption of oxygen via aerobic respiration; a negative flux.

Egestion Loss of unused or undigested material (e.g., feces).

Environmental DNA (eDNA) DNA collected and analyzed from environmental media such as water as compared to collecting DNA from specific tissue.

Environmental optimum The most likely value of an environmental characteristic (e.g., pH or total phosphorus) at which an individual belonging to a particular group (typically species) will be found.

EPT Taxa of the aquatic insect orders Ephemeroptera (mayflies), Plecoptera (stoneflies), and Trichoptera (caddisflies); commonly used in bioassessment.

Excretion Elimination of metabolic waste products, including nitrogenous compounds.

Expected condition Benchmark against which an ecosystem (e.g., stream) is compared.

Fine particulate organic matter (FPOM) Organic particles in the size range of 0.45–1000 μm (1.0 mm).

Gross nutrient uptake Total amount of nutrient taken up by an organism, assemblage, or ecosystem.

Gross primary production (GPP) Total photosynthetic production of oxygen in an ecosystem.

Heterotrophic In the context of an organism, one that only oxidizes organic C to gain energy; in the context of an ecosystem, one that has net ecosystem production (NEP) < 0.

Heterotrophic respiration (R_h) Consumption of oxygen by heterotrophic organisms.

Homeostasis Degree to which an organism's body stoichiometry changes in response to variation in resource stoichiometry, when all else is held equal (e.g., temperature, ontogeny, other aspects of diet quality).

IBI metric calibration The step in the IBI development process by which metrics are selected and evaluated for inclusion in the IBI.

IBI metric validation The final step in the IBI development process, following the calibration step, during which the IBI is tested against data from a reference standard other than that used during the metric calibration step.

In situ bioassay A technique by which organisms are caged or otherwise confined and subsequently exposed to an environmental stressor or stressors in the field; responses to stress may be observed at the time of exposure or at time points following exposure.

Incremental impacts The additive effects of similar stressor events whose combined effect exceeds a critical ecological threshold; measured at specified intervals (e.g., seasons, years).

Index of biotic integrity (IBI) A multimetric index of fish (or other organism) assemblage structure and function that, by integrating information from multiple metrics, can be used to infer the biological integrity of the ecosystem.

Large wood (LW) Functionally refers to wood pieces greater than 1 m in length and 10 cm in diameter.

Lateral litter inputs Leaves, needles, and other organic debris, usually of plant origin, transported overland to the open stream by gravity, wind, or water.

Leaf breakdown The loss of leaf mass through time due to processes including the leaching of solutes, consumption by invertebrates, microbial decomposition, and physical abrasion.

Line intersect (or intercept) method A method for estimating standing vegetation in the riparian zone or the volume or mass of wood in a stream or river channel.

Litterfall/litter inputs Leaves, needles, and other organic debris, usually of plant origin, that fall directly into the open stream.

Macroinvertebrates Multicellular animals without backbones (invertebrates) that are visible without magnification; in streams, mainly consist of aquatic insects, mites, mollusks, crustaceans, and annelids.

Matrix effects Effects on chemical analyses that are due to contents of the sample other than the analyte of interest.

Membrane inlet mass spectrometry (MIMS) A spectrometric process that measures the mass of atoms or molecules with a gas-permeable membrane at the inlet of the machine.

Mesocosm Outdoor or indoor facilities with controlled physicochemical conditions used to simulate natural ecosystems; these systems provide a miniaturized representation of a natural stream contained in a discrete unit which can be experimentally replicated in a manipulative experiment.

Microbial respiration The release of carbon dioxide by microorganisms.

Molecular markers Measurable biologically derived molecules used to indicate the state of a biological process or organism.

Multimetric index A type of bioassessment index that integrates information from multiple measures (metrics) of assemblage structure and function to infer community condition; each component metric in a multimetric index is predictably and reasonably related to specific impacts caused by environmental alterations.

Multiple source impacts Impacts of a single or multiple stressors, arising from spatially separated, discrete sources of stressors whose effects overlap spatially.

Multiple stressor impacts The resultant effect of qualitatively different environmental stress factors interacting in a synergistic or antagonistic fashion to produce a biological response (e.g., mortality); a priori prediction of biotic responses to this class of stressors is challenging.

Net ecosystem production (NEP) GPP plus ER; the balance of gross primary production and ecosystem respiration; can be positive or negative.

Net nutrient uptake Gross nutrient uptake minus the amount of nutrient released from an organism or ecosystem.

Net primary production (NPP) The production of new autotrophic biomass in an ecosystem and equivalent to GPP plus autotrophic respiration (R_a).

Nitrification The conversion of ammonium to nitrate in the presence of O_2 in a chemoautotrophic process that yields cellular energy.

Nitrogen-15 (^{15}N) A natural stable isotope of nitrogen with a nucleus containing one more neutron than the more abundant ^{14}N; naturally present at low levels and often used as a tracer of nitrogen flux.

Nitrogen cycle The sum total of the natural processes of nitrogen conversion in the environment.

Nitrogen fixation The microbial conversion of N_2 gas to ammonium (NH_4) for assimilation into cellular nitrogen.

Nitrogen flux The rate of conversion of nitrogen from one form to another in the nitrogen cycle.

Nutrient diffusing substrata (NDS) In situ tool used to manipulate the chemical environment for colonizing stream biofilms and infer nutrient limitation status of a stream.

Operational taxonomic unit (OTU) Assignable designation used when the identity of an organism is unknown but is known to be unique (e.g., species A, species B).

Organic matter processing The respiration of organic carbon by heterotrophs.

Periphyton A submerged, attached assemblage of algae, bacteria, fungi, and meiofauna held within a mucilaginous, polysaccharide matrix; sometimes referred to as "biofilm."

Phosphatase An enzyme that hydrolyzes phosphomonoesters (nonspecific phosphomonoesterase) into a phosphate ion and a molecule with a free hydroxyl group; alkaline phosphatase has optimum activity at alkaline pH and has been mostly reported on the external surface of algae, but also in the cell wall, external membrane, or periplasmic space.

Phototrophs Organisms that carry out solar radiation (light) capture to acquire energy and carry out various cellular metabolic processes.

Point-centered quarter method (PQ) A method for estimating tree vegetation in the riparian zone based on an area divided into quarters around a randomly selected point.

Potential nitrification The maximum possible rate of nitrification under optimal conditions (i.e., high dissolved oxygen and ammonium concentrations).

Production/biomass ratio (P/B) A metric that can be used in two ways: (1) the ratio of secondary production to mean biomass over a unit of time (e.g., annual P/B), equivalent to the biomass turnover rate with units of inverse time (e.g., $year^{-1}$) or (2) the ratio of secondary production to mean biomass over the life span of a cohort (i.e., cohort P/B), with no time units; annual P/B can vary over orders of magnitude (e.g., <1 to >100 $year^{-1}$, but cohort P/B is typically close to 5.

Quantitative food web A food web in which the feeding connections between species (or higher taxonomic groups such as genus or family) in a community are quantified in some way, such as by measurement of ingestion flows (bottom-up web) or predator impacts (top-down web).

Radioisotope A version of a chemical element that has an unstable nucleus and emits radiation during its decay to a stable form.

Rapid periphyton survey Field-based approaches to quantify the abundance of algal mats and/or macrophytes in a stream by observing mat thickness and/or percent cover.

Rapid turnover compartment Biotic compartment present in a stream that quickly assimilates nutrients into biomass; rapid turnover compartments (e.g., filamentous algae, biofilms) can be sampled following a ^{15}N tracer addition to quantify biotic N assimilation.

Reciprocal food subsidies Resources such as insects, other invertebrates, and fish carcasses that leave or otherwise are transferred from a stream to the land, where they serve as food for an assortment of spiders, mammals, and birds.

Recruitment/delivery/input Introduction of wood or other material into a river channel by processes such as bank erosion, landslides, debris flows, avalanches, blowdown, or fluvial transport.

Redox A measure of the oxidation or reduction capacity of a substance, such as water.

Reference standard A defined set of environmental conditions used to calibrate multimetric indices; reference standards can be "least disturbed/most disturbed" reference sites or can be a set of "composite" reference conditions based on a cumulative pattern of best attainable attributes obtained from multiple sampling sites of comparable condition or from historical or repeated sampling of the same and random sites.

Reference stream A stream used to establish a benchmark to which other steams can be compared.

Remote sensing The acquisition of information about an object or phenomenon without making physical contact with the object and thus in contrast to on-site observation; in stream ecology, remote sensing often refers to image data collected from drones, aircraft, or satellites.

Residence time Time interval during which an atom, molecule, or particulate is stored in a reach of stream before decaying or being transported downstream of the reach.

Retention The process of capturing and storing organic or inorganic matter within the stream channel.

Retentiveness Capacity of the stream channel to capture and store matter.

Riparius The streamside environment/habitat/ecosystem; noun form of the more familiar adjective, riparian.

River invertebrate prediction and classification system (RIVPACS) A multivariate-based assessment method that compares benthic invertebrate taxa observed at a site to the taxa predicted at a site.

Roughness Features of the stream channel, especially the bed that provides heterogeneity and potentially increases retention.

Secondary production (P) The formation of heterotrophic biomass through time [e.g., mg of dry biomass formed by a population within a unit area of habitat (e.g., m^2) over a unit of time (day, week, or year]; secondary production is a flow (or flux) and typical units are g dry mass per m^2 per year.

Seston Fine particulate matter that is suspended in the water column, which includes living and nonliving material.

Shading The extent to which incoming solar radiation to a surface, such as a stream, is prevented from reaching that surface by overhead objects such as shrubs, trees, cliffs, mountains, or buildings.

Solute A material chemically dissolved in water.

Species traits Measurable biological and ecological characteristics that are related to physiology, morphology, behavior, and life history of species; populations; and higher taxa.

Stable isotope Forms of elements that do not decay into other elemental forms (cf. radioisotope); elements may have multiple stable isotopes depending on the number of neutrons in the nucleus.

Standing crop Mass of material of a particular type at a specific time within a specific area, often expressed as dry mass or ash-free dry mass.

Strict homeostasis A common stoichiometric assumption that an organism's body stoichiometry does not change in response to variation in resource stoichiometry, when all else is held equal (e.g., temperature, ontogeny, other aspects of diet quality).

Taxon (pl. taxa) A taxonomic group at any rank (e.g., species, genus, family, order).

Threshold elemental ratio (TER) The dietary ratio of two elements at which limitation of a consumer's growth switches from one element to the other.

Transient storage Water that is traveling in flow paths much more slowly than the main body of water; for example in pools, eddies, or hyporheic flow paths.

Transport Directional movement of material by water flow in the stream channel.

Uptake length Average downstream distance (unit of length) that a reactive (nutrient) solute travels in the stream water column before being transformed (immobilized).

Uptake velocity Mass transfer coefficient that describes a theoretical velocity (units of length per time) at which a nutrient moves toward the location of immobilization.

Weight-of-evidence The use of multiple sources of evidence to define the relative degree of support for candidate causal mechanisms or other conclusions provided by evidence of the underlying cause of observed ecosystem phenomena; the result of weighing the body of evidence.

Weighted average model A mathematical approach to estimate an environmental characteristic using autecological data weighted by the relative abundances of species at that location.

Wood Structural organic matter derived from the secondary xylem of trees and shrubs; wood contains cellulose and lignin, which influence its decomposition rates and food quality.

Index

'Note: Page numbers followed by "f" indicate figures, "t" indicate tables.'